P9-AGG-978

WITHDRAWN
UTSA LIBRAR

SMALL-SCALE FRESHWATER TOXICITY INVESTIGATIONS

Small-scale Freshwater Toxicity Investigations

Volume 2 - Hazard Assessment Schemes

Edited by

Christian Blaise
St. Lawrence Centre, Environment Canada,
QC, Canada

and

Jean-François Férard
Paul Verlaine University, Metz,
Laboratoire Ecotoxicité et Santé Environnementale,
Metz, France

A C.I.P. Catalogue record for this book is available from the Library of Congress.

ISBN-10 1-4020-3543-8 (HB)
ISBN-13 978-1-4020-3543-2 (HB)
ISBN-10 1-4020-3553-5 (e-book)
ISBN-13 978-1-4020-3553-1 (e-book)

Published by Springer,
P.O. Box 17, 3300 AA Dordrecht, The Netherlands.

www.springeronline.com

Cover design: Created by Patrick Bermingham (Montreal, Canada)

Printed on acid-free paper

Printed in the Netherlands.

About the editors

Christian Blaise, D.Sc., is a senior research scientist at the Saint-Lawrence Centre, Environment Canada, Québec Region, where he heads the Aquatic Toxicology Unit (ATU), River Ecosystems Research Section. He also holds an adjunct professor status at UQAR (Université du Québec à Rimouski) where he contributes to teaching and (co)directs graduate students in the field of ecotoxicology. ATU strives to develop, validate, standardize, modernize (and promote the commercialization of) bioanalytical and biomarker techniques, making use of new instrumental technologies whenever possible, in order to determine the potential (geno)toxicity of chemicals and various types of environmental matrices (*e.g.*, effluents, sediments, pore/surface waters). ATU research output provides practical tools and approaches which facilitate decision-making for environmental management of aquatic ecosystems such as the Saint-Lawrence River. ATU also provides (inter)national technology transfer to interested professionals and agencies and promotes graduate student training by co-directing applied research projects with university collaborators.

Dr. Blaise obtained university diplomas from the U. of Montréal (B.A., 1967: biology and chemistry), U. of Ottawa (B.Sc., 1970: cell biology; M.Sc., 1973: environmental microbiology) and U. of Metz (D.Sc., 1984: ecotoxicology). He is a member of the editorial board for two scientific journals (*Environmental Toxicology*; *Ecotoxicology and Environmental Safety*) and holds membership in both the biologists' (Association des Biologistes du Québec) and microbiologists' (Association des Microbiologistes du Québec) associations of the province of Québec. He regularly attends and makes presentations during major venues held in the field of ecotoxicology (SETAC: Society of Environmental Toxicology and Chemistry; SECOTOX: Society of Ecotoxicology and Environmental Safety; ATW-Canada: Aquatic Toxicity Workshop-Canada; ISTA: International Symposium on Toxicity Assessment). Dr. Blaise has (co)authored over 100 scientific articles in internationally refereed journals, as well as having written several book chapters, reviews, and various government technical reports.

He recently co-edited, with Canadian colleagues, a book dedicated to small-scale toxicity testing (Wells, P., K. Lee and C. Blaise (eds.), 1998. *Microscale testing in Aquatic Toxicology Advances, Techniques and Practice*. CRC Lewis Publishers, Boca Raton, Florida, 679 pages). He was scientific organizer of the 10th International Symposium on Toxicity Assessment (ISTA 10), hosted by the Saint-Lawrence Centre where he works, and held in Quebec City, August 26-31, 2001. He further co-edited with another Canadian colleague, a special edition of *Environ. Toxicol.* (Volume 17 [3]: 2002, special issue) highlighting selected papers presented at the ISTA 10 venue.

Jean-François Férard, D.Sc., is a professor at the University of Metz (Lorraine province of France), where he heads a research team (RT) which is part of a C.N.R.S. (Centre National de la Recherche Scientifique) research unit for Ecotoxicity and Environmental Health (E.S.E). He also manages an undergraduate school program dedicated to Environmental Engineering. His teaching duties involve fundamental and applied Ecotoxicology, Physiology and Physiotoxicology, Cell Biology and other related disciplines.

In the field of ecotoxicology, his RT was involved in the behavior of metals, PAHs and complex mixtures in air, water and soil compartments and their effects on different organisms (bacteria, algae, crustaceans, plants, arthropods, etc.). His actual research endeavors are more specifically focused on the development of metal-resistance (*e.g.* phytochelatin) and genotoxic (*e.g.* comet assay) biomarkers. He also promotes knowledge and use of toxicity tests by organizing an annual course entitled "Ecotoxicity and carcinogenicity of chemicals" which provides a theoretical and practical view of numerous toxicity tests to decision-makers, industrialists and consultants. Since 1974, he has markedly contributed to numerous research programs that have successfully lead to i) development and validation of different ecotoxicity tools (*e.g.* new toxicity test methods, trophic chain models, biomarkers), ii) hazard/risk assessment schemes and iii) links between field and laboratory studies. These undertakings were financially supported by the European Economic Community, the French ministry of the Environment, and agencies such as the

French Water Agency of the Rhin-Meuse Watershed, the French Agency for Environment and Energy Ressources.

Professor Férard obtained university diplomas from the U. of Strasbourg (B.A., 1970: biology and chemistry; B.Sc., 1973: biochemistry) and U. of Metz (M.Sc., 1974: chemistry and environmental toxicology; D.Sc., 1978: environmental toxicology; State doctorate, 1986: environmental toxicology). He was European editor for *Environmental Toxicology and Water Quality* from 1992-1996 and holds membership in SETAC (Society of Environmental Toxicology and Chemistry). He regularly makes presentations during major symposia held in the field of ecotoxicology (*e.g.* SETAC meetings, Secotox conferences, International Symposia on Toxicity Assessment, Annual Aquatic Toxicity Workshops in Canada). Professor Férard has (co)authored over 50 scientific articles in (inter)nationally refereed journals, as well as having written several book chapters, reviews, and research reports. He also participates in several OECD (Organization for Economic Cooperation and Development) and AFNOR (*Association française de normalisation* - French standards association) initiatives to standardize and promote the use of biological tests.

Contributors

Inés Ahumada Facultad de Ciencias Químicas y Farmacéuticas Universidad de Chile, Santiago, Chile iahumada@ciq.uchile.cl	Carina Apartin CIMA, Facultad de Ciencias Exactas Universidad Nacional de la Plata La Plata, Argentina apartin@quimica.unlp.edu.ar
Gabriel Bitton Laboratory of Environmental Microbiology and Toxicology University of Florida, Gainesville FL 32611, USA gbitton@ufl.edu	Christian Blaise Centre Saint-Laurent, Env. Canada 105 McGill street, Montréal, Québec Canada, H2Y 2E7 christian.blaise@ec.gc.ca
Manon Bombardier Environmental Technology Centre, Environment Canada, 335 River Road, Ottawa, Canada, K1A 0H3 manon.bombardier@ec.gc.ca	Anne I. Borgmann Environmental Conservation Branch Environment Canada, Ontario Region Burlington, Ontario Canada, L7R 4A6 anne.borgmann@ec.gc.ca
Gustavo Bulus Rossini CIMA, Facultad de Ciencias Exactas Universidad Nacional de la Plata La Plata, Argentina gbulus@quimica.unlp.edu.ar	Gabriela Castillo Departamento de Ingeniería Civil Facultad de Ciencias Físicas y Matemáticas Universidad de Chile Casilla 228-3 Santiago, Chile gcastilo@ing.uchile.cl
Peter M. Chapman EVS Environment Consultants 195 Pemberton Avenue North Vancouver, B.C. Canada V7P 2R4 pchapman@attglobal.net	Roi Dagan Laboratory of Environmental Microbiology and Toxicology Department of Environmental Engineering Sciences University of Florida, Gainesville FL 32611, USA rdagan@ufl.edu
M. Consuelo Díaz-Baez Facultad de Ingeniería Universidad Nacional de Colombia Bogotá, Colombia mcdiazb@unal.edu.co	Adriana Espinosa Ramírez Facultad de Ingeniería Universidad Nacional de Colombia Bogotá, Colombia ajespinosar@unal.edu.co

Contributors

Jean-François Férard Université Paul Verlaine Laboratoire Ecotoxicité et Santé Environnementale CNRS FRE 2635, Campus Bridoux rue du Général Delestraint 57070 METZ, France, ferard@sciences.univ-metz.fr	Benoît Ferrari U. de Genève, Institut F.A. Forel 10, route de Suisse CH-1290 VERSOIX Switzerland benoit.ferrari@terre.unige.ch
Keith E. Holtze Stantec Consulting Ltd. 11B Nicholas Beaver Road Guelph, Ontario Canada, N1H 6H9 kholtze@stantec.com	Falk Krebs German Federal Institute of Hydrology (BfG) Am Mainzer Tor 1 56068 Koblenz, Germany krebs@bafg.de
Blair G. McDonald EVS Environment Consultants 195 Pemberton Avenue North Vancouver, B.C. Canada, V7P 2R4 bmcdonald@golder.com	Jorge Mendoza Facultad de Ciencias Químicas y Farmacéuticas Universidad de Chile Santiago, Chile jmendoza@ciq.uchile.cl
Jennifer A. Miller Miller Environmental Sciences Inc. Innisfil, Ontario L9S 3E9, Canada miller.smith@sympatico.ca	Lesley J. Novak Stantec Consulting Ltd. 11B Nicholas Beaver Road Guelph, Ontario Canada, N1H 6H9 lnovak@stantec.com
Alicia Ronco CIMA, Facultad de Ciencias Exactas Universidad Nacional de la Plata La Plata, Argentina. 47 y 115, (1900) La Plata, Argentina. Cima@quimica.unlp.edu.ar	Rick P. Scroggins Biological Methods Division Environmental Technology Centre Environment Canada Ottawa, Ontario, Canada, K1A 0H3 rick.scroggins@ec.gc.ca
Cecilia Sobrero CIMA, Facultad de Ciencias Exactas Universidad Nacional de la Plata La Plata, Argentina csobrero@quimica.unlp.edu.ar	Eric Vindimian Ministère de l'écologie et du développement durable Service recherche et prospective 20 ave. de Ségur, 75007 Paris, France eric.vindimian@normalesup.org
Marnie Ward Laboratory of Environmental Microbiology and Toxicology Dep. of Environmental. Engineering Sciences University of Florida, Gainesville FL 32611, USA wcward@infionline.net	

Reviewers

Gerald Ankley US Environmental Protection Agency 6201 Congdon Boulevard Duluth, Minnesota 55804 U.S.A.	Larry Ausley Microbiol. & Inorganic Chem. Branch NC Div. of Water Quality Laboratory Section 1623 Mail Service Center Raleigh, North Carolina 27699-1623 U.S.A.
Kristin Becker van Slooten Laboratoire de Chimie environnementale et écotoxicologie (CECOTOX) ENAC-ISTE, Bât GR Ecole polytechnique fédérale de Lausanne 1015 Switzerland	B. Kent Burnison Aquatic Ecosystem Protection Research Branch Environment Canada National Water Research Institute Burlington, Ontario Canada, L7R 4A6
R. Scott Carr USGS, BRD, CERC Marine Ecotoxicology Research Station TAMU-CC, Center for Coastal Studies NRC Suite 3200, 6300 Ocean Drive Corpus Christi, Texas 78412 U.S.A.	Peter M. Chapman EVS Environment Consultants 195 Pemberton Avenue North Vancouver, B.C. Canada, V7P 2R4
Yves Couillard Existing Substances Branch Environnement Canada Place Vincent Massey, 14th floor 351, Bd Saint-Joseph Hull, Quebec Canada, K1A 0H3	Gilles Forget International Development Research Centre B.P. 11007 Peytavin Dakar, Senegal
François Gagné Centre Saint-Laurent, Env. Canada 105 McGill, Montreal, Quebec Canada, H2Y 2E7	Christian Gagnon Centre Saint-Laurent, Env. Canada 105 McGill, Montreal, Quebec Canada, H2Y 2E7
Pierre Gagnon Centre Saint-Laurent, Env. Canada 105 McGill, Montreal, Quebec Canada, H2Y 2E7	Guy Gilron Cantox Environmental Inc. (CEI) 1900 Minnesota Court, Suite 130 Mississauga, Ontario Canada, L5N 3C9

Reviewers

B. Thomas Johnson Environmental Microbiology Columbia Environ. Research Center Columbia, Missouri U.S.A.	Christopher J. Kennedy Dept. of Biological Sciences Simon Fraser University Burnaby, B. C. Canada, V5A 1S6
Guilherme R. Lotufo US Army Engineer Research and Development Center Waterways Experiment Station, EP-R 3909 Halls Ferry Road Vicksburg, MS 39180-6199 U.S.A.	Michael Salazar Applied Biomonitoring 11648 - 72nd Place NE Kirkland, Washington 98034 U.S.A.
Sandra Salazar Applied Biomonitoring 11648 - 72nd Place NE Kirkland, Washington 98034 U.S.A.	Sébastien Sauvé Département de chimie Université de Montréal CP 6128 Centre-ville Montréal, Quebec Canada, H3C 3J7
John B. Sprague Sprague Associates Ltd. Salt Spring Island, B.C., Canada, V8K 2L7	Gladys L. Stephenson Stantec Consulting Ltd. 361 Southgate Dr. Guelph, Ontario Canada, N1G 3M5
Geoffrey I. Sunahara Applied Ecotoxicology Group Biotechnology Research Institute National Research Council - Canada 6100 Royalmount Ave. Montreal, Quebec, Canada H4P 2R2	Paule Vasseur U. de Metz, ESE, Campus Bridoux, rue du Général Delestraint 57070 Metz, France
Sylvia Waara Institutionen för Samhällsteknik Department of Public Technology Mälardalen University Box 883 721 23 Västerås Sweden	Paul A. White Mutagenesis Section, Safe Environments Program, Healthy Environments & Consumer Safety Branch, Health Canada, Tunney's Pasture 0803A, Ottawa, Ontario Canada, K1A 0L2

Preface

Developed, developing and emerging economies worldwide are collectively contributing multiple stresses on aquatic ecosystems by the release of numerous contaminants. This in turn demands that basic toxicological information on their potential to harm living species be available. Hence, environmental protection programs aimed at preserving water quality must have access to comprehensive toxicity screening tools and strategies that can be applied reliably and universally.

While a good number of toxicity testing procedures and hazard assessment approaches have been published in the scientific literature over the past decades, many are wanting in that insufficient detail is available for users to be able to fully understand the test method or scheme and to be able to reproduce it successfully. Even standardized techniques published in recognized international standard organization documents are often lacking in thoroughness and *minutiae*. Paucity of information relating to biological test methods may be consequent and trigger several phenomena including generation of invalid data and resulting toxicity measurements, erroneous interpretation and decision-taking with regards to a particular chemical or environmental issue, or simply abandonment of testing procedures. Clearly, improperly documented toxicity testing methods can be detrimental to their promotion and use, as they open the doorway to unnecessary debate and criticism as to their *raison d'être*. Furthermore, this situation can indirectly contribute to delaying, minimizing or eliminating their application, thereby curtailing the important role toxicity testing plays in the overall protection and conservation of aquatic ecosystems.

The "cry for help" that we have often heard from people having encountered difficulties in properly conducting biological tests was the primary trigger that set off our desire to edit a book on freshwater toxicity testing procedures in the detailed manner described herein. We feel this book is rather unique in that it includes 1) a broad review on toxicity testing applications, 2) comprehensive small-scale toxicity test methods (Volume 1) and hazard assessment schemes (Volume 2) presented in a designated template that was followed by all contributors, and 3) a complete glossary of scientific/technical terms employed by editors/contributors in their respective chapters.

Indeed, the book provides information on the purposes of applying toxicity tests and regroups 15 validated toxicity test methods (Volume 1) and 11 hazard assessment schemes (Volume 2) for the benefit and use of the scientific community at large. Academia (students, professors), government (environmental managers, scientists, regulators) and consulting professionals (biologists, chemists, engineers) should find it of interest, because it encompasses, into a single document, comprehensive information on biological testing which is normally scattered and difficult to find. It should be, for example, very useful for (under)graduate courses in aquatic toxicology involving practical laboratory training. In this respect, it can be attractive, owing to some of its contents, as a laboratory manual for learning purposes or for undertaking applied research to assess chemical hazards. As a further example, it can also prove useful for environmentalists who wish to select the most appropriate test(s) or scheme(s) for future decision-taking with regards to protection of aquatic ecosystems. In short, all groups directly or indirectly involved with the protection and conservation of freshwater environments will find this book appealing, as will those who simply wish to become familiar with the field of toxicity testing.

We are grateful for the financial support given to us in the production of this book by Environment Canada (Centre Saint-Laurent, Québec region, Environmental Conservation), the University of Metz (Metz, France) and IDRC (International Development Research Centre, Ottawa, Ontario, Canada). For their assistance in many dedicated ways which facilitated our tasks and ensured the timely completion of our book, we extend our thanks to the following persons: Mr. Andrés Sanchez and Dr. Jean Lebel (IDRC); Ms. Jacinthe Leclerc, Dr. Alex Vincent and Dr. André Talbot (Centre Saint-Laurent); Ms. Sylvie Bibeau and Dr. Laura Pirastru (University of Québec in Montréal). We are also very appreciative of the dedicated professional help provided us by Anna Besse and Judith Terpos of **Springer Publishers** in guiding us through the editorial process.

Again, how could we not extend our appreciation to all of our devoted colleagues who accepted our invitation to contribute a chapter to this book? They number 54 in total and represent 11 countries including Argentina, Australia, Canada, Chile, Columbia, Denmark, France, Germany, Poland, Switzerland and the U.S.A. Needless to say that it is owing to their outstanding career experience and interest to promote their know-how that ***Small-scale Freshwater Toxicity Investigations, Volumes 1 and 2*** has now become a reality. Last but not least, the ultimate acknowledgment must go to our other estimated colleagues who acted as peer-reviewers for all manuscript contributions and who significantly contributed to their final quality.

We are convinced that this book fills an important scientific gap that will stimulate international use and application of small-scale toxicity tests, whether for research, monitoring, or educational purposes. May the “blue planet” and its aquatic species ultimately profit from such endeavours!

Christian Blaise and Jean-François Férard

January, 2005

Foreword

Much has been said and done since the International Decade for Water and Sanitation of the 1980s to improve access to sufficient and safe drinking water in developing countries. Although we are nowhere near achieving universal access to this basic human need, progress has been accomplished. Technology has played an important role, but another critical legacy of the Decade has been a much better recognition and understanding of the social factors linked to sustainable access to safe drinking water for communities in developing countries.

One of the empowering factors has been the development of simple and affordable technologies for monitoring microbial water quality. Because they are inexpensive and are not dependent of sophisticated laboratories, such technologies have made their way into areas where electrical power has yet to reach and have allowed communities to perform their own water quality monitoring. The identification of specific micro-organisms are less important to rural inhabitants than an alarm system which they can depend on to consistently alert them to fecal contamination of their water supply. With water-borne diarrhea still causing the second highest mortality and morbidity toll in Third World countries (mainly infants and young children) the precautionary principle remains the only responsible strategy for poor communities.

Although fecal contamination of drinking water is still a serious problem in developing countries, it is not the only risk that need concern their populations and ecosystems. Both natural and anthropogenic processes are known to cause another kind, but no less dangerous contamination: recent surveys have shown for example that upwards of 36 million people in the Indian sub-continent are drinking water contaminated by arsenic; such contamination is also known to occur in the Southern Cone of Latin America and in areas of China. In Bangladesh, sadly, this problem has been compounded by altruistic efforts of AID agencies, digging wells to offer an alternative to fecally contaminated surface waters. Alas, the geologic makeup of the region has caused underground water to be heavily laced with Arsenic. Serious pathological manifestations have now been reported in affected areas. Some areas of India have also reported high fluoride concentration in well water leading to severe fluorosis in children and adults alike, with severe skeletal malformations and attendant physiological problems.

Human activity has also exacerbated this problem: Mercury contamination related to gold mining in frontier areas of South America; contamination of both surface and ground water by agricultural inputs such as pesticides and fertilizers; increased chemical pollution by recently implanted industries; global pollution by persistent chemicals used in industrialized countries such as PCBs and bromine-containing fire retardants. Unquestionably, the past and continuing release of toxicants of this nature to receiving waters, one of earth's crucial compartments, by way of numerous (non) point sources of pollution, have equally impaired the health of aquatic biota and even adversely affected the biodiversity of some of its communities (*e.g.*, invertebrates and fish). Indeed, while microbiological pollution poses predominantly a risk to human health, chemical contamination represents a much more global threat to all components of the ecosystem, with a potential for more profound and enduring consequences.

In most cases, laboratory analytical methods exist to detect such chemicals and to quantify them. However, they can be time consuming and very expensive. No one could even propose that screening programs could be set up for routine water testing which would be both timely and affordable. In fact, this would not be feasible for industrialized countries either. How is one to test water for safety from chemicals, then? One approach is to perform routine analysis for specific chemicals in a given area where they are presumed to exist. Therein lies a cautionary tale: in the early nineties, the British Geological Survey carried out a survey of well waters in Bangladesh (in relation to the well digging program discussed earlier), seeking data on iron and phosphorus which were presumed to contaminate the water. No attempts were made to measure other toxic compounds such as arsenic, which we now know constituted a major contaminant. Following the appearance of severe arsenic poisoning in the affected area, Bangladesh sued the agency for failing to warn users that the toxic metal was present in well water. The BGS was cleared by a British court of any wrong doing, since the former had performed the assays for which their services had been retained – and which did not include assays for other contaminants. Could this situation have not been avoided if a test had been applied to evaluate the overall toxicity of water, irrespective of the contaminant present? What about waters which exhibit contamination by multiple chemicals: individual measurements may not give an assessment of the true toxicity if these chemicals act in synergy rather than in an additive fashion.

Thus, some environmental scientists suggest that tests be used that measure "toxicity" rather than individual contaminants. Toxic samples could then be further assayed for specific contaminants if necessary to identify point sources and/or water treatment procedures. Relatively rapid, affordable and dependable assays would be a boon for developing country communities, in the same way as earlier rapid tests were for fecal contamination. The latter have proven to be usable in a sustainable manner in developing country communities, empowering them to monitor water safety and to act appropriately when necessary.

Bioassays appeared to fit the bill to perform this service to monitor chemical contamination. They have been around for a while. Until relatively recently, however, they remained in the realm of the laboratory. Only over the last two decades have they found a niche in testing for toxic chemicals in water and sediment, but not yet specifically as a tool for routine water quality monitoring. As ***Small-scale Freshwater Toxicity Investigations, Volumes 1 and 2*** amply demonstrates, the science has now come of age. Assays based on bacteria, microscopic or multi-cellular algae, protozoa, invertebrates and vertebrates (freshwater fish cell cultures) are discussed in

Volume 1 of this book. Of equal importance to my mind, Volume 2 of the book describes hazard assessment schemes that are based on combinations of the various bioassays, the so-called "battery" of tests. Indeed, all organisms are not similarly sensitive to given toxics. For instance, algae are likely to be very sensitive to herbicides albeit at levels which are unlikely to represent a danger to humans, while vertebrate cells may be less so. Thus, testing the sample on a series of organisms is more likely to reflect an overall toxicity. Whether one is to assess the risk to aquatic organisms or human beings, it is important to monitor the toxicity of samples on more than one trophic level.

Another significant advance is the development of a number of schemes to combine the results of toxicity testing on multiple trophic levels into indices which could be used to standardize results from one sample to another, from one area to another. ***Small-scale Freshwater Toxicity Investigations, Volumes 1 and 2*** presents a number of such schemes, and for this the editors should be congratulated. Only through such approaches can we begin to promote the use of these techniques more generally, especially if we are to encourage their use by field workers who have at best a limited experience of analytic laboratory techniques. Along with the other excellent chapters on hazard assessment schemes described in this book, the paper by Ronco, Castillo and Diaz-Baez *et al.* is significant to my mind because these authors have been working with municipal governments of Latin America (Argentina, Chile and Mexico) to promote WaterTox©. This is a battery of tests which they developed with colleagues elsewhere in Latin America, Canada, India and the Ukraine, with support from the International Development Research Centre (IDRC), the National Water Research Institute (Burlington, Ontario, Environment Canada) and the Saint-Lawrence Centre (Montreal, Quebec, Environment Canada). Results produced by this network of superb scientists have been extremely well received and, in some countries, governments are already incorporating batteries of bioassays in the national water quality testing programs (notably the Ukraine, Mexico and Chile).

All of this bodes very well for the future of bioassays, and for their transfer to poorer communities of the Third World where perhaps they are most needed.

Gilles Forget
Regional Director
In Central and West Africa
International Development Research Centre

Contents

Hazard assessment of solid media

OVERVIEW OF CONTEMPORARY TOXICITY TESTING

CHRISTIAN BLAISE
St. Lawrence Centre, Environment Canada
105 McGill Street, Montreal
Quebec H2Y 2E7, Canada
christian.blaise@ec.gc.ca

JEAN-FRANÇOIS FÉRARD
Université Paul Verlaine
Laboratoire Ecotoxicité et Santé Environnementale
CNRS FRE 2635, Campus Bridoux
rue du Général Delestraint
57070 METZ, France
ferard@sciences.univ-metz.fr

Preamble

In co-editing this book on ***Small-scale Freshwater Toxicity Investigations, Volumes 1 and 2*** we felt it would be of value to bring to light the numerous types of publications which have resulted from the development and use of laboratory bioassays over the past decades. Knowing why toxicity testing has been conducted is obviously crucial knowledge to grasp the importance and breadth of this field.

Our tracking of publications involving toxicity testing was carried out with several databases (Poltox, Current Contents, Medline, Biosis and CISTI: Canada Institute for Scientific and Technical Information) and key words tailored to our objectives. In undertaking our search of the literature, we exclusively circumscribed it to articles or reports dealing with toxicity testing performed in the context of freshwater environments – obviously the focus of this book. Excluded from this review are publications describing sub-cellular bioassays (*e.g.*, assays conducted with sub-mitochondrial particles or where specific enzymes are directly exposed to contaminants) and those carried out with recombinant DNA (micro)organisms (*e.g.*, promoter/reporter bacterial constructs) and biosensors. These essentially newer techniques are unquestionably of interest and will be called upon to play increasingly useful roles in the area of small-scale environmental toxicology in the future, but they are clearly beyond the primary aims of this book.

C. Blaise and J.-F. Férard (eds.), Small-scale Freshwater Toxicity Investigations, Vol. 2, 1-68.

While this review cannot be judged exhaustive, it is nevertheless representative of toxicity tests developed and applied at different levels of biological organization to comprehend toxic effects associated with the discharge of xenobiotics to aquatic environments. In reading this chapter, it is our hope that readers will get a broad sense of the versatile ways in which bioassays have been used by the scientific community at large and of the genuine role they play - along with other tools and approaches in ecotoxicology - in ensuring the protection and conservation of the freshwater aquatic environment.

Introduction

Laboratory toxicity tests have been developed and conducted over the past decades to demonstrate adverse effects that chemicals can have on biological systems. Along with other complementary tools of ecotoxicology available to measure (potential or real) effects on aquatic biota (*e.g.*, microcosm, mesocosm and field study approaches with assessment of a variety of structural and/or functional parameters), they have been, and continue to be, useful to indicate exposure-effect relationships of toxicants under defined, controlled and reproducible conditions (Adams, 2003).

Among their multiple uses, acute and chronic bioassays have served, for example, to rank and screen chemicals in terms of their hazardous potential, to undertake biomonitoring studies, to derive water quality criteria for safe release of single chemicals into aquatic bodies and to assess industrial effluent quality in support of compliance and regulatory statutes.

Because of the pressing contemporary need to assess an ever-growing number of chemicals and complex environmental samples, the development and use of small-scale toxicity tests (also called "micro-scale toxicity tests" or "microbiotests") have increased because of their attractive features. Simply defined as "a test involving the exposure of a unicellular or small multicellular organism to a liquid or solid sample in order to measure a specific effect", small-scale tests are generally simple to execute and characterized by traits which can include small sample volume requirements, rapid turnaround time to results, enhanced sample throughput and hence cost-effectiveness (Blaise et al., 1998a).

Small-scale toxicity tests are numerous and their relative merits (and limitations) for undertaking environmental assessment have been amply documented (Wells et al., 1998; Persoone et al., 2000). The small-scale toxicity tests methods described in this book and the hazard assessment schemes into which they can be incorporated are certainly representative of the field of small-scale aquatic toxicology and of tests and approaches being applied actively in today's world.

Our scrutiny of publications identified in the literature search has enabled us to uncover the various ways in which laboratory toxicity tests have been applied, many of which are small-scale in nature. We have assembled papers based on their application affinities and classified them into specific sections, as shown in Figure 1. This classification scheme essentially comprises the structure of this chapter and each section is subsequently commented hereafter.

Main categories of aquatic bioassay applications based on representative publications involving toxicity testing

1. Liquid media toxicity assessment

- **1.1 Environmental samples**
- **1.2 Chemical contaminants**
- **1.3 Biological contaminants**

2. Sediment toxicity assessment

- **2.1 Assessment of areas of concern**
- **2.2 Critical body residues and links to (sub)lethal toxicity responses**

3. Miscellaneous studies/initiatives linked to aquatic toxicity testing applications (liquid media and sediments)

- **3.1 Endeavors promoting development, validation and refinement of toxicity testing procedures**
  - **3.1.1 Test method development**
  - **3.1.2 Inter-calibration exercises**
  - **3.1.3 Comparative studies**
  - **3.1.4 Factors capable of affecting bioassay responses**
- **3.2 Initiatives promoting the use of toxicity testing procedures**
  - **3.2.1 Review articles, biomonitoring and HAS articles**
  - **3.2.2 Standardized test methods and guidance documents**

Figure 1. Presentation pathway for the overview on toxicity testing exposed in this chapter.

In discussing the developments and applications of bioassays to liquid media and to sediments, we have placed some emphasis on the types of chemicals and environmental samples that have been appraised, on the types and frequency of biotic level(s) employed, as well as on the relative use of single species tests as opposed to test battery approaches.

1. Liquid media toxicity assessment

1.1 ENVIRONMENTAL SAMPLES

Articles related to toxicity testing of waters, wastewaters and other complex media are separated into three groups: studies involving toxicity testing of wastewaters and solid waste leachates (Tab. 1); studies involving toxicity testing of specific receiving media and sometimes including wastewaters (Tab. 2); studies combining toxicity/chemical testing and sometimes integrating other disciplines to assess waters, wastewaters and solid waste leachates (Tab. 3). While some investigations have strictly sought to measure bioassay responses after exposure to (waste)waters (Tables 2 and 3), an equally important number have combined toxicity and chemical testing in an attempt to establish a link between observed effects and putative chemical stressors present in appraised samples (Tab. 3). In both cases, a wide variety of point source effluent wastewaters of diverse industrial and municipal origins, as well as solid matrix leachates and various receiving media have been assessed. On the industrial scene, pulp and paper wastewaters appear to have received more overall attention than other industrial sectors, very likely owing to the fact that the forestry industry is a major enterprise internationally. Historically, also, pulp and paper mills were notorious for their hazardous discharges to aquatic environments (Ali and Sreekrishnan, 2001), although secondary treatment application has greatly reduced their toxicity (Scroggins et al., 2002b).

Table 1. Studies involving toxicity testing of wastewaters and solid waste leachates.

Assessment category	***Type of bioanalytical application***[a]	***Biotic levels employed***[b,c] ***(and reference)***
Industrial effluents		
Dyeing factory	TT	B (Chan et al., 2003)
Electrical utilities	TBA	B,F,I (Rodgers et al., 1996)
Metal plating	TT	P (Roberts and Berk, 1993)
	TBA	B,F,I (Choi and Meier, 2001)
Mining	TT	B,B,B (Gray and O'Neill, 1997); F (Gale et al., 2003)
	TBA	B,B,F,I,I,I,I (CANMET, 1996); A,A,B,F,F,I,L (CANMET, 1997b); I,F (CANMET, 1998); Bi,F,I,I (Milam and Farris, 1998); A,F,I,L (Scroggins et al., 2002a);
Oil refinery	TT	B (Riisberg et al., 1996)
	TBA	A,A,F (Roseth et al., 1996); A,B,F,F,I,I,I,L,S (Sherry et al., 1997)
Pulp and paper	TT	F (Gagné and Blaise, 1993); B (Oanh, 1996); F (Bennett and Farrell, 1998); F (Parrott et al., 2003); F (Sepúlveda et al., 2003); F (van den Heuvel and Ellis, 2002)
	TBA	A,B,F (Blaise et al., 1987); B,B,B,I (Rao et al., 1994); A,B,L (Oanh and Bengtsson, 1995); A,B,B,F,I (Ahtiainen et al., 1996); A,B,F,F (Priha, 1996); B,F,F,I,I,I,I (Côté et al., 1999); A,F,F,I (Scroggins et al., 2002b); B,I (Pintar et al., 2004)
Tannery	TT	B,B (Diaz-Baez and Roldan, 1996)
	TBA	A,B,I,I,I,I,I (Isidori, 2000)
Textile	TT	I (Villegas-Navarro et al., 1999)

Table 1 (cont.). Studies involving toxicity testing of wastewaters and solid waste leachates.

Assessment category	***Type of bioanalytical application[a]***	***Biotic levels employed[b,c] (and reference)***
Industrial effluents		
Various effluents	TT	F (Blaise and Costan, 1987); B (Tarkpea and Hansson, 1989); B (Svenson et al., 1992); I (Seco et al., 2003)
	TBA	B,F,F,F,F,F,I (Williams et al., 1993); B,F,I (Gagné and Blaise, 1997); B,I,I (Jung and Bitton, 1997); B,I (Liu et al., 2002)
Wood industry	TT	F (Rissanen et al., 2003)
Municipal effluents	TT	B,B,B,B,B (Codina et al., 1994); I (Monda et al., 1995); Fc (Gagné and Blaise, 1998a); Fc (Gagné and Blaise, 1999); B (Sánchez-Mata et al., 2001)
	TBA	B,B,I (Arbuckle and Alleman, 1992); A,B,F,P (George et al., 1995); B,B,F,Fc (Dizer et al., 2002); F,I (Gerhardt et al., 2002a)
Municipal and industrial effluents	TT	B (Asami et al., 1996); Fc (Gagné and Blaise, 1998b) ; Fc,Fc,F (Gagné and Blaise, 1998c)
	TBA	F,F,I,I,I (Fisher et al., 1989); F,F,I,I,I (Fisher et al., 1998); B,I (Doherty et al., 1999); B,F,I,I,S (Castillo et al., 2000); A,A,B,I,I,P (Manusadžianas et al., 2003)
WWTP (waste water treatment plants)	TT	B (Hoffmann and Christofi, 2001); B (Paixão and Anselmo, 2002)
	TBA	B,F,I (Sweet et al., 1997)
Solid waste leachates	TT	A (McKnight et al., 1981); B (Bastian and Alleman, 1998); B (Coz et al., 2004)
	TBA	B,B,B,F,F,I,I (Day et al., 1993); A,B,I,I,I,L,P (Clément et al., 1996); A,B,I,I,Pl,Pl,Pl (Ferrari et al., 1999); A,I,I,P (Törökné et al., 2000); A,A,B,B,I,I,P,S (Sekkat et al., 2001)

a) TT (toxicity testing): a study undertaken with test(s) at only one biotic level. TBA (test battery approach): a study involving tests representing two or more biotic levels.
b) Levels of biological organization used in conducting (or describing) TT: A (algae), B (bacteria), Bi (bivalve), F (fish), Fc (fish cells), I (invertebrates), L (*Lemnaceae*, duckweed: small vascular aquatic floating plant), P (protozoans), Pl (plant), and S (seed germination test with various types of seeds, *e.g.*, *Lactuca sativa*).
c) A study reporting the use of more than one toxicity test at the same biotic level is indicated by additional lettering (*e.g.*, use of three different bacterial tests is coded as "B, B, B".

Table 2. Studies involving toxicity testing of specific receiving media and sometimes including wastewaters.

Assessment category	***Type of bioanalytical application***[a]	***Biotic levels employed***[b,c] ***(and reference)***
Groundwater	TBA	A,B,B,I (Dewhurst et al., 2001)
Lake	TT	I (Kungolos et al., 1998)
	TBA	A,B,B,I,S (Okamura et al., 1996); A,I (Angelaki et al., 2000)
River/Stream	TT	I (Viganò et al., 1996); Bi,I (Stuijfzand et al., 1998); I (Jooste and Thirion, 1999); I (Lopes et al. 1999); I,I (Pereira et al., 1999); I (Sakai, 2001); I (Schulz et al., 2001); A (Okamura et al., 2002); I (Sakai, 2002a); I (Williams et al., 2003)
	TBA	A,B,F,I (Wilkes and Beatty-Spence, 1995); B,B,B,I,I (Dutka et al., 1996); A,F,F,I,L (CANMET, 1997c); A,I (Baun et al., 1998); B,B,I (Sabaliunas et al., 2000); A,B,I,I,I (Van der Wielen and Halleux, 2000)
Wetland	TT	B (Dieter et al., 1994)
Specific types of environmental samples		
Packaged water	TT	P (Sauvant et al., 1994)
Pond	TT	I,I,I (Lahr, 1998)
Rainwater	TT	I (Sakai, 2002b)
Rice field	TBA	A,I (Cerejeira et al., 1998)
Runoff water	TT	A (Wong et al., 2001); I (Boulanger and Nikolaidis, 2003)
	TBA	B,B,I (Marsalek et al., 1999); A,B (Heijerick et al., 2002)
Diverse types of environmental samples[d]	TT	B (Coleman and Qureshi, 1985); I (Samaras et al., 1998); I (Lechelt, 2000); A (Graff et al., 2003); Fc (Schweigert et al., 2002)

Table 2 (continued). Studies involving toxicity testing of specific receiving media and sometimes including wastewaters.

Assessment category	Type of bioanalytical application[a]	Biotic levels employed[b,c] (and reference)
Diverse types of environmental samples [d]	TBA	B,B,I (Cortes et al., 1996); B,I (Pardos et al., 1999a); A,I,I,L,P (Blinova, 2000); A,I,I,P (Czerniawska-Kusza and Ebis, 2000); A,I,I,P (Dmitruk and Dojlido, 2000); A,I,I,I (Isidori et al., 2000); B,I,I,P (Stepanova et al., 2000) A,I,I,S,S (Arkhipchuk and Malinovskaya,2002); A,I,I,S (Diaz-Baez et al., 2002); A,I,I (Mandal et al., 2002); A,I,I,S (Ronco et al., 2002)

a) TT (toxicity testing): a study undertaken with test(s) at only one biotic level. TBA test battery approach): a study involving tests representing two or more biotic levels.
b) Levels of biological organization used in conducting (or describing) TT: A (algae), B (bacteria), Bi (bivalve), F (fish), Fc (fish cells), I (invertebrates), L (*Lemnaceae*, duckweed: small vascular aquatic floating plant), P (protozoans), and S (seed germination test with various types of seeds, *e.g.*, *Lactuca sativa*).
c) A study reporting the use of more than one toxicity test at the same biotic level is indicated by additional lettering (*e.g.*, use of three different bacterial tests is coded as "B, B, B".
d) Includes samples such as potable/surface waters, as well as industrial effluents, soil/sediment/sludge extracts, landfill leachates and snow, where individual studies report testing one or more sample type(s).

Table 3. Studies combining toxicity/chemical testing and sometimes integrating other disciplines to assess waters, wastewaters and solid waste leachates.

Assessment category	Type of bioanalytical application[a]	Biotic levels employed[b,c] (and reference)
Industrial effluents		
Chemical plant	TT	B (Chen et al., 1997)
	TBA	B,I,I,I (Guerra, 2001)
Coal industry	TBA	A,I,I,I (Dauble et al., 1982); F,I,I (Becker et al., 1983)
Coke	TBA	A,B (Peter et al., 1995)
Complex munitions	TBA	A,A,A,A,F,F,F,F,I,I,I,I (Liu et al., 1983)
Mining	TT	I,I (Fialkowski et al., 2003)
	TBA	F,I (Erten-Unal et al., 1998); A,B (LeBlond and Duffy, 2001)
Pharmaceutical	TBA	A,B,B,B,F,I (Brorson et al., 1994); B,I (Tišler and Zagorc-Koncan, 1999)

Table 3 (continued). Studies combining toxicity/chemical testing and sometimes integrating other disciplines to assess waters, wastewaters and solid waste leachates.

Assessment category	***Type of bioanalytical application***[a]	***Biotic levels employed***[b,c] ***(and reference)***
Industrial effluents		
Pulp and paper	TBA	B,I,F (Dombroski et al., 1993); B,F,I (Leal et al., 1997); B,F,I (Middaugh et al., 1997); A,B,B,F,I (Ahtiainen et al., 2000); B,I,I,P,P (Michniewicz et al., 2000)
Resin production	TBA	A,B,F,I (Tišler and Zagorc-Koncan, 1997)
Tannery	TT	I,I (Cooman et al., 2003)
	TBA	B,I (Fernández-Sempere et al., 1997); B,I (Font et al., 1998)
Tobacco plant	TBA	A,B,B,B,B,P,P (Sponza, 2001)
Water based drilling muds	TBA	A,I (Terzaghi et al., 1998)
Oily waste		
Olive oil	TBA	B,I,I (Paixão et al., 1999)
Oil refinery	TT	B (Aruldoss and Viraraghavan, 1998)
	TBA	A,B,B,F,F,I,I,I,L,S (Sherry et al., 1994); B,F,I (Bleckmann et al., 1995)
Oil-shale	TT	B,B,B (Kahru et al., 1996)
	TBA	B,B,I,I,I,P (Kahru et al., 1999); A,B,B,B,I,I,I,I,P (Kahru et al., 2000)
Composting oily waste	TBA	B,B,B,B,B,I,I,I,L,S (Juvonen et al., 2000)
Municipal effluents	TT	B (Pérez et al., 2001)
	TBA	B,B,Pl,Pl,S (Monarca et al., 2000)
WWTP (waste water treatment plant)	TT	B (Chen et al., 1999); I (Kosmala et al., 1999); B,B,B (Gilli and Meineri, 2000); B (Svenson et al., 2000); B (Wang et al., 2003)
	TBA	F,I (Fu et al., 1994); A,Fc,I (Pablos et al., 1996); B,B,B,B,P (Ren and Frymier, 2003)
Leachates		
From agricultural production solid waste	TT	B (Redondo et al., 1996)
From industrial solid waste	TT	L (Jenner and Janssen-Mommen, 1989); B (Coya et al., 1996); I,I (Rippon and Riley, 1996); I,I,I,I,I,I (Canivet and Gibert, 2002)

Table 3 (continued). Studies combining toxicity/chemical testing and sometimes integrating other disciplines to assess waters, wastewaters and solid waste leachates.

Assessment category	***Type of bioanalytical application***[a]	***Biotic levels employed***[b,c] ***(and reference)***
Leachates		
From industrial solid waste	TBA	A,B,I (Lambolez et al., 1994); B,B,B,B,L,S,S,S (Joutti et al., 2000); A,B,I (Malá et al., 2000); A,B,B,I (Vaajasaari et al., 2000)
From municipal solid waste	TBA	A,A,B,I,I,S (Latif and Zach, 2000); A,B,B,F,I,I (Rutherford et al., 2000); A,B,I (Ward et al., 2002a)
Miscellaneous types of environmental samples [d]	TT	I (Gasith et al., 1988); I (Doi and Grothe, 1989) B (Bitton et al., 1992); I (Jop et al., 1992); A (Wong et al., 1995); B (Hao et al., 1996); I (Blaise and Kusui, 1997); B,B (Hauser et al., 1997); I (Eleftheriadis et al., 2000); F (Liao et al., 2003); I (Kszos et al., 2004); A,I,I,P,S (Latif and Licek, 2004)
	TBA	F,I,I (Tietge et al., 1997); A,B,I,I,I (Kusui and Blaise, 1999); A,A,I,I,P (Manusadžianas et al., 2000)
Natural waters		
Floodplain	TBA	B,I,I,I,I (de Jonge et al., 1999)
Groundwater	TBA	A,B,I,P,P,P (Helma et al., 1998); B,F,I (Gustavson et al., 2000)
Rivers and streams	TT	A (Guzzella and Mingazzini, 1994); Bi,I,I (Crane et al., 1995); I (Bervoets et al., 1996); A,A (O'Farrell et al., 2002)
Wetland	TT	B (Boluda et al., 2002)

a) TT (toxicity testing): a study undertaken with test(s) at only one biotic level. TBA (test battery approach): a study involving tests representing two or more biotic levels.
b) Levels of biological organization used in conducting (or describing) TT: A (algae), B (bacteria), Bi (bivalve), F (fish), Fc (fish cells), I (invertebrates), L (*Lemnaceae*, duckweed: small vascular aquatic floating plant), P (protozoans), Pl (plant), and S (seed germination test with various types of seeds, *e.g.*, *Lactuca sativa*).
c) A study reporting the use of more than one toxicity test at the same biotic level is indicated by additional lettering (*e.g.*, use of three different bacterial tests is coded as "B, B, B".
d) Includes samples such as storm waters, river waters, as well as industrial/municipal effluents, sludge extracts, where individual studies report testing one or more sample type(s).

While it is beyond our intent to discuss the main purpose(s) that prompted research groups to conduct individual investigations with particular toxicity tests, readers can access this information by consulting references of interest. Others are mentioned hereafter, however, to indicate bioanalytical endeavors that have taken place in past years. For example, Bitton et al. (1992), after developing a metal-specific bacterial toxicity assay, demonstrated its capacity to correctly pinpoint heavy-metal containing industrial wastewaters. In another venture, Roberts and Berk (1993) were motivated to undertake toxicity testing of a metal plating effluent and of a series of (in)organic chemicals in order to further validate a newly-developed protozoan chemo-attraction assay. Again, a test battery approach with chemical support to assess a coke plant effluent identified treatment methods that were superior for decontaminating the wastewater (Peter et al., 1995). In toxicity testing of tannery industry effluent samples, bacterial tests were shown to be sufficiently sensitive to act as screening tools for such wastewaters (Diaz-Baez and Roldan, 1996). In a study conducted on industrial, municipal and sewage treatment plants, toxicity testing identified chlorination as the most important contributor of toxic loading to the receiving environment (Asami et al., 1996). After a comprehensive assessment of pulp and paper mills, toxicity testing proved useful to ameliorate mill process control (Oanh, 1996). Another study conducted with three bacterial toxicity tests showed that oil-shale liquid wastes could be bio-degraded when activated sludge was pre-acclimated to phenolic wastewaters (Kahru et al., 1996). Petrochemical plant assessment using toxicity testing, chemical analysis and a TIE/TRE strategy combined to identify aldehydes as the main agent of effluent toxicity (Chen et al., 1997). Test battery assessment of a mine water discharge, which involved both toxicity testing and in-stream exposure of bivalves, helped to set a no-effect level criterion for a bioavailable form of iron (Milam and Farris, 1998). A comparison of laboratory toxicity testing and *in situ* testing of river sites downstream from an acid mine drainage demonstrated good agreement between the two approaches for the most contaminated stations (Pereira et al., 1999). A similar strategy to assess gold and zinc mining effluents confirmed the reliability of some chronic assays for routine toxicity monitoring (LeBlond and Duffy, 2001). Clearly, there are numerous reasons for conducting toxicity testing and/or chemical analysis of (waste)waters to derive relevant information that have eventually triggered enlightened decisions contributing to their improvement.

Of the 188 studies reported in Tables 1, 2 and 3, more than half (n = 101) were conducted with two or more tests representing at least two biotic levels (*i.e.*, test battery approach or TBA), as opposed to those performed with a single biotic level (n = 87). While test and biotic level selection may be based on a variety of reasons and study objectives (*e.g.*, practicality, cost, personnel availability), preference for TBAs can also be influenced by the need to assess hazard at different levels so as not to underestimate toxicity. Indeed, contaminants can demonstrate "trophic-level specificity" (*e.g.*, phytototoxic effects of herbicides) or they can exert adverse effects at multiple levels (*e.g.*, particular sensitivity of cladocerans toward heavy metals in contrast to bacteria). When TBAs are used, they are mostly conducted with two, three or four trophic levels (Tab. 4).

Whether TT (toxicity testing with single species tests at the same biotic level) or TBAs are performed, some test organisms have been more frequently used than others (Tab. 5). Invertebrates have been the most commonly employed, as had been pointed out in an earlier literature survey conducted between 1979 and 1987 (Maltby and Calow, 1989). Bacteria as well as fish and algal assays come next in frequency of use. Early standardization of invertebrate (*e.g.*, *Daphnia magna*) and bacterial test (*e.g.*, *Vibrio fischeri* luminescence assay) procedures, as well as increased miniaturization and cost-effectiveness, are likely factors explaining their popularity over the past decades. While some groups of small-scale toxicity tests (*i.e.*, fish cell, duckweed and protozoan tests) have thus far received less attention to appraise various environmental samples, recent efforts in test procedure validation and standardisation should effectively promote their use in the future (see Volume 1, Chapters 7, 8, 14 and 15).

Table 4. Frequency of the number of biotic levels employed in test battery approaches (TBAs) for complex liquid media assessment based on the 101 TBA papers classified in Tables 1-3.

TBA studies undertaken with:	***Number and frequency (%)***
Two biotic levels	39/101 (38.6)
Three biotic levels	38/101 (37.6)
Four biotic levels	19/101 (18.8)
Five biotic levels	3/101 (3)
Six biotic levels	2/101 (2)

Table 5. Frequency of use of specific biotic levels employed in toxicity testing (TT) and test battery approaches (TBA) for complex liquid media assessment based on the 188 papers classified in Tables 1-3.

TT and TBA studies undertaken with:	**Number and frequency (%)**
Algae	70/553* (12.7)
Bacteria	152/553 (27.5)
Bivalves	3/553 (< 1)
Fish	68/553 (12.3)
Fish cells	8/553 (1.5)
Invertebrates	199/553 (36.0)
Lemnaceae (duckweed)	10/553 (1.8)
Plants	3/553 (< 1)
Protozoans	23/553 (4.2)
Seeds	15/553 (2.7)

*Total number of single species tests reported in the 188 papers classified in Tables 1-3 (= sum of number of A, B, Bi, F, Fc, I, L, P, Pl, S tests indicated in the "Biotic levels employed" column).

1.2 CHEMICAL CONTAMINANTS

It has been estimated that as many as 250,000 man-made chemicals could possibly enter different compartments of the biosphere and cause adverse effects on ecosystem and human health (OSPAR, 2000). Out of concern for ensuring the protection of aquatic biota, a large number of scientists internationally have turned to bioassays as primary means of assessing the hazard (and risk) posed by these substances. Indeed, the scientific literature abounds with hundreds of publications dealing with toxicity testing of various classes of (in)organic chemicals. While it is beyond the intentions of this chapter to discuss all of these, papers have been selected that reflect the types of chemicals having undergone toxicity assessment. In general, published articles show that test organisms and biotic levels described are the same as those employed for assessing environmental samples.

Representative investigations involving toxicity assessment of metals, ions and oxidizing agents are highlighted in Table 6. Varied toxicological objectives have been pursued to evaluate metals singly or in groups of two or more with one toxicity test or with a test battery. The benefits of these initiatives to enhance our knowledge of undesirable effects that can be directed toward specific biotic levels (*e.g.*, Holdway et al., 2001), to identify useful sentinel species (*e.g.*, Madoni, 2000), or to promote useful (Couture et al., 1989) or potentially safer clean-up technologies (Leynen et al., 1998) should be fairly obvious.

Table 6. Studies involving toxicity assessment of metals, ions and oxidizing agents.

Assessment category	***Type of bioanalytical application***[a]	***Biotic levels employed***[b,c] ***(and reference)***
One metal:		
Aluminium	TT: four species of invertebrates are exposed to Al over a pH range of 3.5 to 6.5.	I,I,I,I (Havas and Likens, 1985)
Cadmium	TT: a simple microcosm experiment associating two biotic levels conducted in a Petri dish allows measurement of reproduction effects on daphnids following Cd contamination of either their food source (algae) or of their water medium.	I (Janati-Idrissi et al., 2001)
Chromium (Cr^{+6})	TT: luminescent bacteria are exposed to assess the influence of pH speciation of chromium on toxicity response.	B (Villaescusa et al., 1997)

Table 6 (continued). Studies involving toxicity assessment of metals, ions and oxidizing agents.

Assessment category	***Type of bioanalytical application***[a]	***Biotic levels employed***[b,c] ***(and reference)***
One metal:		
Copper	TT: comparison of effects occurring at molecular (DNA profiling) and population (ecological fitness parameters including acute and chronic toxicity) levels for *Daphnia magna.*	I (Atienzar et al., 2001)
Gallium	TT: assessment of inter-metallic elements used in making-high speed semiconductors such as gallium arsenic with *Cyprinus carpio.*	F (Yang and Chen, 2003)
Lead	TBA: assessment of toxicity, uptake and depuration of lead in fish and invertebrate species.	F,F,I,I (Oladimeji and Offem, 1989)
Manganese	TT: assessment at three levels of water hardness with *Ceriodaphnia dubia* and *Hyalella azteca.*	I,I (Lasier et al., 2000)
Mercury	TT: assessment of 10 mercury compounds to determine their relative toxicities to luminescent bacteria.	B (Ribo et al., 1989)
Nickel	TT: assessment with 12 species of freshwater ciliates to determine which could become, based on observed sensitivity, a good bio-indicator of waters polluted by heavy metals.	P (Madoni, 2000)
Selenium	TT: assessment of selenium compounds and relationships with uptake in an invertebrate species.	I (Maier and Knight, 1993)
Silver	TBA: assessment of toxicity to fish and invertebrates under a variety of water quality conditions.	F,I (La Point et al., 1996)
Uranium	TT: assessment of depleted uranium on the health and survival of *C. dubia* and *H. azteca.*	I,I (Kuhne et al., 2002)
Zinc	TT: assessment the influence of various ions and pH on phytotoxicity response.	A (Heijerick et al., 2002)

Table 6 (continued). Studies involving toxicity assessment of metals, ions and oxidizing agents.

Assessment category	***Type of bioanalytical application***[a]	***Biotic levels employed***[b,c] ***(and reference)***
One metal:		
Zirconium	TBA: assessment of zirconium ($ZrCl_4$), considered of use as a P-precipitating agent to reduce the eutrophication potential of pig manure wastes to receiving environments.	A,B,F (Couture et al., 1989)
Two metals:		
Cadmium, Zinc	TT: assessment of their acute and chronic toxicity to two *Hydra* species.	I,I (Holdway et al., 2001)
Three metals:		
Arsenic, Cobalt, Copper	TT: assessment of relationships between acute toxicity and various experimental variables (*e.g.*, metal concentration in water, time of exposure, bioconcentration factor) with two fish species.	F,F (Liao and Lin, 2001)
Four metals or more:	TT: assessment of the adequacy of cultured fish cells (Bluegill BF-2) for toxicity testing of aquatic pollutants.	Fc (Babich and Borenfreund, 1987)
Ions:	TT: assessment of the phytotoxicity of high density brines (calcium chloride and calcium bromide) to *L. minor*.	L (Vujevic et al., 2000)
Rare earth elements:	TT: assessment of the aquatic toxicity of rare earth elements (La, Sm, Y, Gd) to a protozoan species.	P (Wang et al., 2000)
Oxidizing agents:	TBA: assessment of the acute toxicity of ozone, an alternative to chlorination to control biofouling in cooling water systems of power plants, to fish larvae of three species and to *D. magna*.	F,F,F,I (Leynen et al., 1998)

a) TT (toxicity testing): a study undertaken with test(s) at only one biotic level. TBA (test battery approach): a study involving tests representing two or more biotic levels.

b) Levels of biological organization used in conducting (or describing) TT: A (algae), B (bacteria), F (fish), Fc (fish cells), I (invertebrates), L (*Lemnaceae*, duckweed: small vascular aquatic floating plant), and P (protozoans).

c) A study reporting the use of more than one toxicity test at the same biotic level is indicated by additional lettering (*e.g.*, use of three different bacterial tests is coded as "B, B, B".

The toxicological properties of chemicals representing various classes and structures of organic substances have also been assessed by a series of bioassays at different levels of biological organization (Tab. 7). Featured in this table is but the tip of the iceberg in terms of the types of studies that have been conducted to further our knowledge about the hazards of anthropogenic molecules. While industrial progress has markedly enhanced the quality of life on this planet through production of countless xenobiotics synthesized for multiple human uses (*e.g.*, diverse household products and pharmaceuticals), it has also increased the risk linked to their discharge and fate in aquatic systems. Understanding their potential for adverse effects through the conduct of bioassays is clearly a first step in the right direction.

Table 7. Examples of studies involving toxicity assessment of organic substances.

Assessment category (and product tested)	***Type of bioanalytical application[a], biotic levels employed[b,c] (and reference)***
Acaricide (Tetradifon)	TT: I (Villarroel et al., 1999)
Adjuvants (several used as surfactants for aquatic herbicide applications)	TT: F (Haller and Stocker, 2003)
Anti-fouling paint (TBT)	TBA: A,I (Miana et al., 1993)
Aromatic hydrocarbon (*para*-methylstyrene)	TBA: A,F,I (Baer et al., 2002)
Cationic fabric softener (DTDMAC)	TBA: A,B,B,I,I,I (Roghair et al., 1992)
Chelator ([S,S]-EDDS)	TBA: A,A,F,I (Jaworska et al., 1999)
Detergents and softeners (26 detergents and 5 softeners)	TT: I (Pettersson et al., 2000)
De-icing / anti-icing fluids	TT: B (Cancilla et al., 1997)
Disinfectant (Mono-chloramine)	TBA: F,I (Farrell et al., 2001)
Dyes (Fluorescein sodium salt, Phloxine B)	TT: I (Walthall and Stark, 1999)
Fatty acids (C_{14} to C_{18})	TT: A (Kamaya et al., 2003)
Fire control substances (Fire-Trol GTS-R and LCG-R, Phos-Chek D75-F and WD-881, Silv-Ex)	TBA: A,I (McDonald et al., 1996)
Flame retardant (Brominated diphenyl ether-99)	TBA: A,I (Evandri et al., 2003)
Fungicide (Ridomil plus 72)	TBA: F,I (Monkiédjé et al., 2000)
Herbicide (Atrazine)	TT: I (Dodson et al.,1999)
Household products (Abrasives, additives, disinfectants)	TBA: A,B,B,F,F,I (Bermingham et al., 1996)
Insecticide (Glyphosate)	TT: L (Lockhart et al., 1989)

Table 7 (cont.). Examples of studies involving toxicity assessment of organic substances.

Assessment category (and product tested)	*Type of bioanalytical application[a], biotic levels employed[b,c] (and reference)*
Lubricant additives (Ashless dispersant A and B, Zinc dialkyldithiophosphate)	TT: A (Ward et al., 2002b)
(Tri *n*-butyl phosphate)	TBA: A,B (Michel et al., 2004)
Nitromusks (Ambrette, Setone, Moskene,Tibetene, Xylene)	TBA: A,B,I (Schramm et al., 1996)
Narcotics (*n*-alkanols)	TT: B (Gustavson et al., 1998)
Organochlorides (PCBs)	TT: B (Chu et al., 1997)
Organosulfur compounds (several benzothiophenes)	TBA: B,I (Seymour et al., 1997)
Pesticide (Cyromazine)	TT: I,I (Robinson and Scott, 1995)
Pharmaceutical compound (β-Blockers)	TBA: F,I,I,I (Huggett et al., 2002)
Phenolic compounds (Pentachlorophenol)	TBA: A,B,I,S (Repetto et al., 2001)
Phtalate esters (several)	TT: I,I,I (Call et al., 2001)
Solvents (Mono-, Di- and Tri PGEs)	TBA: A,B,F,F,F,I,I,L (Staples and Davis, 2002)
Surfactant (Genapol OX-80)	TT: A (Anastácio et al., 2000)
Volatilecompounds (*N*-nitrosodiethylamine, *N*-nitrosodimethylamine)	TBA: A,A,F,I,I (Draper III and Brewer, 1979)
Wood preservative (Bardac 2280)	TBA: F,F,F,F,I,I,I,I (Farrell et al., 1998)

a) TT (toxicity testing): a study undertaken with test(s) at only one biotic level. TBA (test battery approach): a study involving tests representing two or more biotic levels.
b) Levels of biological organization used in conducting (or describing) TT: A (algae), B (bacteria), F (fish), I (invertebrates), L (Lemnaceae, duckweed: small vascular aquatic floating plant) and S (seed germination test with various types of seeds, *e.g.*, Lactuca sativa).
c) A study reporting the use of more than one toxicity test at the same biotic level is indicated by additional lettering (*e.g.*, use of three different bacterial tests is coded as "B, B, B".

Several papers have also reported toxicity data for a variety of metals and organic substances simultaneously. Reasons for conducting such investigations include 1) establishing the concentrations at which chemicals exert their adverse effects (*e.g.*, at the ng/L, μg/L or mg/L levels), 2) estimating environmental risk based on measured toxicity endpoints and predicted environmental concentrations for specific chemicals and 3) defining toxicant concentrations harmful for specific biotic levels and/or assemblages of species within each level.

Studies have assessed the toxicological properties of one or more heavy metal(s) with one or more organic substance(s). Examples include copper and diazinon (van der Geest et al., 2000), cadmium and pentachlorophenol (McDaniel and Snell, 1999), several heavy metals (Cd, Cu, Ni, Pb, Zn) and organic (Chlorpyrifos, DDT, DDD, DDE, Dieldrin) toxicants (Phipps et al., 1995), and two metals (Cu, Zn) and eight surfactants (Dias and Lima, 2002). Again, test organisms employed for toxicity assessment are similar to those discussed previously and investigators make use of one or more biotic levels to undertake their evaluations.

Chemical toxicity assessment should also take into consideration the combined effects that groups of chemicals can have on living organisms. Indeed, contaminants are not discharged singly in aquatic systems but are joined by many others whose composition will depend on the origin of (non)point sources of pollution affecting particular reaches of receiving waters (*e.g.*, industrial, municipal and agricultural sources). The recognition that groups of chemicals can interact together to produce a resulting effect that can reduce (antagonistic effect) or exacerbate (synergistic effect) that of substances tested singularly has prompted scientists to appraise the toxicity characteristics of mixtures.

Published articles indicate that work has focussed on (binary, ternary, etc.) mixtures including metals, organics as well as metal/organic cocktails. For metals, examples include toxicity testing of various mixtures with algae (Chen et al., 1997), bacteria (Mowat and Bundy, 2002) and micro-invertebrates (Burba, 1999). For organics, mixtures have been assessed belonging to groups such as antifouling agents (Fernandés-Alba et al., 2002), herbicides (Hartgers et al., 1998), pesticides (Pape-Lindstrom and Lydy, 1997), and manufactured munitions (Hankenson and Schaeffer, 1991). For (in)organic mixtures, metal/pesticide (Stratton, 1987), metal/composted manure (Ghosal and Kaviraj, 2002), as well as metal/miscellaneous organic (Parrott and Sprague, 1993) combinations offer additional examples of interaction assessments. Because appraising mixtures of compounds (singularly and in binary, ternary or other combinations) is more laborious in time and effort than for single compounds, toxicity testing has, in most cases, been conducted with a single test organism, as opposed to the use of a test battery. Algal, bacterial and micro-invertebrate tests have thus far been favoured in this respect.

Another active field of research intended to estimate the toxic properties of organic compounds lies in the determination of their quantitative structure-activity relationships (QSAR). The rationale for this work is based on the fact that molecules will enter living organisms to exert adverse effects depending on their elemental composition and structure. In brief, QSARs are regression equations relating toxicological endpoints (*e.g.*, LC50s, EC50s, IC50s, NOECs) to physicochemical properties within a class of compounds. A good number of QSARs, for example, are determined with the octanol-water coefficient (K_{ow}), a well-known predictor of the tendency of a compound to be bio-accumulated. QSARs have several potential uses, some of which include 1) predicting the effects of newly-synthesized chemicals, 2) priority ranking of chemicals destined for more elaborate toxicity testing, 3) assistance in deriving water quality guidelines and 4) rapidly estimating toxicity for specific compounds when toxicity test data are unavailable (Environment Canada, 1999).

A quantitative structure-activity relationship (QSAR), for example, has been shown for aliphatic alcohols, where 96h-LC50s for fathead minnows are related to their K_{ow} status (Veith et al., 1983). Other QSARs based on K_{ow} have been reported for several classes of organics with test species including algae, invertebrates and fish (Suter, 1993). Hydrophobicity-based QSARs were also generated for fish and invertebrates with a set of 11 polar narcotics (Ramos et al., 1998) and for bacteria, fish and protozoan test organisms with a large set of (non)polar narcotic classes of chemicals (Schultz et al., 1998). QSARs were also employed to predict the biodegradation, bioconcentration and toxicity potential of more than 5000 xenobiotics (industrial chemicals, pesticides, food additives and pharmaceuticals) having a potential for release into the Great lakes basin (Walker et al., 2004). This study, in particular, illustrates the usefulness of QSARs as a cost-effective pre-screening adjunct to (significantly more expensive) monitoring studies that can then be prioritized towards those chemicals having the potential to persist and bio-accumulate in aquatic species. In these and other recent QSAR-based investigations of chemicals (Junghans et al., 2003; Choi et al., 2004; Schultz et al., 2004), it is noteworthy to mention that small-scale toxicity tests conducted with algae, bacteria, invertebrates and protozoans are used frequently.

1.3 BIOLOGICAL CONTAMINANTS

Besides the many hazards looming on aquatic life owing to the uncontrolled discharge of a myriad of chemicals, exposure to plants or microbes may also place it at risk. Indeed, toxicity tests conducted within the last decade on plant substances/extracts, and on microbes or their products (*e.g.*, metabolites), to investigate their biopesticide or toxicity potential, have indicated that species of different levels of biological organization can be adversely affected by such biological contaminants (Tab. 8). Since undesirable ecological effects to aquatic communities could result from exposure to naturally-produced chemicals or micro-organisms, documenting their toxicity potential via bioassays is fully justified.

As future applications with natural and/or genetically-modified plants and micro-organisms are expected to increase in the future (*e.g.*, for bioremediation treatments of contaminated soils, wastewaters, sediments), so will toxicity assessment programs to insure the protection of aquatic biota. In Canada, for example, information is now required to appraise new microbes (and their products) in terms of their toxicity potential toward aquatic organisms, and standardized toxicity test methods are being developed and recommended for this purpose (Environment Canada, 2004a). Risk assessment of biological contaminants is clearly an area that will receive sustained attention in the coming years.

Table 8. Examples of studies involving toxicity assessment of biological contaminants.

Assessment category and product tested	***Type of bioanalytical application[a], biotic levels employed[b,c] (and reference)***
Biopesticides	
Aquatic plant: essential oils from *Callicarpa americana*	TBA: A,A,A,B,B,B,B,B,S,S (Tellez et al., 2000)
Aquatic plant: phenanthrenoids from *Juncus acutus*	TT: A (DellaGreca et al., 2002)
Aquatic plant: essential oils from *Lepidium meyenii*	TBA: A,A,I,S,S (Tellez et al., 2002)
Aquatic plant: antialgal furano-diterpenes from *Potamogetonaceae*	TT: A (DellaGreca et al., 2001)
Aquatic plant: ent-labdane diterpenes from *Potamogetonaceae*	TBA: A,I,I,I,I (Cangiano et al., 2002)
Bacterium: *Bacillus thuringiensis*	TT: I (Manasherob et al., 1994); TT: I (Kondo et al., 1995)
Fungus: *Metarhizium anisopliae*	TT: B (Milner et al., 2002)
Biotoxins	
Cyanobacteria	
Microcystis aeruginosa	TBA: B,I (Campbell et al., 1994)
Anabaena sp., *M. aeruginosa*, *Microcystis* sp., *P. aghardii*, *P. rubenscens*	TT: I (Törökné, 2000; Törökné et al., 2000)
M. aeruginosa, *M. wesenbergii*	TBA: B,B,B,I,I,I,I,P (Maršálek and Bláha, 2000)
Cyanobacterial blooms	TBA: I,I,P,P (Tarczynska et al., 2000)
Pathogenic bacteria: *Aeromonas hydrophila*, *Flavobacter* spp., *Flexibacter columnaris*	TT: F (Geis et al., 2003)
Odor and taste compounds of microbial origin	
Geosmin, 2-methyliso-borneol	TT: Fc (Gagné et al. 1999)

a) TT (toxicity testing): a study undertaken with test(s) at only one biotic level. TBA (test battery approach): a study involving tests representing two or more biotic levels.

b) Levels of biological organization used in conducting (or describing) TT: A (algae), B (bacteria), F (fish), Fc (fish cells), I (invertebrates), P (protozoans), and S (seed germination test with various types of seeds, *e.g.*, *Lactuca sativa*).

c) A study reporting the use of more than one toxicity test at the same biotic level is indicated by additional lettering (*e.g.*, use of three different bacterial tests is coded as "B, B, B".

2. Sediment toxicity assessment

2.1 ASSESSMENT OF AREAS OF CONCERN

In today's world, sediment contamination continues to be a growing environmental issue. Indeed, the deposition of numerous (in)organic chemicals in aquatic systems stemming from various types of anthropogenic activities (urban, industrial, agricultural) has the potential to adversely affect aquatic biota. Once deposited, resuspension of contaminated sediment *via* both natural (*e.g.*, flood scouring) and man-made (*e.g.*, dredging, navigation, open water deposition) activities can further harm living organisms by increasing their contact with (and uptake of) deleterious chemicals. Integrated strategies to assess the toxic potential of contaminated sediments, such as the sediment quality triad approach (see Volume 2, Chapter 10) continue to favour the presence of a strong bioanalytical component within investigation schemes.

Our literature review has shown that sediment toxicity assessment has received marked attention over the past decades and that bioassays have been largely used for this purpose. Contaminated environments, for instance, have triggered many studies conducted to detect and quantify sediment toxicity, to determine the extent of its impact, and to enhance understanding of its short and long-term effects on aquatic communities.

To give readers a first insight into the ways in which toxicity tests have been applied for sediment assessment, we have regrouped publications dealing with sediments collected from areas of concern (Tab. 9) and those collected from other lotic and lentic environments, also impacted by pollutant discharges, where combined chemical-biological analyses were performed (Tab. 10). Sediments were collected from lakes and rivers to undertake initial assessment of sites, to study effects of diverse (in)organic contamination, as well as to investigate various toxicity aspects linked to oil spills and flooding events (Tab. 9). A number of studies also explored relationships between specific contaminants and observed toxicity effects (Tab. 10).

Table 9. Studies with field-collected sediments: assessment of areas of concern.

Assessment objective, type of bioanalytical application[a] and tested sediment phase(s)		**Biotic levels employed[b,c] (and reference)**
Areas impacted by wastewaters: with sediments potentially contaminated by (in)organic pollution		
Ammonia effects	TT: overlying water, pore water	I (Bartsch et al., 2003)
Initial/preliminary assessment of sites	TT: whole sediment	B (Onorati et al., 1998)
	TT: overlying water	I,I (Rediske et al., 2002)
	TT: whole sediment	I (Bettinetti et al., 2003)
	TT: whole sediment	I,I (Collier and Cieniawski, 2003)
	TBA: elutriate	A,B,I,I,I (Sloterdijk et al., 1989)
	TBA: pore water, whole sediment	B,I (Munawar et al., 2000)
Metal contamination	TT: overlying water	I,I,I (West et al., 1993)
	TT: spiked sediment, whole sediment	I,I (Dave and Dennegard, 1994)
	TT: pore water	I (Besser et al., 1995)
	TT: pore water	I (Deniseger and Kwong, 1996)
	TT: pore water	I (Call et al., 1999)
	TT: pore water	I (Hill and Jooste, 1999)
	TT: overlying water, pore water	I (Bervoets et al., 2004)
	TBA: pore water, whole sediment	B,F,F,I,I,I,I (Kemble et al., 1994)
	TBA: overlying water, pore water, whole sediment	B,I,I,I,I,S (Burton et al., 2001)
Metal and organic contamination	TT: whole sediment	I,I (Nebeker et al., 1988)
	TT: elutriate	A (Lacaze et al., 1989)
	TT: whole sediment	B,B (Kwan and Dutka, 1992)
	TT: whole sediment	I,I (Jackson et al., 1995)
	TT: elutriate	I (Bridges et al., 1996)
	TT: elutriate, pore water, whole sediment	I,I (Ristola et al., 1996)
	TT: whole sediment	B (Svenson et al., 1996)
	TT: pore, elutriate, whole sediment	I,I,I,I (Sibley et al., 1997b)
	TT: whole sediment	A (Blaise and Ménard, 1998)
	TT: OE[d], whole sediment	B (Salizzato et al., 1998)

Table 9 (continued). Studies with field-collected sediments: assessment of areas of concern.

Assessment objective, type of bioanalytical application[a] and tested sediment phase(s)		*Biotic levels employed[b,c] (and reference)*
Areas impacted by wastewaters: with sediments potentially contaminated by (in)organic pollution		
Metal and organic contamination	TT: overlying water	I (Call et al., 1999)
	TT: overlying water	I (Martinez-Madrid, 1999)
	TT: overlying water, whole sediment	I,I,I,I (Munawar et al., 1999)
	TT: overlying water	I,I,I,I (Cheam et al., 2000)
	TT: pore water	I (Kemble et al., 2002)
	TBA: pore water	B,I,I (Giesy et al., 1988)
	TBA: overlying water, whole sediment	A,B,B,B,B,I (Dutka et al., 1989)
	TBA: elutriate, whole sediment	A,I (Gregor and Munawar, 1989)
	TBA: pore water, whole sediment	B,I,I,I (Giesy et al., 1990)
	TBA: elutriate, pore water, whole sediment	A,B,B,F,I(8x) L, Pl (Ross et al., 1992)
	TBA: pore water, whole sediment	B,I,I,I (Hoke et al., 1993)
	TBA: elutriate, OE[d]	B,I,S (Lauten, 1993)
	TBA: elutriate, whole sediment	B,I,I (Moran and Chiles, 1993)
	TBA: elutriate, whole sediment	A,A,B,F,I,I (Naudin et al., 1995)
	TBA: pore water	B,B,I,I (Heida and van der Oost, 1996)
	TBA: overlying water, pore water	F,I,I (Watzin et al., 1997)
	TBA: pore water, whole sediment	A,B,I,I (Carter et al., 1998)
	TBA: pore water, whole sediment	A,B,B,B,I,I,I (Côté et al., 1998a)
	TBA: overlying water, whole sediment	B,I,I,I,S,S,S (Rossi and Beltrami, 1998)
	TBA: elutriate, OE[d]	B,I (Hong et al., 2000)
	TBA: pore water	A,B,I,I,I,I,I,P (Persoone and Vangheluwe, 2000)
	TBA: elutriate, OE[d]	A,B,B,I (Ziehl and Schmitt, 2000)

Table 9 (continued). Studies with field-collected sediments: assessment of areas of concern.

Assessment objective, type of bioanalytical application[a] and tested sediment phase(s)		***Biotic levels employed[b,c] (and reference)***
Areas impacted by wastewaters: with sediments potentially contaminated by (in)organic pollution		
Metal and organic contamination	TBA: whole sediment	B,I,I (Ingersoll et al., 2002)
	TBA: pore water	B,I,I,I,I (Lahr et al., 2003)
	TBA: pore water, whole sediment	B,I,I (Munawar et al., 2003)
Organic contamination	TBA: OE[d]	A,B,I (Santiago et al, 1993)
	TBA: pore water	B,I (Pastorok et al., 1994)
	TBA: elutriate, pore water	B,I (Hyötyläinen and Oikari, 1999)
Areas impacted by oil spill events		
Diesel fuel spill	TT: whole sediment	I,I (Keller et al., 1998)
Oil sands	TT: overlying water	F (Tetreault et al., 2003)
Oil pollution	TT: seepage water, whole sediment	I,I (Wernersson, 2004)
Simulated oil spill experiment	TT: whole sediment	B (Ramirez et al., 1996)
	TT: OE[d]	B (Johnson et al., 2004)
	TBA: whole sediment	B,B,B,I (Mueller et al., 2003)
	TBA: whole sediment	A,B,B,I,I (Blaise et al., 2004)
Areas impacted by flooding events		
Metal and organic contamination	TT: whole sediment	I (Kemble et al., 1998)
	TBA: overlying water, whole sediment	F,I,I (Hatch and Burton, 1999)

a) TT (toxicity testing): a study undertaken with test(s) at only one biotic level. TBA (test battery approach): a study involving tests representing two or more biotic levels.

b) Levels of biological organization used in conducting (or describing) TT: A (algae), B (bacteria), F (fish), I (invertebrates), L (*Lemnaceae*, duckweed: small vascular aquatic floating plant), P (protozoans), Pl (plant), and S (seed germination test with various types of seeds, *e.g.*, *Lactuca sativa*).

c) A study reporting the use of more than one toxicity test at the same biotic level is indicated by additional lettering (e.g., use of three different bacterial tests is coded as "B, B, B".

d) Organic (solvent) extract.

Table 10. Studies with field-collected sediments: assessment of areas of concern where combined toxicity and contaminant analysis studies were undertaken.

Assessment objective, type of bioanalytical application[a], tested sediment phase(s) and type of chemical analysis		***Biotic levels employed[b,c] (and reference)***
Lake sediments	TT: pore water Organic analysis	B (Guzzella et al., 1996)
	TT: elutriate, OE[d] Organic analysis	Fc (Gagné et al., 1999b)
	TT: whole sediment Organic analysis	I,I (Marvin et al., 2002)
River sediments	TT: whole sediment Heavy metal and organic analysis	I, Bc (Canfield et al., 1998)
	TT: overlying water, whole sediment Heavy metal and organic analysis	I,I,I,I (Bonnet, 2000)
	TT: pore water Heavy metal and organic analysis	I (Cataldo et al., 2001)
	TT: overlying water Heavy metal analysis	F (Bervoets and Blust, 2003)
	TT: whole sediment Organic analysis	I,I (Cieniawski and Collier, 2003)
	TBA: elutriate Organic analysis	A,B,F,I (Bradfield et al., 1993)
	TBA: elutriate Organic analysis	B,I (McCarthy et al., 1997)
	TBA: OE[d], pore water, whole sediment NH_3, heavy metal and organic analysis	A,B,B,B,B,Fc,I,I,I,I,I,I (Côté et al., 1998a,b)
	TBA: whole sediment Heavy metals	B,I,I,I (Richardson et al., 1998)

a) TT (toxicity testing): a study undertaken with test(s) at only one biotic level. TBA (test battery approach): a study involving tests representing two or more biotic levels.
b) Levels of biological organization used in conducting (or describing) TT: A (algae), B (bacteria), Bc (various benthic communities), F (fish), Fc (fish cells), and I (invertebrates).
c) A study reporting the use of more than one toxicity test at the same biotic level is indicated by additional lettering (*e.g.*, use of three different bacterial tests is coded as "B, B, B".
d) Organic (solvent) extract.

Of the 75 studies reported in Tables 9 and 10, less than half (n = 34) were conducted with two or more tests representing at least two biotic levels (*i.e.*, test battery approach or TBA), as opposed to those performed with a single biotic level (n = 41). This contrasts somewhat with bioassay applications for liquid media assessment, where TBAs comprised nearly 54% (101/188) of reported studies (Tables 1-3). Again, test and biotic level selection may be based on a variety of reasons and study objectives (*e.g.*, practicality, cost, personnel availability) and have influenced a preference for conducting TT assessments. Another factor may lie in that there were (and still are) less toxicity tests whose use is validated for undertaking sediment appraisals. With the exception of those conducted with several benthic invertebrates, most other tests conducted with other groups (*e.g.*, algae, bacteria, fish) were first developed and intended for liquid media assessment (*e.g.*, chemicals and polluted waters). Unlike invertebrate tests, their use to evaluate different liquid compartments associated with whole sediment (*i.e.*, interstitial waters, elutriates, organic extracts of whole sediment) was generally less frequent until the early 1990's when more small-scale assays were developed and validated for sediment toxicity assessment (Wells et al., 1998). Yet another factor is linked to the fact that sediments, unlike liquid samples, comprise several phases that can be assayed (pore waters, elutriates, whole sediment and organic extracts thereof). Ideally, all of these phases should be assessed with a relevant battery of tests for a comprehensive understanding of the sediment's full toxicity potential. In reality, however, scientists will make choices based on laboratory capability for testing and study objectives. When TBAs are used, they are mostly conducted with two or three trophic levels (Tab. 11), similarly to those TBAs performed to study liquid media (Tab. 4).

Table 11. Frequency of the number of biotic levels employed in test battery approaches (TBA) for sediment assessment based on the 34 TBA papers classified in Tables 9 and 10.

TBA studies undertaken with:	***Number and frequency (%)***
Two biotic levels	18/34 (52.9)
Three biotic levels	11/34 (32.4)
Four biotic levels	4/34 (11.8)
Five biotic levels	0/34 (0)
Six biotic levels	1/34 (2.9)

Whether TT (toxicity testing with single species tests at the same biotic level) or TBAs are performed, some test organisms have been more frequently used than others for sediment assessment (Tab. 12). With an overwhelming majority, invertebrates have unquestionably been the most commonly employed, even more so than for liquid media assessment (Tab. 5). The conduct of solid phase tests on whole sediment with invertebrate species explains their preferential selection as test organisms. Bacterial tests rank second in utilization, likely owing to the frequent use of sediment direct contact bioluminescence inhibition assays whose development began in the early 1990s (Brouwer et al., 1990). Algae and fish have also been used by some workers, in part to study the potential impact of contaminants on water column organisms owing to sediment resuspension.

Several phases associated with sediments are evaluated for their toxic potential as Tables 10 and 11 indicate. Whole sediment and pore water stand out as phases that are most frequently investigated (Tab. 13). Because sediments act as contaminant sinks where both readily-soluble and adsorbed toxicants can be present, it is not surprising that whole sediments should be the compartment to receive marked attention, as the (endo)benthic community lives in intimate contact with this matrix and therefore vulnerable to adverse effects. Man-made activities that cause sediments to move (*e.g.*, dredging) can spread contaminants back into the water column and pose a threat to pelagic organisms. Hence, testing sediment phases including elutriates, interstitial waters and overlying waters are fully justified and these have been amply tested as well. Organic extracts of whole sediment, purported by some to lack environmental relevance because they can extract persistent (lipophilic) compounds that would normally stay sequestered *ad infinitum* in sediments, can nevertheless indicate possible long-term effects for benthic organisms.

Table 12. Frequency of use of specific biotic levels employed in toxicity testing (TT) and test battery approaches (TBA) for sediment assessment based on the 75 papers classified in Tables 9 and 10.

TT and TBA studies undertaken with:	***Number and frequency (%)***
Algae	16/222* (7.2)
Bacteria	53/222 (23.9)
Fish	9/222 (4.1)
Invertebrates	136/222 (61.3)
Lemnaceae (duckweed)	1/222 (< 1)
Plant (*H. verticulata*)	1/222 (< 1)
Protozoans	1/222 (< 1)
Seeds	5/222 (2.3)

*Total number of single species tests reported in the 75 papers classified in Tables 9 and 10 (= sum of number of A,B,F,I,L,P,Pl,S tests indicated in the "Biotic levels employed" column).

Table 13. Testing frequency of specific sediment phases for sediment toxicity assessment based on the 75 papers classified in Tables 9 and 10.

Sediment phase	***Number and frequency (%)***
Elutriate	16/109* (14.7)
Overlying water/seepage water	17/109 (15.6)
Pore water	28/109 (25.7)
Organic extract	7/109 (6.4)
Whole sediment	41/109 (37.6)

*Total number of times different sediment phases have been assayed in the 75 papers classified in Tables 9 and 10 (= sum of number of sediment phases indicated in the "Assessment objective…" column).

2.2 CRITICAL BODY RESIDUE STUDIES AND LINKS TO (SUB)LETHAL TOXICITY RESPONSES

During exposure to contaminated sediments, test organisms can concentrate chemicals in their tissue and exhibit measurable (sub)lethal effects linked to accumulated substances. In the field of sediment toxicity assessment, it is noteworthy to mention that some studies have been conducted to characterize both exposure and biological effects in parallel. Exposure to contaminants can be gauged by measuring their concentrations in water/sediment and tissue, and effects can be estimated with endpoints such as survival and growth. These studies are important, for example, to detect threshold concentrations at which chemicals begin to exert adverse effects. As such, they can be useful to recommend effective chemical quality standards that will be protective of aquatic life.

CBR (critical body residue) studies include research on metals, organics and contaminants in mixtures. For instance, cadmium toxicity was appraised with the midge, *Chironomus tentans*, exposed to spiked-sediments that were stored for different periods of time (Sae-ma et al., 1998). Decreases in toxicity effects (lethality) and Cd accumulation in midge tissue with storage time suggested that decreased bioavailability of this metal had occurred. This work clearly illustrated the influence of sediment storage time on organism toxicity response and the impact it could have on test results. Effects of fluoranthene, a PAH (polycyclic aromatic hydrocarbon) congener, were appraised in benthic copepods exposed to dosed sediments for ten days (Lotufo, 1998). Relationships were found between organism health (survival, reproductive and grazing capacity) and fluoranthene concentration in both sediment and tissue. This study was therefore able to more closely pinpoint the NOEL (no observed effect level) concentration of this chemical for this group of biota. Another initiative in CBR studies sought to find out whether the AVS (acid-volatile sulphide) content of sediments collected in areas impacted by mining activities might influence the bioaccumulation of metals (Zn, Cu) and toxicity to the midge *C. tentans* (Besser et al., 1996). Results indicated differences in metal uptake in organisms based on AVS content and showed that growth inhibition was more markedly linked to Zn than Cu. Recommendations called for considering AVS concentrations in metal-contaminated sediments, because of the importance it can have on uptake by biota and subsequent toxicity responses. These investigations indeed confirm the usefulness of CBR-like approaches for evaluating hazard and risk to sediment-dwelling organisms from metals and organic pollutants.

3. Miscellaneous studies/initiatives linked to aquatic toxicity testing applications (liquid media and sediments)

3.1 ENDEAVOURS PROMOTING THE DEVELOPMENT, VALIDATION AND REFINEMENT OF TOXICITY TESTING PROCEDURES

There are literally hundreds of publications that, directly or indirectly, have contributed to the development, validation and refinement of bioassay techniques both for liquid and solid media assessment. These papers incorporate initiatives that have dealt with 1) test method development, 2) inter-calibration exercises, 3) comparative studies and 4) factors capable of affecting bioassay responses. Anyone familiar with the world of toxicity testing would likely not disagree with the statement that "the perfect bioassay is not of this world" and that developers of these instruments of ecotoxicology simply do their utmost to make each test "as least imperfect as possible". To reach this latter stage, assurance of reproducibility, demonstration of scope of use and understanding confounding factors capable of influencing toxicity responses are some of the issues that must be addressed. Hereunder, examples of such studies are given to reveal some of the ways in which they have contributed to the science of small-scale toxicity testing by enhancing its diagnostic tools.

3.1.1 Test method development

To guarantee that reliable procedures are consistently employed to generate toxicity data, it is first essential that sufficient effort be directed toward the development of reproducible toxicity test methods whose results will remain unchallenged. Those that are featured in this book are representative of dependable micro-assays presently in use internationally. Many other small-scale toxicity test methods have been developed at various levels of biological organization. These include bioassays conducted with **algae** (Daniels et al, 1989*; Radetski et al., 1995; St-Laurent and Blaise, 1995; Chen et al., 1997; Blaise and Ménard, 1998*; Persoone, 1998; Tessier et al., 1999; Geis et al., 2000), **bacteria** (Bitton et al., 1994; Blaise et al., 1994; Bulich and Bailey, 1995; Kwan, 1995*; Bulich et al., 1996; Botsford, 1998; Lappalainen et al., 1999*; Ulitzur et al., 2002; Gabrielson et al., 2003), **fish cells** (Ahne, 1985; Pesonen and Andersson, 1997; Sandbacka et al., 1999), **invertebrates** (Snell and Persoone, 1989; Oris et al., 1991; Kubitz et al., 1996*; Benoit et al., 1997*; Johnson and Delaney, 1998; Chial and Persoone, 2002*; Gerhardt et al., 2002b*; Tran et al., 2003), ***Lemnaceae*** (Bengtsson et al., 1999; Cleuvers and Ratte, 2002a), **protozoans** (Dive et al., 1991; Larsen et al., 1997; Berk and Roberts, 1998; Twagilimana et al., 1998; Gilron et al., 1999) and **yeast** (Ribeiro et al., 2000).

*(tests applying to sediment toxicity testing)

For freshwater solid media investigations, efforts have also been directed towards the development of formulated sediments (also called "artificial" or "synthetic" sediments) to assess their adequacy for conducting contaminant-spiked sediment toxicity studies (Suedel and Rodgers, 1994; Kemble et al., 1999). Among other uses, formulated sediments can be useful to recommend realistic sediment quality criteria for (in)organic substances. Different types of formulated sediments have been employed to evaluate both metal- spiked (Gonzalez, 1996; Harrahy and Clements, 1997; Chapman et al., 1999; Péry et al., 2003) and organic-spiked (Fleming et al., 1998; Besser et al., 2003; Lamy-Enrici et al., 2003) contaminants.

3.1.2 Inter-calibration exercises

Beyond test development and validation, inter-calibration exercises (also known as "round robin" or "inter-laboratory exercises") are mandatory steps that must be undertaken if a toxicity test method is intended for standardization. These exercises further contribute to test validation by insuring reproducibility of results among different laboratories. In most cases, they also contribute to test method improvement and refinement (*e.g.*, Thellen et al., 1989; Dive et al., 1991; Persoone et al., 1993).

For example, inter-calibration exercises have been undertaken with **algae** (Thellen et al., 1989), **bacteria** (Ribo, 1997; Ross et al., 1999*), **fish cells** (Gagné et al., 1999a), **invertebrates** (Cowgill, 1986; Persoone et al., 1993; Burton et al., 1996*; Hayes et al., 1996), **protozoans** (Dive et al., 1990), and **test organisms of several biotic levels** (Rue et al., 1988; Ronco et al., 2002).

*(tests applying to sediment toxicity testing)

If toxicity tests fulfill the scientific criteria set out by inter-calibration exercises, they can then be considered for the standardization process. If this process is followed, an official toxicity test method document is eventually produced that ensures proper conduct of biological tests (see Section 3.2.1).

3.1.3 Comparative studies

Comparative studies involving toxicity tests abound in the scientific literature. There are many reasons compelling ecotoxicologists to conduct work of this nature, some of which are directed 1) to assess the performance, sensitivity and relevance of individual bioassays undertaken on various chemicals and (liquid and solid) media to specify their scope of use, 2) to optimize the diagnostic potential of bioassay batteries to broaden hazard detection (insure that tests in a battery are complementary and not redundant) and 3) to promote the application of novel assays capable of high throughput for cost-effective screening of (complex) environmental samples.

As an overview, **studies carried out with liquid media** have been launched to compare **bioassay responses** (Finger et al., 1985; Blaise et al., 1987; Kaiser and McKinnon, 1993; Ross, 1993; Isomaa et al., 1995; Dodard et al., 1999; Lucivjanskà et al., 2000; Brix et al., 2001a; Nalecz-Jawecki and Sawicki, 2002; Mummert et al., 2003; Sherrard et al., 2003; Tsui and Chu, 2003), **different endpoints** (Dunbar et al., 1983; Fernández-Casalderrey et al., 1993; Pauli and Berger, 1997; Froehner et al., 2000; Snell, 2000; Weyers and Vollmer, 2000; Jos et al., 2003), **responses of laboratory test organism species and endemic species and/or laboratory bioassay responses and field results** (Koivisto and Ketola, 1995; Traunspurger et al., 1996; van Wijngaarden et al., 1996; Jak et al., 1998; Crane et al., 1999; Tchounwou and Reed, 1999; Dyatlov, 2000; Milam et al., 2000; Pascoe et al., 2000; Bérard et al., 2003), **and bioassay and biomarker endpoints** (Gagné and Blaise, 1993; Nyström and Blanck, 1998; Connon et al., 2000; Perkins and Schlenk, 2000; De Coen and Janssen, 1997; Bierkens et al., 1998; Sturm and Hansen, 1999; den Besten and Tuk, 2000; Guilhermino et al., 2000; Maycock et al., 2003; Taylor et al., 2003).

In **studies conducted with sediments**, comparisons have been reported for **artificial (formulated) and natural sediments** (Barrett, 1995; Fleming et al., 1998), **bioassay and biomarker endpoints** (Gillis et al., 2002), **bioassay responses** (Ahlf et al., 1989; Becker et al., 1995; Day et al., 1995a; Kwan and Dutka, 1995; Suedel et al., 1996; Barber et al., 1997; Day et al., 1998; Fuchsman et al., 1998; Guzzella,

1998; Huuskonen et al., 1998; Côté et al., 1998a,b; Vanderbroele et al., 2000; Watts and Pascoe, 2000; Chial et al., 2003; Milani et al., 2003; Mueller et al., 2003; Petänen et al., 2003), **different endpoints** (Suedel et al., 1996; Watts and Pascoe, 1996; Sibley et al., 1997a; Pasteris et al., 2003; Landrum et al., 2004; Vecchi et al., 1999), **different sediment phases** (Harkey et al., 1994), **responses of laboratory test organism species and endemic species** (Conrad et al., 1999) **and/or laboratory bioassay responses and field results** (Reinhold-Dudok et al., 1999; Bombardier and Blaise, 2000; Peeters et al., 2001; den Besten et al., 2003) and **sediment collection techniques** (West et al., 1994).

3.1.4 Factors capable of affecting bioassay responses

Toxicity testing developers and users have also devoted significant energy to the understanding of specific factors capable of confounding (micro-) organism responses and/or interfering with data interpretation (*e.g.*, pH, temperature, light, growth medium, natural contaminants such as NH_3, H_2S, or grain size in case of solid phase tests).

In fact, any aspect of testing likely to impact toxicity results (*e.g.*, stimulatory effects in the case of algal toxicity assays, or sample colour interferences in the case of a toxicity endpoint measured by photometry) have been a focus of concern, as have been ways of minimizing, eliminating or circumventing particular problems or limitations that may be test-specific. In brief, seeking thorough understanding of a test's capabilities and limitations has been considered paramount for proper toxicity assessment (and final data interpretation) and marked efforts have been directed toward this goal.

With this purpose in mind, investigations have explored the influence of such factors as **acid volatile sulfides** (Sibley et al., 1996*; Long et al., 1998*), **alkalinity** (Lasier et al., 1997*), **ammonia** (Besser et al., 1998*; Newton et al., 2003*), **colored samples** (Cleuvers and Weyers, 2003), **equilibration time** (Lee et al., 2004*), **experimental design** (Naylor and Howcroft, 1997*; Bartlett et al., 2004*), **fluid dynamics** (Preston et al., 2001), **food** (Sarma et al., 2001; Gorbi et al., 2002; de Haas et al., 2002*; Antunes et al., 2004; de Haas et al., 2004*); **grain size** (Guerrero et al., 2003*), **genetic variability** (Baird et al., 1991; Barber et al., 1990; Barata et al., 1998), **gut contents** (Sibley et al., 1997c*), **heavy metal speciation** (Gunn et al., 1989*; Ankley et al., 1996*), **humic/fulvic acids** (Ortego and Benson, 1992; Alberts et al., 2001; Guéguen et al., 2003; Koukal et al., 2003; Ma et al., 2003), **intermittent or short exposures to contaminants** (Hickey et al., 1991; Brent and Herricks, 1998; Naddy and Klaine, 2001, Broomhall, 2002), **life-cycle stage/age** (Williams et al., 1986; Stephenson et al., 1991; Watts and Pascoe, 1998*; Hamm et al., 2001), **light regime** (Cleuvers and Ratte, 2002b), **organic matter content** (Ankley et al., 1994*; Lacey et al., 1999*; Besser et al., 2003*; Guerrero et al., 2003*; Lamy-Enrici et al., 2003*; Mäenpää et al., 2003*; VanGenderen et al., 2003), **pH** (Fisher and Wadleigh, 1986; Fu et al., 1991; Svenson and Zhang, 1995; Rousch et al., 1997; Franklin et al., 2000; Peck et al., 2002*; Long et al., 2004), **phosphorus** (Van Donk et al., 1992; Mkandawire et al., 2004), **potassium** (Bervoets et al., 2003*), **pre-exposure to contaminants** (Bearden et al., 1997; Muyssen and Janssen, 2001, 2002; Ristola et al., 2001*; Vidal and Horne, 2003*), **sand** (Thomulka et al., 1997), **sediment**

indigenous animals (Reynoldson et al., 1994*), **sediment processing** (Day et al., 1995b*), **sex** (Sildanchandra and Crane, 2000), **solvents** (Calleja and Persoone, 1993; Fliedner, 1997), **choice of statistical tests** (Isnard et al., 2001), **sulfates** (Brix et al., 2001c), **sulfur** (Jacobs et al., 1992*; Pardos et al., 1999b*), **suspended solids** (Herbrandson et al., 2003a,b), **temperature** (Fisher, 1986; Broomhall, 2002; Buchwalter et al., 2003; Heugens et al., 2003), **test exposure time** (Suedel et al., 1997; Naimo et al., 2000*; Froehner et al., 2002; Feng et al., 2003), **test medium** (Vasseur and Pandard, 1988; Guilhermino et al., 1997; Samel et al., 1999), **test organism inoculum density** (Moreno-Garrido et al., 2000; Franklin et al., 2002), **UV irradiation** (Bonnemoy et al., 2004), **water chemistry/quality** (Persoone et al., 1989; Jop et al., 1991; van Dam et al., 1998; Karen et al., 1999; Clément, 2000; Bury et al., 2002; Graff et al., 2003), **water hardness** (Fu et al., 1991; Baer et al., 1999; Verge et al., 2001; Charles et al., 2002; Gensemer et al., 2002; Naddy et al., 2003; Long et al., 2004), **water-sediment partitioning** (Stewart and Thompson, 1995*).

*(tests applying to sediment toxicity testing)

3.2 INITIATIVES PROMOTING THE USE OF TOXICITY TESTING PROCEDURES

For over three decades, the use of bioassays for toxicity testing has steadily increased and become an indispensable component of aquatic environmental assessment. In this section, specific types of publications are presented as important contributions that have 1) promoted the use of ecotoxicology testing in the biomonitoring, regulatory and compliance arena, 2) disseminated information and understanding relating to toxicity testing issues, 3) favoured technology transfer of test methods internationally and 4) provided overall sound scientific support to facilitate decision-making aimed at environmental protection and conservation.

3.2.1 Review, bio-monitoring and HAS articles

Review articles are particularly useful to synthesize research work that has been undertaken in different spheres relating to toxicity testing. By exposing the state of the art for a selective field, these articles will often circumscribe the limitations, advantages and scope of use of bioassays which then leads to their proper and effective application. Some examples of review articles include papers on **concept/management/policy** (MacGregor and Wells, 1984; U.S. EPA and Environment Canada, 1984; Sergy, 1987; Cairns and Pratt, 1989; Maltby and Callow, 1989; Blaise, 2003), as well as several others on specific trophic groups including **algae** (Blaise, 1993; Lewis, 1995; Sosak-Swiderska and Tyrawska, 1996; Blaise et al., 1998b; Blaise, 2002), **bacteria** (Bennett and Cubbage, 1992b*; Bitton and Koopman, 1992; Kross and Cherryholmes, 1993; Painter, 1993; Bitton and Morel, 1998; Ross, 1998; Doherty, 2001*), **fish cells** (Babich and Borefreund, 1991;Fentem and Balls, 1993; Denizeau, 1998; Fent, 2001; Castaño et al., 2003), **invertebrates** (Burton et al., 1992; Ingersoll et al., 1995*; Snell and Janssen, 1995, 1998; Chapman, 1998*; CANMET, 1999) and **protozoa** (Gilron and Lynn, 1998; Sauvant et al., 1999; Nicolau et al., 2001; Nalecz-Jawecki, 2004).

Other reviews have also encompassed **different levels of toxicity tests** (Giesy and Hoke, 1989*; Bennett and Cubbage, 1992a; CANMET, 1997a; Blaise et al., 1998a; de Vlaming et al., 1999; Blaise et al., 2000; Girling et al., 2000; Janssen et al., 2000; Repetto et al., 2000).

*applying to sediment toxicity assessment

Various papers expounding the value of **biomonitoring, routine and/or regulatory testing** have also advanced the practice of bioassays. Some of these include articles on **drinking water assessment** (Forget et al., 2000), **single chemical or mixture assessment** (Altenburger et al., 1996; Aoyama et al., 2000), **surface water assessment** (Canna-Michaelidou et al., 2000; Marsalek and Rojickova-Padrtova, 2000; Ruck et al., 2000), **wastewater assessment** (OECD, 1987; Blaise et al., 1988; Mackay et al., 1989; Hansen, 1993; Johnson et al., 1993; Stulhfauth, 1995; Kovacs et al., 2002), **sewage treatment plant performance assessment** (Fearnside and Hiley, 1993), and **sediment quality assessment** (Nipper, 1998).

Articles proposing new **hazard assessment schemes** (HAS) for liquid or sediment assessment have equally paved the way for the employment of test batteries in ecotoxicity appraisals. Some describe systems for evaluating **water/wastewater** (Blaise et al., 1985; Heinis et al., 2000; Ronco et al., 2000 ; Persoone et al., 2003), **chemicals** (Fochtman et al., 2000; Garay et al., 2000; Girling et al., 2000; Pica-Granados et al., 2000; Brix et al., 2001a,b,c) and **sediments** (Ingersoll et al., 1997; Côté et al., 1998b). These effects-based indices, varied in their concepts and objectives, demonstrate novel ways of utilizing groups of bioassays to deal with "real-life" environmental situations. As such, they highlight schemes that are complementary to the robust and validated HAS approaches described in Volume 2 of this book.

3.2.2 Standardized test methods and guidance documents

Finally, marked efforts have been undertaken nationally and internationally to publish **standardized toxicity test methods** and several standards organizations (*e.g.*, ASTM, ISO, OECD) have been very active in the production of documents too numerous to reproduce in this chapter. Publishing official test methods is not a simple task and can require a substantial amount of time and energy from dedicated scientists. Again, standardized toxicological method documents are crucial to environmental assessment as they ensure proper use of testing, (inter)national consistency and acceptance, as well as reliability of test results owing to the quality control and assurance components that are integrated in such protocols.

Test method standardization (TMS) calls for several actions that involve 1) preparation of a formal draft test method document for each bioassay intended for standardization, 2) a critical review by an expert subcommittee, 3) the preparation of a final draft test method, 4) an international peer review of each test method, 5) an inter-calibration exercise of the final draft test method, 6) finalization of each test method and 7) the formal publication of the toxicity test method document. Environment Canada (EC) has been particularly active in biological test method standardization and has thus far contributed 18 standardized aquatic and sediment toxicity methods, eight and three of which apply to acute/chronic freshwater liquid (tests with algae, bacteria, fish, invertebrates, and *Lemnaceae*) and solid (tests with bacteria and invertebrates) media assessment, respectively (IGETG, 2004). As a complement to TMS, EC has also produced several **guidance documents** that provide assistance on matters related to choice of reference toxicants (Environment Canada, 1990), sampling and spiking techniques for sediments (Environment Canada, 1994, 1995), interpretation of results (Environment Canada, 1999) and statistical considerations for toxicity tests (Environment Canada, 2004b).

Other **standardized/validated test methods** reported in the literature include acute/chronic tests performed with **algae** (*e.g*, OECD, 2002a; ISO, 2003), **fish cells** (Gagné and Blaise, 2001), **invertebrates** (Borgmann and Munawar, 1989*; Trottier et al., 1997; Pereira et al., 2000*; OECD, 2001*a,b), ***Lemnaceae*** (OECD, 2002b), and with **toxicity tests conducted at different trophic levels** (Nebeker et al., 1984*; U.S. EPA, 2002a,b).

*applying to sediment toxicity assessment

Additionally, **miscellaneous guidance/technical documents** have reported on various aspects linked to ecotoxicity that give advice on:

- choice of bioassays for general contaminant assessment (Calow, 1989);
- criteria to select tests for effluent testing (Grothe et al., 1996; Johnson, 2000);
- choice of species and endpoints for appraising pharmaceuticals (Länge and Deitrich, 2002);
- proper application of algal, bacterial and invertebrate tests (Santiago et al., 2002);
- approaches, design and interpretation of sediment tests (Ross and Leitman, 1995; Ingersoll et al., 2000; Wenning and Ingersoll, 2002; MacDonald and Ingersoll, 2002a,b).

4. Conclusion(s)

Small-scale freshwater toxicity testing is but a modest fraction of a diverse array of scientific activities connected to the field of ecotoxicology. Yet, within this still emerging discipline, few will argue the fact that tools and approaches developed to measure the undesirable effects that countless chemicals (alone or in mixtures) and complex (liquid and solid) media can exert on biota have markedly contributed to aquatic ecosystem preservation. Indeed, the breadth and scope of application of bioassays thus far directed toward obtaining relevant information aimed at problem-solving and prevention of contaminant-based issues has progressed well.

While many developed countries have been effective over past decades in eliminating acute toxicity from point source discharges owing to technological improvement of industrial processes and legislation, chronic effects on aquatic biota are still very much an issue. Furthermore, as the 21rst century unfolds, many emerging and developing countries active in joining the world economy are presently creating new contaminant burdens on aquatic systems that will contribute additional acute and chronic toxicity pressures until, once again, technology and legislation repress pollution. Hence, the techniques and hazard assessment schemes featured in this book can prove to be very relevant for use in all parts of the world. As editors of this book, it is our hope that readers will grasp that an effects-based approach is primordial to deal with hazard and risk assessment of pollutants and that use of toxicity tests is an essential cog in this respect. It is also our hope that many, directly or indirectly involved in ensuring the well-being of aquatic systems, will actually use (or suggest the use of) some of the toxicity testing methods and hazard assessment schemes described in subsequent sections.

Lastly, while acute and chronic (sub)lethal toxicity effects are basic concerns that must be first dealt with and eradicated, new demands will be made on ecotoxicology to address emerging issues. Indeed, several more subtle (and potentially deleterious) effects owing to long-term exposures to low concentrations of contaminants will merit investigation (Eggen et al., 2004). Genotoxicity, teratogenicity, immunotoxicity and endocrine disruption are some of the undesirable consequences of classical (*e.g.*, metals, pesticides, organochlorides) and more recent (*e.g.*, household products and pharmaceuticals) chemical discharges into receiving waters that require urgent comprehensive assessment. Here as well, reliable and relevant standardized tools and approaches will have to be developed and applied.

References

Adams, S.M. (2003) Establishing causality between environmental stressors and effects on aquatic ecosystems, *Human and Ecological Risk Assessment* **19**, 17-35.

Ahlf, W., Calmano, W., Erhard, J. and Förstner, U. (1989) Comparison of rive bioassay techniques for assessing sediment-bound contaminants, in M. Munawar, G. Dixon, C.I. Mayfield, T. Reynoldson and M.H. Sadar (eds.), *Environmental Bioassay Techniques and their Application: Proceedings of the 1st International Conference held in Lancaster, England, 11-14 July 1988*, Kluwer Academic Publishers, Dordrecht, Netherlands, pp. 285-289.

Ahne, W. (1985) Untersuchungen über die Verwendung von Fischzellkulturen fur Toxizitätsbestimmungen zur Einschränkung and Ersatz des Fishtests, *Zentralblatt Fur Bakteriologie, Mikrobiologie Und Hygiene. 1. Abt. Originale B, Hygiene* **180**, 480-504.

Ahtiainen, J., Nakari, T. and Silvonen, J. (1996) Toxicity of TCF and ECF pulp bleaching effluents assessed by biological toxicity tests, in M.R. Servos, K.R. Munkittrick, J.H. Carey and G.J. Van Der Kraak (eds.), *Environmental Fate and Effects of Pulp and Paper Mill Effluents,* St-Lucie Press, FL, pp. 33-40.

Ahtiainen, J., Nakari, T., Ruoppa, M., Verta, M. and Talka, E. (2000) Toxicity screening of novel pulp mill wastewaters in Finnish pulp mills, in G. Persoone, C. Janssen and W.M. De Coen (eds.), *New Microbiotests for Routine Toxicity Screening and Biomonitoring*, Kluwer Academic/Plenum Publishers, New York, pp. 307-317.

Alberts, J.J., Takács, M. and Pattanayek, M. (2001) Influence of IHSS standard and reference materials on copper and mercury toxicity to *Vibrio fischeri*, *Acta Hydrochimica et Hydrobiologica* **28**, 428-435.

Ali, M. and Sreekrishnan, T.R. (2001) Aquatic toxicity from pulp and paper mill effluents: a review, *Advances in Environmental Research* **5**, 175-196.

Altenburger, R., Boedeker, W., Faust, M. and Grimme, L.H. (1996) Regulations for combined effects of pollutants: consequences from risk assessment in aquatic toxicology, *Food and Chemical Toxicology* **34**, 1155-1157.

Anastácio, P.M., Lützhøft, H.C., Halling-Sørensen, B. and Marques, J.C. (2000) Surfactant (Genapol OX-80) toxicity to *Selenastrum capricornutum*, *Chemosphere* **40** (8), 835-838.

Angelaki, A., Sakellariou, M., Pateras, D. and Kungolos, A. (2000) Assessing the quality of natural waters in Magnesia prefecture in Greece using Toxkits, in G. Persoone, C. Janssen and W.M. De Coen (eds.), *New Microbiotests for Routine Toxicity Screening and Biomonitoring*, Kluwer Academic/Plenum Publishers, New York, pp. 281-288.

Ankley, G. T., Benoit, D. A., Balogh, J. C., Reynoldson, T. B., Day, K. E. and Hoke, R. A. (1994) Evaluation of potential confounding factors in sediment toxicity tests with three freshwater benthic invertebrates, *Environmental Toxicology and Chemistry* **13** (4), 627-635.

Ankley, G.T., Liber, K., Call, D.J., Markee, T.P., Canfield, T.J. and Ingersoll, C.G. (1996) A field investigation of the relationship between zinc and acid volatile sulfide concentrations in freshwater sediments, *Journal of Aquatic Ecosystem Health* **5** (4), 255-264.

Antunes, S.C., Castro, B.B. and Gonçalves, F. (2004) Effect of food level on the acute and chronic responses of daphnids to lindane, *Environmental Pollution* **127** (3), 367-375.

Aoyama, I., Okamura, H. and Rong, L. (2000) Toxicity testing in Japan and the use of Toxkit microbiotests, in G. Persoone, C. Janssen and W.M. De Coen (eds.), *New Microbiotests for Routine Toxicity Screening and Biomonitoring*, Kluwer Academic/Plenum Publishers, New York, pp. 123-133.

Arbuckle, W.B. and Alleman, J.E. (1992) Effluent toxicity testing using nitrifiers and Microtox™, *Water Environment Research* **64**, 263-267.

Arkhipchuk, V.V. and Malinovskaya, M.V. (2002) Quality of water types in Ukraine evaluated by WaterTox bioassays, *Environmental Toxicology* **17** (3), 250-257.

Aruldoss, J.A. and Viraraghavan, T. (1998) Toxicity testing of refinery wastewater using Microtox, *Bulletin of Environmental Contamination and Toxicology* **60** (3), 456-463.

Asami, M., Suzuki, N. and Nakanishi, J. (1996) Aquatic toxicity emission from Tokyo: wastewater measured using marine luminescent bacterium, *Photobacterium phosphoreum*, *Water Science and Technology* **33** (6), 121-128.

Atienzar, F.A., Cheung, V.V., Jha, A.N. and Depledge, M.H. (2001) Fitness parameters and DNA effects are sensitive indicators of copper-induced toxicity in *Daphnia magna*, *Toxicological Sciences* **59** (2), 241-250.

Babich, H. and Borenfreund, E. (1987) Cultured fish cells for the ecotoxicity testing of aquatic pollutants, *Toxicity Assessment* **2**, 119-133.

Babich, H. and Borenfreund, E. (1991) Cytotoxicity and genotoxicity assays with cultured fish cells: a review, *Toxicology in Vitro* **5**, 91-100.

Baer, K.N., Ziegenfuss, M.C., Banks, S.D. and Ling, Z. (1999) Suitability of high-hardness COMBO medium for ecotoxicity testing using algae, daphnids, and fish, *Bulletin of Environmental Contamination and Toxicology* **63** (3), 289-296.

Baer, K.N., Boeri, R.L., Ward, T.J. and Dixon, D.W. (2002) Aquatic toxicity evaluation of para-methylstyrene, *Ecotoxicology and Environmental Safety* **53** (3), 432-438.

Baird, D.J., Barber, I., Bradley, M., Soares, A.M.V.M. and Calow, P. (1991) A comparative study of genotype sensitivity to acute toxic stress using clones of *Daphnia magna* Straus, *Ecotoxicology and Environmental Safety* **21**, 257–265.

Barata, C., Baird, D.J. and Markich, S.J. (1998) Influence of genetic and environmental factors on the tolerance of *Daphnia magna* Straus to essential and non-essential metals, *Aquatic Toxicology* **42**(2), 115-137.

Barber, I., Baird, D.J. and Calow, P. (1990) Clonal variation in general responses of *Daphnia magna* Straus to toxic stress. II. Physiological effects, *Functional Ecology* **4**, 409–414.

Barber, T.R., Fuchsman, P.C., Chappie, D.J., Sferra, J.C., Newton, F.C. and Sheehan, P.J. (1997) Toxicity of hexachlorobenzene to *Hyallela azteca* and *Chironomus tentans* in spiked sediment bioassays, *Environmental Toxicology and Chemistry* **16** (8), 1716-1720.

Barrett, K.L. (1995) A comparison of the fate and effects of prochloraz in artificial and natural sediments, *Journal of Aquatic Ecosystem Health* **4** (4), 239-248.

Bartlett, A.J., Borgmann, U., Dixon, D.G., Batchelor, S.P. and Maguire, R.J. (2004) Tributyltin uptake and depuration in *Hyalella azteca*: implications for experimental design, *Environmental Toxicology and Chemistry* **23** (2), 426-434.

Bartsch, M.R., Newton, T.J., Allran, J.W., O'Donnell, J.A. and Richardson, W.B. (2003) Effects of pore-water ammonia on *in situ* survival and growth of juvenile mussels (*Lampsilis cardium*) in the St. Croix Riverway, Wisconsin, USA, *Environmental Toxicology and Chemistry* **22** (11), 2561-2568.

Bastian, K.C. and Alleman, J.E. (1998) Microtox characterization of foundry sand residuals, *Waste Management* **18** (4), 227-234.

Baun, A., Bussarawit, N. and Nyholm, N. (1998) Screening of pesticide toxicity in surface water from an agricultural area at Phuket Island (Thailand), *Environmental Pollution* **102** (2-3), 185-190.

Bearden, A.P., Gregory, B.W. and Schultz, T.W. (1997) Growth kinetics of preexposed and naive populations of *Tetrahymena pyriformis* to 2-decanone and acetone, *Ecotoxicology and Environmental Safety* **37** (3), 245-250.

Becker, C.D., Fallon, W.E., Crass, D.W. and Scott, A.J. (1983) Acute toxicity of water soluble fractions derived from a coal liquid (SRC-II) to three aquatic organisms, *Water, Air, and Soil Pollution* **19**, 171-184.

Becker, D.S., Rose, C.D. and Bigham, G.N. (1995) Comparison of the 10-day freshwater sediment toxicity tests using *Hyallela azteca* and *Chironomus tentans*, *Environmental Toxicology and Chemistry* **14** (12), 2089-2094.

Bengtsson, B.-E., Bongo, J.P. and Eklund, B. (1999) Assessment of duckweed *Lemna aequinoctialis* as a toxicological bioassay for tropical environments in developing countries, *Ambio* **28** (2), 152-155.

Bennett, J. and Cubbage, J. (1992a) *Evaluation of bioassay organisms for freshwater sediment toxicity testing*, Environmental Investigations and Laboratory Services, Washington State Department of Ecology, Washington, DC (December 19, 2003); http://www.nic.edu/library/superfund/refdocs%5Ccda0159.pdf.

Bennett, J. and Cubbage, J. (1992b) *Review and evaluation of Microtox test for freshwater sediments*, Washington State Department of Ecology, Washington, 28 pp.

Bennett, W.R. and Farrell, A.P. (1998) Acute toxicity testing with juvenile white sturgeon (*Acipenser transmontanus*), *Water Quality Research Journal of Canada* **33** (1), 95-110.

Benoit, D.A., Sibley, P.K., Juenemann, J.L. and Ankley, G.T. (1997) *Chironomus tentans* life-cycle test: design and evaluation for use in assessing toxicity of contaminated sediments, *Environmental Toxicology and Chemistry* **16** (6), 1165-1176.

Bérard, A., Dorigo, U., Mercier, I., Becker-van Slooten, K., Grandjean, D. and Leboulanger, C. (2003) Comparison of the ecotoxicological impact of the triazines Irgarol 1051 and atrazine on microalgal cultures and natural microalgal communities in Lake Geneva, *Chemosphere* **53** (8), 935-944.

Berk, S.G. and Roberts, R.O. (1998) Development of a protozoan chemoattraction inhibition assay for evaluating toxicity of aquatic pollutants, in P.G. Wells, K. Lee and C. Blaise (eds.) *Microscale Testing in Aquatic Toxicology: Advances, Techniques, and Practice*, CRC Press, Boca Raton, FL, pp. 337-348.

Bermingham, N., Costan, G., Blaise, C. and Patenaude, L. (1996) Use of micro-scale aquatic toxicity tests in ecolabelling guidelines for general purpose cleaners, in M. Richardson (ed.), *Environmental Xenobiotics*, Taylor & Francis Books Ltd, London, England, pp. 195-212.

Bervoets, L. and Blust, R. (2003) Metal concentrations in water, sediment and gudgeon (*Gobio gobio*) from a pollution gradient: relationship with fish condition factor, *Environmental Pollution* **126** (1), 9-19.

Bervoets, L., Baillieul, M., Blust, R. and Verheyen, R. (1996) Evaluation of effluent toxicity and ambient toxicity in a polluted lowland river, *Environmental Pollution* **91** (3), 333-341.

Bervoets, L., De Bruyn, L., Van Ginneken, L. and Blust, R. (2003) Accumulation of ^{137}Cs by larvae of the midge *Chironomus riparius* from sediment: effect of potassium, *Environmental Toxicology and Chemistry* **22** (7), 1589-1596.

Bervoets, L., Meregalli, G., De Cooman, W., Goddeeris, B. and Blust, R. (2004) Caged midge larvae (*Chironomus riparius*) for the assessment of metal bioaccumulation from sediments *in situ*, *Environmental Toxicology and Chemistry* **23** (2), 443-454.

Besser, J.M., Kubitz, J.A., Ingersoll, C.G., Braselton, W.E. and Giesy, J.P. (1995) Influences on copper bioaccumulation, growth, and survival of the midge, *Chironomus tentans*, in metal contaminated sediments, *Journal of Aquatic Ecosystem Health* **4** (3), 157-168.

Besser, J.M., Ingersoll, C.G. and Giesy, J.P. (1996) Effects of spatial and temporal variation of acid-volatile sulfide on the bioavailability of copper and zinc in freshwater sediments, *Environmental Toxicology and Chemistry* **15** (3), 286-293.

Besser, J.M., Ingersoll, C.G., Leonard, E.N. and Mount, D.R. (1998) Effect of zeolite on toxicity of ammonia in freshwater sediments: implications for toxicity identification evaluation procedures, *Environmental Toxicology and Chemistry* **17** (11), 2310-2317.

Besser, J.M., Brumbaugh, W.G., May, T.W. and Ingersoll, C.G. (2003) Effects of organic amendments on the toxicity and bioavailability of cadmium and copper in spiked formulated sediments, *Environmental Toxicology and Chemistry* **22** (4), 805-815.

Bettinetti, R., Giarei, C. and Provini, A. (2003) Chemical analysis and sediment toxicity bioassays to assess the contamination of the River Lambro (Northern Italy), *Archives of Environmental Contamination and Toxicology* **45** (1), 72-78.

Bierkens, J., Maes, J. and Plaetse, F.V. (1998) Dose-dependent induction of heat shock protein 70 synthesis in *Raphidocelis subcapitata* following exposure to different classes of environmental pollutants, *Environmental Pollution* **101** (1), 91-97.

Bitton, G. and Koopman, B. (1992) Bacterial and enzymatic bioassays for toxicity testing in the environment, *Reviews of Environmental Contamination and Toxicology* **125**, 1-22.

Bitton, G. and Morel, J.L. (1998) Microbial enzyme assays for the detection of heavy metal toxicity, in P.G. Wells, K. Lee and C. Blaise (eds.), *Microscale Testing in Aquatic Toxicology: Advances, Techniques, and Practice*, CRC Press, Boca Raton, FL, pp. 143-152.

Bitton, G., Koopman, B. and Agami, O. (1992) MetPAD™: a bioassay for rapid assessment of heavy metal toxicity in wastewater, *Water Environment Research* **64** (6), 834-836.

Bitton, G., Jung, K. and Koopman, B. (1994) Evaluation of a microplate assay specific for heavy metal toxicity, *Archives of Environmental Contamination and Toxicology* **27** (1), 25-28.

Blaise, C. (1993) Practical laboratory applications with micro-algae for hazard assessment of aquatic contaminants, in M. Richardson (ed.), *Ecotoxicology Monitoring*, VCH Publishers, Weinheim, Germany, pp. 83-107.

Blaise, C. (2002) Use of microscopic algae in toxicity testing, in G. Bitton (ed.), *Encyclopedia of Environmental Microbiology*, John Wiley & Sons Inc., New York, pp. 3219-3230.

Blaise, C. (2003) Canadian application of bioassays for environmental management: a review, in M. Munawar (ed.), *Sediment Quality Assessment and Management: Insight and Progress*, Aquatic Ecosystem Health and Management Society, Canada, pp. 39-58.

Blaise, C. and Costan, G. (1987) La toxicité létale aiguë des effluents industriels au Québec vis-à-vis de la truite arc-en-ciel, *Water Pollution Research Journal of Canada* **22** (3), 385-402.

Blaise, C. and Kusui, T. (1997) Acute toxicity assessment of industrial effluents with a microplate-based *Hydra attenuata* assay, *Environmental Toxicology and Water Quality* **12** (1), 53-60.

Blaise, C. and Ménard, L. (1998) A micro-algal solid-phase test to assess the toxic potential of freshwater sediments, *Water Quality Research Journal of Canada* **33** (1), 133-151.

Blaise, C., Bermingham, N. and Van Collie, R. (1985) The integrated ecotoxicological approach to assessment of ecotoxicity, *Water Quality Bulletin* **10** (1), 3-10.

Blaise, C., Van Coillie, R., Bermingham, N. and Coulombe, G. (1987) Comparaison des réponses toxiques de trois indicateurs biologiques (bactéries, algues, poissons) exposés à des effluents de fabriques de pâtes et papiers, *Revue Internationale des Sciences de l'Eau* **3** (1), 9-17.

Blaise, C., Sergy, G., Wells, P., Bermingham, N. and Van Coillie, R. (1988) Biological Testing-Development, Application, and Trends in Canadian Environmental Protection Laboratories, *Toxicity Assessment* **3**, 385-406.

Blaise, C., Forghani, R., Legault, R., Guzzo, J. and Dubow, M.S. (1994) A bacterial toxicity assay performed with microplates, microluminometry and Microtox reagent, *Biotechniques* **16** (5), 932-937.

Blaise, C., Wells, P.G. and Lee, K. (1998a) Microscale testing in aquatic toxicology: introduction, historical perspective, and context, in P.G. Wells, K. Lee and C. Blaise (eds.), *Microscale Testing in Aquatic Toxicology: Advances, Techniques, and Practice*, CRC Press, Boca Raton, FL, pp. 1-9.

Blaise, C., Férard, J.-F. and Vasseur, P. (1998b) Microplate toxicity tests with microalgae: a review, in P.G. Wells, K. Lee and C. Blaise (eds.), *Microscale Testing in Aquatic Toxicology: Advances, Techniques, and Practice*, CRC Press, Boca Raton, FL, pp. 269-288.

Blaise, C., Gagné, F. and Bombardier, M. (2000) Recent developments in microbiotesting and early millennium prospects, *Water, Air, and Soil Pollution* **123** (1-4), 11-23.

Blaise, C, Gagné, F., Chèvre, N., Harwood, M., Lee, K., Lappalainen, J, Chial, B., Persoone, G. and Doe, K. (2004) Toxicity assessment of oil-contaminated freshwater sediments, *Environmental Toxicology* **19**, 329-335.

Bleckmann, C.A., Rabe, B., Edgmon, S.J. and Fillingame, D. (1995) Aquatic toxicity variability for fresh- and saltwater species in refinery wastewater effluent, *Environmental Toxicology and Chemistry* **14** (7), 1219-1223.

Blinova, I. (2000) Comparison of the sensitivity of aquatic test species for toxicity evaluation of various environmental samples, in G. Persoone, C. Janssen and W.M. De Coen (eds.), *New Microbiotests for Routine Toxicity Screening and Biomonitoring*, Kluwer Academic/Plenum Publishers, New York, pp. 217-220.

Boluda, R., Quintanilla, J.F., Bonilla, J.A., Saez, E. and Gamon, M. (2002) Application of the Microtox test and pollution indices to the study of water toxicity in the Albufera Natural Park (Valencia, Spain), *Chemosphere* **46** (2), 355-369.

Bombardier, M. and Blaise, C. (2000) Comparative study of the sediment-toxicity index, benthic community metrics and contaminant concentrations, *Water Quality Research Journal of Canada* **35** (4), 753-780.

Bonnemoy, F., Lavédrine, B. and Boulkamh, A. (2004) Influence of UV irradiation on the toxicity of phenylurea herbicides using Microtox® test, *Chemosphere* **54** (8), 1183-1187.

Bonnet, C. (2000) Développement de bioessais sur sédiments et applications à l'étude, en laboratoire, de la toxicité de sédiments dulçaquicoles contaminés, UFR Sciences Fondamentales et Appliquées, Université de METZ, Metz, France, 326 pages.

Borgmann, U. and Munawar, M. (1989) A new standardized sediment bioassay protocol using the amphipod *Hyalella azteca* (Saussure), in M. Munawar, G. Dixon, C.I. Mayfield, T. Reynoldson and M.H. Sadar (eds.), *Environmental Bioassay Techniques and their Application: Proceedings of the 1st International Conference held in Lancaster, England, 11-14 July 1988,* Kluwer Academic Publishers, Dordrecht, Netherlands, pp. 425-531.

Botsford, J.L. (1998) A simple assay for toxic chemicals using a bacterial indicator, *World Journal of Microbiology and Biotechnology* **14** (3), 369-376.

Boulanger, B. and Nikolaidis, N.P. (2003) Mobility and aquatic toxicity of copper in an urban watershed, *Journal of the American Water Resources Association* **39** (2), 325-336.

Bradfield, A.D., Flexner, N.M. and Webster, D.A. (1993) *Water quality, organic chemistry of sediment, and biological conditions of streams near an abandoned wood-preserving plant site at Jackson, Tennessee*, Water-resources investigations report; 93-4148, USGS, Earth Science Information Center, Denver, CO, pp. 1-50.

Brent, R.N. and Herricks, E.E. (1998) Postexposure effects of brief cadmium, zinc, and phenol exposures on freshwater organisms, *Environmental Toxicology and Chemistry* **17** (10), 2091-2099.

Bridges, T.S., Wright, R.B., Gray, B.R., Gibson, A.B. and Dillon, T.M. (1996) Chronic toxicity of Great Lakes sediments to *Daphnia magna*: elutriate effects on survival, reproduction and population growth, *Ecotoxicology* **5**, 83-102.

Brix, K.V., DeForest, D.K. and Adams, W.J. (2001a) Assessing acute and chronic copper risks to freshwater aquatic life using species sensitivity distributions for different taxonomic groups, *Environmental Toxicology and Chemistry* **20** (8), 1846-1856.

Brix, K.V., Henderson, D.G., Adams, W.J., Reash, R.J., Carlton, R.G. and McIntyre, D.O. (2001b) Acute toxicity of sodium selenate to two daphnids and three amphipods, *Environmental Toxicology* **16** (2), 142-150.

Brix, K.V., Volosin, J.S., Adams, W.J., Reash, R.J., Carlton, R.G. and McIntyre, D.O. (2001c) Effects of sulfate on the acute toxicity of selenate to freshwater organisms, *Environmental Toxicology and Chemistry* **20** (5), 1037-1045.

Broomhall, S. (2002) The effects of endosulfan and variable water temperature on survivorship and subsequent vulnerability to predation in *Litoria citropa* tadpoles, *Aquatic Toxicology* **61** (3-4), 243-250.

Brorson, T., Björklund, I., Svenstam, G. and Lantz, R. (1994) Comparison of two strategies for assessing ecotoxicological aspects of complex wastewater from a chemical-pharmaceutical plant, *Environmental Toxicology and Chemistry* **13** (4), 543-552.

Brouwer, H., Murphy T. and McArdle, L. (1990) A sediment contact assay with *Photobacterium phosphoreum*, *Environmental Toxicology and Chemistry* **9**, 1353-1358.

Buchwalter, D.B., Jenkins, J.J. and Curtis, L.R. (2003) Temperature influences on water permeability and chlorpyrifos uptake in aquatic insects with differing respiratory strategies, *Environmental Toxicology and Chemistry* **22** (11), 2806-2812.

Bulich, A.A. and Bailey, G. (1995) Environmental toxicity assessment using luminescent bacteria, in M. Richardson (ed.), *Environmental Toxicology Assessment*, Taylor & Francis Ltd., London, England, pp. 29-40.

Bulich, A.A., Huynh, H. and Ulitzur, S. (1996) The use of luminescent bacteria for measuring chronic toxicity, in G.K. Ostrander (ed.), *Techniques in Aquatic Toxicology*, CRC Press, Boca Raton, FL, pp. 3-12.

Burba, A. (1999) The design of an experimental system of estimation methods for effects of heavy metals and their mixtures on *Daphnia magna*, *Acta Zoologica Lituanica, Hydrobiologia* **9** (2), 21-29.

Burton Jr, G.A., Nelson, M.K. and Ingersoll, C.G. (1992) Freshwater benthic toxicity tests, in G.A. Burton Jr. (ed.), *Sediment Toxicity Assessment*, Lewis Publishers, Boca Raton, FL, pp. 213-240.

Burton Jr, G.A., Norberg-King, T.J., Ingersoll, C.G., Benoit, D.A., Ankley, G.T., Winger, P.V., Kubitz, J.A., Lazorchak, J.M., Smith, M.E., Greer, E., Dwyer, F.J., Call, D.J., Day, K.E., Kennedy, P. and Stinson, M. (1996) Interlaboratory study of precision: *Hyallela azteca* and *Chironomus tentans* freshwater sediment toxicity assays, *Environmental Toxicology and Chemistry* **15** (8), 1335-1343.

Burton Jr, G.A., Baudo, R., Beltrami, M. and Rowland, C. (2001) Assessing sediment contamination using six toxicity assays, *Journal of Limnology* **60** (2), 263-267.

Bury, N.R., Shaw, J., Glover, C. and Hogstrand, C. (2002) Derivation of a toxicity-based model to predict how water chemistry influences silver toxicity to invertebrates, *Comparative Biochemistry and Physiology, Part C* **133** (1-2), 259-270.

Cairns Jr, J. and Pratt, J.R. (1989) The scientific basis of bioassays, in M. Munawar, G. Dixon, C.I. Mayfeld, T. Reynoldson and M.H. Sadar (eds.), *Environmental Bioassay Techniques and their Application: Proceedings of the 1st International Conference held in Lancaster, England, 11-14 July 1988*, Kluwer Academic Publishers, Dordrecht, Netherlands, pp. 5-20.

Call, D.J., Liber, K., Whiteman, F.W., Dawson, T.D. and Brooke, L.T. (1999) Observations on the 10-day *Chironomus tentans* survival and growth bioassay in evaluating Great Lakes sediments, *Journal of the Great Lakes Research* **25**, 171-178.

Call, D.J., Markee, T.P., Geiger, D.L., Brooke, L.T., VandeVenter, F.A., Cox, D.A., Genisot, K.I., Robillard, K.A., Gorsuch, J.W., Parkerton, T.F., Reiley, M.C., Ankley, G.T. and Mount, D.R. (2001) An assessment of the toxicity of phthalate esters to freshwater benthos. 1. Aqueous exposures, *Environmental Toxicology and Chemistry* **20** (8), 1798-1804.

Calleja, M.C. and Persoone, G. (1993) The influence of solvents on the acute toxicity of some lipophilic chemicals to aquatic invertebrates, *Chemosphere* **26** (11), 2007-2022.

Calow, P. (1989) The choice and implementation of environmental bioassays, in M. Munawar, G. Dixon, C.I. Mayfeld, T. Reynoldson and M.H. Sadar (eds.), *Environmental Bioassay Techniques and their Application: Proceedings of the 1st International Conference held in Lancaster, England, 11-14 July 1988*, Kluwer Academic Publishers, Dordrecht, Netherlands, pp. 61-64.

Campbell, D.L., Lawton, L.A., Beattie, K.A. and Codd, G.A. (1994) Comparative assessment of the specificity of the brine shrimp and Microtox assays to hepatotoxic (microcystin-LR-containing) cyanobacteria, *Environmental Toxicology and Water Quality* **9** (1), 71-77.

Cancilla, D.A., Holtkamp, A., Matassa, L. and Fang, X. (1997) Isolation and characterization of Microtox®-active components from aircraft de-icing/anti-icing fluids, *Environmental Toxicology and Chemistry* **16**(3), 430-434.

Canfield, T.J., Brunson, E.L., Dwyer, F.J., Ingersoll, C.G. and Kemble, N.E. (1998) Assessing sediments from Upper Mississippi River navigational pools using a benthic invertebrate community evaluation and the sediment quality triad approach, *Archives of Environmental Contamination and Toxicology* **35** (2), 202-212.

Cangiano, T., Dellagreca, M., Fiorentino, A., Isidori, M., Monaco, P. and Zarrelli, A. (2002) Effect of ent-labdane diterpenes from *Potamogetonaceae* on *Selenastrum capricornutum* and other aquatic organisms, *Journal of Chemical Ecology* **28** (6), 1091-1102.

Canivet, V. and Gibert, J. (2002) Sensitivity of epigean and hypogean freshwater macroinvertebrates to complex mixtures. Part I: Laboratory experiments, *Chemosphere* **46** (7), 999-1009.

CANMET (1996) Comparison of results from alternative acute toxicity tests with rainbow trout for selected mine effluents, *Aquatic Effects Technology Evaluation (AETE) Program*, Project 1.1.4, Canada Centre for Mineral and Energy Technology (CANMET), Mining Association of Canada (MAC), Ottawa, Ontario, pp. 1-228.

CANMET (1997a) Review of methods for sublethal aquatic toxicity tests relevant to the Canadian metal-mining industry, *Aquatic Effects Technology Evaluation (AETE) Program*, Project 1.2.1, Canada Centre for Mineral and Energy Technology (CANMET), Mining Association of Canada (MAC), Ottawa, Ontario, pp. 1-132.

CANMET (1997b) Laboratory screening of sublethal toxicity tests for selected mine effluents, *Aquatic Effects Technology Evaluation (AETE) Program*, Project 1.2.2, Canada Center for Mineral and Energy Technology (CANMET), Mining Association of Canada (MAC), Ottawa, Ontario, pp. 1-69.

CANMET (1997c) Toxicity assessment of highly mineralized waters from potential mine sites, *Aquatic Effects Technology Evaluation (AETE) Program*, Project 1.2.4, Canada Centre for Mineral and Energy Technology (CANMET), Mining Association of Canada (MAC), Ottawa, Ontario, 38 pp.

CANMET (1998) Toxicity assessment of mining effluents using up-stream or reference site waters and test organism acclimation techniques, *Aquatic Effects Technology Evaluation (AETE) Program*, Project 4.1.2a, Canada Centre for Mineral and Energy Technology (CANMET), Mining Association of Canada (MAC), Ottawa, Ontario, 81 pp.

CANMET (1999) Technical evaluation of determining mining related impacts utilizing benthos macroinvertebrate fitness parameters, *Aquatic Effects Technology Evaluation (AETE) Program*, Project 2.1.5, Canada Centre for Mineral and Energy Technology (CANMET), Mining Association of Canada (MAC), Ottawa, Ontario, 81 pp.

Canna-Michaelidou, S., Nicolaou, A.S., Neopfytou, E. and Christodoulidou, M., (2000) The use of a battery of microbiotests as a tool for integrated pollution control evaluation and perspectives in Cyprus, in G. Persoone, C. Janssen and W.M. De Coen (eds.), *New Microbiotests for Routine Toxicity Screening and Biomonitoring,* Kluwer Academic / Plenum Publishers, New York, pp. 39-48.

Carter, J.A., Mroz, R.E., Tay, K.L. and Doe, K.G. (1998) An evaluation of the use of soil and sediment bioassays in the assessment of three contaminated sites in Atlantic Canada, *Water Quality Research Journal of Canada* **33** (2), 295-317.

Castaño, A., Bols, N.C., Braunbeck, T., Dierickx, P.J., Halder, M., Isomaa, B., Kawahara, K., Lee, L.E.J., Mothersill, C., Pärt, P., Repetto, G., Sintes, J.R., Rufli, H., Smith, R., Wood, C. and Segner, H. (2003) The use of fish cells in ecotoxicology. The report and recommendations of ECVAM Workshop 47, *ATLA (Alternatives To Laboratory Animals)* **31** (3), 317-351.

Castillo, G.C., Vila, I.C. and Neild, E. (2000) Ecotoxicity assessment of metals and wastewater using multitrophic assays, *Environmental Toxicology* **15** (5), 370-375.

Cataldo, D., Colombo, J.C., Boltovskoy, D., Bilos, C. and Landoni, P. (2001) Environmental toxicity assessment in the Paraná river delta (Argentina): simultaneous evaluation of selected pollutants and mortality rates of *Corbicula fluminea* (Bivalvia) early juveniles, *Environmental Pollution* **112** (3), 379-389.

Cerejeira, M.J., Pereira, T. and Silva-Fernandes, A. (1998) Use of new microbiotests with *Daphnia magna* and *Selenastrum capricornutum* immobilized forms, *Chemosphere* **37** (14-15), 2949-2955.

Chan, Y.K., Wong, C.K., Hsieh, D.P.H., Ng, S.P., Lau, T.K. and Wong, P.K. (2003) Application of a toxicity identification evaluation for a sample of effluent discharged from a dyeing factory in Hong Kong, *Environmental Toxicology* **18** (5), 312-316.

Chapman, P.M. (1998) Death by mud: amphipod sediment toxicity tests, in P.G. Wells, K. Lee and C. Blaise (eds.), *Microscale Testing in Aquatic Toxicology: Advances, Techniques, and Practice*, CRC Press, Boca Raton, FL, pp. 451-463.

Chapman, K.K., Benton, M.J., Brinkhurst, R.O. and Scheuerman, P.R. (1999) Use of the aquatic oligochaetes *Lumbriculus variegatus* and *Tubifex tubifex* for assessing the toxicity of copper and cadmium in a spiked-artificial-sediment toxicity test, *Environmental Toxicology* **14** (2), 271-278.

Charles, A.L., Markich, S.J., Stauber, J.L. and De Filippis, L.F. (2002) The effect of water hardness on the toxicity of uranium to a tropical freshwater alga (*Chlorella* sp.), *Aquatic Toxicology* **60** (1-2), 61-73.

Cheam, V., Reynoldson, T., Garbai, G., Rajkumar, J. and Milani, D. (2000) Local impacts of coal mines and power plants across Canada. II. Metals, organics and toxicity in sediments, *Water Quality Research Journal of Canada* **35** (4), 609-631.

Chen, C.-Y. and Lin, K.-C. (1997) Optimization and performance evaluation of the continuous algal toxicity test, *Environmental Toxicology and Chemistry* **16** (7), 1337-1344.

Chen, C.-Y., Huang, J.-B. and Chen, S.-D. (1997) Assessment of the microbial toxicity test and its application for industrial wastewaters, *Water Science and Technology* **36** (12), 375-382.

Chen, C.-Y., Chen, J.-N. and Chen, S.-D. (1999) Toxicity assessment of industrial wastewater by microbial testing method, *Water Science and Technology* **39** (10-11), 139-143.

Chial, B.Z. and Persoone, G. (2002) Cyst-based toxicity tests XIII - Development of a short chronic sediment toxicity test with the ostracod crustacean *Heterocypris incongruens*: Methodology and precision, *Environmental Toxicology* **17** (6), 528-532.

Chial, B.Z., Persoone, G. and Blaise, C. (2003) Cyst-based toxicity tests. XVIII. Application of ostracodtoxkit microbiotest in a bioremediation project of oil-contaminated sediments: Sensitivity comparison with *Hyalella azteca* solid-phase assay, *Environmental Toxicology* **18** (5), 279 - 283.

Choi, K. and Meier, P.G. (2001) Toxicity evaluation of metal plating wastewater employing the Microtox® assay: a comparison with cladocerans and fish, *Environmental Toxicology* **16** (2), 136-141.

Choi, K., Sweet, L.I., Meier, P.G. and Kim, P.G. (2004) Aquatic toxicity of four alkylphenols (3-tert-butylphenol, 2-isopropylphenol, 3-isopropylphenol, and 4-iso-propylphenol) and their binary mixtures to microbes, invertebrates and fish, *Environmental Toxicology* **19**, 45-50.

Chu, S., He, Y. and Xu, X. (1997) Determination of acute toxicity of polychlorinated biphenyls to *Photobacterium phosphoreum*, *Bulletin of Environmental Contamination and Toxicology* **58** (2), 263-267.

Cieniawski, S. and Collier, D. (2003) *Post-Remediation Sediment Sampling on the Raisin River Near Monroe, Michigan*, U.S. Environmental Protection Agency, Great Lakes National Programm Office, Chicago, Illinois, 52 pp.

Clément, B. (2000) The use of microbiotests for assessing the influence of the dilution medium quality on the acute toxicity of chemicals and effluents, in G. Persoone, C. Janssen and W.M. De Coen (eds.), *New Microbiotests for Routine Toxicity Screening and Biomonitoring*, Kluwer Academic/Plenum Publishers, New York, pp. 221-228.

Clément, B., Persoone, G., Janssen, C. and Le Dû-Delepierre, A. (1996) Estimation of the hazard of landfills through toxicity testing of leachates - I. Determination of leachate toxicity with a battery of acute tests, *Chemosphere* **33** (11), 2303-2320.

Cleuvers, M. and Ratte, H.T. (2002a) Phytotoxicity of coloured substances: is *Lemna* duckweed an alternative to the algal growth inhibition test?, *Chemosphere* **49** (1), 9-15.

Cleuvers, M. and Ratte, H.T. (2002b) The importance of light intensity in algal tests with coloured substances, *Water Research* **36** (9), 2173-2178.

Cleuvers, M. and Weyers, A. (2003) Algal growth inhibition test: does shading of coloured substances really matter?, *Water Research* **37** (11), 2718-2722.

Codina, J.C., Pérez-García, A. and de Vicente, A. (1994) Detection of heavy metal toxicity and genotoxicity in wastewaters by microbial assay, *Water Science and Technology* **30** (10), 145-151.

Coleman, R.N. and Qureshi, A.A. (1985) Microtox and *Spirillum volutans* tests for assessing toxicity of environmental samples, *Bulletin of Environmental Contamination and Toxicology* **35** (4), 443-451.

Collier, D. and Cieniawski, S. (2003) *Survey of sediment contamination in the Chicago River, Chicago, Illinois*, U.S. Environmental Protection Agency, Great Lakes National Program Office, Chicago, IL, (2003-12-30), 46 pp.; http://www.epa.gov/glnpo/sediment/ChgoRvr/chgorvrpt.pdf.

Connon, R., Printes, L.B., Dewhurst, R.E., Crane, M. and Callaghan, A. (2000) *Groundwater Pollution: development of biomarkers for the assessment of sublethal toxicity*, URGENT Annual Meeting 2000 Proceedings of the NERC URGENT Thematic Programme, Cardiff University, Wales, UK (2003-12-22); http://urgent.nerc.ac.uk/Meetings/2000/2000Proc/water/connon.htm.

Conrad, A.U., Fleming, R.J. and Crane, M. (1999) Laboratory and field response of *Chironomus riparius* to a pyrethroid insecticide, *Water Research* **33**(7), 1603-1610.

Cooman, K., Gajardo, M., Nieto, J., Bornhardt, C. and Vidal, G. (2003) Tannery wastewater characterization and toxicity effects on *Daphnia* spp., *Environmental Toxicology* **18** (1), 45-51.

Cortes, G., Mendoza, A. and Muñoz, D. (1996) Toxicity evaluation using bioassays in rural developing district 063 Hidalgo, Mexico, *Environmental Toxicology and Water Quality* **11** (2), 137-143.

Côté, C., Blaise, C., Michaud, J.-R., Ménard, L., Trottier, S., Gagné, F. and Lifshitz, R. (1998a) Comparisons between microscale and whole-sediment assays for freshwater sediment toxicity assessment, *Environmental Toxicology and Water Quality* **13** (1), 93-110.

Côté, C., Blaise, C., Schroeder, J., Douville, M. and Michaud, J.-R. (1998b) Investigating the adequacy of selected micro-scale bioassays to predict the toxic potential of freshwater sediments through a tier process, *Water Quality Research Journal of Canada* **33** (2), 253-277.

Côté, C., Douville, M. and Michaud, J.-R. (1999) *Eaux usées industrielles : évaluation de micro-bioessais pour la surveillance et l'identification de la toxicité des effluents de l'industrie papetière*, Saint-Laurent Vision 2000, Environnement Canada, Québec.

Couture, P., Blaise, C., Cluis, D. and Bastien, C. (1989) Zirconium toxicity assessment using bacteria, algae and fish assay, *Water, Air, and Soil Pollution* **47** (1-2), 87-100.

Cowgill, U.M. (1986) Why round-robin testing with zooplankton often fails to provide acceptable results, in T.M. Poston and R. Purdy (eds.), *Aquatic Toxicology and Environmental Fate: 9th Volume, ASTM STP 921*, American Society for Testing and Materials, Philadelphia, PA, pp. 349-356.

Coya, B., Marañón, E. and Sastre, H. (1996) Evaluation of the ecotoxicity of industrial wastes by microtox bioassay, *Toxicology Letters* **88** (Supplement 1), 79-79.

Coz, A., Andrés, A. and Irabien, A. (2004) Ecotoxicity assessment of stabilized/solifidied foundry sludge, *Environmental Science and Technology* **38**, 1897-1900.

Crane, M., Delaney, P., Mainstone, C. and Clarke, S. (1995) Measurement by *in situ* bioassay of water quality in an agricultural catchment, *Water Research* **29** (11), 2441-2448.

Crane, M., Attwood, C., Sheahan, D. and Morris, S. (1999) Toxicity and bioavailability of the organophosphorus insecticide pirimiphos methyl to the freshwater amphipod *Gammarus pulex* L. in laboratory and mesocosm systems, *Environmental Toxicology and Chemistry* **18** (7), 1456-1461.

Czerniawska-Kusza, I. and Ebis, M. (2000) Toxicity of waste dump leachates and sugar factory effluents and their impact on groundwater and surface water quality in the Opole Province in Poland, in G. Persoone, C. Janssen and W.M. De Coen (eds.), *New Microbiotests for Routine Toxicity Screening and Biomonitoring*, Kluwer Academic/Plenum Publishers, New York, pp. 319-322.

Daniels, S.A., Munawar, M. and Mayfield, C.I. (1989) An improved elutriation technique for the bioassessment of sediment contaminants, in M. Munawar, G. Dixon, C.I. Mayfield, T. Reynoldson and M.H. Sadar (eds.), *Environmental Bioassay Techniques and their Application: Proceedings of the 1st International Conference held in Lancaster, England, 11-14 July 1988*, Kluwer Academic Publishers, Dordrecht, Netherlands, pp. 619-631.

Dauble, D.D., Fallon, W.E., Gray, R.H. and Bean, R.M. (1982) Effects of coal liquid water-soluble fractions on growth and survival of four aquatic organisms, *Archives of Environmental Contamination and Toxicology* **11** (5), 553-560.

Dave, G. and Dennegard, B. (1994) Sediment toxicity and heavy metals in the Kattegat and Skaggerak, *Journal of Aquatic Ecosystem Health* **3** (3), 207-219.

Day, K.E., Holtze, K.E., Metcalfe-Smith, J.L., Bishop, C.T. and Dutka, B.J. (1993) Toxicity of leachate from automobile tires to aquatic biota, *Chemosphere* **27** (4), 665-675.

Day, K.E., Dutka, B.J., Kwan, K.K., Batista, N., Reynoldson, T.B. and Metcalfe-Smith, J.L. (1995a) Correlations between solid-phase microbial screening assays, whole-sediment toxicity tests with macroinvertebrates and *in situ* benthic community structure, *Journal of the Great Lakes Research* **21** (2), 192-206.

Day, K.E., Kirby, R.S. and Reynoldson, T.B. (1995b) The effect of manipulations of freshwater sediments on responses of benthic invertebrates in whole-sediment toxicity tests, *Environmental Toxicology and Chemistry* **14** (8), 1333-1343.

Day, K.E., Maguire, R.J., Milani, D. and Batchelor, S.P. (1998) Toxicity of tributyltin to four species of freshwater benthic invertebrates using spiked sediment bioassays, *Water Quality Research Journal of Canada* **33** (1), 111-132.

De Coen, W.M. and Janssen, C.R. (1997) The use of biomarkers in *Daphnia magna* toxicity testing II. Digestive enzyme activity in *Daphnia magna* exposed to sublethal concentrations of cadmium, chromium and mercury, *Chemosphere* **35** (5), 1053-1067.

de Haas, E.M., Reuvers, B., Moermond, C.T.A., Koelmans, A.A. and Kraak, M.H.S. (2002) Responses of benthic invertebrates to combined toxicant and food input in floodplain lake sediments, *Environmental Toxicology and Chemistry* **21** (10), 2165-2171.

de Haas, E.M., Paumen, M.L., Koelman, A.A. and Kraak, M.H.S. (2004) Combined effects of copper and food on the midge *Chironomus riparius* in whole-sediment bioassays, *Environmental Pollution* **127** (1), 99-107.

de Jonge, J., Brils, J.M., Hendriks, A.J. and Ma, C. (1999) Ecological and ecotoxicological surveys of moderately contaminated floodplain ecosystems in the Netherlands, *Aquatic Ecosystem Health and Management* **2** (1), 9-18.

de Vlaming, V. and Norberg-King, T.J. (1999) *A review of single species toxicity tests: are the tests reliable predictors of aquatic ecosystem community reponses?*, EPA 600/R-97/114, Office of Research and Development, U.S. Environmental Protection Agency, Duluth, MN.

DellaGreca, M., Fiorentino, A., Isidori, M., Monaco, P., Temussi, F. and Zarrelli, A. (2001) Antialgal furano-diterpenes from *Potamogeton natans* L., *Phytochemistry* **58** (2), 299-304.

DellaGreca, M., Fiorentino, A., Isidori, M., Lavorgna, M., Monaco, P., Previtera, L. and Zarrelli, A. (2002) Phenanthrenoids from the wetland *Juncus acutus*, *Phytochemistry* **60** (6), 633-638.

DellaGreca, M., Fiorentino, A., Isidori, M., Lavorgna, M., Previtera, L., Rubino, M. and Temussi, F. (2004) Toxicity of prednisolone, dexamethasone and their photochemical derivatives on aquatic organisms, *Chemosphere* **54** (5), 629-637.

den Besten, P.J. and Tuk, C.W. (2000) Relation between responses in the neutral red retention test and the comet assay and life history parameters of *Daphnia magna*, *Marine Environment Research* **50** (1-5), 513-516.

den Besten, P.J., Naber, A., Grootelaar, E.M.M. and van de Guchte, C. (2003) *In situ* bioassays with *Chironomus riparius*: laboratory-field comparisons of sediment toxicity and effects during wintering, *Aquatic Ecosystem Health and management* **6** (2), 217-228.

Deniseger, J. and Kwong, Y.T.J. (1996) Risk Assessment of Copper-Contaminated Sediments in the Tsolum River Near Courtenay, British Columbia, *Water Quality Research Journal of Canada* **31** (4), 725-740.

Denizeau, F. (1998) The use of fish cells in the toxicological evaluation of environmental contaminants, in P.G. Wells, K. Lee and C. Blaise (eds.), *Microscale Testing in Aquatic Toxicology: Advances, Techniques, and Practice*, CRC Press, Boca Raton, FL, pp. 113-128.

Dewhurst, R.E., Connon, R., Crane, M., Callaghan, A. and Mather, J.D. (2001) *URGENT in Hounslow and Heathrow. The application of acute and sub-lethal ecotoxicity tests to groundwater quality assessment*, URGENT Annual Meeting 2000 Proceedings of the NERC URGENT Thematic Programme, Cardiff University, Wales, UK (2004-01-02); http://urgent.nerc.ac.uk/Meetings/2001/Abstracts/mather.htm.

Dias, N. and Lima, N. (2002) A comparative study using a fluorescence-based and a direct-count assay to determine cytotoxicity in *Tetrahymena pyriformis*, *Research in Microbiology* **153** (5), 313-322.

Diaz-Baez, M.C. and Roldan, F. (1996) Evaluation of the agar plate method for rapid toxicity assessment with some heavy metals and environmental samples, *Environmental Toxicology and Water Quality* **11** (3), 259-263.

Diaz-Baez, M.C., Sanchez, W.A., Dutka, B.J., Ronco, A., Castillo, G., Pica-Granados, Y., Castillo, L.E., Ridal, J., Arkhipchuk, V. and Srivastava, R.C. (2002) Overview of results from the WaterTox intercalibration and environmental testing phase II program: part 2, ecotoxicological evaluation of drinking water supplies, *Environmental Toxicology* **17** (3), 241-249.

Dieter, C.D., Hamilton, S.J., Duffy, W.G. and Flake, L.D. (1994) Evaluation of the Microtox test to detect phorate contamination in wetlands, *Journal of Freshwater Ecology* **9** (4), 271-280.

Dive, D., Blaise, C., Robert, S., Le Du, A., Bermingham, N., Cardin, R., Kwan, A., Legault, R., Mac Carthy, L., Moul, D. and Veilleux, L. (1990) Canadian workshop on the *Colpidium campylum* ciliate protozoan growth inhibition test, *Zeitschrift für angewandte Zoologie* **76** (1), 49-63.

Dive, D., Blaise, C. and Le Du, A. (1991) Standard protocol proposal for undertaking the *Colpidium campylum* ciliate protozoan growth inhibition test, *Zeitschrift für angewandte Zoologie* **78** (1), 79-90.

Dizer, H., Wittekindt, E., Fischer, B. and Hansen, P.-D. (2002) The cytotoxic and genotoxic potential of surface water and wastewater effluents as determined by bioluminescence, umu-assays and selected biomarkers, *Chemosphere* **46** (2), 225-233.

Dmitruk, U. and Dojlido, J. (2000) Application of Toxkit microbiotests for toxicity evaluation of river waters and waste waters in the region of Warsaw in Poland, in G. Persoone, C. Janssen and W.M. De Coen (eds.), *New Microbiotests for Routine Toxicity Screening and Biomonitoring*, Kluwer Academic/Plenum Publishers, New York, pp. 323-325.

Dodard, S.G., Renoux, A.Y., Hawari, J., Ampleman, G., Thiboutot, S. and Sunahara, G.I. (1999) Ecotoxicity characterization of dinitrotoluenes and some of their reduced metabolites, *Chemosphere* **38** (9), 2071-2079.

Dodson, S.I., Merritt, C.M., Shannahan, J.-P. and Shults, C.M. (1999) Low exposure concentrations of atrazine increase male production in *Daphnia pulicaria*, *Environmental Toxicology and Chemistry* **18** (7), 1568-1573.

Doherty, F.G. (2001) A review of the Microtox toxicity test system for assessing the toxicity of sediments and soils, *Water Quality Research Journal of Canada* **36** (3), 475-518.

Doherty, F.G., Qureshi, A.A. and Razza, J.B. (1999) Comparison of the *Ceriodaphnia dubia* and Microtox® inhibition tests for toxicity assessment of industrial and municipal wastewaters, *Environmental Toxicology* **14** (4), 375-382.

Doi, J. and Grothe, D.R. (1989) Use of fractionation and chemical analysis schemes for plant effluent toxicity evaluations, in G.W. Suter II and M.A. Lewis (eds.), *Aquatic Toxicology and Environmental Fate: Eleventh Volume, ASTM STP 1007*, American Society for Testing and Materials, Philadelphia, PA, pp. 204-215.

Dombroski, E.C., Smiley, K.L., Johnson, C.I., Florence, L.Z. and Dieken, F.P. (1993) A comparison of bioassay results from untreated CTMP effluent, *Canadian Technical Report of Fisheries and Aquatic Sciences* **1942**, 96-104.

Draper III, A.C. and Brewer, W.S. (1979) Measurement of the aquatic toxicity of volatile nitrosamines, *Journal of Toxicology and Environmental Health* **5** (6), 985-993.

Dunbar, A.M., Lazorchak, J.M. and Waller, W.T. (1983) Acute and chronic toxicity of sodium selenate to *Daphnia magna* Straus, *Environmental Toxicology and Chemistry* **2** (2), 239-244.

Dutka, B.J., Tuominen, T., Churchland, L. and Kwan, K.K. (1989) Fraser river sediments and waters evaluated by the battery of screening tests technique, in M. Munawar, G. Dixon, C.I. Mayfield, T. Reynoldson and M.H. Sadar (eds.), *Environmental Bioassay Techniques and their Application: Proceedings of the 1st International Conference held in Lancaster, England, 11-14 July 1988*, Kluwer Academic Publishers, Dordrecht, Netherlands, pp. 301-315.

Dutka, B.J., McInnis, R., Jurkovic, A., Liu, D. and Castillo, G. (1996) Water and sediment ecotoxicity studies in Temuco and Rapel River Basin, Chile, *Environmental Toxicology and Water Quality* **11** (3), 237-247.

Dyatlov, S. (2000) Comparison of Ukrainian standard methods and new microbiotests for water toxicity assessment, in G. Persoone, C. Janssen and W.M. De Coen (eds.), *New Microbiotests for Routine Toxicity Screening and Biomonitoring*, Kluwer Academic/Plenum Publishers, New York, pp. 229-232.

Eggen, R.I.L., Behra, R., Burkhardt-Holm, P., Escher, B.I. and Schweigert, N. (2004) Challenges in ecotoxicology, *Environmental Science and Technology*, February 1, 2004, pp. 59A-64A.

Eleftheriadis, K., Angelaki, A., Kungolos, A., Nalbandian, L. and Sakellaropoulos, G.P. (2000) Assessing the impact of atmospheric wet and dry deposition using chemical and toxicological analysis, in G. Persoone, C. Janssen and W.M. De Coen (eds.), *New Microbiotests for Routine Toxicity Screening and Biomonitoring*, Kluwer Academic/Plenum Publishers, New York, pp. 469-473.

Environment Canada (1990) Guidance document on control of toxicity test precision using reference toxicants, Report EPS 1/RM/12, Environment Canada, Ottawa, 85 pp.

Environment Canada (1994) Guidance document on collection and preparation of sediment for physicochemical characterization and biological testing, Report EPS 1/RM/29, Environment Canada, Ottawa, 144 pp.

Environment Canada (1995) Guidance document on measurement of toxicity test precision using control sediments spiked with a reference toxicant, Report EPS 1/RM/30, Environment Canada, Ottawa, 56 pp.

Environment Canada (1999) Guidance document on application and interpretation of single-species tests in environmental toxicology, Report EPS 1/RM/34, Environment Canada, Ottawa, 203 pp.

Environment Canada (2004a) Guidance document for testing the pathogenicity and toxicity of new microbial substances to aquatic and terrestrial organisms, Report EPS 1/RM/44, Environment Canada, Ottawa, 171 pp.

Environment Canada (2004b) Guidance document on statistical methods to determine endpoints to toxicity tests, Report EPS 1/RM/46, Environment Canada, Ottawa, 265 pp.

Erten-Unal, M., Wixson, B.G., Gale, N. and Pitt, J.L. (1998) Evaluation of toxicity, bioavailability and speciation of lead, zinc and cadmium in mine/mill wastewaters, *Chemical Speciation and Bioavailability* **10** (2), 37-46.

Evandri, M.G., Costa, L.G. and Bolle, P. (2003) Evaluation of brominated diphenyl ether-99 toxicity with *Raphidocelis subcapitata* and *Daphnia magna*, *Environmental Toxicology and Chemistry* **22** (9), 2167-2172.

Farrell, A.P., Kennedy, C.J., Wood, A., Johnston, B.D. and Bennett, W.R. (1998) Acute toxicity of a didecyldimethylammonium chloride-based wood preservative, bardac 2280, to aquatic species, *Environmental Toxicology and Chemistry* **17** (8), 1552-1557.

Farrell, A.P., Kennedy, C., Cheng, W. and Lemke, M.A. (2001) Acute toxicity of monochloramine to juvenile chinook salmon (*Oncorhynchus tshawytscha* Walbaum) and *Ceriodaphnia dubia*, *Water Quality Research Journal of Canada* **36** (1), 133-149.

Fearnside, D. and Hiley, P.D. (1993) The role of Microtox® in the detection and control of toxic trade effluents and spillages, in M. Richardson (ed.), *Ecotoxicology Monitoring*, VCH Publishers, Weinheim, Germany, pp. 319-332.

Feng, Q., Boone, A.N. and Vijayan, M.M. (2003) Copper impact on heat shock protein 70 expression and apoptosis in rainbow trout hepatocytes, *Comparative Biochemistry and Physiology, Part C* **135** (3), 345-355.

Fent, K. (2001) Fish cell lines as versatile tools in ecotoxicology: assessment of cytotoxicity, cytochrome P4501A induction potential and estrogenic activity of chemicals and environmental samples, *Toxicology In Vitro* **15** (4-5), 477-488.

Fentem, J. and Balls, M. (1993) Replacement of fish in ecotoxicology testing: use of bacteria, other lower organisms and fish cells *in vitro*, in M. Richardson (ed.), *Ecotoxicology Monitoring*, VCH Publishers, Weinheim, Germany, pp. 71-81.

Fernández-Alba, A.R., Hernando, M.D., Piedra, L. and Chisti, Y. (2002) Toxicity evaluation of single and mixed antifouling biocides measured with acute toxicity bioassays, *Analytica Chimica Acta* **456** (2), 303-312.

Fernández-Casalderrey, A., Ferrando, M.D. and Andreu-Moliner, E. (1993) Chronic toxicity of methylparathion to the rotifer *Brachionus calyciflorus* fed on *Nannochloris oculata* and *Chlorella pyrenoidosa*, *Hydrobiologia* **255/256**, 41-49.

Fernández-Sempere, J., Barrueso-Martínez, M.L., Font-Montesinos, R. and Sabater-Lillo, M.C. (1997) Characterization of tannery wastes. Comparison of three leachability tests, *Journal of Hazardous Materials* **54** (1-2), 31-45.

Ferrari, B., Radetski, C.M., Veber, A.-M. and Férard, J.-F. (1999) Ecotoxicological assessment of solid wastes: a combined liquid- and solid-phase testing approach using a battery of bioassays and biomarkers, *Environmental Toxicology and Chemistry* **18** (6), 1195-1202.

Fialkowski, W., Klonowska-Olejnik, M., Smith, B.D. and Rainbow, P.S. (2003) Mayfly larvae (*Baetis rhodani* and *B. vernus*) as biomonitors of trace metal pollution in streams of a catchment draining a zinc and lead mining area of Upper Silesia, Poland, *Environmental Pollution* **121** (2), 253-267.

Finger, S.E., Little, E.F., Henry, M.G., Fairchild, J.F. and Boyle, T.P. (1985) Comparison of laboratory and field assessment of fluorene - Part I: Effects of fluorene on the survival, growth, reproduction, and behavior of aquatic organisms in laboratory tests, in T.P. Boyle (ed.), *Validation and Predictability of Laboratory Methods for Assessing the Fate and Effects of Contaminants in Aquatic Ecosystems, ASTM STP 865*, American Society for Testing and Materials, Philadelphia, PA, pp. 120-133.

Fisher, D.J., Hersh, C.M., Paulson, R.L., Burton, D.T. and Hall Jr., L.W. (1989) Acute toxicity of industrial and municipal effluents in the state of Maryland, USA: results from one year of toxicity testing, in M. Munawar, G. Dixon, C.I. Mayfield, T. Reynoldson and M.H. Sadar (eds.), *Environmental Bioassay Techniques and their Application: Proceedings of the 1st International Conference held in Lancaster, England, 11-14 July 1988*, Kluwer Academic Publishers, Dordrecht, Netherlands, pp. 641-648.

Fisher, D.J., Knott, M.H., Turley, B.S., Yonkos, L.T. and Ziegler, G.P. (1998) Acute and chronic toxicity of industrial and municipal effluents in Maryland, U.S., *Water Environment Research* **10** (1), 101-107.

Fisher, S.W. (1986) Effects of temperature on the acute toxicity of PCP in the midge *Chironomus riparius* Meigen, *Bulletin of Environmental Contamination and Toxicology* **36** (5), 744-748.

Fisher, S.W. and Wadleigh, R.W. (1986) Effects of pH on the acute toxicity and uptake of [^{14}C]pentachlorophenol in the midge, *Chironomus riparius*, *Ecotoxicology and Environmental Safety* **11** (1), 1-8.

Fleming, R.J., Holmes, D. and Nixon, S.J. (1998) Toxicity of permethrin to *Chironomus riparius* in artificial and natural sediments, *Environmental Toxicology and Chemistry* **17** (7), 1332-1337.

Fliedner, A. (1997) Ecotoxicity of poorly water-soluble substances, *Chemosphere* **35** (1-2), 295-305.

Fochtman, P., Raszka, A. and Nierzedska, E. (2000) The use of conventional bioassays, microbiotests, and some rapid methods in the selection of an optimal test battery for the assessment of pesticides toxicity, *Environmental Toxicology* **15** (5), 376-384.

Font, R., Gomis, V., Fernandez, J. and Sabater, M.C. (1998) Physico-chemical characterization and leaching of tannery wastes, *Waste Management and Research* **16** (2), 139-149.

Forget, G., Gagnon, P., Sanchez, W.A. and Dutka, B.J. (2000) Overview of methods and results of the eight country International Development Research Centre (IDRC) WaterTox project, *Environmental Toxicology* **15** (4), 264-276.

Franklin, N.M., Stauber, J.L., Markich, S.J. and Lim, R.P. (2000) pH-dependent toxicity of copper and uranium to a tropical freshwater alga (*Chlorella* sp.), *Aquatic Toxicology* **48** (2-3), 275-289.

Franklin, N.M., Stauber, J.L., Apte, S.C. and Lim, R.P. (2002) Effect of initial cell density on the bioavailability and toxicity of copper in microalgal bioassays, *Environmental Toxicology and Chemistry* **21** (4), 742-751.

Froehner, K., Backhaus, T. and Grimme, L.H. (2000) Bioassays with *Vibrio fischeri* for the assessment of delayed toxicity, *Chemosphere* **40** (8), 821-828.

Froehner, K., Meyer, W. and Grimme, L.H. (2002) Time-dependent toxicity in the long-term inhibition assay with *Vibrio fischeri*, *Chemosphere* **46** (7), 987-997.

Fu, L.-J., Staples, R.E. and Stahl Jr, R.G. (1991) Application of the *Hydra attenuata* assay for identifying developmental hazards among natural waters and wastewaters, *Ecotoxicology and Environmental Safety* **22** (3), 309-319.

Fu, L.-J., Staples, C.A. and Stahl Jr, R.G. (1994) Assessing acute toxicities of pre- and post-treatment industrial wastewaters with *Hydra attenuata*: a comparative study of acute toxicity with the fathead minnow, *Pimephales promelas*, *Environmental Toxicology and Chemistry* **13** (4), 563-569.

Fuchsman, P.C., Barber, T.R. and Sheehan, P.J. (1998) Sediment toxicity evaluation for hexachlorobenzene: spiked sediment tests with *Leptocheirus plumulosus*, *Hyalella azteca*, and *Chironomus tentans*, *Archives of Environmental Contamination and Toxicology* **35** (4), 573-579.

Gabrielson, J., Kühn, I., Colque-Navarro, P., Hart, M., Iversen, A., McKenzie, D. and Möllby, R. (2003) Microplate-based microbial assay for risk assessment and (eco)toxic fingerprinting of chemicals, *Analytica Chimica Acta* 485, 121-130.

Gagné, F. and Blaise, C. (1993) Hepatic metallothionein level and mixed function oxidase activity in fingerling rainbow trout (*Oncorhynchus mykiss*) after acute exposure to pulp and paper mill effluents, *Water Research* **27** (11), 1669-1682.

Gagné, F. and Blaise, C. (1997) Evaluation of industrial wastewater quality with a chemiluminescent peroxidase activity assay, *Environmental Toxicology and Water Quality* **12** (4), 315-320.

Gagné, F. and Blaise, C. (1998a) Toxicological evaluation of municipal wastewaters to rainbow trout hepatocytes, *Toxicology Letters* **95** (Supplement 1), 194-194.

Gagné, F. and Blaise, C. (1998b) Estrogenic properties of municipal and industrial wastewaters evaluated with a rapid and sensitive chemoluminescent in situ hybridization assay (CISH) in rainbow trout hepatocytes, *Aquatic Toxicology* **44** (1), 83-91.

Gagné, F. and Blaise, C. (1998c) Differences in the measurement of cytotoxicity of complex mixtures with rainbow trout hepatocytes and fibroblasts, *Chemosphere* **37** (4), 753-769.

Gagné, F. and Blaise, C. (1999) Toxicological effects of municipal wastewaters to rainbow trout hepatocytes, *Bulletin of Environmental Contamination and Toxicology* **63** (4), 503-510.

Gagné, F. and Blaise, C. (2001) Acute cytotoxicity assessment of liquid samples using rainbow trout (*Oncorhynchus mykiss*) hepatocytes, *Environmental Toxicology* **16** (1), 104-109.

Gagné, F., Blaise, C., van Aggelen, G., Boivin, P., Martel, P., Chong-Kit, R., Jonczyk, E., Marion, M., Kennedy, S.W., Legault, R. and Goudreault, J. (1999a) Intercalibration study in the evaluation of toxicity with rainbow trout hepatocytes, *Environmental Toxicology* **14** (4), 429-437.

Gagné, F., Pardos, M., Blaise, C., Turcotte, P., Quémerais, B. and Fouquet, A. (1999b) Toxicity evaluation of organic sediment extracts resolved by size exclusion chromatography using rainbow trout hepatocytes, *Chemosphere* **39** (9), 1545-1570.

Gagné, F., Ridal, J., Blaise, C. and Brownlee, B. (1999c) Toxicological effects of geosmin and 2-methylisoborneol on rainbow trout hepatocytes, *Bulletin of Environmental Contamination and Toxicology* **63** (2), 174-180.

Gale, S.A., Smith, S.V., Lim, R.P., Jeffree, R.A. and Petocz, P. (2003) Insights into the mechanisms of copper tolerance of a population of black-banded rainbowfish (*Melanotaenia nigrans*) (Richardson) exposed to mine leachate, using 64/67Cu, *Aquatic Toxicology* **62** (2), 135-153.

Garay, V., Roman, G. and Isnard, P. (2000) Evaluation of PNEC values: extrapolation from Microtox®, algae, daphnid, and fish data to HC5, *Chemosphere* **40** (3), 267-273.

Gasith, A., Jop, K.M., Dickson, K.L., Parkerton, T.F. and Kaczmarek, S.A. (1988) Protocol for the identification of toxic fractions in industrial wastewater effluents, in W.J. Adams, G.A. Chapman and W.G. Landis (eds.), *Aquatic Toxicology and Hazard Assessment: 10th volume, ASTM STP 971*, American Society for Testing and Materials, Philadelphia, PA, pp. 204-215.

Geis, S.W., Fleming, K.L., Korthals, E.T., Searle, G., Reynolds, L. and Karner, D.A. (2000) Modifications to the algal growth inhibition test for use a regulatory assay, *Environmental Toxicology and Chemistry* **19** (1), 36-41.

Geis, S.W., Fleming, K.L., Mager, A. and Reynolds, L. (2003) Modifications to the fathead minnow (*Pimephales promelas*) chronic test method to remove mortality due to pathogenic organisms, *Environmental Toxicology and Chemistry* **22** (10), 2400-2404.

Gensemer, R.W., Naddy, R.B., Stubblefield, W.A., Hockett, J.R., Santore, R. and Paquin, P. (2002) Evaluating the role of ion composition on the toxicity of copper to *Ceriodaphnia dubia* in very hard waters, *Comparative Biochemistry and Physiology, Part C* **133** (1-2), 87-97.

George, D.B., Berk, S.G., Adams, V.D., Ting, R.S., Roberts, R.O., Parks, L.H. and Lott, R.C. (1995) Toxicity of alum sludge extracts to a freshwater alga, protozoan, fish, and marine bacterium, *Archives of Environmental Contamination and Toxicology* **29** (2), 149-158.

Gerhardt, A., Janssens de Bisthoven, L., Mo, Z., Wang, C., Yang, M. and Wang, Z. (2002a) Short-term responses of *Oryzias latipes* (Pisces: Adrianichthyidae) and *Macrobrachium nipponense* (Crustacea: Palaemonidae) to municipal and pharmaceutical waste water in Beijing, China: survival, behaviour, biochemical biomarkers, *Chemosphere* **47** (1), 35-47.

Gerhardt, A., Schmidt, S. and Höss, S. (2002b) Measurement of movement patterns of *Caenorhabditis elegans* (Nematoda) with the Multispecies Freshwater Biomonitor® (MFB) - a potential new method to study a behavioral toxicity parameter of nematodes in sediments, *Environmental Pollution* **120** (3), 513-516.

Ghosal, T.K. and Kaviraj, A. (2002) Combined effects of cadmium and composted manure to aquatic organisms, *Chemosphere* **46**(7), 1099-1105.

Giesy, J.P. and Hoke, R.A. (1989) Freshwater sediment toxicity bioassessment: rationale for species selection and test design, *Journal of the Great Lakes Research* **15** (4), 539-569.

Giesy, J.P., Graney, R.L., Newsted, J.L., Rosiu, C.J., Benda, A., Kreis, J.R.G. and Horvath, F.J. (1988) Comparison of three sediment bioassay methods using Detroit River sediments, *Environmental Toxicology and Chemistry* **7**, 483-498.

Giesy, J.P., Rosiu, C.J., Graney, R.L. and Henry, M.G. (1990) Benthic invertebrate bioassays with toxic sediment and pore water, *Environmental Toxicology and Chemistry* **9** (2), 233-248.

Gilli, G. and Meineri, V. (2000) Assessment of the toxicity and genotoxicity of wastewaters treated in a municipal plant, in G. Persoone, C. Janssen and W.M. De Coen (eds.), *New Microbiotests for Routine Toxicity Screening and Biomonitoring*, Kluwer Academic/Plenum Publishers, New York, pp. 327-338.

Gillis, P.L., Diener, L.C., Reynoldson, T.B. and Dixon, D.G. (2002) Cadmium-induced production of a metallothioneinlike protein in *Tubifex tubifex* (oligochaeta) and *Chironomus riparius* (diptera): correlation with reproduction and growth, *Environmental Toxicology and Chemistry* **21** (9), 1836-1844.

Gilron, G.L. and Lynn, D.H. (1998) Ciliated protozoa as test organisms in toxicity assessments, in P.G. Wells, K. Lee and C. Blaise (eds.), *Microscale Testing in Aquatic Toxicology: Advances, Techniques, and Practice*, CRC Press, Boca Raton, FL, pp. 323-336.

Gilron, G.L., Gransden, S.G., Lynn, D.H., Broadfoot, J. and Scroggins, R. (1999) A behavioral toxicity test using the ciliated protozoan *Tetrahymena thermophila*. I. Method description, *Environmental Toxicology and Chemistry* **18** (8), 1813-1816.

Girling, A.E., Pascoe, D., Janssen, C.R., Peither, A., Wenzel, A., Schäfer, H., Neumeier, B., Mitchell, G.C., Taylor, E.J., Maund, S.J., Lay, J.P., Jüttner, I., Crossland, N.O., Stephenson, R.R. and Persoone, G. (2000) Development of methods for evaluating toxicity to freshwater ecosystems, *Ecotoxicology and Environmental Safety* **45** (2), 148-176.

Gonzalez, A.M. (1996) A laboratory formulated sediment incorporating synthetic acid-volatile sulfide, *Environmental Toxicology and Chemistry* **15** (12), 2209-2220.

Gorbi, G., Corradi, M.G., Invidia, M., Rivara, L. and Bassi, M. (2002) Is Cr(VI) toxicity to *Daphnia magna* modified by food availability or algal exudates? The hypothesis of a specific chromium/algae/exudates interaction, *Water Research* **36** (8), 1917-1926.

Graff, L., Isnard, P., Cellier, P., Bastide, J., Cambon, J.-P., Narbonne, J.-F., Budzinski, H. and Vasseur, P. (2003) Toxicity of chemicals to microalgae in river and in standard waters, *Environmental Toxicology and Chemistry* **22** (6), 1368-1379.

Gray, N.F. and O'Neill, C. (1997) Acid mine-drainage toxicity testing, *Environmental Geochemistry and Health* **19** (4), 165-171.

Gregor, D.J. and Munawar, M. (1989) Assessing toxicity of Lake Diefenbaker (Saskatchewan, Canada) sediments using algal and nematode bioassays, in M. Munawar, G. Dixon, C.I. Mayfield, T. Reynoldson and M.H. Sadar (eds.), *Environmental Bioassay Techniques and their Application: Proceedings of the 1st International Conference held in Lancaster, England, 11-14 July 1988*, Kluwer Academic Publishers, Dordrecht, Netherlands, pp. 291-300.

Grothe, D.R., Dickson, K.L. and Reed-Judkins, D.K. (eds.) (1996) *Whole effluent toxicity testing: an evaluation of methods and prediction of receiving system impacts*, Proceedings from a SETAC - sponsored Pellston Workshop, Society of Environmental Toxicology and Chemistry, Pensacola, FL, 346 pp.

Guéguen, C., Koukal, B., Dominik, J. and Pardos, M. (2003) Competition between alga (*Pseudokirchneriella subcapitata*), humic substances and EDTA for Cd and Zn control in the algal assay procedure (AAP) medium, *Chemosphere* **53** (8), 927-934.

Guerra, R. (2001) Ecotoxicological and chemical evaluation of phenolic compounds in industrial effluents, *Chemosphere* **44** (8), 1737-1747.

Guerrero, N.R.V., Taylor, M.G., Wider, E.A. and Simkiss, K. (2003) Influence of particle characteristics and organic matter content on the bioavailability and bioaccumulation of pyrene by clams, *Environmental Pollution* **121** (1), 115-122.

Guilhermino, L., Diamantino, T.C., Ribeiro, R., Gonçalves, F. and Soares, A.M. (1997) Suitability of test media containing EDTA for the evaluation of acute metal toxicity to *Daphnia magna* Straus, *Ecotoxicology and Environmental Safety* **38** (3), 292-295.

Guilhermino, L., Lacerda, M.N., Nogueira, A.J.A. and Soares, A.M.V.M. (2000) *In vitro* and *in vivo* inhibition of *Daphnia magna* acetylcholinesterase by surfactant agents: possible implications for contamination biomonitoring, *The Science of The Total Environment* **247** (2-3), 137-141.

Gunn, A.M., Hunt, D.T.E. and Winnard, D.A. (1989) The effect of heavy metal speciation in sediment on bioavailability to tubificid worms, in M. Munawar, G. Dixon, C.I. Mayfield, T. Reynoldson and M.H. Sadar (eds.), *Environmental Bioassay Techniques and their Application: Proceedings of the 1st International Conference held in Lancaster, England, 11-14 July 1988*, Kluwer Academic Publishers, Dordrecht, Netherlands, pp. 487-496).

Gustavson, K.E., Svenson, A. and Harkin, J.M. (1998) Comparison of toxicities and mechanism of action of *n*-alkanols in the submitochondrial particle and the *Vibrio fischeri* bioluminescence (Microtox®) bioassay, *Environmental Toxicology and Chemistry* **17** (10), 1917-1921.

Gustavson, K.E., Sonsthagen, S.A., Crunkilton, R.A. and Harkin, J.M. (2000) Groundwater toxicity assessment using bioassay, chemical, and toxicity identification evaluation analyses, *Environmental Toxicology* **15** (5), 421-430.

Guzzella, L. (1998) Comparison of test procedures for sediment toxicity evaluation with *Vibrio fischeri* bacteria, *Chemosphere* **37** (14-15), 2895-2909.

Guzzella, L. and Mingazzini, M. (1994) Biological assaying of organic compounds in surface waters, *Water Science and Technology* **30** (10), 113-124.

Guzzella, L., Bartone, C., Ross, P., Tartari, G. and Muntau, H. (1996) Toxicity identification evaluation of Lake Orta (Northern Italy) sediments using the Microtox system, *Ecotoxicology and Environmental Safety* **35** (3), 231-235.

Haller, W.T. and Stocker, R.K. (2003) Toxicity of 19 adjuvants to juvenile *Lepomis macrochirus* (bluegill sunfish), *Environmental Toxicology and Chemistry* **22**(3), 615-619.

Hamm, J.T., Wilson, B.W. and Hinton, D.E. (2001) Increasing uptake and bioactivation with development positively modulate diazinon toxicity in early life stage medaka (*Oryzias latipes*), *Toxicological Sciences* **61** (2), 304-313.

Hankenson, K. and Schaeffer, D.J. (1991) Microtox assay of trinitrotoluene, diaminonitrotoluene, and dinitromethylaniline mixtures, *Bulletin of Environmental Contamination and Toxicology* **46**(4), 550-553.

Hansen, P.D. (1993) Regulatory significance of toxicological monitoring by and summarizing effect parameters, in M. Richardson (ed.), *Ecotoxicology Monitoring*, VCH Publishers, Weinheim, Germany, pp. 273-286.

Hao, O.J., Shin, C.-J., Lin, C.-F., Jeng, F.-T. and Chen, Z.-C. (1996) Use of microtox tests for screening industrial wastewater toxicity, *Water Science and Technology* **34** (10), 43-50.

Harkey, G.A., Landrum, P.F. and Klaine, S.J. (1994) Comparison of whole-sediment, elutriate and pore-water exposures for use in assessing sediment-associated organic contaminants in bioassays, *Environmental Toxicology and Chemistry* **13** (8), 1315-1329.

Harrahy, E.A. and Clements, W.H. (1997) Toxocity and bioaccumulation of a mixture of heavy metals in *Chironomus tentans* (Diptera: Chironomidae) in synthetic sediment, *Environmental Toxicology and Chemistry* **16** (2), 317-327.

Hartgers, E.M., Aalderink, G.H.R., Van den Brink, P.J., Gylstra, R., Wiegman, J.W.F. and Brock, T.C.M. (1998) Ecotoxicological threshold levels of a mixture of herbicides (atrazine, diuron and metolachlor) in freshwater microcosms, *Aquatic Ecology* **32** (2), 135-152.

Hatch, A.C. and Burton Jr, G.A. (1999) Sediment toxicity and stormwater runoff in a contaminated receiving system: consideration of different bioassays in the laboratory and field, *Chemosphere* **39** (6), 1001-1017.

Hauser, B., Schrader, G. and Bahadir, M. (1997) Comparison of acute toxicity and genotoxic concentrations of single compounds and waste elutriates using the Microtox/Mutatox test system, *Ecotoxicology and Environmental Safety* **38** (3), 227-231.

Havas, M. and Likens, G.E. (1985) Toxicity of aluminum and hydrogen ions to *Daphnia catawba*, *Holopedium gibberum*, *Chaoborus punctipennis*, and *Chironomus anthrocinus* from Mirror Lake, New Hampshire, *Canadian Journal of Zoology* **63** (5), 1114-1119.

Hayes, K.R., Douglas, W.S. and Fischer, J. (1996) Inter- and intra-laboratory testing of the *Daphnia magna* IQ toxicity test, *Bulletin of Environmental Contamination and Toxicology* **57** (4), 660-666.

Heida, H. and van der Oost, R. (1996) Sediment pore water toxicity testing, *Water Science and Technology* **34** (7-8), 109-116.

Heijerick, D.G., Janssen, C.R., Karlèn, C., Wallinder, I.O. and Leygraf, C. (2002) Bioavailability of zinc in runoff water from roofing materials, *Chemosphere* **47** (10), 1073-1080.

Heinis, F., Brils, J.M., Klapwijk, S.P. and De Poorter, L.R.M. (2000) From microbiotest to decision support system: an assessment framework for surface water toxicity, in G. Persoone, C. Janssen and W.M. De Coen (eds.), *New Microbiotests for Routine Toxicity Screening and Biomonitoring*, Kluwer Academic / Plenum Publishers, New York, pp. 65-72.

Helma, C., Eckl, P., Gottmann, E., Kassie, F., Rodinger, W., Steinkellner, H., Windpassinger, C., Schulte-Hermann, R. and Knasmüller, S. (1998) Genotoxic and ecotoxic effects of groundwaters and their relation to routinely measured chemical parameters, *Environmental Science and Technology* **32** (12), 1799-1805.

Herbrandson, C., Bradbury, S.P. and Swackhamer, D.L. (2003a) Influence of suspended solids on acute toxicity of carbofuran to *Daphnia magna*: I. Interactive effects, *Aquatic Toxicology* **63** (4), 333-342.

Herbrandson, C., Bradbury, S.P. and Swackhamer, D.L. (2003b) Influence of suspended solids on acute toxicity of carbofuran to *Daphnia magna*: II. An evaluation of potential interactive mechanisms, *Aquatic Toxicology* **63**(4), 343-355.

Heugens, E.H., Jager, T., Creyghton, R., Kraak, M.H., Hendriks, A.J., Van Straalen, N.M. and Admiraal, W. (2003) Temperature-dependent effects of cadmium on *Daphnia magna*: accumulation versus sensitivity, *Environmental Science and Technology* **37** (10), 2145-2151.

Hickey, C.W., Blaise, C. and Costan, G. (1991) Microtesting appraisal of ATP and cell recovery toxicity end points after acute exposure of *Selenastrum capricornutum* to selected chemicals, *Environmental Toxicology and Water Quality* **6**, 383-403.

Hill, L. and Jooste, S. (1999) The effects of contaminated sediments of the Blesbok Spruit near Witbank on water quality and the toxicity thereof to *Daphnia pulex*, *Water Science and Technology* **39** (10-11), 173-176.

Hoffmann, C. and Christofi, N. (2001) Testing the toxicity of influents to activated sludge plants with the *Vibrio fischeri* bioassay utilising a sludge matrix, *Environmental Toxicology* **16** (5), 422-427.

Hoke, R.A., Giesy, J.P., Zabik, M. and Ungers, M. (1993) Toxicity of sediments and sediment pore waters from the Grand Calumet River - Indiana Harbor, Indiana area of concern, *Ecotoxicology and Environmental Safety* **26** (1), 86-112.

Holdway, D.A., Lok, K. and Semaan, M. (2001) The acute and chronic toxicity of cadmium and zinc to two *Hydra* species, *Environmental Toxicology* **16** (6), 557-565.

Hong, L.C.D., Becker-van Slooten, K., Sauvain, J.-J., Minh, T.L. and Tarradellas, J. (2000) Toxicity of sediments from the Ho Chi Minh City canals and Saigon River, Viet Nam, *Environmental Toxicology* **15** (5), 469-475.

Huggett, D.B., Brooks, B.W., Peterson, B., Foran, C.M. and Schlenk, D. (2002) Toxicity of select beta adrenergic receptor-blocking pharmaceuticals (B-blockers) on aquatic organisms, *Archives of Environmental Contamination and Toxicology* **43** (2), 229-235.

Huuskonen, S.E., Ristola, T.E., Tuvikene, A., Hahn, M.E., Kukkonen, J.V.K. and Lindström-Seppä, P. (1998) Comparison of two bioassays, a fish liver cell line (PLHC-1) and a midge (*Chironomus riparius*), in monitoring freshwater sediments, *Aquatic Toxicology* **44** (1-2), 47-67.

Hyötyläinen, T. and Oikari, A. (1999) The toxicity and concentrations of PAHs in creosote-contaminated lake sediment, *Chemosphere* **38** (5), 1135-1144.

IGETG (Inter-Governmental Ecotoxicological Testing Group) (2004) The evolution of toxicological testing in Canada, Environment Canada, Environmental Technology Centre Report, January 2004, Ottawa, Ontario, K1A 0H3, 19 pp.

Ingersoll, C.G., Ankley, G.T., Benoit, D.A., Brunson, E.L., Burton, G.A., Dwyer, F.J., Hoke, R.A., Landrum, P.F., Norberg-King, T.J. and Winger, P.V. (1995) Toxicity and bioaccumulation of sediment-associated contaminants using freshwater invertebrates: a review of methods and applications, *Environmental Toxicology and Chemistry* **14** (11), 1885-1894.

Ingersoll, C., Besser, J. and Dwyer, J. (1997) *Development and application of methods for assessing the bioavailability of contaminants associated with sediments: I. Toxicity and the sediment quality triad*, Proceedings of the U.S. Geological Survey (USGS) Sediment Workshop, U.S. Geological Survey, Columbia, Missouri (2003-12-22); http://water.usgs.gov/osw/techniques/workshop/ingersoll.html.

Ingersoll, C.G., MacDonald, D.D., Wang, N., Crane, J.L., Field, L.J., Haverland, P.S., Kemble, N.E., Lindskoog, R.A., Severn, C. and Smorong, D.E. (2000) *Prediction of sediment toxicity using consensus-based freshwater sediment quality guidelines*, EPA 905/R-00/007, U.S. Enviromnental Protection Agency, Great Lakes National Program Office, Chicago, IL (2004-02-25); http://www.cerc.usgs.gov/pubs/center/pdfdocs/91126.pdf.

Ingersoll, C.G., MacDonald, D.D., Brumbaugh, W.G., Johnson, B.T., Kemble, N.E., Kunz, J.L., May, T.W., Wang, N., Smith, J.R., Sparks, D.W. and Ireland, D.S. (2002) Toxicity assessment of sediments from the Grand Calumet River and Indiana Harbor Canal in Northwest Indiana, USA., *Archives of Environmental Contamination and Toxicology* **43** (2), 156-167.

Isidori, M. (2000) Toxicity monitoring of waste waters from tanneries with microbiotests, in G. Persoone, C. Janssen and W.M. De Coen (eds.), *New Microbiotests for Routine Toxicity Screening and Biomonitoring*, Kluwer Academic / Plenum Publishers, New York, pp. 339-345.

Isidori, M., Parrella, A., Piazza, C.M.L. and Strada, R. (2000) Toxicity screening of surface waters in southern Italy with Toxkit microbiotests, in G. Persoone, C. Janssen and W.M. De Coen (eds.), *New Microbiotests for Routine Toxicity Screening and Biomonitoring*, Kluwer Academic/Plenum Publishers, New York, pp. 289-293.

Isnard, P., Flammarion, P., Roman, G., Babut, M., Bastien, P., Bintein, S., Esserméant, L., Férard, J.F., Gallotti-Schmitt, S., Saouter, E., Saroli, M., Thiébaud, H., Tomassone, R. and Vindimian, E. (2001) Statitical analysis of regulatory ecotoxicity tests, *Chemosphere* **45**, 659-669.

ISO (2003) Water quality - Freshwater algal growth inhibition test with unicellular green algae, (ISO/FDIS 8692:2004), International Standard (under development). Water quality - Freshwater algal growth inhibition test with unicellular green algae.

Isomaa, B., Lilius, H., Sandbacka, M. and Holmström, T. (1995) The use of freshly isolated rainbow trout hepatocytes and gill epithelial cells in toxicity testing, *Toxicology Letters* **78** (1), 42-42.

Jackson, M., Milne, J., Johnston, H. and Dermott, R. (1995) Assays of Hamilton Harbour sediments using *Diporeia hoyi* (Amphipoda) and *Chironomus plumosus* (Diptera), *Canadian Technical Report of Fisheries and Aquatic Sciences* **2039**, 1-21.

Jacobs, M.W., Delfino, J.J. and Bitton, G. (1992) The toxicity of sulfur to Microtox® from acetonitrile extracts of contaminated sediments, *Environmental Toxicology and Chemistry* **11** (8), 1137-1143.

Jak, R.G., Maas, J.L. and Scholten, M.C.T. (1998) Ecotoxicity of 3,4-dichloroaniline in enclosed freshwater plankton communities at different nutrient levels, *Ecotoxicology* **7** (1), 49-60.

Janati-Idrissi, M., Guerbet, M. and Jouany, J.M. (2001) Effect of cadmium on reproduction of daphnids in a small aquatic microcosm, *Environmental Toxicology* **16**(4), 361-364.

Janssen, C.R., Vangheluwe, M. and Van Sprang, P. (2000) A brief review and critical evaluation of the status of microbiotests, in G. Persoone, C. Janssen and W. M. De Coen (eds.), *New Microbiotests for Routine Toxicity Screening and Biomonitoring*, Kluwer Academic / Plenum Publishers, New York, pp. 27-37.

Jaworska, J.S., Schowanek, D. and Feijtel, T.C. (1999) Environmental risk assessment for trisodium [S,S]-ethylene diamine disuccinate, a biodegradable chelator used in detergent applications, *Chemosphere* **38** (15), 3597-3625.

Jenner, H.A. and Janssen-Mommen, J.P.M. (1989) Phytomonitoring of pulverized fuel ash leachates by the duckweed *Lemna minor*, in M. Munawar, G. Dixon, C.I. Mayfield, T. Reynoldson and M.H. Sadar (eds.), *Environmental Bioassay Techniques and their Application: Proceedings of the 1st International Conference held in Lancaster, England, 11-14 July 1988,* Kluwer Academic Publishers, Dordrecht, Netherlands, pp. 361-366.

Johnson, B.T., Petty, J.D., Huckins, J.N., Lee, K. and Gauthier, J. (2004) Hazard assessment of a simulated oil spill on intertidal areas of the St-Lawrence River with SPMP-TOX, *Environmental Toxicology* **19**, 329-335.

Johnson, I. (2000) Criteria-based procedure for selecting test methods for effluent testing and its application to Toxkit microbiotests, in G. Persoone, C. Janssen and W.M. De Coen (eds.), *New Microbiotests for Routine Toxicity Screening and Biomonitoring,* Kluwer Academic / Plenum Publishers, New York, pp. 73-94.

Johnson, I. and Delaney, P. (1998) Development of a 7-day *Daphnia magna* growth test using image analysis, *Bulletin of Environmental Contamination and Toxicology* **61** (3), 355-362.

Johnson, I., Butler, R., Milne, R. and Redshaw, C.J. (1993) The role of Microtox® in the monitoring and control of effluents, in M. Richardson (ed.), *Ecotoxicology Monitoring*, VCH Publishers, Weinheim, Germany, pp. 309-317.

Jooste, S. and Thirion, C. (1999) An ecological risk assessment for a South African acid mine drainage, *Water Science and Technology* **39** (10-11), 297-303.

Jop, K.M., Foster, R.B. and Askew, A.M. (1991) Factors affecting toxicity identification evaluation: the role of source water use in industrial processes, in M.A. Mayes and M.G. Barron (eds.), *Aquatic Toxicology and Risk Assessment: Fourteenth Volume, ASTM STP 1124*, American Society for Testing and Materials, Philadelphia, PA, pp. 84-93.

Jop, K.M., Askew, A.M., Terrio, K.F. and Simoes, A.T. (1992) Application of the short-term chronic test with *Ceriodaphnia dubia* in identifying sources of toxicity in industrial wastewaters, *Bulletin of Environmental Contamination and Toxicology* **49** (5), 765-771.

Jos, A., Repetto, G., Rios, J.C., Hazen, M.J., Molero, M.L., del Peso, A., Salguero, M., Fernández-Freire, P., Pérez-Martín, M. and Cameán, A. (2003) Ecotoxicological evaluation of carbamazepine using six different model systems with eighteen endpoints, *Toxicology in Vitro* **17** (5-6), 525-532.

Joutti, A., Schultz, E., Tuukkanen, E. and Vaajasaari, K. (2000) Industrial waste leachates toxicity detection with microbiotests and biochemical tests, in G. Persoone, C. Janssen and W.M. De Coen (eds.), *New Microbiotests for Routine Toxicity Screening and Biomonitoring*, Kluwer Academic/Plenum Publishers, New York, pp. 347-355.

Jung, K. and Bitton, G. (1997) Use of Ceriofast™ for monitoring the toxicity of industrial effluents: comparison with the 48-h acute *Ceriodaphnia* toxicity test and Microtox®, *Environmental Toxicology and Chemistry* **16** (11), 2264-2267.

Junghans, M., Backhaus T., Faust M., Scholze M., Grimme L.H. (2003) Predictability of combined effects of eight chloroacetanilide herbicides on algal reproduction, *Pest Management Science* **59**, 1101-1110.

Juvonen, R., Martikainen, E., Schultz, E., Joutti, A., Ahtiainen, J. and Lehtokari, M. (2000) A battery of toxicity tests as indicators of decontamination in composting oily waste, *Ecotoxicology and Environmental Safety* **47** (2), 156-166.

Kahru, A., Kurvet, M. and Külm, I. (1996) Toxicity of phenolic wastewater to luminescent bacteria *Photobacterium phosphoreum* and activated sludges, *Water Science and Technology* **33**(6), 139-146.

Kahru, A., Põllumaa, L., Reiman, R. and Rätsep, A. (1999) Predicting the toxicity of oil-shale industry wastewater by its phenolic composition, *ATLA (Alternatives To Laboratory Animals)* **27** (3), 359-366.

Kahru, A., Põllumaa, L., Reiman, R. and Rätsep, A. (2000) Microbiotests for the evaluation of the pollution from the oil shale industry, in G. Persoone, C. Janssen and W.M. De Coen (eds.), *New Microbiotests for Routine Toxicity Screening and Biomonitoring* Kluwer Academic/Plenum Publishers, New York, pp. 357-365.

Kaiser, K.L.E. and McKinnon, M.B. (1993) Qualitative and quantitative relationships of Microtox data with acute and subchronic toxicity data for other aquatic species, *Canadian Technical Report of Fisheries and Aquatic Sciences* **1942**, 1-24.

Kamaya, Y., Kurogi, Y. and Suzuki, K. (2003) Acute toxicity of fatty acids to the freshwater green alga *Selenastrum capricornutum*, *Environmental Toxicology* **18** (5), 289-294.

Karen, D.J., Ownby, D.R., Forsythe, B.L., Bills, T.P., La Point, T.W., Cobb, G.B. and Klaine, S.J. (1999) Influence of water quality on silver toxicity to rainbow trout (*Onchorhynchus mykiss*), fathead minnow (*Pimephales promelas*), and water fleas (*Daphnia magna*), *Environmental Toxicology and Chemistry* **18** (1), 63-70.

Keller, A.E., Ruessler, D.S. and Chaffee, C.M. (1998) Testing the toxicity of sediments contaminated with diesel fuel using glochidia and juvenile mussels (Bivalvia, Unionidae), *Aquatic Ecosystem Health and Management* **1** (1), 37-47.

Kemble, N.E., Brumbaugh, W.G., Brunson, E.L., Dwyer, F.J., Ingersoll, C.G., Monda, D.P. and Woodward, D.F. (1994) Toxicity of metal-contaminated sediments from the Upper Clark Fork River, Montana, to aquatic invertebrates and fish in laboratory exposures, *Environmental Toxicology and Chemistry* **13**, 1985-1997.

Kemble, N.E., Brunson, E.L., Canfield, T.J., Dwyer, F.J. and Ingersoll, C.G. (1998) Assessing sediment toxicity from navigational pools of the Upper Mississippi River using a 28-D *Hyalella azteca* test., *Archives of Environmental Contamination and Toxicology* **35** (2), 181-190.

Kemble, N.E., Dwyer, F.J., Ingersoll, C.G., Dawson, T.D. and Norberg-King, T.J. (1999) Tolerance of freshwater test organisms to formulated sediments for use as control materials in whole-sediment toxicity tests, *Environmental Toxicology and Chemistry* **18** (2), 222-230.

Kemble, N.E., Ingersoll, C.G. and Kunz, J.L. (2002) Toxicity assessment of sediment samples collected from North Carolina streams, U.S. Geological Survey, Columbia, Missouri, Columbia Environmental Research Center, Columbia, MO, Final Report CERC-8335-FY03-20-01, 69 pages.

Koivisto, S. and Ketola, M. (1995) Effects of copper on life-history traits of *Daphnia pulex* and *Bosmina longirostris*, *Aquatic Toxicology* **32** (2-3), 255-269.

Kondo, S., Fujiwara, M., Ohba, M. and Ishii, T. (1995) Comparative larvicidal activities of the four *Bacillus thuringiensis* serovars against a chironomid midge, *Paratanytarsus grimmii* (Diptera: Chironomidae), *Microbiological Research* **150** (4), 425-428.

Kosmala, A., Charvet, S., Roger, M.-C. and Faessel, B. (1999) Impact assessment of a wastewater treatment plant effluent using instream invertebrates and the *Ceriodaphnia dubia* chronic toxicity test, *Water Research* **33** (1), 266-278.

Koukal, B., Guéguen, C., Pardos, M. and Dominik, J. (2003) Influence of humic substances on the toxic effects of cadmium and zinc to the green alga *Pseudokirchneriella subcapitata*, *Chemosphere* **53** (8), 953-961.

Kovacs, T., Gibbons, J.S., Naish, V. and Voss, R. (2002) Complying with effluent toxicity regulation in Canada, *Water Quality Research Journal of Canada* **37** (4), 671-679.

Kross, B.C. and Cherryholmes, K. (1993) Toxicity screening of sanitary landfill leachates: a comparative evaluation with Microtox® analyses, chemical, and other toxicity screening methods, in M. Richardson (ed.), *Ecotoxicology Monitoring*, VCH Publishers, Weinheim, Germany, pp. 225-249.

Kszos, L.A., Morris, G.W. and Konetsky, B.K. (2004) Source of toxicity in storm water: zinc from commonly used paint, *Environmental Toxicology and Chemistry* **23** (1), 12-16.

Kubitz, J.A., Besser, J.M. and Giesy, J.P. (1996) A two-step experimental design for a sediment bioassay using growth of the amphipod *Hyallela azteca* for the test end point, *Environmental Toxicology and Chemistry* **15** (10), 1783-1792.

Kuhne, W.W., Caldwell, C.A., Gould, W.R., Fresquez, P.R. and Finger, S.E. (2002) Effects of depleted uranium on the health and survival of *Ceriodaphnia dubia* and *Hyallela azteca*, *Environmental Toxicology and Chemistry* **21** (10), 2198-2203.

Kungolos, A., Samaras, P., Kimeroglu, V., Dabou, X. and Sakellaropoulos, G.P. (1998) Water quality and toxicity assessment in Koronia Lake, Greece, *Fresenius Environmental Bulletin* **7** (7A-8A, Sp.), 615-622.

Kusui, T. and Blaise, C. (1999) Ecotoxicological assessment of japanese industrial effluents using a battery of small-scale toxicity tests, in S.S. Rao (ed.), *Impact Assessment of Hazardous Aquatic Contaminants: Concepts and Approaches*, Lewis Publishers, Boca Raton, Florida, pp. 161-181.

Kwan, K.K. (1995) Direct sediment toxicity testing procedure using sediment-chromotest kit, *Environmental Toxicology and Water Quality* **10**, 193-196.

Kwan, K.K. and Dutka, B.J. (1992) Evaluation of Toxi-Chromotest direct sediment toxicity testing procedure and Microtox solid-phase testing procedure, *Bulletin of Environmental Contamination and Toxicology* **49** (5), 656-662.

Kwan, K.K. and Dutka, B.J. (1995) Comparative assessment of two solid-phase toxicity bioassays: the direct sediment toxicity testing procedure (DSTTP) and the Microtox® solid-phase test (SPT), *Bulletin of Environmental Contamination and Toxicology* **55** (3), 338-346.

La Point, T.W., Cobb, G.P., Klaine, S.J., Bills, T., Forsythe, B., Jeffers, R., Waldrop, V.C. and Wenholz, M. (1996) Water quality components affecting silver toxicity in *Daphnia magna* and *Pimephales promelas*, in A.W. Andren and T.W. Bober (eds.), *Proceedings of the Fourth International Conference on Transport, Fate and Effects of Silver in the Environment*, International Argentum Conference, Madison, Wisconsin, USA, pp. 121-124.

Lacaze, J.C., Chesterikoff, A. and Garban, B. (1989) Bioévaluation de la pollution des sédiments de la Seine (région parisienne) par l'emploi d'un bioessai basé sur la croissance à court terme de la micro-algue *Selenastrum capricornutum* Printz, *Revue des Sciences de l'Eau* **2**, 405-427.

Lacey, R., Watzin, M.C. and McIntosh, A.W. (1999) Sediment organic matter content as a confounding factor in toxicity tests with *Chironomus tentans*, *Environmental Toxicology and Chemistry* **18** (2), 231-236.

Lahr, J. (1998) An ecological assessment of the hazard of eight insecticides used in Desert Locust control, to invertebrates in temporary ponds in the Sahel, *Aquatic Ecology* **32** (2), 153-162.

Lahr, J., Maas-Diepeveen, J.L., Stuijfzand, S.C., Leonards, P.E.G., Drüke, J.M., Lücker, S., Espeldoorn, A., Kerkum, L.C.M., van Stee, L.L.P. and Hendriks, A.J., (2003) Responses in sediment bioassays used in the Netherlands: can observed toxicity be explained by routinely monitored priority pollutants?, *Water Research* **37** (8), 1691-1710.

Lambolez, L., Vasseur, P., Férard, J.-F. and Gisbert, T. (1994) The environmental risks of industrial waste disposal: an experimental approach including acute and chronic toxicity studies, *Ecotoxicology and Environmental Safety* **28** (3), 317-328.

Lamy-Enrici, M.-H., Dondeyne, A. and Thybaud, E. (2003) Influence of the organic matter on the bioavailability of phenanthrene for benthic organisms, *Aquatic Ecosystem Health and Management* **6** (4), 391-396.

Landrum, P.F., Leppänen, M.T., Robinson, S.D., Gossiaux, D.C., Burton, G.A., Greenberg, M., Kukkonen, J.V.K., Eadie, B.J. and Lansing, M.B. (2004) Comparing behavioral and chronic endpoints to evaluate the response of *Lumbriculus variegatus* to 3,4,3',4'-tetrachlorobiphenyl sediment exposures, *Environmental Toxicology and Chemistry* **23** (1), 187-194.

Länge, R. and Dietrich, D. (2002) Environmental risk assessment of pharmaceutical drug substances - conceptual considerations, *Toxicology Letters* **131** (1-2), 97-104.

Lappalainen, J., Juvonen, R., Vaajasaari, K. and Karp, M. (1999) A new flash method for measuring the toxicity of solid and colored samples, *Chemosphere* **38** (5), 1069-1083.

Larsen, J., Schultz, T.W., Rasmussen, L., Hooftman, R. and Pauli, W. (1997) Progress in an ecotoxicological standard protocol with protozoa: results from a pilot ring test with *Tetrahymena pyriformis*, *Chemosphere* **35** (5), 1023-1041.

Lasier, P.J., Winger, P.V. and Reinert, R.E. (1997) Toxicity of alkalinity to *Hyalella azteca*, *Bulletin of Environmental Contamination and Toxicology* **59** (5), 807-814.

Lasier, P.J., Winger, P.V. and Bogenrieder, K.J. (2000) Toxicity of manganese to *Ceriodaphnia dubia* and *Hyalella azteca*, *Archives of Environmental Contamination and Toxicology* **38** (3), 298-304.

Latif, M. and Zach, A. (2000) Toxicity studies of treated residual wastes in Austria using different types of conventional assays and cost-effective microbiotests, in G. Persoone, C. Janssen and W.M. De Coen (eds.), *New Microbiotests for Routine Toxicity Screening and Biomonitoring*, Kluwer Academic/Plenum Publishers, New York, pp. 367-383.

Latif, M. and Licek, E. (2004) Toxicity assessment of wastewaters, river waters, and sediments in Austria using cost-effective microbiotests, *Environmental Toxicology* **19**, 302-309.

Lauten, K.P. (1993) Sediment toxicity assessment - North Saskatchewan River, *Canadian Technical Report of Fisheries and Aquatic Sciences* **1942**, 360-367.

Leal, H.E., Rocha, H.A. and Lema, J.M. (1997) Acute toxicity of hardboard mill effluents to different bioindicators, *Environmental Toxicology and Water Quality* **12** (1), 39-42.

LeBlond, J.B. and Duffy, L.K. (2001) Toxicity assessment of total dissolved solids in effluent of Alaskan mines using 22-h chronic Microtox® and *Selenastrum capricornatum* assays, *The Science of The Total Environment* **271** (1-3), 49-59.

Lechelt, M., Blohm, W., Kirschneit, B., Pfeiffer, M., Gresens, E., Liley, J., Holz, R., Lüring, C. and Moldaenke, C. (2000) Monitoring of surface water by ultrasensitive *Daphnia* toximeter, *Environmental Toxicology* **15** (5), 390-400.

Lee, J.-S., Lee, B.-G., Luoma, S.N. and Yoo, H. (2004) Importance of equilibration time in the partitioning and toxicity of zinc in spiked sediment bioassays, *Environmental Toxicology and Chemistry* **23** (1), 65-71.

Lewis, M.A. (1995) Use of freshwater plants for phytotoxicity testing: a review, *Environmental Pollution* **87** (3), 319-336.

Leynen, M., Duvivier, L., Girboux, P. and Ollevier, F. (1998) Toxicity of ozone to fish larvae and *Daphnia magna*, *Ecotoxicology and Environmental Safety* **41** (2), 176-179.

Liao, C.M. and Lin, M.C. (2001) Acute toxicity modeling of rainbow trout and silver sea bream exposed to waterborne metals, *Environmental Toxicology* **16** (4), 349-360.

Liao, C.-M., Chen, B.-C., Singh, S., Li, M.-C., Liu, C.-W. and Han, B.-C. (2003) Acute toxicity and bioaccumulation of arsenic in tilapia (*Oreochromis mossambicus*) from a blackfoot disease area in Taiwan, *Environmental Toxicology* **18** (4), 252-259.

Liu, D.H.W., Bailey, H.C. and Pearson, J.G. (1983) Toxicity of a complex munitions wastewater to aquatic organisms, in W.E. Bishop, R.D. Cardwell and B.B. Heidolph (eds.), *Aquatic Toxicology and Hazard Assessment: Sixth Symposium, ASTM STP 802*, American Society for Testing and Materials, Philadelphia, PA, pp. 135-150.

Liu, M.C., Chen, C.M., Cheng, H.Y., Chen, H.Y., Su, Y.C. and Hung, T.Y. (2002) Toxicity of different industrial effluents in Taiwan: a comparison of the sensitivity of *Daphnia similis* and Microtox®, *Environmental Toxicology* **17** (2), 93-97.

Lockhart, W.L., Billeck, B.N. and Baron, C.L. (1989) Bioassays with a floating aquatic plant (*Lemna minor*) for effects of sprayed and dissolved glyphosate, in M. Munawar, G. Dixon, C.I. Mayfield, T. Reynoldson and M.H. Sadar (eds.), *Environmental Bioassay Techniques and their Application: Proceedings of the 1st International Conference held in Lancaster, England, 11-14 July 1988*, Kluwer Academic Publishers, Dordrecht, Netherlands, pp. 353-359.

Long, E.R., MacDonald, D.D., Cubbage, J.C. and Ingersoll, C.G. (1998) Predicting the toxicity of sediment associated trace metals with simultaneously extracted trace metal: acid-volatile sulfide concentrations and dry weight-normalized concentrations: a critical comparison, *Environmental Toxicology and Chemistry* **17** (5), 972-974.

Long, K.E., Van Genderen, E.J. and Klaine, S.J. (2004) The effects of low hardness and pH on copper toxicity to *Daphnia magna*, *Environmental Toxicology and Chemistry* **23** (1), 72-75.

Lopes, I., Gonçalves, F., Soares, A.M.V.M. and Ribeiro, R. (1999) Discriminating the ecotoxicity due to metals and to low pH in acid mine drainage, *Ecotoxicology and Environmental Safety* **44** (2), 207-214.

Lotufo, G.R. (1998) Lethal and sublethal toxicity of sediment-associated fluoranthene to benthic copepods: application of the critical-body-residue approach, *Aquatic Toxicology* **44** (1-2), 17-30.

Lucivjanská, V., Lucivjanská, M. and Cízek, V. (2000) Sensitivity comparison of the ISO *Daphnia* and algal test procedures with Toxkit microbiotests, in G. Persoone, C. Janssen and W.M. De Coen (eds.), *New Microbiotests for Routine Toxicity Screening and Biomonitoring,* Kluwer Academic/Plenum Publishers, New York, pp. 243-246.

Ma, M., Zhu, W., Wang, Z. and Witkamp, G.J. (2003) Accumulation, assimilation and growth inhibition of copper on freshwater alga (*Scenedesmus subspicatus* 86.81 SAG) in the presence of EDTA and fulvic acid, *Aquatic Toxicology* **63** (3), 221-228.

MacDonald, D.D. and Ingersoll, C.G. (2002a) A guidance manual to support the assessment of contaminated sediments in freshwater ecosystems. Volume II - Design and implementation of sediment quality investigations, *EPA-905-B02-001-B*, U.S. Environmental Protection Agency, Great Lakes National Program Office, Chicago, IL, 136 pp.

MacDonald, D.D. and Ingersoll, C.G. (2002b) A guidance manual to support the assessment of contaminated sediments in freshwater ecosystems. Volume III - Interpretation of the results of sediment quality investigations, *EPA-905-B02-001-C,* U.S. Environmental Protection Agency, Great Lakes National Program Office, Chicago, Il, 232 pp.

MacGregor, D.J. and Wells, P.G. (1984) The role of ecotoxicological testing of effluents and chemicals in the Environmental Protection Service, A working paper for E.P.S., Environment Canada, Ottawa, Ontario, November 1984, 56 pp.

Mackay, D.W., Holmes, P.J. and Redshaw, C.J. (1989) The application of bioassay techniques to water pollution problems - The United Kingdom experience, in M. Munawar, G. Dixon, C.I. Mayfield, T. Reynoldson and M.H. Sadar (eds.), *Environmental Bioassay Techniques and their Application: Proceedings of the 1st International Conference held in Lancaster, England, 11-14 July 1988*, Kluwer Academic Publishers, Dordrecht, Netherlands, pp. 77-86.

Madoni, P. (2000) The acute toxicity of nickel to freshwater ciliates, *Environmental Pollution* **109** (1), 53-59.

Mäenpää, K.A., Sormunen, A.J. and Kukkonen, J.V.K. (2003) Bioaccumulation and toxicity of sediment associated herbicides (ioxynil, pendimethalin, and bentazone) in *Lumbriculus variegatus* (Oligochaeta) and *Chironomus riparius* (Insecta), *Ecotoxicology and Environmental Safety* **56** (3), 398-410.

Maier, K.J. and Knight, A.W. (1993) Comparative acute toxicity and bioconcentration of selenium by the midge *Chironomus decorus* exposed to selenate, selenite, and seleno-DL-methionine, *Archives of Environmental Contamination and Toxicology* **25** (3), 365-370.

Malá, J., Maršálková, E. and Rovnaníková, P. (2000) Toxicity testing of solidified waste leachates with microbiotests, in G. Persoone, C. Janssen and W.M. De Coen (eds.), *New Microbiotests for Routine Toxicity Screening and Biomonitoring*, Kluwer Academic/Plenum Publishers, New York, pp. 385-390.

Maltby, L. and Calow, P. (1989) The application of bioassays in the resolution of environmental problems; past, present and future, in M. Munawar, G. Dixon, C.I. Mayfield, T. Reynoldson and M.H. Sadar (eds.), *Environmental Bioassay Techniques and their Application: Proceedings of the 1st International Conference held in Lancaster, England, 11-14 July 1988*, Kluwer Academic Publishers, Dordrecht, Netherlands, pp. 65-76.

Manasherob, R., Ben-Dov, E., Zaritsky, A. and Barak, Z. (1994) Protozoan-enhanced toxicity of *Bacillus thuringiensis* var. israelensis - Endotoxin against *Aedes aegypti* larvae, *Journal of Invertebrate Pathology* **63** (3), 244-248.

Mandal, R., Hassan, N.M., Murimboh, J., Chakrabarti, C.L., Back, M.H., Rahayu, U. and Lean, D.R.S. (2002) Chemical speciation and toxicity of nickel species in natural waters from the Sudbury area (Canada), *Environmental Science and Technology* **36** (7), 1477-1484.

Manusadžianas, L., Balkelyte, L., Sadauskas, K. and Stoškus, L. (2000) Microbiotests for the toxicity assessment of various types of water samples, in G. Persoone, C. Janssen and W.M. De Coen (eds.), *New Microbiotests for Routine Toxicity Screening and Biomonitoring*, Kluwer Academic/Plenum Publishers, New York, pp. 391-399.

Manusadžianas, L., Balkelyte, L., Sadauskas, K., Blinova, I., Põllumaa, L. and Kahru, A. (2003) Ecotoxicological study of Lithuanian and Estonian wastewaters: selection of the biotests, and correspondence between toxicity and chemical-based indices, *Aquatic Toxicology* **63** (1), 27-41.

Maršálek, B. and Bláha, L. (2000) Microbiotests for cyanobacterial toxins screening, in G. Persoone, C. Janssen and W.M. De Coen (eds.), *New Microbiotests for Routine Toxicity Screening and Biomonitoring*, Kluwer Academic/Plenum Publishers, New York, pp. 519-525.

Maršálek, B. and Rojícková-Padrtová, R. (2000) Selection of a battery of microbiotests for various purposes - the Czech experience, in G. Persoone, C. Janssen and W.M. De Coen (eds.), *New Microbiotests for Routine Toxicity Screening and Biomonitoring*, Kluwer Academic/Plenum Publishers, New York, pp. 95-101.

Marsalek, J., Rochfort, Q., Brownlee, B., Mayer, T. and Servos, M. (1999) An exploratory study of urban runoff toxicity, *Water Science and Technology* **39** (12), 33-39.

Martinez-Madrid, M., Rodriguez, P. and Perez-Iglesias, J.I. (1999) Sediment toxicity bioassays for assessment of contaminated sites in the Nervion River (Northern Spain). I. Three-brood sediment chronic bioassay of *Daphnia magna* Straus, *Ecotoxicology* **8**, 97-109.

Marvin, C.H., Howell, E.T., Kolic, T.M. and Reiner, E.J. (2002) Polychlorinated dibenzo-p-dioxins and dibenzofurans and dioxinlike polychlorianted biphenyls in sediments and mussels at three sites in the lower Great Lakes, North America, *Environmental Toxicology and Chemistry* **21** (9), 1908-1921.

Maycock, D.S., Prenner, M.M., Kheir, R., Morris, S., Callaghan, A., Whitehouse, P., Morritt, D. and Crane, M. (2003) Incorporation of *in situ* and biomarker assays in higher-tier assessment of the aquatic toxicity of insecticides, *Water Research* **37** (17), 4180-4190.

McCarthy, L.H., Williams, T.G., Stephens, G.R., Peddle, J., Robertson, K. and Gregor, D.J. (1997) Baseline studies in the Slave River, NWT, 1990-1994: Part I. Evaluation of the chemical quality of water and suspended sediment from the Slave River (NWT), *The Science of The Total Environment* **197** (1-3), 21-53.

McDaniel, M. and Snell, T.W. (1999) Probability distributions of toxicant sensitivity for freshwater rotifer species, *Environmental Toxicology* **14** (3), 361-366.

McDonald, S.F., Hamilton, S.J., Buhl, K.J. and Heisinger, J.F. (1996) Acute toxicity of fire control chemicals to *Daphnia magna* (Straus) and *Selenastrum capricornutum* (Printz), *Ecotoxicology and Environmental Safety* **33** (1), 62-72.

McKnight, D.M., Feder, G.L. and Stiles, E.A. (1981) Toxicity of volcanic-ash leachate to a blue-green alga. Results of a preliminary bioassay experiment, *Environmental Science and Technology* **15** (3), 362-364.

Miana, P., Scotto, S., Perin, G. and Argese, E. (1993) Sensitivity of *Selenastrum capricornutum*, *Daphnia magna* and submitochondrial particles to tributyltin, *Environmental Technology (Letters) ETLEDB* **14** (2), 175-181.

Michel, K., Brinkmann, C., Hahn, S., Dott, W. and Eisentraeger, A. (2004) Acute toxicity investigations of ester-based lubricants by using biotests with algae and bacteria, *Environmental Toxicology* **19**, 445-448.

Michniewicz, M., Nalecz-Jawecki, G., Stufka-Olczyk, J. and Sawicki, J. (2000) Comparison of chemical composition and toxicity of wastewaters from pulp industry, in G. Persoone, C. Janssen and W.M. De Coen (eds.), *New Microbiotests for Routine Toxicity Screening and Biomonitoring*, Kluwer Academic/Plenum Publishers, New York, pp. 401-411.

Middaugh, D.P., Beckham, N., Fournie, J.W. and Deardorff, T.L. (1997) Evaluation of bleached kraft mill process water using Microtox®, *Ceriodaphnia dubia*, and *Menidia beryllina* toxicity tests, *Archives of Environmental Contamination and Toxicology* **32** (4), 367-375.

Milam, C.D. and Farris, J.L. (1998) Risk identification associated with iron-dominated mine discharges and their effect upon freshwater bivalves, *Environmental Toxicology and Chemistry* **17** (8), 1611-1619.

Milam, C.D., Farris, J.L. and Wilhide, J.D. (2000) Evaluating mosquito control pesticides for effect on target and nontarget organisms, *Archives of Environmental Contamination and Toxicology* **39** (3), 324-328.

Milani, D., Reynoldson, T.B., Borgmann, U. and Kolasa, J. (2003) The relative sensitivity of four benthic invertebrates to metals in spiked-sediment exposures and application to contaminated field sediment, *Environmental Toxicology and Chemistry* **22** (4), 845-854.

Milner, R.J., Lim, R.P. and Hunter, D.M. (2002) Risks to the aquatic ecosystem from the application of *Metarhizium anisopliae* for locust control in Australia, *Pest Management Science* **58** (7), 718-723.

Mkandawire, M., Lyubun, Y.V., Kosterin, P.V. and Dudel, E.G. (2004) Toxicity of arsenic species to *Lemna gibba* L. and the influence of phosphate on arsenic bioavailability, *Environmental Toxicology* **19** (1), 26-34.

Monarca, S., Feretti, D., Collivignarelli, C., Guzzella, L., Zerbini, I., Bertanza, G. and Pedrazzani, R. (2000) The influence of different disinfectants on mutagenicity and toxicity of urban wastewater, *Water Research* **34** (17), 4261-4269.

Monda, D.P., Galat, D.L., Finger, S.E. and Kaiser, M.S. (1995) Acute toxicity of ammonia (NH3-N) in sewage effluent to *Chironomus riparius*: II. Using a generalized linear model, *Archives of Environmental Contamination and Toxicology* **28** (3), 385-390.

Monkiédjé, A., Njiné, T., Tamatcho, B. and Démanou, J. (2000) Assessment of the acute toxic effects of the fungicide Ridomil plus 72 on aquatic organisms and soil micro-organisms, *Environmental Toxicology* **15** (1), 65-70.

Moran, T. and Chiles, C. (1993) Multi-species toxicity assessment of sediments from the St-Clair River using *Hyalella azteca*, *Daphnia magna* and Microtox (*Photobacterium phosphoreum*) as test organisms, *Canadian Technical Report of Fisheries and Aquatic Sciences* **1942**, 447-456.

Moreno-Garrido, I., Lubián, L.M. and Soares, A.M.V.M. (2000) Influence of cellular density on determination of EC(50) in microalgal growth inhibition tests, *Ecotoxicology and Environmental Safety* **47** (2), 112-116.

Mowat, F.S. and Bundy, J.G. (2002) Experimental and mathematical/computational assessment of the acute toxicity of chemical mixtures from the Microtox® assay, *Advances in Environmental Research* **6** (4), 547-558.

Mueller, D.C., Bonner, J.S., McDonald, S.J., Autenrieth, R.L., Donnelly, K.C., Lee, K., Doe, K. and Anderson, J. (2003) The use of toxicity bioassays to monitor the recovery of oiled wetland sediments, *Environmental Toxicology and Chemistry* **22** (9), 1945-1955.

Mummert, A.K., Neves, R.J., Newcomb, T.J. and Cherry, D.S. (2003) Sensitivity of juvenile freshwater mussels (*Lampsilis fasciola, Villora iris*) to total and un-ionized ammonia, *Environmental Toxicology and Chemistry* **22** (11), 2545-2553.

Munawar, M., Dermott, R., McCarthy, L.H., Munawar, S.F. and van Stam, H.A. (1999) A comparative bioassessment of sediment toxicity in lentic and lotic ecosystems of the North American Great Lakes, *Aquatic Ecosystem Health and Management* **2** (4), 367-378.

Munawar, M., Munawar, I.F., Sergeant, D. and Wenghofer, C. (2000) A preliminary bioassessment of Lake Baikal sediment toxicity in the vicinity of a pulp and paper mill, *Aquatic Ecosystem Health and Management* **3** (2), 249-257.

Munawar, M., Munawar, I.F., Burley, M., Carou, S. and Niblock, H. (2003) Multi-trophic bioassessment of stressed "Areas of Concern" of the Lake Erie watershed, in M. Munawar (ed.), *Sediment Quality Assessment and Management: Insight and Progress*, Aquatic Ecosystem Health and Management Society, Canada, pp. 169-192.

Muyssen, B.T. and Janssen, C.R. (2001) Zinc acclimation and its effect on the zinc tolerance of *Raphidocelis subcapitata* and *Chlorella vulgaris* in laboratory experiments, *Chemosphere* **45** (4-5), 507-514.

Muyssen, B.T. and Janssen, C.R. (2002) Tolerance and acclimation to zinc of *Ceriodaphnia dubia*, *Environmental Pollution* **117**(2), 301-306.

Naddy, R.B. and Klaine, S.J. (2001) Effect of pulse frequency and interval on the toxicity of chlorpyrifos to *Daphnia magna*, *Chemosphere* **45** (4-5), 497-506.

Naddy, R.B., Stern, G.R. and Gensemer, R.W. (2003) Effect of culture water hardness on the sensitivity of *Ceriodaphnia dubia* to copper toxicity, *Environmental Toxicology and Chemistry* **22** (6), 1269-1271.

Naimo, T.J., Cope, W.G. and Bartsch, M.R. (2000) Sediment-contact and survival of fingernail clams: implications for conducting short-term laboratory tests, *Environmental Toxicology* **15** (1), 23 - 27.

Nalecz-Jawecki, G. (2004) Spirotox – *Spirostomum ambiguum* acute toxicity test – 10 years of experience, *Environmental Toxicology* **19**, 359-364.

Nalecz-Jawecki, G. and Sawicki, J. (2002) A comparison of sensitivity of spirotox biotest with standard toxicity tests, *Archives of Environmental Contamination and Toxicology* **42** (4), 389-395.

Naudin, S., Pardos, M. and Quiniou, F. (1995) Toxicité des sédiments du bassin versant du Stang Alar (Brest) déterminée par une batterie de bio-essais, Cemagref, France, *La revue Ingénieries - EAT no spécial Rade de Brest*, 67-74.

Naylor, C. and Howcroft, J. (1997) Sediment bioassays with *Chironomus riparius*: understanding the influence of experimental design on test sensitivity, *Chemosphere* **35** (8), 1831-1845.

Nebeker, A.V., Cairns, M.A., Gakstatter, J.H., Malueg, K.W., Schuytema, G.S. and Krawczyk, D.F. (1984) Biological methods for determining toxicity of contaminated freshwater sediments to invertebrates, *Environmental Toxicology and Chemistry* **3** (4), 617-630.

Nebeker, A.V., Onjukka, S.T. and Cairns, M.A. (1988) Chronic effects of contaminated sediment on *Daphnia magna* and *Chironomus tentans*, *Bulletin of Environmental Contamination and Toxicology* **41**, 574-581.

Newton, T.J., Allran, J.W., O'Donnell, J.A., Bartsch, M.R. and Richardson, W.B. (2003) Effects of ammonia on juvenile unionid mussels (*Lampsilis cardium*) in laboratory sediment toxicity tests, *Environmental Toxicology and Chemistry* **22** (11), 2554-2560.

Nicolau, A., Dias, N., Mota, M. and Lima, N. (2001) Trends in the use of protozoa in the assessment of wastewater treatment, *Research in Microbiology* **152** (7), 621-630.

Nipper, M.G. (1998) The development and application of sediment toxicity tests for regulatory purposes, in P.G. Wells, K. Lee and C. Blaise (eds.), *Microscale Testing in Aquatic Toxicology: Advances, Techniques, and Practice*, CRC Press, Boca Raton, FL, pp. 631-643.

Nyström, B. and Blanck, H. (1998) Effects of the sulfonylurea herbicide metsulfuron methyl on growth and macromolecular synthesis in the green alga *Selenastrum capricornutum*, *Aquatic Toxicology* **43** (1), 25-39.

Oanh, N.T.K. (1996) A comparative study of effluent toxicity for three chlorine-bleached pulp and paper mills in Southeast Asia, *Resources, Conservation and Recycling* **18** (1-4), 87-105.

Oanh, N.T.K. and Bengtsson, B.-E. (1995) Toxicity to Microtox, micro-algae and duckweed of effluents from the Bai Bang paper company (BAPACO), a Vietnamese bleached kraft pulp and paper mill, *Environmental Pollution* **90** (3), 391-399.

OECD (1987) The use of biological tests for water pollution assessment and control, Environment Monograph No. 11, 70 pp.

OECD (2001a) Proposal for a new guideline 218: Sediment-water Chironomid toxicity test using spiked sediment, OECD Guidelines for the Testing of Chemicals, Organisation for Economic Co-operation and Development (OECD), Washington, DC (2003-12-23); http://www.oecd.org/dataoecd/40/3/2739721.pdf.

OECD (2001b) Proposal for a new guideline 219: Sediment-water Chironomid toxicity test using spiked water, OECD Guidelines for the Testing of Chemicals, Organisation for Economic Co-operation and Development (OECD), Washington, DC (2004-02-25); http://www.oecd.org/dataoecd/40/45/2739742.pdf.

OECD (2002a) Proposal for updating guideline 201: Freshwater alga and cyanobacteria, growth inhibition test, OECD Guidelines for the Testing of Chemicals, Organisation for Economic Co-operation and Development (OECD), Washington, DC (2004-02-25); http://www.oecd.org/dataoecd/58/60/1946914.pdf.

OECD (2002b) Revised proposal for a new guideline 221: Lemna sp. growth inhibition test, OECD Guidelines for the Testing of Chemicals, Organisation for Economic Co-operation and Development (OECD), Washington, DC (2004-02-25); http://www.oecd.org/dataoecd/16/51/1948054.pdf.

O'Farrell, I., Lombardo, R.J., de Tezanos Pinto, P. and Loez, C. (2002) The assessment of water quality in the Lower Lujan River (Buenos Aires, Argentina): phytoplankton and algal bioassays, *Environmental Pollution* **120** (2), 207-218.

Okamura, H., Luo, R., Aoyama, I. and Liu, D. (1996) Ecotoxicity assessment of the aquatic environment around Lake Kojima, Japan, *Environmental Toxicology and Water Quality* **11** (3), 213-221.

Okamura, H., Piao, M., Aoyama, I., Sudo, M., Okubo, T. and Nakamura, M. (2002) Algal growth inhibition by river water pollutants in the agricultural area around Lake Biwa, Japan, *Environmental Pollution* **117** (3), 411-419.

Oladimeji, A.A. and Offem, B.O. (1989) Toxicity of lead to *Clarias lazera*, *Oreochromis niloticus*, *Chironomus tentans* and *Benacus* sp., *Water, Air, and Soil Pollution* **44** (3-4), 191-201.

Onorati, F., Pellegrini, D. and Ausili, A. (1998) Sediment toxicity assessment with *Photobacterium phosphoreum*: a preliminary evaluation of natural matrix effect, *Fresenius Environmental Bulletin* **7** (Special), 596-604.

Oris, J.T., Winner, R.W. and Moore, M.V. (1991) A four day survival and reproduction toxicity test for *Ceriodaphnia dubia*, *Environmental Toxicology and Chemistry* **10** (2), 217-224.

Ortego, L.S. and Benson, W.H. (1992) Effects of dissolved humic material on the toxicity of selected pyrethroid insecticides, *Environmental Toxicology and Chemistry* **11** (2), 261-265.

OSPAR (2000) Briefing document on the work of DYNAMEC and the DYNAMEC mechanism for the selection and prioritisation of hazardous substances. OSPAR Commission PRAM 2000. Summary Record (PRAM 00/12/1, Annex 5).

Pablos, V., Fernández, C., Valdovinos, C., Castaño, A., Muñoz, M.J. and Tarazona, J.V. (1996) Use of ecotoxicity tests as biological detectors of toxic chemicals in the environmental analysis of complex sewages, *Toxicology Letters* **88** (Supplement 1), 82-82.

Painter, H.A. (1993) A review of tests for inhibition of bacteria (especially those agreed internationally), in M. Richardson (ed.), *Ecotoxicology Monitoring*, VCH Publishers, Weinheim, Germany, pp. 17-36.

Paixão, S.M. and Anselmo, A.M. (2002) Effect of olive mill wastewaters on the oxygen consumption by activated sludge microorganisms: an acute toxicity test method, *Journal of Applied Toxicology* **22** (3), 173-176.

Paixão, S.M., Mendonça, E., Picado, A. and Anselmo, A.M. (1999) Acute toxicity evaluation of olive oil mill wastewaters: A comparative study of three aquatic organisms, *Environmental Toxicology* **14** (2), 263-269.

Pape-Lindstrom, P.A. and Lydy, M.J. (1997) Synergistic toxicity of atrazine and organophosphate insecticides contravenes the response addition mixture model, *Environmental Toxicology and Chemistry* **16** (11), 2415-2420.

Pardos, M., Benninghoff, C., Guéguen, C., Thomas, R., Dobrowolski, J. and Dominik, J. (1999a) Acute toxicity assessment of Polish (waste)water with a microplate-based *Hydra attenuata* assay: a comparison with the Microtox® test, *The Science of The Total Environment* **243-244**, 141-148.

Pardos, M., Benninghoff, C., Thomas, R.L. and Khim-Heang, S. (1999b) Confirmation of elemental sulfur toxicity in the Microtox® assay during organic extracts assessment of freshwater sediments, *Environmental Toxicology and Chemistry* **18** (2), 188-193.

Parrott, J.L. and Sprague, J.B. (1993) Patterns in toxicity of sublethal mixtures of metals and organic chemicals determined by Microtox® and by DNA, RNA, and protein content of fathead minnows (*Pimephales promelas*), *Canadian Journal of Fisheries and Aquatic Sciences* **50** (10), 2245-2253.

Parrott, J.L., Wood, C.S., Boutot, P. and Dunn, S. (2003) Changes in growth and secondary sex characteristics of fathead minnows exposed to bleached sulfite mill effluent, *Environmental Toxicology and Chemistry* **22** (12), 2908-2915.

Pascoe, D., Wenzel, A., Janssen, C.R., Girling, A.E., Jüttner, I., Fliedner, A., Blockwell, S.J., Maund, S.J., Taylor, E.J., Diedrich, M., Persoone, G., Verhelst, P., Stephenson, R.R., Crossland, N.O., Mitchell, G.C., Pearson, N., Tattersfield, L., Lay, J.P., Peither, A., Neumeier, B. and Velletti, A.R. (2000) The development of toxicity tests for freshwater pollutants and their validation in stream and pond mesocosms, *Water Research* **34** (8), 2323-2329.

Pasteris, A., Vecchi, M., Reynoldson, T.B. and Bonomi, G. (2003) Toxicity of copper-spiked sediments to *Tubifex tubifex* (Oligochaeta, Tubificidae): a comparison of the 28-day reproductive bioassay with a 6-month cohort experiment, *Aquatic Toxicology* **65** (3), 253-265.

Pastorok, R.A., Peek, D.C., Sampson, J.R. and Jacobson, M.A. (1994) Ecological risk assessment for river sediments contaminated by creosote, *Environmental Toxicology and Chemistry* **13** (12), 1929-1941.

Pauli, W. and Berger, S. (1997) Toxicological comparisons of *Tetrahymena* species, end points and growth media: supplementary investigations to the pilot ring test, *Chemosphere* **35** (5), 1043-1052.

Peck, M.R., Klessa, D.A. and Baird, D.J. (2002) A tropical sediment toxicity test using the dipteran *Chironomus crassiforceps* to test metal bioavailability with sediment pH change in tropical acid-sulfate sediments, *Environmental Toxicology and Chemistry* **21** (4), 720-728.

Peeters, E.T.H.M., Dewitte, A., Koelmans, A.A., van der Velden, J.A. and den Besten, P.J. (2001) Evaluation of bioassays versus contaminant concentrations in explaining the macroinvertebrate community structure in the Rhine-Meuse delta, the Netherlands, *Environmental Toxicology and Chemistry* **20** (12), 2883-2891.

Pereira, A.M.M., Soares, A.M.V.M., Gonçalves, F. and Ribeiro, R. (1999) Test chambers and test procedures for *in situ* toxicity testing with zooplankton, *Environmental Toxicology and Chemistry* **18** (9), 1956-1964.

Pereira, A.M.M., Soares, A.M.V.M., Gonçalves, F. and Ribeiro, R. (2000) Water-column, sediment, and *in situ* chronic bioassays with cladocerans, *Ecotoxicology and Environmental Safety* **47** (1), 27-38.

Pérez, S., Farré, M., García, M.J. and Barceló, D. (2001) Occurrence of polycyclic aromatic hydrocarbons in sewage sludge and their contribution to its toxicity in the toxalert 100 bioassay, *Chemosphere* **45** (6-7), 705-712.

Perkins Jr, E.J. and Schlenk, D. (2000) *In vivo* acetylcholinesterase inhibition, metabolism, and toxicokinetics of aldicarb in channel catfish: role of biotransformation in acute toxicity, *Toxicological Sciences* **53** (2), 308-315.

Persoone, G. (1998) Development and first validation of a "stock-culture free" algal microbiotest: the Algaltoxkit, in P.G. Wells, K. Lee and C. Blaise (eds.), *Microscale Testing in Aquatic Toxicology: Advances, Techniques, and Practice*, CRC Press, Boca Raton, FL, pp. 311-320.

Persoone, G. and Vangheluwe, M.L. (2000) Toxicity determination of the sediments of the river Seine in France by application of a battery of microbiotests, in G. Persoone, C. Janssen and W.M. De Coen (eds.), *New Microbiotests for Routine Toxicity Screening and Biomonitoring*, Kluwer Academic / Plenum Publishers, New York, pp. 427-439.

Persoone, G., Van de Vel, A., Van Steertegem, M. and De Nayer, B. (1989) Predictive value of laboratory tests with aquatic invertebrates: influence of experimental conditions, *Aquatic Toxicology* **14** (2), 149-167.

Persoone, G., Blaise, C., Snell, T., Janssen, C. and Van Steertegem, M. (1993) Cyst-based toxicity tests: II. - Report on an international intercalibration exercise with three cost-effective Toxkits, *Zeitschrift für Angewandte Zoologie* **79** (1), 17-36.

Persoone, G. Janssen, C. and De Coen, W. (2000) New microbiotests for routine toxicity screening and biomonitoring, Kluwer Academic / Plenum Publishers, New York, 550 pp.

Persoone, G., Marsalek, B., Blinova, I., Törökné, A., Zarina, D., Manusadžianas, L., Nalecz-Jawecki, G., Tofan, L., Stepanova, N., Tothova, L. and Kolar, B. (2003) A practical and user-friendly toxicity classification system with microbiotests for natural waters and wastewaters, *Environmental Toxicology* **18** (6), 395-402.

Péry, A.R.R., Ducrot, V., Mons, R. and Garric, J. (2003) Modelling toxicity and mode of action of chemicals to analyse growth and emergence tests with the midge *Chironomus riparius*, *Aquatic Toxicology* **65** (3), 281-292.

Pesonen, M. and Andersson, T.B. (1997) Fish primary hepatocyte culture; an important model for xenobiotic metabolism and toxicity studies, *Aquatic Toxicology* **37** (2-3), 253-267.

Petänen, T., Lyytikäinen, M., Lappalainen, J., Romantschuk, M. and Kukkonen, J.V. K. (2003) Assessing sediment toxicity and arsenite concentration with bacterial and traditional methods, *Environmental Pollution* **122** (3), 407-415.

Peter, S., Siersdorfer, C., Kaltwasser, H. and Geiger, M. (1995) Toxicity estimation of treated coke plant wastewater using the luminescent bacteria assay and the algal growth inhibition test, *Environmental Toxicology and Water Quality* **10**, 179-184.

Pettersson, A., Adamsson, M. and Dave, G. (2000) Toxicity and detoxification of Swedish detergents and softener products, *Chemosphere* **41** (10), 1611-1620.

Phipps, J.L., Mattson, V.R. and Ankley, G.T. (1995) Relative sensitivity of three freshwater benthic macroinvertebrates to ten contaminants, *Archives of Environmental Contamination and Toxicology* **28** (3), 281-286.

Pica-Granados, Y., Trujillo, G.D. and Hernández, H.S. (2000) Bioassay standardization for water quality monitoring in Mexico, *Environmental Toxicology* **15** (4), 322-330.

Pintar, A., Besson, M., Gallezot, P., Gibert, J.J. and Martin, D. (2004) Toxicity to *Daphnia magna* and *Vibrio fischeri* of Kraft bleach plant effluents treated by catalytic wet-air oxidation, *Water Research* **38** (2), 289-300.

Preston, B.L., Snell, T.W., Fields, D.M. and Weissburg, M.J. (2001) The effects of fluid motion on toxicant sensitivity of the rotifer *Brachionus calyciflorus*, *Aquatic Toxicology* **52** (2), 117-131.

Priha, M.H. (1996) Ecotoxicological impacts of pulp mill effluents in Finland, in M.R. Servos, K.R. Munkittrick, J.H. Carey and G.J. Van Der Kraak (eds.), *Environmental Fate and Effects of Pulp and Paper Mill Effluents*, St- Lucie Press, Delray Beach, FL, pp. 637-650.

Radetski, C.M, Férard, J.F. and Blaise, C. (1995) A semi-static microplate-based phytotoxicity test, *Environmental Toxicology and Chemistry* **14**, 299-302.

Ramirez, N.E., Vargas, M.C. and Sanchez, F.N. (1996) Use of the sediment Chromotest for monitoring simulated hydrocarbon biodegradation processes, *Environmental Toxicology and Water Quality* **11**, 223-230.

Ramos, E.U., Vermeer, C., Vaes, W.H.J. and Hermens, J.L.M. (1998) Acute toxicity of polar narcotics to three aquatic species (*Daphnia magna*, *Poecilia reticulata* and *Lymnaea stagnalis*) and its relation to hydrophobicity, *Chemosphere* **37** (4), 633-650.

Rao, S.S., Burnison, B.K., Rokosh, D.A. and Taylor, C.M. (1994) Mutagenicity and toxicity assessment of pulp mill effluent, *Chemosphere* **28** (10), 1859-1870.

Rediske, R., Thompson, C., Schelske, C., Gabrosek, J., Nalepa, T. and Peaslee, G. (2002) *Preliminary investigation of the extent of sediment contamination in Muskegon Lake*, U.S. Environmental Protection Agency, Great Lakes National Program Office, Chicago, IL, #GL-97520701, (2003-12-30), 112 pp.; http://www.epa.gov/glnpo/sediment/muskegon/MuskRpt8.pdf.

Redondo, M.J., López-Jaramillo, L., Ruiz, M.J. and Font, G. (1996) Toxicity assessment using the microtox test and determination of pesticides in soil and water samples by chromatographic techniques, *Toxicology Letters* **88** (Supplement 1), 30-30.

Reinhold-Dudok van Heel, H.C. and den Besten, P.J. (1999) The relation between macroinvertebrate assemblages in the Rhine–Meuse delta (The Netherlands) and sediment quality, *Aquatic Ecosystem Health and Management* **2** (1), 19-38.

Ren, S. and Frymier, P.D. (2003) Use of multidimensional scaling in the selection of wastewater toxicity test battery components, *Water Research* **37** (7), 1655-1661.

Repetto, G., del Peso, A. and Repetto, M. (2000) Alternative ecotoxicological methods for the evaluation, control and monitoring of environmental pollutants, *Ecotoxicology and Environmental Restoration* **3** (1), 47-51.

Repetto, G., Jos, A., Hazen, M.J., Molero, M.L., del Peso, A., Salguero, M., del Castillo, P.D., Rodríguez-Vicente, M.C. and Repetto, M. (2001) A test battery for the ecotoxicological evaluation of pentachlorophenol, *Toxicology In Vitro* **15** (4-5), 503-509.

Reynoldson, T.B., Day, K.E., Clarke, C. and Milani, D. (1994) Effect of indigenous animals on chronic end points in freshwater sediment toxicity tests, *Environmental Toxicology and Chemistry* **13** (6), 973-977.

Ribeiro, I.C., Veríssimo, I., Moniz, L., Cardoso, H., Sousa, M.J., Soares, A.M. and Leão, C. (2000) Yeasts as a model for assessing the toxicity of the fungicides penconazol, cymoxanil and dichlofluanid, *Chemosphere* **41** (10), 1637-1642.

Ribo, J.M. (1997) Interlaboratory comparison studies of the luminescent bacteria toxicity bioassay, *Environmental Toxicology and Water Quality* **12** (4), 283-294.

Ribo, J.M., Yang, J.E. and Huang, P.M. (1989) Luminescent bacteria toxicity assay in the study of mercury speciation, in M. Munawar, G. Dixon, C.I. Mayfield, T. Reynoldson and M.H. Sadar (eds.), *Environmental Bioassay Techniques and their Application: Proceedings of the 1st International Conference held in Lancaster, England, 11-14 July 1988*, Kluwer Academic Publishers, Dordrecht, Netherlands, pp. 155-162.

Richardson, J.S., Hall, K.J., Kiffney, P.M., Smith, J.A. and Keen, P. (1998) *Ecological impacts of contaminants in an urban watershed*, DOE FRAP 1998-25, Environment Canada, Environmental Conservation Branch, Aquatic and Atmospheric Sciences Division, Vancouver, BC, 22 pp.

Riisberg, M., Bratlie, E. and Stenersen, J. (1996) Comparison of the response of bacterial luminescence and mitochondrial respiration to the effluent of an oil refinery, *Environmental Toxicology and Chemistry* **15** (4), 501-502.

Rippon, G.D. and Riley, S.J. (19960 Environmental impact assessment of tailings dispersal from a uranium mine using toxicity testing protocols, *Water Resources Bulletin* **32** (6), 1167-1175.

Rissanen, E., Krumschnabel, G. and Nikinmaa, M. (2003) Dehydroabietic acid, a major component of wood industry effluents, interferes with cellular energetics in rainbow trout hepatocytes, *Aquatic Toxicology* **62** (1), 45-53.

Ristola, T., Pellinen, J., Leppänen, M. and Kukkonen, J. (1996) Characterization of Lake Ladoga sediments. I. Toxicity to *Chironomus riparius* and *Daphnia magna*, *Chemosphere* **32** (8), 1165-1178.

Ristola, T., Parker, D. and Kukkonen, J.V.K. (2001) Life-cycle effects of sediment-associated 2,4,5-trichlorophenol on two groups of the midge *Chironomus riparius* with different exposure histories, *Environmental Toxicology and Chemistry* **20** (8), 1772-1777.

Roberts, R.O. and Berk, S.G. (1993) Effect of copper, herbicides, and a mixed effluent on chemoattraction of *Tetrahymena pyriformis*, *Environmental Toxicology and Water Quality* **8** (1), 73-85.

Robinson, P.W. and Scott, R.R. (1995) The toxicity of cyromazine to *Chironomus zealandicus* (Chironomidae) and *Deleatidium* sp. (Leptophlebiidae), *Pesticide Science* **44** (3), 283-292.

Rodgers, D.W., Schröder, J. and Sheehan, L.V. (1996) Comparison of *Daphnia magna*, rainbow trout and bacterial-based toxicity tests of Ontario Hydro aquatic effluents, *Water, Air, and Soil Pollution* **90** (1-2), 105-112.

Roghair, C.J., Buijze, A. and Schoon, H.N.P. (1992) Ecotoxicological risk evaluation of the cationic fabric softener DTDMAC. I. Ecotoxicological effects, *Chemosphere* **24** (5), 599-609.

Ronco, A.E., Castillo, G. and Díaz-Baez, M.C. (2000) Development and application of microbioassays for routine testing and biomonitoring in Argentina, Chile and Colombia, in G. Persoone, C. Janssen and W.M. De Coen (eds.), *New Microbiotests for Routine Toxicity Screening and Biomonitoring*, Kluwer Academic / Plenum Publishers, New York, pp. 49-61.

Ronco, A., Gagnon, P., Díaz-Baez, M.C., Arkhipchuk, V., Castillo, G., Castillo, L.E., Dutka, B.J., Pica-Granados, Y., Ridal, J., Srivastava, R.C. and Sanchez, A. (2002) Overview of results from the WaterTox intercalibration and environmental testing phase II program: Part 1, statistical analysis of blind sample testing, *Environmental Toxicology* **17** (3), 232-240.

Roseth, S., Edvardsson, T., Botten, T.M., Fuglestad, J., Fonnum, F. and Stenersen, J. (1996) Comparison of acute toxicity of process chemicals used in the oil refinery industry, tested with the diatom *Chaetoceros gracilis*, the flagellate *Isochrysis galbana*, and the zebra fish, *Brachydanio rerio*, *Environmental Toxicology and Chemistry* **15** (7), 1211-1217.

Ross, P. (1993) The use of bacterial luminescence systems in aquatic toxicity testing, in M. Richardson (ed.), *Ecotoxicology Monitoring*, VCH Publishers, Weinheim, Germany, pp. 185-195.

Ross, P. (1998) Role of microbiotests in contaminated sediment assessment batteries, in P.G. Wells, K. Lee and C. Blaise (eds.), *Microscale Testing in Aquatic Toxicology: Advances, Techniques, and Practice*, CRC Press, Boca Raton, FL, pp. 549-556.

Ross, P. and Leitman, P.A. (1995) Solid phase testing of aquatic sediments using *Vibrio fischeri*: test design and data interpretation, in M. Richardson (ed.), *Environmental Toxicology Assessment*, Taylor & Francis Ltd., London, England, pp. 65-76.

Ross, P.E., Burton Jr, G.A., Crecelius, E.A., Filkins, J.C., Giesy, J.P., Ingersoll, C.G., Landrum, P.F., Mac, M.J., Murphy, T.J., Rathbun, J.E., Smith, V.E., Tatem, H.E. and Taylor, R.W. (1992) Assessment of sediment contamination at Great Lakes areas of concern: the ARCS Program Toxicity-Chemistry Work Group strategy, *Journal of Aquatic Ecosystem Health* **1** (3), 193-200.

Ross, P., Burton Jr, G.A., Greene, M., Ho, K., Meier, P.G., Sweet, L.I., Auwarter, A., Bispo, A., Doe, K., Erstfeld, K., Goudey, S., Goyvaerts, M., Henderson, D.G., Jourdain, M., Lenon, M., Pandard, P., Qureshi, A., Rowland, C., Schipper, C., Schreurs, W., Trottier, S. and Van Aggelen, G. (1999) Interlaboratory precision study of a whole sediment toxicity test with the bioluminescent bacterium *Vibrio fischeri*, *Environmental Toxicology* **14** (3), 339-345.

Rossi, D. and Beltrami, M. (1998) Sediment ecological risk assessment: *in situ* and laboratory toxicity testing of Lake Orta sediments, *Chemosphere* **37** (14-15), 2885-2894.

Rousch, J.M., Simmons, T.W., Kerans, B.L. and Smith, B.P. (1997) Relative acute effects of low pH and high iron on the hatching and survival of the water mite (*Arrenurus manubriator*) and the aquatic insect (*Chironomus riparius*), *Environmental Toxicology and Chemistry* **16** (10), 2144-2150.

Ruck, J.G., Martin, M. and Mabon, M. (2000) Evaluation of Toxkits as methods for monitoring water quality in New Zealand, in G. Persoone, C. Janssen and W.M. De Coen (eds.), *New Microbiotests for Routine Toxicity Screening and Biomonitoring*, Kluwer Academic / Plenum Publishers, New York, pp. 103-119.

Rue, W.J., Fava, J.A. and Grothe, D.R. (1988) A review of inter- and intralaboratory effluent toxicity test method variablility, in M.S. Adams, G.A. Chapman and W.G. Landis (eds.), *Aquatic Toxicology and Hazard Assessment: 10th Volume, ASTM STP 971*, American Society for Testing and Materials, Philadelphia, PA, pp. 190-203.

Rutherford, L.A., Matthews, S.L., Doe, K.G. and Julien, G.R.J. (2000) Aquatic toxicity and environmental impact of leachate discharges from a municipal landfill, *Water Quality Research Journal of Canada* **35** (1), 39-57.

Sabaliunas, D., Lazutka, J. and Sabaliuniené, I. (2000) Acute toxicity and genotoxicity of aquatic hydrophobic pollutants sampled with semipermeable membrane devices, *Environmental Pollution* **109** (2), 251-265.

Sae-Ma, B., Meier, P.G. and Landrum, P.F. (1998) Effect of extended storage time on the toxicity of sediment-associated cadmium on midge larvae (*Chironomus tentans*), *Ecotoxicology* **7** (3), 133-139.

Sakai, M. (2001) Chronic toxicity tests with *Daphnia magna* for examination of river water quality, *Journal of Environmental Science and Health, Part B* **36** (1), 67-74.

Sakai, M. (2002a) Use of chronic tests with *Daphnia magna* for examination of diluted river water, *Ecotoxicology and Environmental Safety* **53** (3), 376-381.

Sakai, M. (2002b) Determination of pesticides and chronic test with *Daphnia magna* for rainwater samples, *Journal of Environmental Science and Health, Part B,* **37**(3), 247-254.

Salizzato, M., Pavoni, B., Ghirardini, A.V. and Ghetti, P.F. (1998) Sediment toxicity measured using *Vibrio fischeri* as related to the concentrations of organic (PCBs, PAHs) and inorganic (metals, sulfur) pollutants, *Chemosphere* **36** (14), 2949-2968.

Samaras, P., Sakellaropoulos, G.P., Kungolos, A. and Dermissi, S. (1998) Toxicity assessment assays in Greece, *Fresenius Environmental Bulletin* **7** (7-8, Special), 623-630.

Samel, A., Ziegenfuss, M., Goulden, C.E., Banks, S. and Baer, K.N. (1999) Culturing and bioassay testing of *Daphnia magna* using Elendt M4, Elendt M7, and COMBO media, *Ecotoxicology and Environmental Safety* **43** (1), 103-110.

Sánchez-Mata, J.D., Fernández, V., Chordi, A. and Tejedor, C. (2001) Toxicity and mutagenecity of urban wastewater treated with different purifying processes, *Aquatic Ecosystem Health and Management* **4** (1), 61-72.

Sandbacka, M., Pärt, P. and Isomaa, B. (1999) Gill epithelial cells as tools for toxicity screening - comparison between primary cultures, cells in suspension and epithelia on filters, *Aquatic Toxicology* **46** (1), 23-32.

Santiago, S., Thomas, R.L., Larbaigt, G., Rossel, D., Echeverria, M.A., Tarradellas, J., Loizeau, J.L., McCarthy, L., Mayfield, C.I. and Corvi, C. (1993) Comparative ecotoxicity of suspended sediment in the lower Rhone River using algal fractionation, Microtox and *Daphia magna* bioassays, *Hydrobiologia* **252** (3), 231-244.

Santiago, S., van Slooten, K.B., Chèvre, N., Pardos, M., Benninghoff, C., Dumas, M., Thybaud, E. and Garrivier, F. (2002) *Guide pour l'Utilisation des Tests Ecotoxicologiques avec les Daphnies, les Bactéries Luminescentes et les Algues Vertes, Appliqués aux Echantillons de l'Environnement*, Soluval Institut Forel, Genève, 56 pp.

Sarma, S.S.S., Nandini, S. and Flores, J.L.G. (2001) Effect of methyl parathion on the population growth of the rotifer *Brachionus patulus* (O.F. Muller) under different algal food (*Chlorella vulgaris*) densities, *Ecotoxicology and Environmental Safety* **48** (2), 190-195.

Sauvant, M.P., Pépin, D. and Piccinni, E. (1999) *Tetrahymena pyriformis*: a tool for toxicological studies. A review, *Chemosphere* **38** (7), 1631-1669.

Sauvant, M.P., Pépin, D., Bohatier, J., Grolière, C.A. and Veyre, A. (1994) Comparative study of two in vitro models ($_L$-929 fibroblasts and *Tetrahymena pyriformis* GL) for the cytotoxicological evaluation of packaged water, *The Science of The Total Environment* **156** (2), 159-167.

Schramm, K.-W., Kaune, A., Beck, B., Thumm, W., Behechti, A., Kettrup, A. and Nickolova, P. (1996) Acute toxicities of five nitromusk compounds in *Daphnia*, algae and photoluminescent bacteria, *Water Research* **30** (10), 2247-2250.

Schultz, T.W., Sinks, G.D. and Bearden, A.P. (1998) QSAR in aquatic toxicology: a mechanism of action approach comparing toxic potency to *Pimephales promelas*, *Tetrahymena pyriformis*, and *Vibrio fischeri*, in J. Devillers (ed.), *Comparative QSAR*, Taylor & Francis, New York, pp. 51-109.

Schulz, R., Peall, S.K., Dabrowski, J.M. and Reinecke, A.J. (2001) Spray deposition of two insecticides into surface waters in a South African orchard area, *Journal of Environmental Quality* **30** (3), 814-822.

Schultz, T.W., Seward-Nagel, J., Foster, K.A. and Tucker, V.A. (2004) Population growth impairment of aliphatic alcohols to *Tetrahymena*, *Environmental Toxicology* **19**, 1-10.

Schweigert, N., Eggen, R.I., Escher, B.I., Burkhardt-Holm, P. and Behra, R. (2002) Ecotoxicological assessment of surface waters: a modular approach integrating in vitro methods, *ALTEX (Alternatives to Animal Experiments)* **19** (Suppl 1), 30-37.

Scroggins, R., van Aggelen, G. and Schroeder, J. (2002a, Monitoring sublethal toxicity in effluent under the metal mining EEM Program, *Water Quality Research Journal of Canada* **37** (1), 279-294.

Scroggins, R.P., Miller, J.A., Borgmann, A.I. and Sprague, J.B. (2002b) Sublethal toxicity findings by the pulp and paper industry for cycles 1 and 2 of the environmental effects monitoring program, *Water Quality Research Journal of Canada* **37** (1), 21-48.

Seco, J.I., Fernández-Pereira, C. and Vale, J. (2003) A study of the leachate toxicity of metal-containing solid wastes using *Daphnia magna*, *Ecotoxicology and Environmental Safety* **56** (3), 339-350.

Sekkat, N., Guerbet, M. and Jouany, J.-M. (2001) Étude comparative de huit bioessais à court terme pour l'évaluation de la toxicité de lixiviats de déchets urbains et industriels, *Revue des Sciences de l'Eau* **14** (1), 63-72.

Sepúlveda, M.S., Quinn, B.P., Denslow, N.D., Holm, S.E. and Gross, T.S. (2003) Effects of pulp and paper mill effluent on reproductive success of largemouth bass, *Environmental Toxicology and Chemistry* **22** (1), 205-213.

Sergy, G. (1987) Recommendations on aquatic biological tests and procedures for environmental protection, Conservation and Protection, Department of Environment Manuscript Report, Environment Canada, Ottawa, Ontario, 102 pp.

Seymour, D.T., Verbeek, A.G., Hrudey, S.E. and Fedorak, P.M. (1997) Acute toxicity and aqueous solubility of some condensed thiophenes and their microbial metabolites, *Environmental Toxicology and Chemistry* **16** (4), 658-665.

Sherrard, R.M., Murray-Gulde, C.L., Rodgers Jr, J.H. and Shah, Y.T. (2003) Comparative toxicity of chlorothalonil: *Ceriodaphnia dubia* and *Pimephales promelas*, *Ecotoxicology and Environmental Safety* **56** (3), 327-333.

Sherry, J.P., Scott, B.F., Nagy, V. and Dutka, B.J. (1994) Investigation of the sublethal effects of some petroleum refinery effluents, *Journal of Aquatic Ecosystem Health* **3** (2), 129-137.

Sherry, J.P, Scott, B.F and Dutka, B. (1997) Use of various acute, sublethal and early life-stage tests to evaluate the toxicity of refinery effluents, *Environmental Toxicology and Chemistry* **16** (11), 2249-2257.

Sibley, P.K., Ankley, G.T., Cotter, A.M. and Leonard, E.N. (1996) Predicting chronic toxicity of sediments spiked with zinc: an evaluation of the acid-volatile sulfide model using a life-cycle test with the midge *Chironomus tentans*, *Environmental Toxicology and Chemistry* **15** (12), 2102–2112.

Sibley, P.K., Benoit, D.A. and Ankley, G.T. (1997a) The significance of growth in *Chironomus tentans* sediment toxicity tests: relationship to reproduction and demographic endpoints, *Environmental Toxicology and Chemistry* **16** (2), 336–345.

Sibley, P.K., Legler, J., Dixon, D.G. and Barton, D.R. (1997b) Environmental health assessment of the benthic habitat adjacent to a pulp mill discharge. I. Acute and chronic toxicity of sediments to benthic macroinvertebrates, *Archives of Environmental Contamination and Toxicology* **32** (3), 274-284.

Sibley, P.K., Monson, P.D. and Ankley, G.T. (1997c) The effect of gut contents on dry weight estimates of *Chironomus tentans* larvae: implications for interpreting toxicity in freshwater sediment toxicity tests, *Environmental Toxicology and Chemistry* **16** (8), 1721–1726.

Sildanchandra, W. and Crane, M. (2000) Influence of sexual dimorphism in *Chironomus riparius* Meigen on toxic effects of cadmium, *Environmental Toxicology and Chemistry* **19** (9), 2309-2313.

Sloterdijk, H., Champoux, L., Jarry, V., Couillard, Y. and Ross, P. (1989) Bioassay responses of micro-organisms to sediment elutriates from the St. Lawrence River (Lake St. Louis), in M. Munawar, G. Dixon, C.I. Mayfield, T. Reynoldson and M.H. Sadar (eds.), *Environmental Bioassay Techniques and their Application: Proceedings of the 1st International Conference held in Lancaster, England, 11-14 July 1988,* Kluwer Academic Publishers, Dordrecht, Netherlands, pp. 317-335.

Snell, T.W. (2000) The distribution of endpoint chronic value, for freshwater rotifer, in G. Persoone, C. Janssen and W.M. De Coen (eds.), *New Microbiotests for Routine Toxicity Screening and Biomonitoring*, Kluwer Academic/Plenum Publishers, New York, pp. 185-190.

Snell, T. W. and Persoone, G. (1989) Acute toxicity bioassays using rotifers. II. A freshwater test with *Brachionus rubens*, *Aquatic Toxicology* **14** (1), 81-91.

Snell, T.W. and Janssen, C.R. (1995) Rotifers in ecotoxicology: a review, *Hydrobiologia* **313/314**, 231-247.

Snell, T.W. and Janssen, C.R. (1998) Microscale toxicity testing with rotifers, in P.G. Wells, K. Lee and C. Blaise (eds.), *Microscale Testing in Aquatic Toxicology: Advances, Techniques, and Practice*, CRC Press, Boca Raton, FL, pp. 409-422.

Sosak-Swiderska, B. and Tyrawska, D. (1996) The role of algae in ecotoxicological tests, in M. Richardson (ed.), *Environmental Xenobiotics*, Taylor & Francis Books Ltd, London, England, pp. 179-193.

Sponza, D.T. (2001) Toxicity studies of tobacco wastewater, *Aquatic Ecosystem Health and Management* **4**(4), 479-492.

Staples, C.A. and Davis, J.W. (2002) An examination of the physical properties, fate, ecotoxicity and potential environmental risks for a series of propylene glycol ethers, *Chemosphere* **49** (1), 61-73.

Stepanova, N.J., Petrov, A.M., Gabaydullin, A.G. and Shagidullin, R.R. (2000) Toxicity of snow cover for the assessment of air pollution as determined with microbiotests, in G. Persoone, C. Janssen and W.M. De Coen (eds.), *New Microbiotests for Routine Toxicity Screening and Biomonitoring*, Kluwer Academic/Plenum Publishers, New York, pp. 475-478.

Stephenson, G.L., Kaushik, N.K. and Solomon, K.R. (1991) Acute toxicity of pure pentachlorophenol and a technical formulation to three species of *Daphnia*, *Archives of Environmental Contamination and Toxicology* **20** (1), 73-80.

Stewart, K.M. and Thompson, R.S. (1995) Fluoranthene as a model toxicant in sediment studies with *Chironomus riparius*, *Journal of Aquatic Ecosystem Health* **4** (4), 231-238.

St-Laurent, D. and Blaise, C. (1995) Validation of a microplate-based algal lethality test developed with the help of flow cytometry, in M. Richardson (ed.), *Environmental Toxicology Assessment*, Taylor & Francis Ltd., London, England, pp. 137-155.

Stratton, G.W. (1987) The effects of pesticides and heavy metals towards phototrophic microorganisms. In: E. Hodgson, Editor, Reviews in Environmental Toxicology vol. 3, Elsevier, NY (1987), pp. 71–147.

Stuhlfauth, T. (1995) Ecotoxicological monitoring of industrial effluents, in M. Richardson (ed.), *Environmental Toxicology Assessment*, Taylor & Francis Ltd., London, England, pp. 187-198.

Stuijfzand, S.C., Drenth, A., Helms, M. and Kraak, M.H. (1998) Bioassays using the midge *Chironomus riparius* and the zebra mussel *Dreissena polymorpha* for evaluation of river water quality, *Archives of Environmental Contamination and Toxicology* **34** (4), 357-363.

Sturm, A. and Hansen, P. (1999) Altered cholinesterase and monooxygenase levels in *Daphnia magna* and *Chironomus riparius* exposed to environmental pollutants, *Ecotoxicology and Environmental Safety* **42** (1), 9-15.

Suedel, B.C. and Rodgers Jr, J.H. (1994) Development of formulated reference sediments for freshwater and estuarine sediment testing, *Environmental Toxicology and Chemistry* **13** (7), 1163-1175.

Suedel, B.C. and Rodgers Jr, J.H. (1996) Toxicity of fluoranthene to *Daphnia magna*, *Hyalella azteca*, *Chironomus tentans*, and *Stylaria lacustris* in water-only and whole sediment exposures, *Bulletin of Environmental Contamination and Toxicology* **57** (1), 132-138.

Suedel, B.C., Deaver, E. and Rodgers Jr, J.H. (1996) Experimental factors that may affect toxicity of aqueous and sediment-bound copper to freshwater organisms, *Archives of Environmental Contamination and Toxicology* **30** (1), 40-46.

Suedel, B.C., Rodgers Jr, J.H. and Deaver, E. (1997) Experimental factors that may affect toxicity of cadmium to freshwater organisms, *Archives of Environmental Contamination and Toxicology* **33** (2), 188-193.

Suter, G.W. II. (1993) Ecological risk assessment, Lewis Publishers, Boca Raton, Fl., U.S.A., 538 pp.

Svenson, A. and Zhang, L. (1995) Acute aquatic toxicity of protolyzing substances studied as the Microtox effect, *Ecotoxicology and Environmental Safety* **30** (3), 283-288.

Svenson, A., Linlin, Z. and Kaj, L. (1992) Primary chemical and physical characterization of acute toxic components in wastewaters, *Ecotoxicology and Environmental Safety* **24** (2), 234-242.

Svenson, A., Edsholt, E., Ricking, M., Remberger, M. and Röttorp, J. (1996) Sediment contaminants and Microtox toxicity tested in a direct contact exposure test, *Environmental Toxicology and Water Quality* **11** (4), 293-300.

Svenson, A., Sandén, B., Dalhammar, G., Remberger, M. and Kaj, L. (2000) Toxicity identification and evaluation of nitrification inhibitors in wastewaters, *Environmental Toxicology* **15** (5), 527-532.

Sweet, L.I., Travers, D.F. and Meier, P.G. (1997) Short Communication-chronic toxicity evaluation of wastewater treatment plant effluents with bioluminescent bacteria: a comparison with invertebrates and fish, *Environmental Toxicology and Chemistry* **16** (10), 2187-2189.

Tarczynska, M., Nalecz-Jawecki, G., Brzychcy, B., Zalewski, M. and Sawicki, J. (2000) The toxicity of cyanobacterial blooms as determined by microbiotests and mouse assays, in G. Persoone, C. Janssen and W.M. De Coen (eds.), *New Microbiotests for Routine Toxicity Screening and Biomonitoring*, Kluwer Academic/Plenum Publishers, New York, pp. 527-532.

Tarkpea, M. and Hansson, M. (1989) Comparison between two Microtox test procedures, *Ecotoxicology and Environmental Safety* **18** (2), 204-210.

Taylor, L.N., Wood, C.M. and McDonald, D.G. (2003) An evaluation of sodium loss and gill metal binding properties in rainbow trout and yellow perch to explain species differences in copper tolerance, *Environmental Toxicology and Chemistry* **22** (9), 2159-2166.

Tchounwou, P.B. and Reed, L. (1999) Assessment of lead toxicity to the marine bacterium, *Vibrio fischeri*, and to a heterogeneous population of microorganisms derived from the Pearl River in Jackson, Mississippi, USA, *Review of Environmental Health* **14** (2), 51-61.

Tellez, M.R., Dayan, F.E., Schrader, K.K., Wedge, D.E. and Duke, S.O. (2000) Composition and some biological activities of the essential oil of *Callicarpa americana* (L.), *Journal of Agricultural and Food Chemistry* **48** (7), 3008-3012.

Tellez, M.R., Khan, I.A., Kobaisy, M., Schrader, K.K., Dayan, F.E. and Osbrink, W. (2002) Composition of the essential oil of *Lepidium meyenii* (Walp), *Phytochemistry* **61** (2), 149-155.

Terzaghi, C., Buffagni, M., Cantelli, D., Bonfanti, P. and Camatini, M. (1998) Physical-chemical and ecotoxicological evaluation of water based drilling fluids used in Italian off-shore, *Chemosphere* **37** (14-15), 2859-2871.

Tessier, L., Unfer, S., Férard, J.F., Loiseau, C., Richard, E. and Brumas, V. (1999) Potential of acoustic wave microsensors for aquatic ecotoxicity assessment based on microplates. *Sensors and Actuators* **B 59**: 177-179.

Tetreault, G.R., McMaster, M.E., Dixon, D.G. and Parrott, J.L. (2003) Physiological and biochemical responses of Ontario slimy sculpin (*Cottus cognatus*) to sediment from the Athabasca oil sands area, *Water Quality Research Journal of Canada* **38** (2), 361-377.

Thellen, C., Blaise, C., Roy, Y. and Hickey, C. (1989) Round Robin testing with the *Selenastrum capricornutum* microplate toxicity assay, in M. Munawar, G. Dixon, C.I. Mayfield, T. Reynoldson and M.H. Sadar (eds.), *Environmental Bioassay Techniques and their Application: Proceedings of the 1st International Conference held in Lancaster, England, 11-14 July 1988*, Kluwer Academic Publishers, Dordrecht, Netherlands, pp. 259-268.

Thomulka, K.W., Schroeder, J.A. and Lange, J.H. (1997) Use of *Vibrio harveyi* in an aquatic bioluminescent toxicity test to assess the effects of metal toxicity: Treatment of sand and water-buffer, with and without EDTA, *Environmental Toxicology and Water Quality* **12** (4), 343-348.

Tietge, J.E., Hockett, J.R. and Evans, J.M. (1997) Major ion toxicity of six produced waters to three freshwater species: application of ion toxicity models and TIE procedures, *Environmental Toxicology and Chemistry* **16** (10), 2002-2008.

Tišler, T. and Zagorc-Koncan, J. (1997) Comparative assessment of toxicity of phenol, formalhehyde, and industrial wastewater to aquatic organisms, *Water, Air, and Soil Pollution* **97** (3-4), 315-322.

Tišler, T. and Zagorc-Koncan, J. (1999) Toxicity evaluation of wastewater from the pharmaceutical industry to aquatic organisms, *Water Science and Technology* **39** (10-11), 71-76.

Törökné, A.K. (2000) The potential of the Thamnotoxkit microbiotest for routine detection of cyanobacterial toxins, in G. Persoone, C. Janssen and W.M. De Coen (eds.), *New Microbiotests for Routine Toxicity Screening and Biomonitoring*, Kluwer Academic / Plenum Publishers, New York, pp. 533-539.

Törökné, A., Oláh, B., Reskóné, M., Báskay, I. and Bérciné, J. (2000) Utilization of microbiotests to assess the contamination of water-bases, *Central European Journal of Public Health* **8** (8), 97-99.

Tran, D., Ciret, P., Ciutat, A., Durrieu, G. and Massabuau, J.-C. (2003) Estimation of potential and limits of bivalve closure response to detect contaminants: application to cadmium, *Environmental Toxicology and Chemistry* **22** (4), 914-920.

Traunspurger, W., Schäfer, H. and Remde, A. (1996) Comparative investigation on the effect of a herbicide on aquatic organisms in single species tests and aquatic microcosms, *Chemosphere* **33** (6), 1129-1141.

Trottier, S., Blaise, C., Kusui, T. and Johnson, E.M. (1997) Acute toxicity assessment of aqueous samples using a microplate-based *Hydra attenuata* assay, *Environmental Toxicology and Water Quality* **12** (3), 265-271.

Tsui, M.T.K. and Chu, L.M. (2003) Aquatic toxicity of glyphosate-based formulations: comparison between different organisms and the effects of environmental factors, *Chemosphere* **52** (7), 1189-1197.

Twagilimana, L., Bohatier, J., Grolière, C.A., Bonnemoy, F. and Sargos, D. (1998) A new low-cost microbiotest with the Protozoan *Spirostomum teres*: culture conditions and assessment of sensitivity of the ciliate to 14 pure chemicals, *Ecotoxicology and Environmental Safety* **41** (3), 231-244.

Ulitzur, S., Lahav, T. and Ulitzur, N. (2002) A novel and sensitive test for rapid determination of water toxicity, *Environmental Toxicology* **17** (3), 291-296.

U.S. EPA (2002a) Methods for measuring the acute toxicity of effluents and receiving waters to freshwater and marine organisms, EPA-821-R-02-012, United States Environmental Protection Agency, Washington, DC, pp. 1-275.

U.S. EPA (2002b) Short-term methods for estimating the chronic toxicity of effluents and receiving waters to freshwater organisms, EPA-821-R-02-013, U.S. Environmental Protection Agency, Washington, DC, pp. 1-350.

U.S. EPA (United States Environmental Protection Agency) and Environment Canada (1984) Proceedings of the International OECD workshop on biological testing of effluents (and related receiving waters), September 10 through 14, 1984, Duluth, Minnesota, USA, 367 pp.

Vaajasaari, K., Ahtiainen, J., Nakari, T. and Dahlbo, H. (2000) Hazard assessment of industrial waste leachability: chemical characterization and biotesting by routine effluent tests, in G. Persoone, C. Janssen and W. M. De Coen (eds.), *New Microbiotests for Routine Toxicity Screening and Biomonitoring*, Kluwer Academic / Plenum Publishers, New York, pp. 413-423.

van Dam, R.A., Barry, M.J., Ahokas, J.T. and Holdway, D.A. (1998) Effects of water-borne iron and calcium on the toxicity of diethylenetriamine pentaacetic acid (DTPA) to *Daphnia carinata*, *Aquatic Toxicology* **42** (1), 49-66.

van den Heuvel, M.R. and Ellis, R.J. (2002) Timing of exposure to a pulp and paper effluent influences the manifestation of reproductive effects in rainbow trout, *Environmental Toxicology and Chemistry* **21** (11), 2338-2347.

van der Geest, H.G., Greve, G.D., Kroon, A., Kuijl, S., Kraak, M.H.S. and Admiraal, W. (2000) Sensitivity of characteristic riverine insects, the caddisfly *Cyrnus trimaculatus* and the mayfly *Ephoron virgo*, to copper and diazinon, *Environmental Pollution* **109** (2), 177-182.

Van der Wielen, C. and Halleux, I. (2000) Toxicity monitoring of the Scheldt and Meuse rivers in Wallonia (Belgium) by conventional tests and microbiotests, in G. Persoone, C. Janssen and W.M. De Coen (eds.), *New Microbiotests for Routine Toxicity Screening and Biomonitoring*, Kluwer Academic/Plenum Publishers, New York, pp. 295-303.

Van Donk, E., Abdel-Hamid, M.I., Faafeng, B.A. and Källqvist, T. (1992) Effects of Dursban® 4E and its carrier on three algal species during exponential and P-limited growth, *Aquatic Toxicology* **23** (3-4), 181-191.

van Wijngaarden, R.P.A., van den Brink, P.J., Crum, S.J.H., Oude, V.J.H., Brock, T.C.M. and Leeuwangh, P. (1996) Effects of the insecticide Dursban® 4E (active ingredient chlorpyrifos) in outdoor experimental ditches: I. Comparison of short-term toxicity between the laboratory and the field, *Environmental Toxicology and Chemistry* **15** (7), 1133-1142.

VanGenderen, E.J., Ryan, A.C., Tomasso, J.R. and Klaine, S.J. (2003) Influence of dissolved organic matter source on silver toxicity to *Pimephales promelas*, *Environmental Toxicology and Chemistry* **22** (11), 2746-2751.

Vanderbroele, M.C., Heijerick, D.G., Vangheluwe, M.L. and Janssen, C.R. (2000) Comparison of the conventional algal assay and the Algaltoxkit F™ microbiotest for toxicity evaluation of sediment pore waters, in G. Persoone, C. Janssen and W. M. De Coen (eds.), *New Microbiotests for Routine Toxicity Screening and Biomonitoring*, Kluwer Academic / Plenum Publishers, New York, pp. 261-268.

Vasseur, P. and Pandard, P. (1988) Influence of some experimental factors on metal toxicity to *Selenastrum capricornutum*, *Toxicity Assessment* **3**: 331-343.

Vecchi, M., Reynoldson, T.B., Pasteris, A. and Bonomi, G. (1999) Toxicity of copper-spiked sediments to *Tubifex tubifex* (Oligochaeta, Tubificidae): comparison of the 28-day reproduction bioassay with an early-life stage bioassay, *Environmental Toxicology and Chemistry* **18** (6), 1173-1179.

Veith, G.D., Call, D.J. and Brooke, L.T. (1983) Estimating the acute toxicity of narcotic industrial chemicals to fathead minnows, in W.E. Bishop, R.D. Cardwell and B.B. Heidolph (eds.), *Aquatic Toxicology and Hazard Assessment: Sixth Symposium, ASTM STP 802*, American Society for Testing and Materials, Philadelphia, PA, pp. 90-97.

Verge, C., Moreno, A., Bravo, J. and Berna, J.L. (2001) Influence of water hardness on the bioavailability and toxicity of linear alkylbenzene sulphonate (LAS), *Chemosphere* **44** (8), 1749-1757.

Vidal, D.E. and Horne, A.J. (2003) Inheritance of mercury tolerance in the aquatic oligochaete *Tubifex tubifex*, *Environmental Toxicology and Chemistry* **22**(9), 2130-2135.

Viganò, L., Bassi, A. and Garino, A. (1996) Toxicity evaluation of waters from a tributary of the River Po using the 7-Day *Ceriodaphnia dubia* test, *Ecotoxicology and Environmental Safety* **35** (3), 199-208.

Villaescusa, I., Martí, S., Matas, C., Martínez, M. and Ribó, J.M. (1997) Chromium(VI) toxicity to luminescent bacteria, *Environmental Toxicology and Chemistry* **16** (5), 871-874.

Villarroel, M.J., Sancho, E., Ferrando, M.D. and Andreu-Moliner, E. (1999) Effect of an acaricide on the reproduction and survival of *Daphnia magna*, *Bulletin of Environmental Contamination and Toxicology* **63** (2), 167-173.

Villegas-Navarro, A., González, M.C.R., López, E.R., Aguilar, R.D. and Marçal, W.S. (1999) Evaluation of *Daphnia magna* as an indicator of toxicity and treatment efficacy of textile wastewaters, *Environment International* **25** (5), 619-624.

Vujevic, M., Vidakovic-Cifrek, Z., Tkalec, M., Tomic, M. and Regula, I. (2000) Calcium chloride and calcium bromide aqueous solutions of technical and analytical grade in *Lemna* bioassay, *Chemosphere* **41** (10), 1535-1542.

Walker, J.D., Knaebel, D., Mayo, K., Tunkel, J. and Gray, D.A. (2004) Use of QSARs to promote more cost-effective use of chemical monitoring resources. 1. Screening industrial chemicals and pesticides, direct food additives, indirect food additives and pharmaceuticals for biodegradation, bioconcentration and aquatic toxicity potential, *Water Quality Research Journal of Canada* **39**, 35-39.

Walthall, W.K. and Stark, J.D. (1999) The acute and chronic toxicity of two xanthene dyes, fluorescein sodium salt and phloxine B, to *Daphnia pulex*, *Environmental Pollution* **104** (2), 207-215.

Wang, C., Wang, Y., Kiefer, F., Yediler, A., Wang, Z. and Kettrup, A. (2003) Ecotoxicological and chemical characterization of selected treatment process effluents of municipal sewage treatment plant, *Ecotoxicology and Environmental Safety* **56** (2), 211-217.

Wang, Y., Zhang, M. and Wang, X. (2000) Population growth responses of *Tetrahymena shanghaiensis* in exposure to rare earth elements, *Biological Trace Element Research* **75** (1-3), 265-275.

Ward, M.L., Bitton, G., Townsend, T. and Booth, M. (2002a) Determining toxicity of leachates from Florida municipal solid waste landfills using a battery-of-tests approach, *Environmental Toxicology* **17** (3), 258-266.

Ward, T.J., Rausina, G.A., Stonebraker, P.M. and Robinson, W.E. (2002b) Apparent toxicity resulting from the sequestering of nutrient trace metals during standard *Selenastrum capricornutum* toxicity tests, *Aquatic Toxicology* **60** (1-2), 1-16.

Watts, M.M. and Pascoe, D. (1996) Use of the freshwater macroinvertebrate *Chironomus riparius* (diptera: chironomidae) in the assessment of sediment toxicity, *Water Science and Technology* **34** (7-8), 101-107.

Watts, M.M. and Pascoe, D. (1998) Selection of an appropriate life-cycle stage of *Chironomus riparius* Meigen for use in chronic sediment toxicity testing, *Chemosphere* **36** (6), 1405-1413.

Watts, M.M. and Pascoe, D. (2000) Comparison of *Chironomus riparius* Meigen and *Chironomus tentans* Fabricius (Diptera: Chironomidae) for assessing the toxicity of sediments, *Environmental Toxicology and Chemistry* **19** (7), 1885-1892.

Watzin, M.C., McIntosh, A.W., Brown, E.A., Lacey, R., Lester, D.C., Newbrough, K.L. and Williams, A.R. (1997) Assessing sediment quality in heterogeneous environments: a case study of a small urban harbor in Lake Champlain, Vermont, USA, *Environmental Toxicology and Chemistry* **16** (10), 2125-2135.

Wells, P.,K. Lee and C. Blaise (eds.) (1998) *Microscale testing in Aquatic Toxicology Advances, Techniques and Practice*, CRC Lewis Publishers, Boca Raton, Florida, 679 pp.

Wenning, R.J. and Ingersoll, C.G. (eds.) (2002) *Use of sediment quality guidelines and related tools for the assessment of contaminated sediments*, Executive Summary Booklet of a SETAC Pellston Workshop, Society of Environmental Toxiclogy and Chemistry, Pensacola, FL, 48 pp.

Wernersson, A.S. (2004) Aquatic ecotoxicity due to oil pollution in the Ecuadorian Amazon, *Aquatic ecosystem Health and Management* **7**, 127-136.

West, C.W., Mattson, V.R., Leonard, E.N., Phipps, G.L. and Ankley, G.T. (1993) Comparison of the relative sensitivity of three benthic invertebrates to copper-contaminated sediments from the Keweenaw Waterway, *Hydrobiologia* **262**, 57-63.

West, C.W., Phipps, G.L., Hoke, R.A., Goldenstein, T.A., Vandermeiden, F.M., Kosian, P.A. and Ankley, G.T. (1994) Sediment core versus grab samples: evaluation of contamination and toxicity at a DDT-contaminated site, *Ecotoxicology and Environmental Safety* **28** (2), 208-220.

Weyers, A. and Vollmer, G. (2000) Algal growth inhibition: effect of the choice of growth rate or biomass as endpoint on the classification and labelling of new substances notified in the EU, *Chemosphere* **41** (7), 1007-1010.

Wilkes, B.D. and Beatty Spence, J.M. (1995) Assessing the toxicity of surface waters downstream from a gold mine using a battery of bioassays, *Canadian Technical Report of Fisheries and Aquatic Sciences* **2050**, 38-44.

Williams, K.A., Green, D.W.J., Pascoe, D. and Gower, D.E. (1986) The acute toxicity of cadmium to different larval stages of *Chironomus riparius* (Diptera : Chironomidae) and its ecological significance for pollution regulation, *Oecologia (Berlin)* **70** (3), 362-366.

Williams, M.L., Palmer, C.G. and Gordon, A.K. (2003) Riverine macroinvertebrate responses to chlorine and chlorinated sewage effluents - Acute chlorine tolerances of *Baetis harrisoni* (Ephemeroptera) from two rivers in KwaZulu-Natal, South Africa, *Water SA* **29** (4), 483-488.

Williams, T.D., Hutchinson, T.H., Roberts, G.C. and Coleman, C.A. (1993) The assessment of industrial effluent toxicity using aquatic microorganisms, invertebrates and fish, *The Science of The Total Environment* **Supplement**, 1129-1141.

Wong, M.-Y., Sauser, K.R., Chung, K.-T., Wong, T.-Y. and Liu, J.-K. (2001) Response of the ascorbate-peroxidase of *Selenastrum capricornutum* to copper and lead in stormwaters, *Environmental Monitoring and Assessment* **67** (3), 361-378.

Wong, S.L., Wainwright, J.F. and Pimenta, J. (1995) Quantification of total and metal toxicity in wastewater using algal bioassays, *Aquatic Toxicology* **31** (1), 57-75.

Yang, J.-L. and Chen, H.-C. (2003) Effects of gallium on common carp (*Cyprinus carpio*): acute test, serum biochemistry, and erythrocyte morphology, *Chemosphere* **53** (8), 877-882.

Ziehl, T.A. and Schmitt, A. (2000) Sediment quality assessment of flowing waters in South-West Germany using acute and chronic bioassays, *Aquatic Ecosystem Health and Management* **3** (3), 347-357.

Abbreviations

ASTM	American Society for Testing and Materials
AVS	Acid-Volatile Sulphide
CANMET	Canada Center for Mineral and Energy Technology

CISTI	Canada Institute for Scientific and Technical Information
CBR	Critical Body Residue
DDD	dichlorodiphenyldichloroethane
DDE	dichlorodiphenyldichloroethylene
DDT	dichlorodiphenyltrichloroethane
EC	Environment Canada
HAS	Hazard Assessment Schemes
IGETG	Inter-Governmental Ecotoxicological Testing Group
ISO	International Standard Organisation
Kow	octanol-water coefficient
NOEC	no observed effect concentration
NOEL	no observed effect level
PCBs	polychlorinated biphenyls
PGE	Propylene glycol ether
QSAR	Quantitative Structure-Activity Relationships
OECD	Organization for Economic Cooperation and Development
PAH	Polycyclic Aromatic Hydrocarbon
[S,S]-EDDS	trisodium[S,S]-ethylene diamine disuccinate
TBA	Test Battery Approach
TMS	Test Method Standardization
TT	Toxicity Testing
U.S. EPA	U.S. Environmental Protection Agency.

1. EFFLUENT ASSESSMENT WITH THE PEEP (POTENTIAL ECOTOXIC EFFECTS PROBE) INDEX

CHRISTIAN BLAISE
Saint-Lawrence Centre
Environment Canada
105 McGill street, Montreal
Quebec H2Y 2E7, Canada
christian.blaise@ec.gc.ca

JEAN-FRANÇOIS FÉRARD
Université de Metz, EBSE
Campus Bridoux, rue du Général Delestraint
57070 Metz, France
ferard@sciences.univ-metz.fr

1. Objective and scope of the PEEP index

The PEEP index (or PEEP scale) was originally-developed as a simple effects-based hazard assessment scheme to compare the toxic potential of a series of wastewaters discharging to a common receiving aquatic ecosystem. Within the group of point source emissions being investigated, the PEEP index expresses the toxic loading of each as a single numerical value which integrates both its toxic potential (determined with a battery of small-scale bioassays representing different biological levels and types of toxic effects) and its flow. The PEEP index formula with its corresponding units has been formulated such that resulting values generally vary from 0 to 10, thereby simulating a type of "environmental Richter scale" that readily points out effluent samples that are more problematic than others in terms of toxic loading to an aquatic environment (*i.e.*, the higher an effluent's PEEP index value, the more potentially hazardous it is toward aquatic biota). Hence, the PEEP index is useful as a cost-effective aid to decision-making aimed at environmental protection because it allows prioritizing curative action of wastewaters displaying the highest toxic loadings.

Originally, the PEEP index was designed to assess industrial and municipal effluent toxicity (Section 3). Because the PEEP index formula can accommodate any number and types of bioassays, it could also be applied in other versatile ways (Section 8).

C. Blaise and J.-F. Férard (eds.), Small-scale Freshwater Toxicity Investigations, Vol. 2, 69-87.

2. Summary of the PEEP index

The PEEP (Potential Ecotoxic Effects Probe) index enables the assessment and comparison of the toxic potential of industrial effluents. It is one example of an integrated bioassay battery approach developed to serve the purposes of environmental management. This effluent assessment index relies on the use of an appropriate suite of multitrophic bioassays (decomposers, primary producers and consumers) allowing the measurement of various types (acute, chronic) and levels (lethal, sublethal) of toxicity. At the time of its conception, this index integrated the results of selected small-scale screening bioassays (*Vibrio fischeri* Microtox® test, *Selenastrum capricornutum* growth inhibition microtest, *Ceriodaphnia dubia* lethality and reproduction inhibition tests, *Escherichia coli* genotoxicity SOS Chromotest), and took into account the persistence of toxicity (meaning that biotests were performed on an effluent before and after a five-day biodegradability procedure), (multi)specificity of toxic impact (number of biological responses affected by an effluent), as well as toxic loading (effluent flow in m^3/h). The resulting Potential Ecotoxic Effects Probe (PEEP) index number is reflected by a $\log_{10}$ value that can normally vary from 0 to 10. The structure of the mathematical formula generating PEEP values (illustrated further on) is simple and "user-friendly" in that it can accommodate any number and type of bioassays to fit particular needs.

Table 1. Summary of the PEEP index hazard assessment scheme for effluents.

PEEP index : *Potential Ecotoxic Effects Probe* index
Purpose
• This index, which integrates bioassays as aids to decision-making, was developed as a management tool to assess and compare the relative toxic hazard of a series of industrial effluents discharging to a common aquatic receiving system. Once PEEP index values have been determined, enlightened decisions can take place to proceed with corrective action on those effluents (or classes thereof) which have been identified as the most potentially harmful to a receiving aquatic system (*i.e.*, those with the highest PEEP values). Application of the PEEP scale can thus contribute cost-effectively to reducing toxic input of industrial pollutants to water bodies.
Principle
• Determination of the toxic loading of each effluent by measuring its toxic potential with an appropriate suite of bioassays 1) taking different levels of biological organization into account, 2) taking persistence of toxicity into account and 3) taking effluent flow into account.
Bioassays employed in effluent studies
• *Vibrio fischeri* (Microtox® light inhibition test); *Escherichia coli* (SOS Chromotest); *Selenastrum capricornutum* (micro-algal growth inhibition assay); *Ceriodaphnia dubia* acute immobilization test; *Ceriodaphnia dubia* chronic reproduction test (Costan et al., 1993).

Table 1 (continued). Summary of the PEEP index hazard assessment scheme for effluents.

PEEP index : *Potential Ecotoxic Effects Probe* index
Bioassays employed in effluent studies
• *Vibrio fischeri* (Microtox® light inhibition test); *Selenastrum capricornutum* (micro-algal growth inhibition assay); *Daphnia magna* acute immobilization test; *Thamnocephalus platyurus* (ThamnoToxkit® lethality assay); *Hydra attenuata* (sub)lethality assay (Kusui and Blaise, 1999).
Determination of effluent hazard potential
• Hazard potential for each effluent, calculated with a mathematical formula integrating the product of effluent toxicity (= summation of toxic units generated by all bioassays) and effluent flow (expressed in m^3/h), yields a toxic loading value (= toxic units discharged per cubic meter per hour). The $\log_{10}$ value of the latter is the resulting PEEP index for an effluent.
Notes of interest
• In theory, the PEEP index can vary from 0 to infinity. In practice, it has been shown to produce values ranging from 0 to about 10, thereby simulating a readily-understandable "environmetal Richter scale" indicative of point source industrial toxicity. • Because of the mathematical formula employed, the PEEP index can be determined from any appropriate number and type of tests depending on laboratory expertise and means.
Documented applications with the PEEP index
• PEEP values were generated for 77 Canadian-based industrial effluents targeted for ecotoxicological studies under two Saint-Lawrence Action Plans (Costan et al., 1993; Kusui and Blaise, 1999). • PEEP values were generated for 20 industrial effluents discharging into Toyama Bay (Toyama Prefecture, Japan) to identify those possessing the highest toxic loads (Kusui and Blaise, 1999).

3. Historical overview and applications reported with the PEEP index

The development of the PEEP index was intimately linked with the Saint-Lawrence River Action Plan (SLAP), initiated in 1988 by the Government of Canada as part of a national commitment to sustain the biodiversity of its major aquatic environments. Originating in the Great Lakes, the 1600 km long Saint-Lawrence River slices through the province of Québec, where its freshwater portion first flows through major industrialized cities (especially the greater Montreal area and Quebec City) and then makes its way toward the Saint-Lawrence estuary, east of Québec City, where it later joins the Gulf of Saint-Lawrence (Fig. 1). During the first two five-year plans running from 1988 to 1998, SLAP's basic goals were to protect, conserve and restore this economically-significant and biologically-rich fluvial system which, over the years, had suffered ecosystem setbacks owing to environmental negligence. Indeed, industrial pollution stemming from several important sectors (Pulp and Paper, Metallurgy, Chemical production, Mining, Oil refinery, Metal finishing) constituted at the time a major source of toxic wastes to the Saint-Lawrence River. Reduction of such toxic input, therefore, comprised a major objective of these Action Plans.

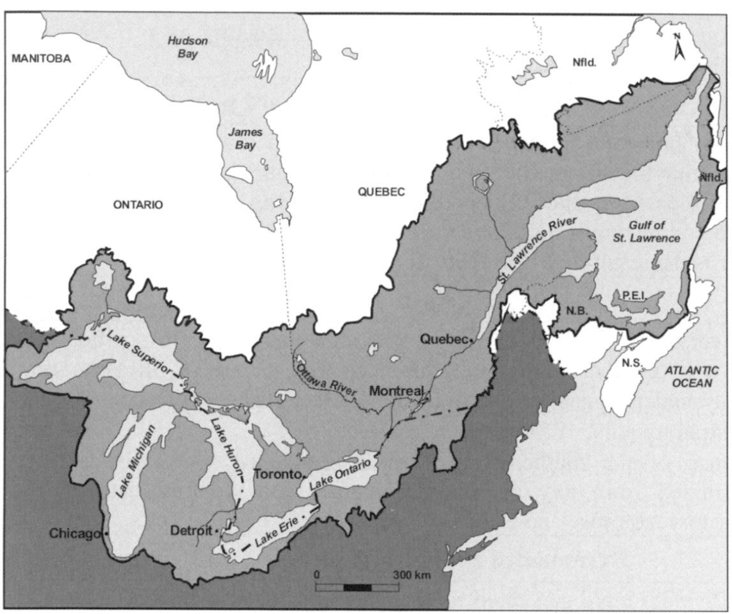

Figure 1. The Saint-Lawrence River watershed including the Great Lakes, the freshwater and estuarine portions of the Saint-Lawrence River and the Gulf of Saint-Lawrence.

The question then arose as to how to best make use of ecotoxicological tools and approaches to determine the toxic contribution of each of a series of industrial effluents prioritized for possible curative actions, based on knowledge of their chemical emission characteristics. While bioassays were clearly sought to identify effluent ecotoxic effects, a second underlying question dealt with their cost-effectiveness, owing to budgetary considerations. This issue was essentially resolved thanks to the emergence of small-scale aquatic toxicology which had made marked progress during the 1980's in contributing to the development of several attractive small-scale toxicity assays characterized by features including simplicity, sensitivity and low sample volume requirements (Blaise, 1991).

Recognition of aquatic species diversity and different modes of actions of contaminants next dictated that a suite of small-scale tests should be employed to properly evaluate the hazards of wastewaters, as pointed out by studies of that time period (Blaise et al., 1988; Dutka, 1988; Garric et al., 1993). A further request of

SLAP environmental managers pleaded for an effluent assessment system that would be user-friendly in that it would unambiguously identify problematic liquid discharges and readily enable decision-making with respect to clean-up actions that should follow. To serve the purposes of environmental management, therefore, what eventually would become the PEEP index had to be both a scientific tool based on sound ecotoxicological principles (Section 5), as well as a simple-to-use and simple-to-interpret managerial tool capable of discriminating effluents based on their toxic loading to the Saint-Lawrence River.

As seen further on in this chapter, individual PEEP index values express a condensed portrait of an effluent's hazard potential which takes into account several important ecotoxicological notions (toxic intensity and scope in terms of biotic levels impacted, bioavailability, persistence of toxicity and effluent flow). Unlike wastewater investigations limited to chemical characterization, this bioassay-based scale reflects the integrated responses of several representative toxicity tests to all interaction phenomena (antagonistic, additive and/or synergistic effects) that can be present in effluent samples.

Under the first two Saint-Lawrence Action Plans (1988-98), the PEEP index was employed to determine and compare the relative toxicity of 106 priority industrial sites all of which discharged their wastewaters to the Saint-Lawrence River. Several publications have reported on various aspects of this initiative (Costan et al., 1993; Environment Canada, 1996; Blaise, 1996; Blaise et al., 2000). Presented at several scientific venues since 1993, the PEEP index concept has generated interest among the international scientific community. We are aware that it has been employed in Australia, France, Lithuania, Japan and South Africa. In Toyama Prefecture, Japan, PEEP index values were recently determined for 20 industrial sites and sewage plants, whose effluents discharged to Toyama Bay, an important commercial fisheries resource area, to identify those responsible for the highest toxic loadings (Kusui and Blaise, 1999; Kusui, 2002). This index was also employed to assess the toxic loading of wastewaters discharging to the Bogotá River in Columbia (see Chapter 7 of this volume).

4. Advantages of the PEEP index scheme for effluent assessment

Outstanding PEEP index characteristics can be summarized as follows:

- The PEEP scale is a cost-effective tool to determine the toxic loading potential of a series of point source liquid wastes discharging to a common receiving environment, owing to the use of small-scale bioassays.
- The PEEP index formula, is easy to use and interpret, and capable of accommodating any number and types of toxicity tests, thereby enabling its application internationally.
- Application of the PEEP index provides unambiguous capacity to discriminate between effluents having low and high toxic loading potentials.
- Numerical PEEP index values are the $\log_{10}$ expression of an effluent's toxic loading (= toxic potential of effluent generated with a relevant battery of toxicity tests multiplied by effluent flow) and normally vary between 0 and 10. The PEEP scale can thus be considered as a type of "environmental Richter scale" for

wastewaters that describes an effluent's hazardous potential toward a receiving aquatic ecosystem.

- Because they are readily understandable by environmental managers, industrialists and the general public, PEEP values enable rapid and enlightened decision-making to circumscribe specific effluents which should be targeted for clean-up actions in order to reduce toxic loading to aquatic environments.

5. Description of the PEEP index scheme

While the PEEP index can theoretically be employed to assess the toxic potential of varied liquid media and groups of specific chemical products of interest (Section 8.3), it was originally conceived to appraise that of liquid wastes of point source discharges to aquatic environments. The index is thus described for this purpose in this section.

5.1 EFFLUENT SAMPLES

Samples are taken from the final effluent of each industrial plant investigated with an automatic sampler (*e.g.*, Manning sampler) that collects a set volume of wastewater after a designated time interval (*e.g.*, 400 mL every 15 min). A 24-h sample (in this case, 38,400 mL or 38.4 L) can then be stored in an appropriate glass container (40-L container in this case). If additional volumes are required (*e.g.*, for subsequent biological and chemical analyses), the automatic sampler can be programmed to collect larger effluent quantities after a 24-h time period <u>or</u> sampling can proceed for up to three consecutive days to collect 3×40 L samples (as in this example). Composite 24-h or 72-h samples are then truly representative of the liquid emissions of each industrial site being assessed. Ideally, and during the sampling period, effluent flow can be determined on site with a Parshall flume (see glossary) or this information can be obtained directly from plant authorities. For plants having more than one effluent, each composite sample can be combined as a function of flow volume. All samples should be kept in coolers (~ 4°C) after sampling and during transport and storage. Biological testing should commence as soon as possible afterwards, but no more than 5 days after each composite sample has been prepared.

Prior to toxicity testing, an appropriate volume of each composite sample is vacuum-filtered (0.45 μ membrane) to remove suspended material which would otherwise interfere with the conduct of some of the bioassays employed. In removing particulates from effluent samples, it is important to note that the PEEP scale only evaluates their soluble toxicity. This issue is further discussed in Section 6.3.

5.2 TYPES OF BIOASSAYS EMPLOYED

At the time of its conception, the PEEP index integrated the results of a selection of practical small-scale screening bioassays which included the *Vibrio fischeri* bioluminescence inhibition test, the *Selenastrum capricornutum* growth inhibition microtest, the *Ceriodaphnia dubia* lethality and reproduction inhibition tests and the *Escherichia coli* genotoxicity SOS Chromotest (Costan et al., 1993). When later applied toward the assessment of Japanese industrial and municipal effluents, PEEP scale values were calculated with a different suite of small-scale toxicity tests that reflected those in use in a particular laboratory at the time (Kusui and Blaise, 1999). The characteristics of all tests are reported in Table 2.

Table 1. Characteristics of small-scale bioassays used in PEEP scale studies of wastewaters.

Trophic level	***Toxicity test***	***Assessment endpoint***	***Reference***
Decomposer	Bacterial test [a,b] *Vibrio fischeri* Microtox® toxicity	Acute sublethal light inhibition (after a 15-min exposure)	Environment Canada, 1992a
Decomposer	Bacterial test[a] *Escherichia coli SOS* Chromotest genotoxicity assay	SOS gene DNA repair induction with and without metabolic activation (after a 2-h exposure)	Quillardet et al., 1982; Legault et al., 1996
Primary producer	Algal test[a,b] *Selenastrum capricornutum* microplate assay	Chronic sublethal growth inhibition (after a 96-h exposure)[c]	Environment Canada, 1992b
Primary consumer	Cladoceran test[b] *Ceriodaphnia dubia* assay	Acute motility inhibition (after a 48-h exposure)	Environment Canada, 1990
	Cladoceran test[a] *Ceriodaphnia dubia* assay	Chronic lethality (after a 7-d exposure)	Environment Canada, 1992c
		Chronic sublethal reproduction inhibition (after a 7-d exposure)	Environment Canada, 1992c
Secondary consumer	Cnidarian test[b] *Hydra attenuata* assay	Acute lethality (after a 96-h exposure)	Blaise and Kusui, 1997
		Acute sublethality indicated by morphology changes (after a 96-h exposure)	Blaise and Kusui, 1997

a) Test performed during the Saint-Lawrence Action Plan effluent study (Costan et al., 1993).
b) Test performed during the Toyama Bay Japanese effluent study (Kusui and Blaise, 1999).
c) The *S. capricornutum* chronic growth inhibition assay is now a 72-h exposure test.

For both the Saint-Lawrence River Action Plan (Costan et al., 1993) and the Toyama Bay Japanese (Kusui and Blaise, 1999) studies, the two suites of bioassays employed represented three trophic levels (decomposers, primary producers and primary or secondary consumers), and sought to measure both acute and chronic toxicity. Toxicity tests were selected on the basis of practical and scientific criteria including low sample volume requirement, sensitivity, simplicity of undertaking the assay, ease in maintaining laboratory cultures, cost-effectiveness, procedural reliability and/or frequency of use internationally.

5.3 MEASUREMENT ENDPOINTS

Because 50% effect effluent concentrations cannot always be determined owing to several factors (*e.g.*, nonlinear or lack of an apparent concentration-response, effect below 50% in the case of slightly toxic effluents, other confounding factors such as the co-presence of both toxic and stimulatory contaminants), the common measurement endpoint selected for all bioassays employed in the Saint-Lawrence River Action investigation was the (geno)toxic threshold concentration (TC), calculated from NOEC (no observed effect concentration) and LOEC (lowest observed effect concentration) values as indicated in the formula below (U.S. EPA, 1989). A second reason for choosing a TC endpoint over a median effect endpoint is its increased sensitivity over the latter (*i.e.*, a TC endpoint will signal the onset of an adverse effect at a lower effluent concentration than will a EC50 endpoint, for example, which reflects an effluent concentration at which 50% of exposed organisms are affected).

Hence, hypothesis testing (ANOVA analysis followed by multiple comparison analysis) was used to determine NOEC and LOEC values expressed as % v/v of effluent. In order to satisfy statistical analysis requirements enabling NOEC and LOEC determinations, some bioassay protocols were adjusted to make sure that there were at least three replicates per effluent concentration and at least five effluent concentrations tested. TC % effluent values were then determined as follows:

$$TC = (NOEC \times LOEC)^{1/2} \tag{1}$$

So that toxic effects of all bioassays could be later integrated in the PEEP scale formula (Section 5.5), TC values were again transformed into toxic unit (TU) values by means of the following formula (Sprague and Ramsay, 1965):

$$TU = 100\%\ v/v\ effluent \div TC \tag{2}$$

In the Toyama Bay Japanese effluent study, 20% endpoint effect values (*e.g.*, LC20s for the *D. magna* assay and IC20s for the *S. capricornutum* assay), which are close approximations of TC values determined from NOEC and LOEC data (as in the Canadian study), were transformed into TU values and integrated into the PEEP formula. In applying the PEEP index concept to a designated series of wastewaters discharging to a common aquatic environment, it is paramount, of course, to use the same battery of bioassays and to report all of their toxicity responses with the same measurement endpoint and statistical analysis system (*i.e.*, TC values for all effluents in the case of the Canadian study and 20% effect responses for all effluents in the case of the Japanese study). Only in this way can it be justified to appraise the relative toxicity contribution of a set of wastewaters on a common comparative basis.

5.4 BIODEGRADABILITY OF EFFLUENT TOXICITY

Persistence of toxicity is an important notion to consider in assessing the hazards of liquid wastes that are discharged to aquatic environments. Clearly, effluents whose toxic components are persistent stand to harm aquatic biota more severely than those that are not. The PEEP index incorporates the notion of "persistence of toxicity" in its wastewater assessment and employs a simple procedure to determine this aspect. Essentially, the selected suite of bioassays is first conducted on the neat effluent sample that has been collected. The same suite of tests is then undertaken on a sub-sample after it has been subjected to a five-day biodegradability test. This test calls for incorporating a micro-volume of buffered solution of inorganic salts to a 1-L volume of effluent, adding a commercial bacterial seed solution, followed by a 5-d room temperature incubation (in darkness) under continuous low-bubbling aeration (Costan et al., 1993). An identical cocktail added to a 1-L sample of deionized water and processed as described was run in parallel and bioassays confirmed its non toxic characteristics. This type of biodegradation step simulates aerobic biological treatment (*e.g.*, biodegradation activities of aquatic microbes) and allows effluent toxicity potential to be determined on pre- and post-biodegradation effluent samples.

5.5 PEEP INDEX FORMULA

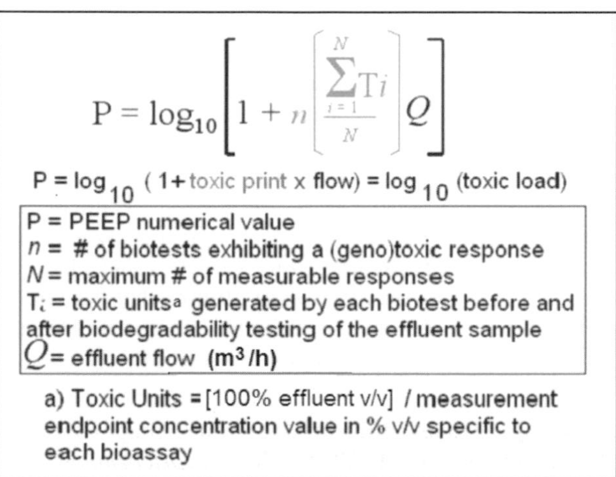

Figure 2. PEEP index formula.

The formula enabling the determination of effluent PEEP index values with the description of its various components is shown in Fig. 2. It takes into account persistence of toxicity (meaning that biotests are performed on an effluent before and after a five-day biodegradability procedure described above), (multi)specificity of toxic impact (number of biological responses affected by an effluent), as well as toxic loading (effluent flow in m^3/h). The resulting Potential Ecotoxic Effects Probe (PEEP) index number is reflected by a $\log_{10}$ value that will normally vary from 0 to 10. The structure of the mathematical formula generating PEEP values is simple and "user-friendly" in that it can accommodate any number and type of bioassays to fit particular needs and/or specific laboratory capabilities.

The PEEP formula can best be understood by means of an example showing genuine toxicity data generated with an effluent sample collected in 1991 from a Pulp & Paper mill located in the province of Quebec, Canada (Fig. 3). Measurement endpoint effluent concentration data calculated in % v/v (*i.e.*, TC values as calculated from formula 1, Section 5.3) are first transformed into toxic units (formula 2, Section 5.3). Taken on their own, these values are clearly informative in terms of the toxicity characteristics of the effluent under investigation. For example, it is clear that this wastewater sample is highly phytotoxic (183.8 toxic units observed following micro-algal testing of the neat effluent sample), but that most of this toxicity is not persistent (5.7 toxic units remaining after the effluent sample has been subjected to a biodegradation procedure). Again, while the effluent appears to contain one (or more) directly-acting genotoxicant(s), because it is initially genotoxic in the absence of metabolic activation (*i.e.*, SOS Chromotest performed without rat liver S-9 enzyme mix), this potentially adverse effect is not persistent (no genotoxic units measured after the biodegradability test). The effluent is also devoid of pro-genotoxicants (*i.e.*, those displaying genotoxic activity after S-9 activation). Furthermore, this example points out that effluent toxicity is trophic-level dependent and that bioassays should be conducted with different organisms to properly circumscribe the full hazard potential that complex discharges can represent with respect to aquatic biota. Additional details on various conceptual aspects and information on the significance of bioassay data generated with this managerial tool can be obtained by consulting the original PEEP article (Costan et al., 1993).

Once toxic units are calculated for all bioassays, they are integrated in the ***toxic print*** portion of the PEEP formula, which is multiplied by effluent ***flow*** datum ($Q = 3213\ m^3/h$). The product of ***toxic print*** and ***flow*** yields the ***toxic loading*** of the effluent. The resulting PEEP index value of 5.8 is then simply the $\log_{10}$ of the calculated effluent sample ***toxic loading*** (plus 1). The value of « ***1*** », inserted into the PEEP formula just ahead of the ***toxic print***, insures that the inferior scale of the PEEP index will commence at « ***0*** » for effluents which are non toxic (*i.e.*, those where toxicity responses are absent for all of the bioassays and which yield a ΣTi value = 0).

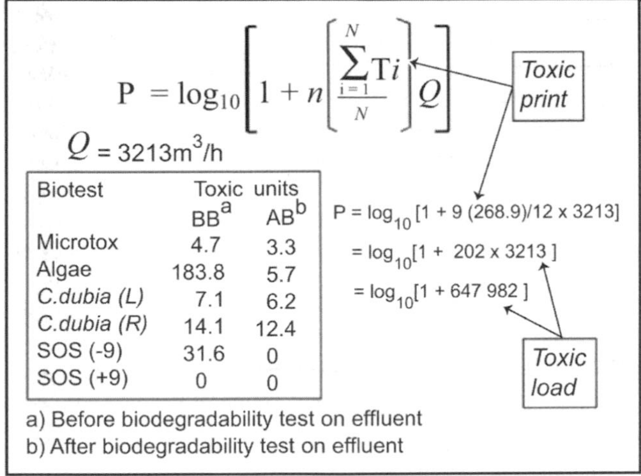

Figure 3. PEEP index calculation for actual toxicity data obtained for a Pulp & Paper effluent.

6. Factors capable of influencing effluent PEEP scale interpretation

6.1 EFFLUENT SAMPLING

Collection of a reliable composite sample of effluent taken over time (Section 5.1) is paramount to ensure that its determined PEEP value will reflect reality. Unless an industrial process discharges wastewaters that are constant in their physical-chemical constituents and flow characteristics, a grab (*i.e.*, instantaneous) sample cannot pretend to be representative of an effluent's hazard/risk potential in terms of toxic loading. More often than not, and for a variety of reasons, process waters are seldom uniform and show variation over time. Again, grab and/or even composite samples cannot hope to take account of unpredictable discharges that can be accidental (*e.g.*, uncontrollable slug of pollution released owing to plant technical problem) or illegal (*e.g.*, wilful nocturnal release of contaminated wastes) in nature.

6.2 TYPES AND NUMBERS OF BIOASSAYS

The same types and numbers of bioassays must always be applied to the same series of effluents discharging to a common aquatic environment. Failure to do so invalidates appraising the relative toxicity contribution of each wastewater in relation to others, as one would clearly be comparing "apples and oranges" in such an event. It is also important to employ bioassays (within a designated battery) that are not redundant in the toxicity information they yield (*e.g.*, two bioassays significantly correlated to one another in terms of their effluent measurement endpoint values). While this does not invalidate PEEP results, it would signify that the test battery has likely not been optimally designed in terms of cost-effectiveness and also possibly in terms of circumscribing the full toxic potential of the set of effluents being investigated.

6.3 EFFLUENT TOXICITY POTENTIAL

In applying the PEEP index concept to sets of industrial effluents thus far, wastewater samples have been filtered prior to bio-analysis (see Section 5.1). Hence, only their soluble toxicity potential is taken into consideration. This is certainly a drawback at this time as toxic and genotoxic potential linked to suspended matter of some industrial plant effluents, for example, have been shown to be important (White et al., 1996; Pardos and Blaise, 1999). Particulate toxicity in effluent samples should certainly be addressed in future PEEP applications, as soon as reliable small-scale toxicity tests are developed and available to estimate it. Indeed, the issue of soluble and particulate toxicity is especially relevant in relation to technology-based reduction of hazardous liquid emissions.

7. Application of the PEEP index: a case study

We now demonstrate how the potential hazards of 50 industrial effluents discharging to the Saint-Lawrence River were effectively appraised during the first five-year (1988-93) Saint-Lawrence Action Plan (SLAP I). This is one example of how small-scale testing, integrated within the PEEP scale concept, can be advantageously employed as a helpful management tool for decision-making. Industrial effluents identified for priority assessment under SLAP I are shown in Fig. 4. Most were situated in the more industrialized part of the freshwater portion of the Saint-Lawrence River lying between Montreal and Quebec City. Taken together, the 50 effluents represented four major industrial sectors, namely Pulp and Paper, Inorganic, Organic and Metallurgic plants.

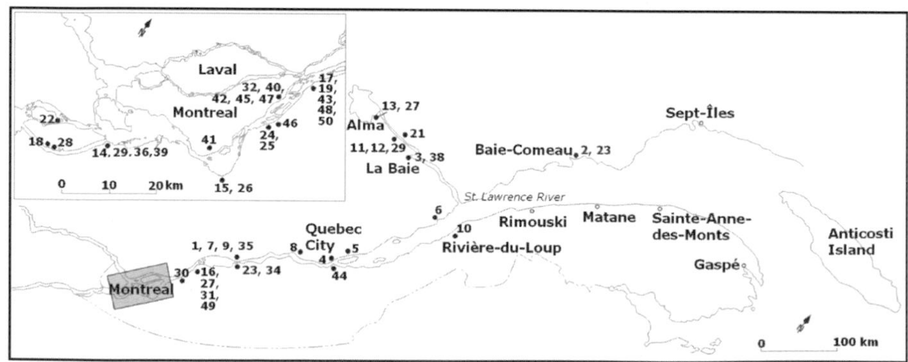

Figure 4. Emplacement of the 50 priority industrial effluents investigated during the first Saint-Lawrence River Action Plan (1988-93).

By observing the range of PEEP values determined for these 50 industrial effluents, it is evident that some were irrefutably more harmful than others (Fig. 5). For example, at the time at which sampling and bioassays were conducted for these effluents (*circa* 1990-92), the 15 Pulp and Paper sector effluents discharging to the Saint-Lawrence River unmistakably showed a high range of PEEP values (dark bars in Fig. 5). Furthermore, if the toxic loading of all effluents (*i.e.*, total toxic units discharged per m^3 for the 50 effluents) is separated into the four industrial sectors they represent, Pulp and Paper mills and the Inorganic chemical producing plants unquestionably stand out as the worst polluters (Fig. 6), contributing to 57 and 39% of toxic emissions to the Saint-Lawrence river, respectively. In quantifying the toxicity of industrial discharges, the PEEP index unambiguously points a finger at the most problematic ones requiring priority attention in terms of clean-up action, such that environmental protection effectiveness can be achieved. Clearly, curative action in terms of toxic effects is first required at sites belonging to the two aforementioned sectors in order to maximize reduction of toxic input to the Saint-Lawrence River ecosystem, a top priority at the time. Along with the chemical characterization conducted on these 50 effluents, their PEEP index values provided environmental managers with valuable information leading to effective decision-taking, based on focused clean-up initiatives, establishment of site-specific wastewater standards and enforcement regulations. A few years after the completion of the first SLAP plan (1988-93), toxic loading reduction to the Saint-Lawrence River was estimated to be 96%, based on a comparison of the 1988 and 1995 chemical loading of these 50 effluents (Thériault, 1996).

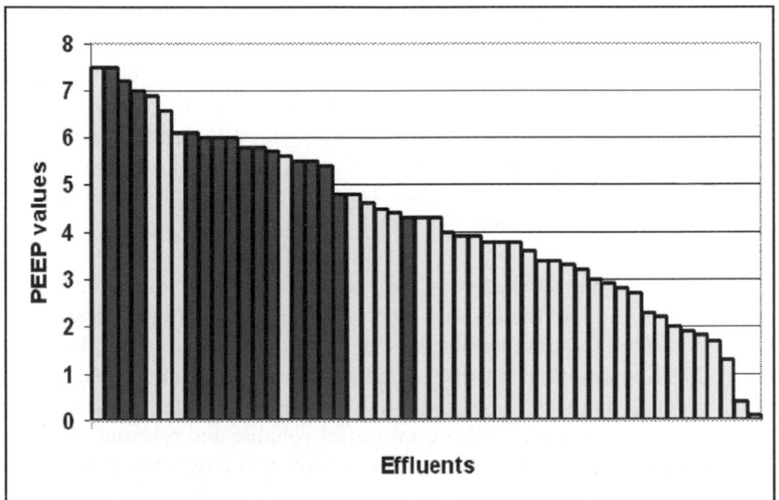

Figure 5. PEEP values for the 50 priority industrial effluents investigated during the first Saint-Laurence River Action Plan (1988-93). Dark bars are PEEP values for the 15 Pulp & Paper effluents belonging to this set of wastewaters.

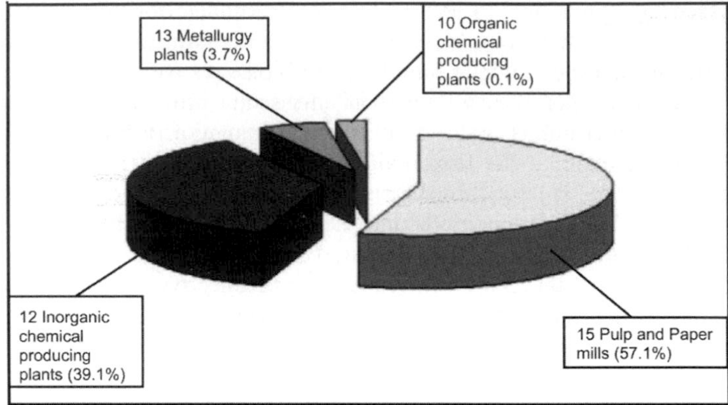

Figure 6. Relative toxic loading to the Saint-Lawrence River of the four industrial sectors made up by the 50 priority effluents identified for study under the first Saint-Lawrence River Action Plan (1988-93).

8. Accessory/miscellaneous PEEP scale procedure information

8.1 INCLUSION OF A FISH TEST IN FUTURE PEEP BIOASSAY BATTERIES

Requiring low-sample volume micro-scale tests for its cost-effective application, the PEEP index has thus far employed bioassays with bacteria, algae and micro-invertebrates. While well-standardized toxicity tests using freshwater fish existed at the time of the PEEP's conception in the early 1990's (*e.g.*, the Environment Canada fingerling rainbow trout 96-h lethality test to assess industrial wastewaters), they were excluded because of their large sample volume needs (*e.g.*, close to 400 L of effluent sample required to undertake a multiple dilution 96-h LC50 bioassay in the case of the trout test). In addition to effluent sample volume, the cost of carrying out salmonid fish acute lethality bioassays for the 50 priority industrial effluents identified under SLAP I (the first 1988-93 Saint-Lawrence River Action Plan) was prohibitive.

Inclusion of a test representative of the fish level of organization in future PEEP bioassay batteries is nevertheless highly advisable owing to the specific adverse effects that liquid wastes can manifest on this trophic level. To offset the constraints mentioned above, appropriate surrogates can now be found with tests conducted with fish cells. Indeed, fish cell bioassays such as those reported in this book (see Chapters 14 and 15, volume 1 of this book) offer reliable and relevant alternatives to whole organism testing that alleviate sample volume and budgetary considerations.

8.2 REPRODUCIBILITY OF PEEP EFFLUENT VALUES

Some of the 50 industrial effluents investigated under SLAP I had their composite samples collected more than once at different time periods. Subsequent determination of their PEEP values gives some estimate of reproducibility for this index (Tab. 3). In general, it appears that effluent toxic loading potential, as reflected by two to four PEEP values reported for each effluent, is fairly constant and reflects, at the very least, similar orders of magnitude over time. Some temporal variations in PEEP index values (*e.g.*, PEEP values of 3.0 and 3.7, Inorganic plants, Effluent # 4) can possibly be explained by various factors (*e.g.*, change in effluent flow characteristics, plant treatment modifications). In such cases, information exchanges between PEEP index value producers and plant officials may prove useful to explain fluctuations in toxic loading.

Table 2. Reproducibility of effluent PEEP index values.

Industrial sector and effluent number	***PEEP values (and effluent sample date: month-year)***
Inorganic chemical production plants:	
Effluent # 1	7.6 (10-91), 7.6 (11-91), 7.4 (12-91), 7.5 (01-92)
Effluent # 2	7.1 (11a-91)[a], 6.7 (11b-91)[a], 6.3 (11c-91)[a], 6.9 (12-91)
Effluent # 3	6.1 (01-93), 5.9 (02-93), 5.7 (03-93)
Effluent # 4	3.0 (01-91), 3.7 (10-92)
Pulp and Paper mills:	
Effluent # 1	5.9 (02-89), 6.0 (07-90)
Organic chemical plants:	
Effluent # 1	3.9 (08-90), 3.9 (12-92)
Effluent # 2	2.8 (09-90), 3.7 (02-91)
Effluent # 3	2.9 (08-90), 3.6 (12-92)

a) Effluent # 2 was sampled on three consecutive days in November (11a, 11b, 11c) of 1991.

8.3 OTHER POSSIBLE USES OF THE PEEP SCALE

While initially developed to compare the toxic loading of a series of industrial effluents discharging to a common receiving water body, the PEEP scale could theoretically serve several other useful purposes (Box 1). Toxicity monitoring and/or regulatory activities associated with specific emissions or industrial sectors comprise one set of potential applications (Items 1-3, Box 1). Again, determining the relative toxic loading contribution of individual liquid wastes discharging to a common wastewater treatment plant (WTP) may prove useful (Item 4, Box 1), particularly to identify those whose toxic input may inhibit the performance of secondary treatment processes (*e.g.*, by intoxicating activated sludge micro-organisms). Assuming that a battery of sufficiently sensitive bioassays and endpoints can be exploited, comparing the toxic loading of a series of tributaries (*e.g.*, rivers) discharging to a common receiving water body (*e.g.*, a larger river such as the St-Lawrence River or estuary), would also be worthwhile so as to focus future studies on those tributaries presenting the highest toxic charges (Item 5, Box 1). Finally, the PEEP scale concept does not have to be confined to liquid wastes, but its application can be extended to solid media as well (Item 6, Box 1). Chapters 8 (SED-TOX) and 11 (WASTOXHAS) of this volume offer examples in this respect.

Box 1. Other possible applications of the PEEP index concept.

1. Toxicity management of process effluents within particular industrial plants.
2. Toxicity reduction assessment of new treatment technology for specific industrial sectors
3. Effluent regulatory control by setting common emission objectives.
4. Toxicity management of wastewater treatment plants dealing with multiple effluent source situations.
5. PEEP index mapping of major waterways and their associated tributaries (employing highly sensitive bioassays) to identify environmental hot spots and pollution sources.
6. Toxicity management of various environmental media (contaminated sediments/soils, land-based solid waste sites, solid waste disposal, etc.) with suites of bioassays appropriate for each case.

Besides actual or potential applications linked to liquid wastewater discharges, the PEEP index concept has additional value in being able to appraise the ecotoxicity of chemical products. In this sense, it was successfully used to evaluate the relative toxicity of nine general purpose cleaners to assist Environment Canada's Environmental Choice Program in authorizing the labelling of EcoLogos on formulations deemed less harmful to the environment (Bermingham et al., 1996). While the battery of bioassays employed unsurprisingly demonstrated that all products were toxic, some were clearly more toxic than others. The relative toxicity of this set of general purpose cleaners was then ranked by comparing the summation of toxic units of each product generated with the toxic print portion of the PEEP formula (Fig. 2). On the basis of this PEEP index information, followed by expert judgement considerations, three out of the nine products were recommended as qualifiers for the Ecologo labels.

Finally, a recent study was undertaken to assess the human and environmental hazard of recycled tire crumb as ground covering in playgrounds (Birkholz et al., 2003). Here, the PEEP scale was called upon to estimate hazard associated with aquatic exposure to water-soluble extracts of tire crumbs. Based on an initially-determined PEEP value of 3.2 for projected volumes of tire crumb leachates to the aquatic environment and a documented decrease in toxicity three months after tire crumb cover had been in place, the study concluded that tires recycled in this fashion would not present a significant risk of contamination for either receiving surface or groundwaters.

9. Conclusions/prospects

The PEEP index was originally designed to be a scientifically-sound management tool, integrating bioassays as an aid to decision-making, capable of assessing the relative toxic loading (expressed by a single numerical value) of each of a series of industrial effluents discharging to a common receiving aquatic environment. To be effective, this effluent assessment index is dependent on the use of an appropriate suite of bioassays undertaken at several biological levels (*e.g.*, decomposers, primary producers and consumers) enabling the measurement of various types (acute, chronic) and levels (lethal, sublethal) of toxicity. Its effectiveness in predicting the overall hazard potential of wastewaters was further revealed when the selected panel of bioassays featured non redundancy in toxic responses generated with effluents representative of four different industrial sectors (Costan et al., 1993). The approach is novel in that it combines information on 1) the biodegradability/persistence of effluent toxicity (indicative of its possible fate in receiving waters), 2) the trophic levels targeted by effluent toxicity (indicative of the ecological scope of impact) and on 3) the flow characteristics of the effluent (indicative of toxic loading released to the environment). The integration of these concepts into a PEEP scale or index is clearly an unparalleled attempt to bring together factors of relevant ecotoxicological importance into a simple, practical and useful management tool to literally "peep" into the hazardous potential of industrial effluents *via* an initial bioanalytical screening strategy. Once PEEP index values have been generated for a designated set of point source pollution emissions, enlightened control efforts can then be directed toward the most problematic ones to optimize toxicity reduction.

Beyond its capability to identify generic toxicity hazards linked to complex liquid media and classes of chemical products (recalled in Section 8.3), the PEEP concept might in future yet unfold in different ways. For one, as the discipline of aquatic toxicology and instrumental technology evolve, so likely will the choice of bioassays based on cost-efficiency and improved sample throughput considerations. For another, PEEP batteries of bioassays might be later designed to focus on specific environmental issues of concern (*e.g.*, genotoxicity, immunotoxicity or endocrine disruption) or designed to target individual trophic levels (*e.g.*, a bioassay battery composed of a suite of phytotoxicity tests and endpoints, if primary producers are deemed of importance in a particular ecosystem). As the demand for improved information on long-term effects of low level of pollutants increases, we can also expect biomarker measurements to complement sensitive bioassays in PEEP scales designed to investigate the potential chronic effects of wastewaters devoid of acute toxicity effects and/or of lotic/lentic receiving systems. What will not change is the genuine usefulness of applying PEEP-based strategies to provide relevant hazard assessment information that should prove to be beneficial for protection and conservation of the aquatic environment.

Acknowledgements

The assistance of François Boudreault and Denise Séguin, Saint-Lawrence Centre, is appreciated for providing assistance in graphics preparation (Figures 1 and 3, respectively). We are also grateful to John Wiley & Sons, Inc. for permission to reproduce Figures 2 and 3 taken from:

- Figure 9 (a, b) on page 3226 of the following publication: Blaise, C., 2002. Use of microscopic algae in toxicity testing, in G. Bitton (ed.), Encyclopedia of Environmental Microbiology, Vol. 6, Wiley Publishers, New York, NY, USA, pp. 3219-3230.

References

Bermingham, N., Costan, G., Blaise, C. and Patenaude, L. (1996) Use of micro-scale aquatic toxicity tests in ecolabelling guidelines for general purpose cleaners, in M. Richardson (ed.), *Environmental Xenobiotics*, Taylor & Francis, 195-212.

Blaise, C., Sergy, G., Wells, P., Bermingham, N. and van Coillie, R. (1988) Biological testing - Development, application and trends in Canadian environmental protection laboratories, *Toxicity Assessment* **3**, 385-406.

Blaise, C. (1991) Microbiotests in aquatic ecotoxicology: characteristics, utility and prospects, *Toxicity Assessment* **6**, 145-155.

Blaise, C. (1996) A micro-scale bioassay approach to industrial effluent assessment and management. Water Report (ISSN 0917-0456), pp. 53-56.

Blaise, C. and Kusui, T. (1997) Acute toxicity assessment of industrial effluents with a microplate-based *Hydra attenuata* assay, *Environmental Toxicology and Water Quality* **12**, 53-60.

Blaise, C., Gagné, F. and Bombardier, M. (2000) Recent developments in microbiotesting and early millennium prospects, in S. Belkin and S. Gabbay (eds.), *Environmental Challenges*, Kluwer Academic/Plenum Publishers, pp. 11-23.

Birkholz, D., Belton, K. and Guidotti, T. (2003) Toxicological evaluation for the hazard assessment of tire crumb for use in public playgrounds, *Journal of Air & Waste Management Association* **53**, 903-907.

Costan, G., Bermingham, N., Blaise, C. and Férard, J.F. (1993) Potential ecotoxic effects probe (PEEP): a novel index to assess and compare the toxic potential of industrial effluents, *Environmental Toxicology and Water Quality* **8**, 115-140.

Dutka, B. (1988) Priority setting of hazards in waters and sediments by proposed ranking scheme and battery of tests approach, *German Journal of Applied Zoology* **75**, 303-316.

Environment Canada (1990) Biological test method: reference method for determining acute lethality of effluents to *D. magna*, Environmental Protection Publications, Environment Canada, Ottawa, EPS Report 1/RM/14, 18 pp.

Environment Canada (1992a) Biological test method: toxicity test using luminescent bacteria (*Vibrio fisheri*), Environmental Protection Publications, Environment Canada, Ottawa, EPS Report 1/RM/24, 61 pp.

Environment Canada (1992b) Biological test method: growth inhibition test using the freshwater alga *Selenastrum capricornutum*, Environmental Protection Publications, Environment Canada, Ottawa, EPS Report 1/RM/25, 41 pp.

Environment Canada (1992c) Biological test method: test of reproduction and survival using the cladoceran *Ceriodaphnia dubia*, Environmental Protection Publications, Environment Canada, Ottawa, EPS Report 1/RM/21, 71 pp.

Environment Canada (1996) Industrial Plants: highlights of effluent files 1-106, Governments of Canada and Québec, Saint-Lawrence Action Plan Vision 2000, ISBN 0-662-80860-6.

Garric, J, Vindimian, E. and Férard, J.F. (1993) Ecotoxicology and wastewater: some practical applications, *Science of The Total Environment* **Suppl. (Part 2)**, 1085-1103.

Kusui, T. and Blaise, C. (1999) Ecotoxicological assessment of Japanese industrial effluents using a battery of small-scale toxicity tests, in S.S. Rao (ed.), *Impact Assessment of Hazardous Aquatic Contaminants*, Ann Arbor Press, Michigan, USA, pp. 161-181.

Kusui, T. (2002) Japanese application of bioassays for environmental management, in *The International Conference on Environmental Concerns and Emerging Abatement Technologies 2001*: Collection of Short Communications, *The Scientific World Journal* **2**, 537-541.

Legault, R., Blaise, C., Trottier, S. and White, P. (1996) Detecting the genotoxic activity of industrial effluents with the SOS Chromotest microtitration procedure, *Environmental Toxicology and Water Quality* **11**, 151-165.

Pardos, M. and Blaise, C. (1999) Assessment of toxicity and genotoxicity of hydrophobic organic compounds in wastewater, *Environmental Toxicology* **14**, 241-247.

Quillardet, P. Huisman, O., D'Ari, R. and Hofnung, M. (1982) SOS Chromotest, a direct assay of induction of an SOS function in *Escherichia coli* K-12 to measure genotoxicity, *Proceedings of National Academy of Science of USA* **79**, 5971-5975.

Sprague, J.B. and Ramsay, B.A. (1965) Lethal levels of mixed copper-zinc solutions for juvenile salmon, *Journal of the Fisheries Research Board of Canada* **22**, 425-432.

Thériault, F. (1996) La Réduction des rejets liquides toxiques des 50 établissements industriels prioritaires du Plan d'action Saint-Laurent, Rapport-synthèse 1988-1995.- Montréal: Environnement Canada - région de Québec, Direction de la protection de l'environnement; Longueuil : Ministère de l'Environnement et de la Faune du Québec, Direction régionale de la Montérégie, 1996, Saint-Laurent Vision 2000 – volet Protection - 12 pp., tabl. + annexes.

U.S. EPA (U.S. Environmental Protection Agency) (1989) Short-term methods for estimating the chronic toxicity of effluents and receiving waters to freshwater organisms, EPA/600/4-89/001, Office of Research and Development, Cincinnati, OH, 248 pp.

White, P., Rasmussen, J. and Blaise, C. (1996) Sorption of organic genotoxins to particulate matter in industrial effluents, *Environmental. Molecular Mutagenesis* **27**, 140-151.

Abbreviations

EC50	Effective concentration at which 50% of exposed organisms are affected
LOEC	Lowest Observed Effect Concentration
NOEC	No Observed Effect Concentration
PEEP	Potential Ecotoxic Effects Probe
SLAP	Saint-Lawrence Action Plan
TC	Toxic threshold Concentration
TU	Toxicity Unit.

2. A MULTITEST INDEX OF EFFLUENT TOXICITY BY PLS REGRESSION

ÉRIC VINDIMIAN
Ministère de l'écologie et du développement durable
Direction des études économiques et
de l'évaluation environnementale
Service de la recherche et de la prospective
20 avenue de Ségur, 75007 Paris, France
eric.vindimian@normalesup.org

1. Objective and scope

This method makes use of a test battery to derive a toxicity index that can be employed to classify effluents as a function of their overall toxicity. A formula is given as an example and a procedure to calculate the index using expert judgements and a PLS (Partial Least Square) regression procedure is described using data on 30 effluents.

2. Summary of the multitest index

2.1 PURPOSE

The use of a common battery of toxicity tests for effluents is a key feature for the management of the aquatic environment. The choice of the tests and the expression of results should be made as objective as possible to maximize the information on potential dangers and minimize cost. This text describes a procedure that was used in order to choose a test battery by means of a series of expert judgments and the use of modeling.

2.2 PRINCIPLE

The steps involved in designing the index of toxicity are as follows:

(1) Evaluation of a series of effluents using different acute and chronic toxicity test procedures.
(2) Collation of expert judgments from a panel of volunteers.
(3) Modeling an index of effluent toxicity with different batteries of tests by PLS regression.
(4) Choice of an optimum test battery and subsequent index that ranks the effluents as closely to the expert judgments as possible.

C. Blaise and J.-F. Férard (eds.), Small-scale Freshwater Toxicity Investigations, Vol. 2, 89-113.

This was carried out by suppressing tests and calculating the resulting sum of squares of deviation of the index from expert judgements. The more the sum of squares increases after suppression of one test, the more important the test is.

2.3 BIOASSAYS EMPLOYED TO DEVELOP THE INDEX

Vibrio fischeri, Microtox® light inhibition test; *Pseudokirchneriella subcapitata,* micro-algal growth inhibition assay; *Daphnia magna,* acute immobilization test; *Ceriodaphnia dubia,* chronic reproduction and survival test; *Thamnocephalus platyurus,* Thamnotoxkit® lethality assay.

2.4 FINAL TEST BATTERY

ALG: *Pseudokirchneriella subcapitata* (micro-algal growth inhibition assay); DM: *Daphnia magna* acute immobilization test; CER: *Ceriodaphnia dubia* chronic reproduction and CES: *Ceriodaphnia dubia* survival test.

2.5 CALCULATION OF TOXICITY PARAMETERS

The EC10 was calculated using a non linear regression procedure that is available as an Excel macro at the following link: http://eric.vindimian.9online.fr

2.6 CALCULATION OF THE INDEX (See Section 7.5)

$$I = 1 + LogQ \times \left[0.25Log\frac{100}{EC_{10}^{DM}} + 0.3Log\frac{100}{EC_{10}^{ALG}} + 0.35\left(Log\frac{100}{EC_{10}^{CES}} + Log\frac{100}{EC_{10}^{CER}} \right) \right]$$

3. Historical overview and applications

The use of toxicity tests for assessing water quality is often suggested as an appropriate tool to integrate several chemical impacts within a single relevant biological endpoint. Toxicity tests belong to the set of tools used for biomonitoring the environment with biomarkers (biochemical or physiological variables) and bioindicators (ecological or biological variables) (Vindimian, 2001). The advantage of biological testing allows a quantification of chemical stressors on the environment and may be the basis for regulatory action since sources of pollution are more clearly characterized. Regulation of environmental impacts using discharge permits and limitations or toxicity-based taxation tools may use the information driven from biotests according to national regulatory strategies.

For more than 30 years French water agencies have controlled effluents discharged in waters using a taxation principle based on several parameters. Most of the parameters are chemically-defined characteristics of water quality like chemical oxygen demand, ammonia, heavy metals and organic chemicals. However, a single biological parameter has been used for about 20 years. The potential effect of effluents is based on motility inhibition of *D. magna* Strauss after 48h of exposure and reported in TUs (toxic units). The TU is determined with the ratio "100%/EC50" where effluent dilution is expressed as a percentage, and considered proportional to the toxicity of the effluent. Taxation is based on the amount of inhibiting material disposed. It is calculated as the product of the effluent in TUs and the volume of discharged effluent. This procedure has proven to be useful in a span of twenty years and has contributed to a substantial decrease in the number of TUs released to local waters (Fig. 1).

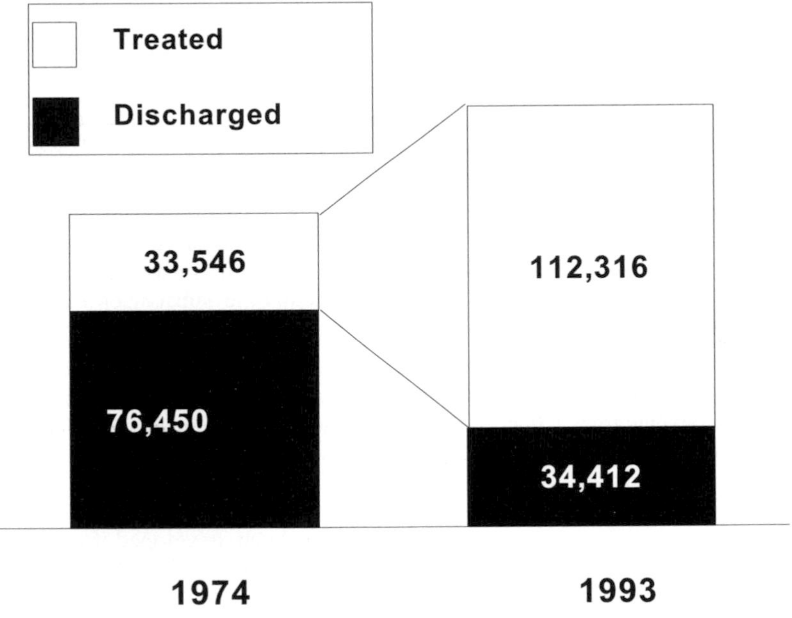

Figure 1. Evolution of effluent toxicity in France from 1974 to 1993 in terms of actual (discharged toxic units) and avoided (treated toxic units) toxic loading reaching surface waters.

However, the use of a single species acute toxicity test for such an assessment is questionable. The permanent release of chemicals in waters may cause long-term chronic toxicity that is not detectable in short-term laboratory tests. Long-term effects resulting in mortality or inhibition of growth and reproduction can only be assessed by means of chronic tests lasting at least for several reproduction cycles of each species of interest. Moreover, the impact of toxic substances on aquatic ecosystems cannot be adequately assessed by a single species and uncertainty can only be reduced by testing several species. Ideally, it would be helpful to have information on the effects of many species from different trophic levels and from diverse phyla. When such data are available, ecotoxicologists fit species sensitivity distributions to the available data and calculate the concentration that affects only a small percentage of species. The level of 5% is often used but this is purely conventional.

In most cases, limited information is available regarding the toxic effects of chemicals. Empirical guidelines are then used in an attempt to protect most of the aquatic ecosystem's biota. The regulation of chemicals, for instance, generally uses safety factors from 10 to 1000 depending on the number of species tested. Mesocosm studies or comparisons with real field situations are accepted with lower safety factors on a case by case basis, since these studies reduce the uncertainty linked to the relevance of laboratory models in terms of site-specific data.

It is even more difficult to protect aquatic ecosystems from potentially toxic effluents because of the uncertainty associated with complex mixtures of chemicals. Effluents often contain waste chemicals with an unknown composition and toxicity. The mixture of chemicals might also have an effect that cannot be predicted from the composition of the effluent. Moreover, effluents are highly variable in chemical composition and concentrations over time. However, it is important to regulate the release of potentially toxic chemicals in effluents even though it will be difficult to achieve. It would be preferable to know the chronic effects of effluents and each chemical substance within the effluent on several species. Chronic toxicity testing is the best way to get this information, including potential synergism or antagonism, with a direct assessment method. It would also be useful to assess effluents for genotoxicity. This is even more difficult since genotoxic substances might cause deleterious effects that would only be manifested in long-term chronic tests.

Once accepted, the utility of chronic toxicity tests in regulating toxic effluents could be enhanced by selecting only the most appropriate tests and calculating a summary toxicity parameter that could be used as a regulatory tool. The choice of tests should be theoretically made according to the level of biological diversity that needs to be protected. This could be achieved by the use of species endemic to the receiving waters or surrogate species with biological traits most closely resembling species of concern. However, methods have only been developed for a limited number of species relative to those found in the wild. The choice of test species is mainly made from considerations such as ease in laboratory rearing, availability, speed of the test and the ability to measure sensitive chronic endpoints such as reproduction. A test battery is therefore commonly used to account for species diversity but is also based on testing logistics (Vasseur et al., 1991; Latif et al., 1995; Kusui and Blaise, 1999).

The work presented here started from a matrix of test results on 30 effluents using seven common tests, including two chronic tests and two genotoxicity tests. The methodology used for this work has been published (Vindimian et al., 1999). Data from all toxicity tests were statistically interpreted using a common procedure able to calculate an ECx (where x = 10, 20 or 50) by non linear regression on a simple model (Hill, 1910). The genotoxicity tests were interpreted using a qualitative procedure since no concentration-response relationship was obtained with these tests. Then, all test results were sent to a panel of experts in the field of ecotoxicology. Each expert was requested to rank the effluents for their overall toxicity based on the information available from test results. Partial least square regression was used to determine the optimal test battery and an index of toxicity as a summary parameter. This procedure was developed to reproduce the average expert ranking as closely as possible. The index is a linear combination of each toxic unit TU value defined by the ratio “100%/EC10”.

4. Data set used

4.1 THE SERIES OF EFFLUENTS AND THE TESTS USED

Table 1. List of the 30 effluents that were sent to three laboratories. The second column shows the type of industrial activity, the last column the acidity of the effluent upon arrival at the laboratory.

N°	*Activity*	*Laboratory*	*pH*
1	Dye-works factory	1	8.75
2	Chemistry	1	7.95
3	Organic chemistry	2	7.00
4	Paper mill	2	7.30
5	Paper mill	1	8.40
6	Surface treatment	1	8.10
7	Surface treatment	1	9.75
8	Paper mill	2	-
9	Paper mill	2	-
10	Chemistry	1	10.30
11	Printing	2	7.60
12	Paper mill	2	9.40
13	Paper mill	2	5.00
14	Organic chemistry	2	6.00
15	Paper mill	2	6.00
16	Wool	2	7.00
17	Paper mill	2	4.40
18	Tannery	3	9.00
19	Surface treatment	3	7.00
20	Pharmaceuticals	3	7.00
21	Surface treatment	3	9.50
22	Slaughter	3	7.00
23	Inorganic	3	8.00
24	Dye-works factory	2	7.00
25	Unspecified	2	8.30
26	Organic chemistry	2	7.75
27	Organic chemistry	2	10.45
28	Organic chemistry	2	7.85
29	Paper mill	2	6.50
30	Coal	2	7.60

The 30 effluents are representative of various industrial sectors and were provided by different French water agencies. Their origin was kept confidential. Care was taken to use a set of effluents from diverse origin in order to ensure that the results from this study could be extrapolated to a large number of effluents. Table 1 lists the effluents tested, the type of industry and activity they represent, their pH and the laboratory code indicating where individual effluents were processed and analyzed. Three laboratories were involved in the testing exercise for toxicity and genotoxicity testing. Only one laboratory conducted micronucleus testing.

Table 2. List of tests used for the present study. The third column gives the code used for each test. Note that the test using Ceriodaphnia uses two codes which are endpoint related. The last column gives the ISO standard (or ISO project) number for an assay that is standardized or considered for standardization by (www.iso.org); those with no reference are commercialized tests.

Test	***Duration***	***Code***	***ISO standard***
Daphnia magna, acute toxicity test	24 h	DM	6341
Thamnocephalus platyurus, acute THAMNOTOX® kit	24 h	TM	-
Vibrio fisheri, luminescence test MICROTOX®	15 min	MT	11348
Ceriodaphnia dubia, chronic reproduction test	7 d	CER	20665
Ceriodaphnia dubia, chronic survival test	7 d	CES	20665
Pseudokirchneriella subcapitata, algal growth test	72 h	ALG	8692
MUTATOX® without S9	24 h	MU	-
MUTATOX® with S9	24 h	MUS9	-
Micronucleus *Xenopus laevis*	12 d	XE	21427-1

Table 3. Endpoints of the different tests. Endpoints of survival and reproduction were recorded for the 7-d Ceriodaphnia dubia test and population growth was also included since it is possible to calculate a growth rate using reproduction and survival across the age structure.

Endpoints	***Toxicity tests***
Survival or immobilization	*Daphnia magna* THAMNOTOX® kit *Ceriodaphnia dubia*
Bioluminescence	MICROTOX®
Population growth	Algae
Reproduction	*Ceriodaphnia dubia*
Genotoxicity	MUTATOX® Micronucleus

Seven different tests were used including two genotoxicity tests. Tables 2 and 3 summarize these tests. A more precise description is beyond the scope of this chapter. Among these toxicity tests, most are well known and routinely performed in laboratories. A more difficult one is that using *C. dubia* since it implies the manipulation of very small animals and is highly sensitive to laboratory conditions. However, this procedure has been used for many years around the world especially in North America where it is routinely used for the testing of effluents.

Among the two genotoxicity tests, the Mutatox® is a commercial test using a bacterial strain and is no more difficult to perform than the well known Microtox® test. The test on induction of micronuclei in erythrocytes of *Xenopus laevis* larvae developed by van Hummelen et al. (1989) requires specialized skills and training and is routinely performed only in a few laboratories. However, this test is currently being standardized by ISO (International Standardization Organisation).

5. Calculation of ECx

5.1 TOXICITY TESTS

Since NOECs and LOECs have been severely criticized, toxicity parameters were estimated by regression analysis (Chapman et al., 1996). The effects observed at different effluent concentrations were fitted to a Hill equation. The recommended equation (Hill, 1910) for the hemoglobin dissociation sigmoid curve is similar to a logistic equation. In 1983, we demonstrated the possibility of using non linear regression for estimation of the parameters of the Hill equation (Vindimian et al., 1983). It was later used with a "bootstrap" method to estimate confidence intervals for each parameter (Garric et al., 1990). For this study, we used Excel® software to estimate these metrics. This approach has also been used previously using various equations (Caux and Moore, 1997). Subsequently, we have developed mathematical solutions using this software to calculate ECx (ECx means a concentration with x% level of effect compared to control) with their corresponding confidence intervals. It is a macro using Excel spreadsheet and can be downloaded from the following web site http://eric.vindimian.9online.fr/. The equation is the following:

$$Y = Y_{\max} \times \left(1 - \frac{C^{nH}}{EC_{50}{}^{nH} + C^{nH}}\right) \tag{1}$$

where: Y is the observed biological variable (*e.g.*, number of juveniles, population growth, number of survivors); Y_{max} is the adjusted value of Y for control; EC50 is the median effective concentration, C is a designated concentration of interest and nH is the Hill number.

With such a program and spreadsheet, it is simple to compute the ECx values for different x levels. ECx means a concentration with x% level of effect compared to control. The confidence intervals of ECx were estimated using a "bootstrap" method as previously described (Efron, 1981).

The confidence intervals were used as guidance parameters to choose an optimal value of x. We found 10% to be a good compromise between sensitivity and precision for all tests used in this study. EC10 is also close to the range of NOEC values which makes it convenient for qualitative comparisons with former tests for which no raw data are available for recalculation (Isnard et al., 2001).

5.2 GENOTOXICITY TESTS

5.2.1 Mutatox®

This test estimates genotoxicity as the induction of reverse mutations in luminescent bacteria of a modified strain of *V. fischeri*. The genotoxicity tests were interpreted using a different strategy since no clear concentration-response was generated with this assay. The MUTATOX® responses were non monotonous for most tests. It appears that the effluent toxicity concentration range impacting bacterial growth or survival may have been similar to that causing genotoxic effects, thereby confounding results and making the test results more difficult to interpret. Such interactions precluded the use of a regression model that would have led to non interpretable results. We therefore used the following algorithm to help derive useful qualitative information from Mutatox® test results.

An effluent was considered:

- Non genotoxic (-1) if, at all exposure times, no more than one effluent concentration indicated a light induction factor greater than 2.
- Genotoxic (+1) if, for at least one exposure time, there are at least 3 effluent concentrations where the light induction factor is greater than 3.
- Suspect (0) in all other cases.

The induction factor is defined as the ratio of the treated batch luminescence to the average luminescence of controls.

5.2.2 *Xenopus micronucleus test*

This test is based on the observation and counting of micronuclei within the tadpole erythrocytes of the African toad *X. laevis*. The level of spontaneous micronuclei is low (< 5‰). The calculation method is as follows:

For each batch of results, erythrocytes with micronuclei per thousand observed are ranked in increasing values. The median and quartiles are determined.

The confidence interval on the median is given as

$$\mathrm{M} \pm 1.57\ IQ/\sqrt{n} \tag{2}$$

where:

- M = median of the sample
- IQ = inter-quartile difference
- n = sample size

Two alternative methods are used to interpret test results: 1) the lowest effective concentration or 2) the induction factor. The lowest effective concentration is the lowest at which the median number of micronucleated erythrocytes is significantly different from that of control animals. IF (induction factor) is the ratio between the treated group median and that of controls. In some cases, such parameters were not applicable and a qualitative assessment was employed.

An effluent was considered:

- Non genotoxic, when no effluent concentration gave a significant increase of micronuclei.
- Genotoxic, when at least one effluent concentration demonstrated that the median # of micronucleated cells was significantly different from that of controls.
- Suspect, when the median # of micronucleated cells was double that of control animals although not significantly different.

6. Design of an optimal battery and calculation of an index of toxicity

6.1 COMPARISON OF THE DIFFERENT EFFLUENTS AND TESTS BY PCA

Once the toxicity parameters were computed to a spreadsheet yielding a table of 30 rows (effluents) and 9 columns (bioassays), we ran a principal component analysis (PCA) to check the diversity patterns of effluents and the correlation between tests. The PCA calculations were carried out using the ADE 3.6 statistical package on a Macintosh computer. ADE was developed by the University of Lyon II and by the French National Centre of Scientific Research (CNRS) common biometry laboratory. The new version ADE version 4 running on Mac and PC computers is now available on this university's internet site at http://pbil.univ-lyon1.fr/ADE-4/

6.2 ASSOCIATION BETWEEN INDUSTRIAL ACTIVITY AND TOXICITY PATTERN

Having a set of effluents of different industrial origins allowed us to verify whether any relationships could be found between the activity and the pattern of toxicological responses. In this event, identifying specific test batteries adapted to the monitoring of different industrial types would then be an objective of further interest. To verify this, a discriminant analysis was used where industry type was entered as a categorical variable. Again, the analysis was performed using the ADE software mentioned above.

6.3 EXPERTS CONSULTATION

In order to get an expert judgement on the toxicity of the different effluents, we sent a questionnaire to 58 experts world-wide and asked them to classify the effluents on a 1 (least toxicity) to 5 (highest toxicity) scale. The only information the experts were given was the type of industrial activity, the pH, and the biological test responses. The effluents were randomly assigned a letter code. The experts were also asked for their opinion on bioassay performance and adequacy in order to guide the choice of tests for future application of a test battery approach. Numbers of experts by country who answered our call out of those solicited were the following: France 7/23, Portugal 1/1, United Kingdom 1/6, Finland 1/1, Canada 1/2, Sweden 0/3, United States of America 0/6, The Netherlands 0/5, Italy 0/2, Denmark 0/3, Germany 0/4, Belgium 0/2.

6.4 CALCULATION OF AN INDEX BY PLS REGRESSION

The average values of expert judgements were used as a vector of dependent variables in a PLS regression procedure where the results of each toxicity test became the independent variables. PLS is analogous to a multiple regression method but allows a more stable and coherent set of coefficients for the variables (Geladi and Kowalski, 1986). The principle of PLS is to calculate factors as latent structures being linear combinations of the X and Y tables in order to find a relation between the two tables. It begins by a co-inertia analysis. PLS can be used when Y is a single vector of a table. The latent structures corresponding to the highest eigen values represent the highest proportion of the table variance. These factors are used to set up the linear relationship between the two tables. Then the coefficient of each variable is calculated from its contribution to the latent structures.

We ran PLS using ADE 3.6 where no automatic PLS procedure was implemented. Thus, we had to run PCA on the table of variables, find the covariance factors and then run linear regression on those components. Due to the high variance explained by the first factor, we only used one component for the PLS regression. No improvement was obtained with a second component.

7. Application of the multitest index in a case study

7.1 TOXICITY DATA

The wide range in EC10 values obtained for the different effluents is shown in Table 4. Some effluents were found non toxic with some tests. Since a numerical value for computations was needed, the value of 100% as a "virtual" EC10 was used. It only served for computation but not for the expert judgements.

The table sent to the experts showed a non toxic (NT) sign when the effluent sample was not found to be toxic for a specific test. A first investigation sought to check the differences between different tests. This was carried out using a correlation analysis. The correlation coefficients from all the tests are reported in Table 5. It shows that the tests are generally correlated to each other. However, this may be misleading since high correlations are expected in such comparisons. When an effluent contains several toxic substances, it is likely that many tests will show a positive correlated response. Conversely, effluents containing no (or low concentrations of) toxic substances would lead to a similar no-response with all the tests. With that expected correlation in mind, the correlation matrix seems to show quite a good diversity of associations among the tests. The highest correlation is obtained with two acute crustacean bioassays (DM and TM), for which the coefficient is 0.8. The algal test generally shows a very weak correlation with all other tests. This is not unexpected in view of the evident phylogenetic differences that algae have with animals.

Table 4. Effluent toxicity data for each bioassay. Effluents with no toxicity were assigned a numerical value of 100 (in bold numbers).

Effluent	***Toxicity tests[a] (EC10s in percent)***						***Genotoxicity tests[a,b]***		
code	***DM***	***TM***	***MT***	***ALG***	***CER***	***CES***	***MU***	***MUS9***	***XE***
A	85	78	10.2	2.2	0.049	0.48	-1	0	0
B	22	0.15	6.9	19	5.3	14	-1	-1	0.02
C	**100**	**100**	0	8.5	64	**100**	+1	-1	0
D	0.76	0.57	0.03	0.4	0.05	0.14	0	0	0
E	**100**	83	5.4	17	10.2	7.1	+1	+1	0
F	0.067	0.22	0.027	0.0066	0.0022	0.0032	+1	+1	35.5
G	10.7	13	0.27	20	3.2	**100**	-1	-1	0
H	1.5	0.058	3.4	0.13	0.016	0.057	-1	-1	0
I	27	43	3.2	41	3.5	10.3	0	-1	0
J	**100**	**100**	**100**	7.9	20	**100**	+1	-1	0
K	40	7.4	**100**	79	1.945	4.5	-1	-1	0.008
L	3.4	0.4	0.13	1.05	0.29	0.70	-1	-1	4
M	**100**	84	88	**100**	36	81	0	+1	0
N	57	91	9.4	13.8	1.6	32	0	0	0.036
O	**100**	34	2.3	2.1	10.1	7.9	-1	-1	0.01
P	34	11.5	2.8	0.56	2.07	3.8	0	+1	0.44
Q	37	**100**	3.8	8.6	6.4	8.6	0	+1	0.06
R	23	11.3	2.8	0.56	0.046	0.0071	0	+1	0
S	0.89	0.077	2.2	0.48	0.36	0.46	-1	-1	0
T	**100**	75	**100**	28	**100**	0.51	+1	+1	0
U	**100**	88	17.8	1.1	4.3	7.4	+1	0	0
V	90	53	**100**	24	17	15.5	-1	+1	0.008
W	3.7	2.8	0.18	3.1	0.057	0.62	-1	0	0
X	1.16	13.6	1.06	2.2	0.13	0.11	-1	-1	0
Y	4.4	8.4	25	9.1	1.16	3.37	-1	+1	0
Z	**100**	22.9	8.2	24	22	**100**	-1	-1	0
AA	**100**	97	**100**	**100**	**100**	**100**	-1	0	0
AB	14	19.2	2. 9	1.9	1.20	8.96	+1	0	0
AC	4.2	5.2	0.23	8.6	0.73	1.5	+1	-1	0.6
AD	0.23	0.17	0.013	0.066	0.11	0.24	+1	+1	18

[a] Refer to Table 2 for test codes.
[b] MU (Mutatox® test): genotoxic (+1), non genotoxic (-1), suspect (0); *Xenopus* test (XE): induction factor as defined in Section 5.2.2.

Noteworthy as well is the low correlation observed between the two *C. dubia* chronic toxicity endpoints, where the correlation coefficient is only 0.5. One explanation may lie in the fact that survival was not affected at the highest concentration for several effluents shown to be toxic towards crustacean reproduction. In the case of one effluent, however, no effect on reproduction could be seen at sublethal concentrations of exposure, although survival was highly impacted above these. Figure 2 illustrates the observed relationship between survival and reproduction and argues in favor of using both endpoints for wastewater toxicity assessment.

Table 5. Correlation matrix for all sets of tests used in the study. The correlation coefficients between each set of tests are indicated.

Tests	TM	ALG	MT	CES	CER	MU	MUS9	XE
DM	0.806	0.397	0.549	0.508	0.602	0.168	0.102	-0.294
TM		0.328	0.441	0.461	0.547	0.331	0.199	-0.269
ALG			0.696	0.453	0.531	-0.178	0.039	-0.176
MT				0.354	0.608	-0.009	0.166	-0.172
CES					0.498	-0.004	-0.256	-0.175
CER						0.157	0.118	-0.142
MU							0.334	-0.330
MUS9								0.301

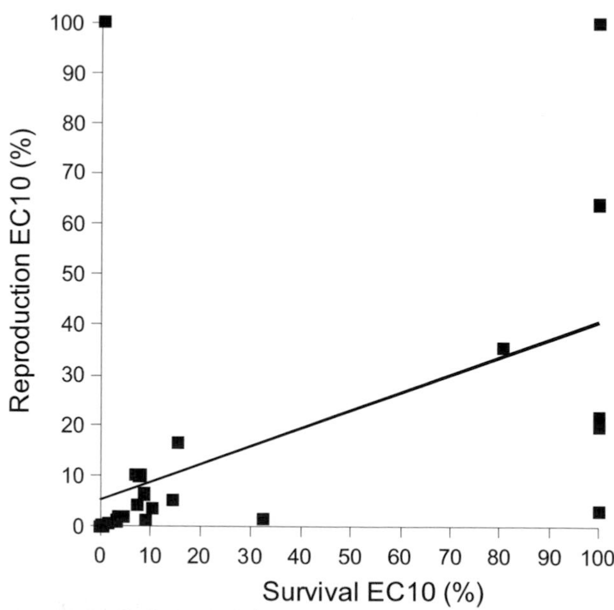

Figure 2. Correlation graph between the two endpoints of the Ceriodaphnia test. The apparent correlation coefficient is 0.5. Note that the values for which the EC10 is 100% are not true values but correspond to effluents where no toxicity was observed at 100% concentration.

7.2 PRINCIPAL COMPONENT ANALYSIS

The coordinates of effluents on the first factorial plane are represented in Fig. 3. The first factor encompasses 41% of the overall variance. It clearly distinguishes effluents from the most toxic to the least toxic. The second factor is related to genotoxicity. Thus the two main characteristics of effect (toxicity and genotoxicity) can be represented on the same graph. For instance, effluent F is not only the most toxic effluent but it is also genotoxic (see Tab. 4). Other effluents may be essentially genotoxic (*e.g.*, effluent T, Tab. 4), while others only display toxic effects (*e.g.*, effluent S, Tab. 4).

The correlation circle of the tests is shown in Fig. 4. This graph is visually useful to check the correlations of each test with PCA factors. The coordinates of each test are their correlation coefficients with each factor, the radius of the circle representing a coefficient of one. This analysis confirms the orthogonal relationships of genotoxicity and toxicity (Fig. 4). Toxicity tests conducted with unicellular organisms are essentially differentiated from those carried out with multicellular species based on factor 3 (Tab. 6). PCA analysis also shows that toxicity tests are only weakly correlated despite the effect of contrasting effluents which are either not toxic or toxic for most bioassays (Fig. 3). This is noteworthy as one drawback of correlation analysis is that correlation coefficients are strongly influenced by extreme values. In this case, correlation coefficients obtained between tests should be seen as low values owing to the range of effluent toxicity.

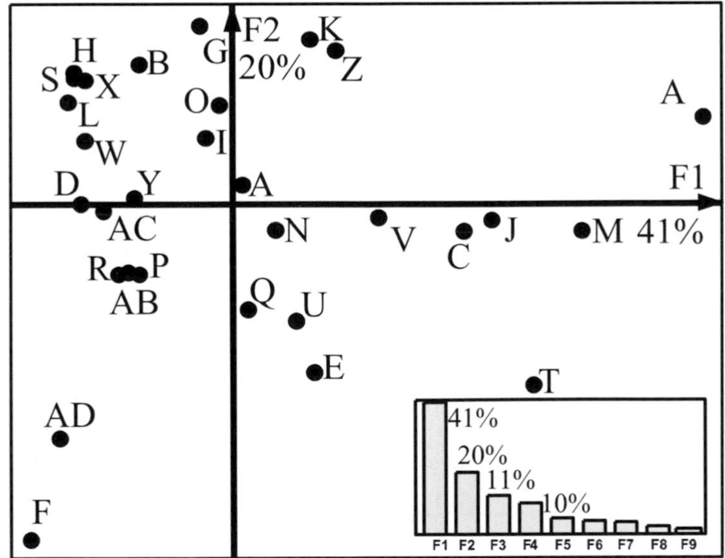

Figure 3. First factorial plane of the principal component analysis showing the effluent coordinates in this plane. The inner graph shows the percentages of the eigen values of this analysis, corresponding to the part of the global variance for each factor.

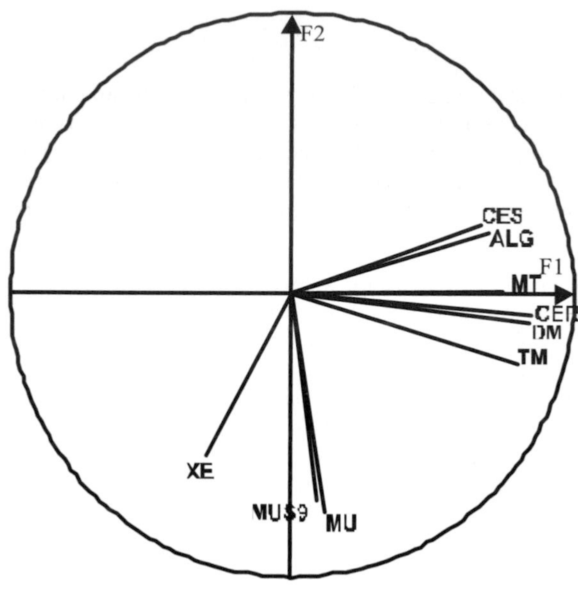

Figure 4. Correlation circle showing all tests according to their correlation with the two first factors.

Table 6. Correlation coefficients of each test with each of the three first factors of PCA. These values are the coordinates that were used to draw the correlation circle.

Test	***F1***	***F2***	***F3***
DM	0.85	-0.1	0.27
TM	0.80	-0.25	0.38
ALG	0.71	0.22	-0.53
MT	0.78	0.007	-0.44
CES	0.67	0.25	0.21
CER	0.81	-0.1	-0.09
MU	0.12	-0.8	0.35
MUS9	0.09	-0.76	-0.35
XE	-0.33	-0.61	-0.32

Only the two acute tests with *D. magna* and *T. platyurus* showed a high correlation coefficient. This confirms the need for using a wide range of different species covering different phyla to increase the power of a test battery to detect toxic effects associated with effluents. It is also worth emphasizing once more that the two *C. dubia* (survival and reproduction) endpoints are weakly correlated (Tab. 5) and that they may be equally relevant to estimate effects of effluents on populations.

The data suggesting that genotoxicity is orthogonal to toxicity (Fig. 4) supports the argument for separating these two variables. An index using both endpoints would imply taking the sum of two independent terms, one dealing with genotoxicity and the other with toxicity. However, the experts consulted agreed that the genotoxicity tests employed may not have represented the best choices and suggested, for instance, that an alternative such as the SOS Chromotest should be included in future test batteries (Legault et al., 1996). Based on this rationale, we decided to focus the approach strictly on a toxicity index and discarded genotoxicity in this effluent investigation.

7.3 DISCRIMINANT ANALYSIS

The result of the discriminant analysis is shown in Fig. 5. The industrial activities are represented by ellipses of inertia containing the coordinates of industrial categories on the first factorial plane. All the categories largely overlap each other and show that this set of ecotoxicological data could not allow any linkage between the type of industry and the toxicological pattern.

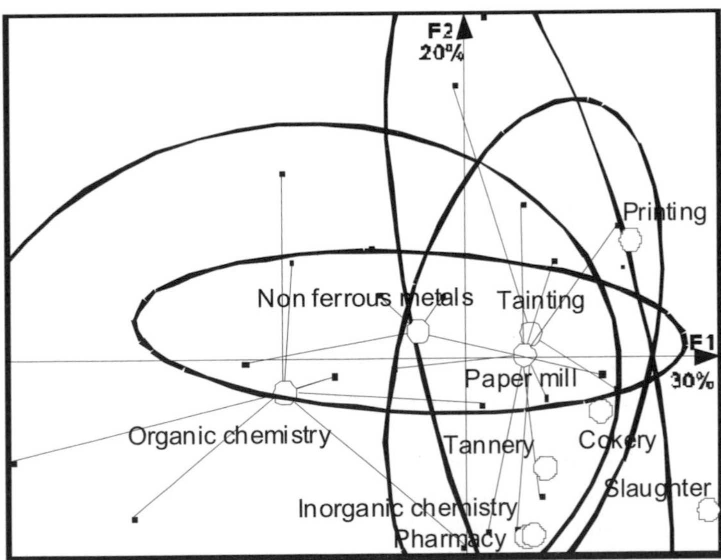

Figure 5. Graph showing the result of a discriminant analysis on the effluents as a function of the type of activity. The circles represent the centers of gravity of each effluent category linked with the individual effluents lying within an ellipse showing 90% of the variance for each category.

7.4 EXPERTS JUDGEMENTS

In light of the PCA results, it was fitting to ask the panel of experts to render their judgement on toxicity and genotoxicity, the two categories of bioassays considered in this study. The average values of experts judgements and their standard deviations are summarized in Table 7. The dispersion of these values is reported in Fig. 7. The toxicity quotations were coherent between the experts since the standard deviations are low. The overall average is well centered within the 1-5 (not toxic to highly toxic) classification ranking scheme, so we did not need to centre the quotations as would have been required with a non-symmetric distribution.

The experts were also requested to give some rationale for the (bio)analytical parameters that they routinely use to classify effluents (Tab. 8). Their responses clearly favored the most commonly used chronic toxicity tests and the *D. magna* test. Opinions on the use of commercial toxicity kits were mixed, probably due to a lack of information or experience with these tests at that time. The genotoxicity tests, especially Mutatox, were not appreciated by everyone. Moreover, chemical analysis of effluents was found a very useful parameter to include in effluent assessment. Despite the fact that the group of experts belonged mainly to the field of biological testing, they still considered chemical characterization of effluents to be a necessary complement to toxicity testing to ensure adequate protection of aquatic environments.

Whole effluent toxicity testing is in fact a good tool to assess the global toxicity of mixtures of chemicals susceptible to reach the aquatic environment. However, it does not take into account some chemicals that are not disposed at toxic concentrations but might create ecological problems on the long term because of their persistence and bio-accumulative properties. The loading of such identified chemicals in the environment needs to be assessed directly using a set of chemical analyses. Priority lists should also be setup for these chemicals in order to enforce good environment protection policies. This is the case for the water framework directive in Europe that identified 33 substances that need to be monitored (European Union, 2001).

Other unsolicited comments of the experts included the following:

- the *Ceriodaphnia* test was found expensive and insufficiently standardized at the time of this exercise (this problem is now being resolved, as a standard protocol is being drafted by ISO: ISO 20665);
- two experts recommended the use of a fish test;
- two experts suggested the use of other tests, *Hydra attenuata, Lemna minor* and the chronic test using *D. magna;*
- one expert suggested the use of the SOS Chromotest for genotoxicity;
- the *T. platyurus* micro-crustacean test was well correlated with the *D. magna* acute test (Fig. 4 and Tab. 5), suggesting that either could be used within a battery of test approach to assess effluent toxicity;
- the Mutatox® test was severely criticized, essentially owing to the fact that the results were found difficult to interpret;
- although appreciated because of its relevance, the *in vivo* test conducted with *X. laevis* was found difficult to interpret.

Table 7. Average value of the experts judgements on the toxicity and genotoxicity of the effluents. The experts had to rank their relative hazard as an integer varying from 1 to 5 (not toxic = 1, highly toxic = 5). σ stands for standard deviation.

Code	Toxicity		Genotoxicity		Code	Toxicity		Genotoxicity	
	Average	σ	Average	σ		Average	σ	Average	σ
A	3.5	0.9	1.3	0.5	P	3.4	0.5	2.3	0.8
B	2.9	0.8	2.0	1.5	Q	2.8	0.7	2.3	1.0
C	1.9	0.4	1.5	0.5	R	3.9	1.0	1.7	0.8
D	4.6	0.5	2.0	1.5	S	4.1	0.6	1.0	0.0
E	2.3	0.5	2.0	1.1	T	2.3	0.7	2.0	1.3
F	5.0	0.0	4.2	1.0	U	2.9	0.6	1.7	0.8
G	2.4	0.5	1.0	0.0	V	2.0	0.5	2.3	1.0
H	4.5	0.5	1.0	0.0	W	3.8	0.7	1.2	0.4
I	2.3	0.5	1.3	0.5	X	3.5	0.8	1.0	0.0
J	2.0	0.9	1.5	0.5	Y	2.9	0.6	1.5	0.5
K	2.3	0.7	1.8	1.2	Z	2.0	0.5	1.0	0.0
L	3.9	0.8	2.2	0.8	AA	1.0	0.0	1.2	0.4
M	1.4	0.5	1.8	1.2	AB	3.3	0.9	1.7	0.8
N	2.0	0.5	2.5	1.2	AC	3.3	1.0	2.2	0.8
O	2.6	0.7	1.8	1.2	AD	4.6	0.5	3.5	0.8
				General average		**3.0**	**0.4**	**1.8**	**0.5**

Other miscellaneous remarks questioned the use of the EC10 whose precision as an endpoint of toxicity was judged to be less reliable than other metrics. Again, some seemingly ambiguous results relating to *Ceriodaphnia* survival suggesting that it could sometimes be as sensitive as reproduction surprised some experts.

Collectively, PCA results combined with the experts judgements provided a justification to run the PLS regression only using toxicity tests. Genotoxicity was clearly found to be a different phenomenon related to a different mode of action and only the *Xenopus* test appeared interesting to evaluate the genotoxic potential of effluents. Clearly, the design of an index based solely on effluent genotoxicity merits further investigation and should comprise a variety of tests that were not available within this study.

7.5 PLS REGRESSION

The PLS regression was first run on the global data set using all the toxicity tests with the exception of the genotoxicity tests. For each effluent the regression gives a calculated index value that is intended to most closely represent the average of the experts judgements. Graphical comparisons appear to be the most convenient way of examining the calculated index values with the survey results of experts judgements.

Table 8. Average experts judgements on the adequacy of bioassays and other parameters in term of their usefulness for effluent toxicity assessment. Each expert had to rank usefulness as an integer value between 1 (least useful) to 5 (most useful) and they were also allowed to suggest other tests or parameters deemed useful. Some experts recommended tests without numerical ranking (Lemna minor and chronic Daphnia magna).

Test or parameters	***Average***	***σ***	***Number of responses***
Acute *Daphnia magna*	4.0	1.2	8
Thamnotoxkit	2.6	1.1	5
Microtox	3.0	1.2	8
Algal growth	4.4	0.7	8
Ceriodaphnia	4.0	0.9	8
Mutatox	2.4	0.9	5
Xenopus	3.3	1.0	4
Industrial activity	3.1	1.9	7
Effluent chemistry	4.7	1.3	6
Fish	5	-	2
SOS chromotest	(4)	-	1
Hydra attenuata	(4)	-	1
Chronic *Daphnia magna*	-	-	1
Lemna minor	-	-	1

First, a PLS was run on variables which were either the original EC10 or transformed variables. Experts' judgements were clearly correlated to the logarithm of EC10s rather than to the actual values. Hence, a PLS was run on a transformed variable. Such a variable is similar to the " Equitox " parameter which is currently used by French water agencies to classify effluents using *D. magna* 24h LC50s.

The result of this first analysis is shown in Fig. 6. The graph shows a fairly good linear relationship between the experts judgement and the calculated index. This can be visualized by the position of the circles compared to the line y = x. This comparative analysis can also be used to pass judgment on the proposed battery of tests, with the degree of departure from the line being a criterion to estimate its adequacy.

In an initial step, different toxicity test batteries were derived from the original data set by virtually suppressing one or more tests and then making the appropriate comparisons. A comprehensive view of the influence of each test on the goodness of fit is shown in Table 9 where the sum of squares deviations from the experts judgements is presented for a series of data sets with missing tests. The sum of squares of residues is higher when the fit is weaker, indicating that the suppressed test is important in revealing the observed effect. The more the sum of squares increases after suppression of one test, the more important the test is. This in turn drove our approach to find the "best" test battery. The value obtained with all tests was 2.2. The *C. dubia* test was the most influential on the experts judgement followed by the *P. subcapitata* test. The influence of the *Ceriodaphnia* test is likely strongest because of its double endpoint including survival and reproduction.

The PLS regression run on data sets where only a few acute tests were missing was found to give only slightly modified indices. This shows that the experts were strongly influenced by the chronic toxicity tests. We chose to keep only the *D. magna* acute test because it has been used for two decades as a tool for regulating French effluents and considerable data have already been amassed with this bioassay. Hence, the *D. magna* test is only kept because of historical links with previous French legislation and is not driven here by any ecotoxicological considerations.

The last step in calculating an index was to simplify the numerical values of the coefficients from each test in order to have an easier way to calculate the model. Fig. 7 compares the calculated results with the experts judgements. We also considered the possibility of including effluent flow in the index formula such that the quantity of toxic material (*i.e.*, toxic loading) discharged to the receiving environment can be estimated.

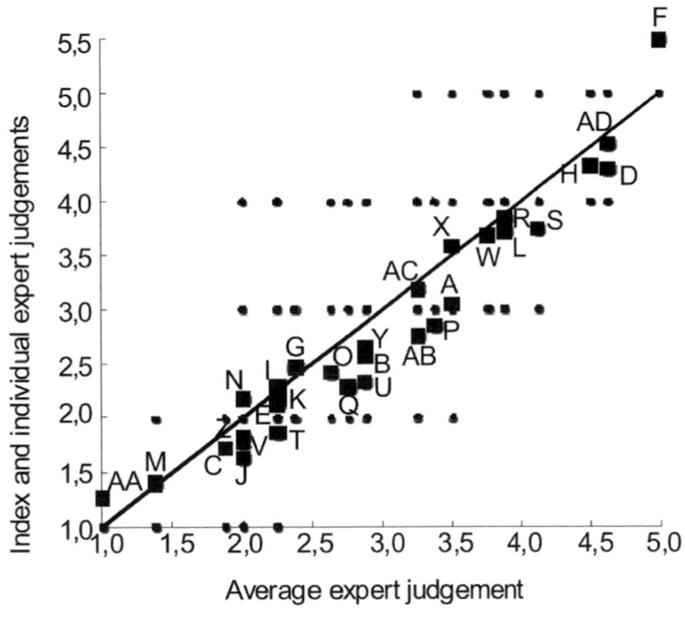

Figure 6. Calculated index using all tests versus the expert judgement. The small dots represent the individual experts judgements; large squares show the average of experts judgements as an abscissa and the index value as an ordinate for each effluent. The line drawn shows the identity y = x.

Table 9. Coefficient values of the PLS regression using different test batteries (B0 to B6) with missing tests. The 1st set of rows indicates the results of the regression with no constraint on the coefficients, the 2nd set of rows were obtained using the regression with the constraint that B0=1, the 3rd set of rows used rounded coefficients. Σ sq stands for the sum of square residues.

		Index without								
TB[a]	***PLS-R[b]***	***DM***	***TM***	***ALG***	***MT***	***CER***	***CES***	***CE***	***DM & TM***	***TM & MT***
			Index calculated by PLS regression							
B0	1.27	1.44	1.47	1.54	1.54	1.51	1.56	1.67	1.43	1.55
B1	0.12	**0**	0.13	0.13	0.13	0.14	0.15	0.20	**0**	0.19
B2	0.15	0.15	**0**	0.16	0.16	0.17	0.18	0.23	**0**	**0**
B3	0.2	0.20	0.20	**0**	0.21	0.22	0.24	0.31	0.22	0.23
B4	0.2	0.20	0.20	0.21	**0**	0.23	0.24	0.31	0.20	**0**
B5	0.22	0.22	0.22	0.23	0.24	**0**	0.26	**0**	0.27	0.28
B6	0.25	0.25	0.26	0.27	0.27	0.28	**0**	**0**	0.27	0.28
Σ sq	2.15	1.53	2.09	2.42	1.94	2.42	2.00	4.04	2.14	2.52
			Optimization using B0 =1							
B0	1	1	1	1	1	1	1	1	1	1
B1	0.14	**0**	0.153	0.164	0.166	0.173	0.189	0.260	**0**	0.239
B2	0.17	0.179	**0**	0.197	0.199	0.208	0.227	0.313	**0**	**0**
B3	0.22	0.235	0.242	**0**	0.263	0.274	0.299	0.412	0.266	0.286
B4	0.22	0.240	0.246	0.264	**0**	0.279	0.305	0.420	0.242	**0**
B5	0.25	0.264	0.271	0.291	0.295	**0**	0.336	**0**	0.328	0.352
B6	0.28	0.303	0.312	0.335	0.339	0.354	**0**	**0**	0.327	0.351
Σ sq	3.90	3.22	4.09	5.15	4.69	4.83	5.09	8.59	3.74	5.37
			Simplified index by rounding off of coefficients							
B0		1	1	1	1	1	1	1	1	1
B1		**0**	0.15	0.15	0.15	0.15	0.2	0.25	**0**	0.25
B2		0.2	**0**	0.2	0.2	0.2	0.25	0.3	**0**	**0**
B3		0.25	0.25	**0**	0.25	0.25	0.3	0.4	0.25	0.3
B4		0.25	0.25	0.25	**0**	0.25	0.3	0.4	0.25	**0**
B5		0.25	0.25	0.3	0.3	**0**	0.35	**0**	0.35	0.35
B6		0.3	0.3	0.35	0.35	0.35	**0**	**0**	0.35	0.35
Σ sq		3.24	4.11	5.06	4.68	5.27	5.30	8.75	4.03	5.35

[a] TB: test battery.
[b] PLS-R: PLS regression.

The final index formula is as follows:

$$I = 1 + LogQ \times \left[0.25Log\frac{100}{EC_{10}^{DM}} + 0.3Log\frac{100}{EC_{10}^{ALG}} + 0.35\left(Log\frac{100}{EC_{10}^{CES}} + Log\frac{100}{EC_{10}^{CER}} \right) \right] \quad (3)$$

where: I is the index value and Q the flow of the effluent ($m^3.s^{-1}$) which may also be a ratio of the effluent flow to a percentile of the median river flow. The coefficient values are those rounded off from Table 9.

Fig. 7 also compares the values of the experts judgement index with the corresponding Canadian PEEP index values (see Chapter 1 of this volume). It shows that the value differences for both indexes are minor and suggests that the index defined in this work was a good estimator of an expert consensus.

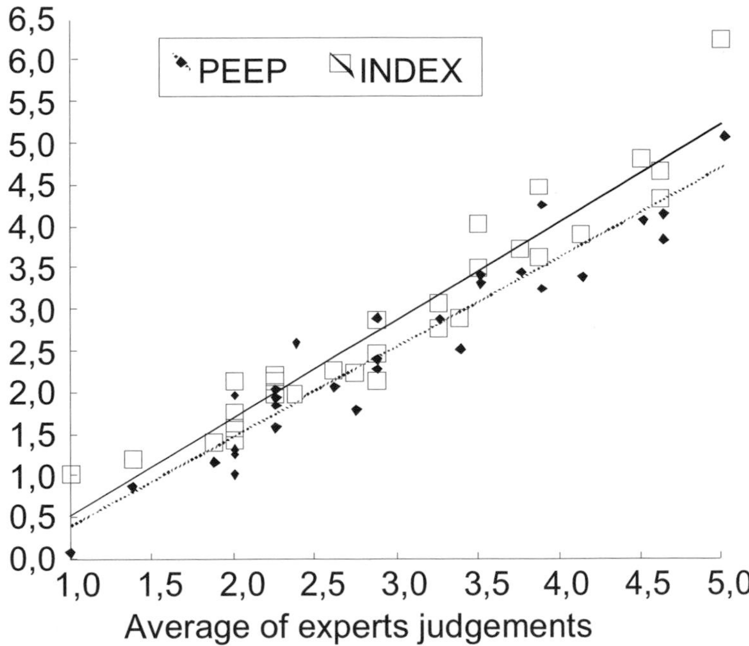

Figure 7. Comparison of the index that we developed (called "index") and the Canadian PEEP index. The correlation coefficient between the two indexes is 0.95. In both cases we draw a regression line to show the good agreement between the experts judgements and the indexes.

8. Discussion

The PCA results show that toxicity and genotoxicity are orthogonal. This confirms the independence of these two biological effects and suggests that they could be addressed separately by two different indexes. These results also emphasize the importance of assessing genotoxicity in order to consider different and more subtle environmental hazards to aquatic ecosystems. The available genotoxicity data from this study were not sufficient to be recommended as the basis for calculating a genotoxicity index. Other tests such as the SOS Chromotest (Legault et al., 1996), or the UMU test (Oda et al., 1985) that have been validated for assessing complex mixtures (Wong et al., 1986), or the well known Ames test (Ames et al., 1973), could be added to the genotoxicity battery. However, these tests are all microbial tests, which may not prove sufficiently protective of higher level species in the environment. Additionally, there is a need to undertake tests with eucaryotic cells, which can be achieved by incorporating bioassays such as the *X. laevis* micronucleus test used in this study. Ideally, a set of different tests conducted with eucaryotic organisms representing different trophic levels would increase the environmental relevance of this approach. *In vitro* tests on eukaryotic cell cultures would also be very useful (Chung et al., 1977).

Correlation between the toxicity tests could be invoked as an argument for using a single test to derive a cost effective index. However, the correlations obtained in the present study were not very high. This may be explained by the fact that the effluents originated from different industrial processes and thus contained different chemical substances. This clearly shows that the response of different species to toxicants is highly variable and stresses the importance of using a test battery to ensure that most of the potentially toxic chemical effects are circumscribed by the bioanalytical strategy in place.

Despite having investigated a set of effluents of diverse origins, there was no clear relationship between toxicity pattern and industrial activity. Indeed, discriminant analysis showed that most effluents could not be distinguished from one another.

PLS regression proved useful as a tool to design the index because it gave coherent results that can easily be applied in a regulatory context. Use of multiple regression would have led to coefficients with opposite signs in the event that tests would have shown correlation, even with low correlation coefficients. This cannot occur with PLS regression since the latent variables employed are linear combinations of all the variables dependent on the correlation with those variables. This approach facilitates calculating different index values with different test batteries and helps in determining which tests were most useful in matching the experts judgements.

In the end, the proposed battery is a compromise between the most efficient battery of tests and historical considerations. Results indicated that discarding acute invertebrate tests and including the Microtox® test would have been slightly more efficient. However, the improvement would have been negligible. The main reason for keeping the *D. magna* test within the battery is because a huge collection of data exists with this test, which has been used for over two decades by regulatory agencies around the world to monitor effluent discharges. For instance the *D. magna* acute toxicity test is recommended by the U.S. EPA as part of the set of acute toxicity tests for use in the National Pollutant Discharge Elimination System (NPDES) Permits Program to identify effluents and receiving waters containing toxic materials in acutely toxic concentrations (U.S. EPA, 2002). The French water agencies recommend this test (Agence de bassin Rhône-Mediterranée-Corse, 2002) that they have been using since 1972 for the control of effluents. In the event that this index should be accepted by regulatory authorities, its implementation would facilitate comparisons based on the *D. magna* test. This is also very important for the French situation, which was the original reason for performing this study.

It now seems possible to go one step further and to include chronic toxicity testing for several species into the environmental protection equation. Chronic bioassays could be utilized in several ways to:

- establish effluent discharge limits,
- quantify effluent toxic loadings entering wastewater treatment plants and enforce effluent quality based on toxicity emission,
- monitor effects-based stresses of point source discharges on the aquatic environment.

Implementing the last two approaches can best be carried out with a test battery. The battery and index described herein can be used for such purposes. When results are expressed with an index, hazard may be more accurately defined than with a single endpoint, even if it proves to be the most sensitive. The index designed, based on a judgement from experts, seems to validate this assumption since they did not rely on the most sensitive result as a unique classification tool. However, it was also shown that the experts did not use a simple average of the test results but tended to assign greater weight to chronic tests than to acute tests.

The index was derived from only 11 experts who accepted to answer our questionnaire and this might be considered as a weak point. However, the standard errors of the experts judgements are rather small suggesting that an increase of the number of experts would probably not have changed the results of this study. It might also be thought undesirable to decrease the amount of information included in a battery of test in a single synthesizing parameter. An index is clearly not a substitute for a more refined analysis using substantial data. It is rather a value that can be used for routine pollution control. The use of such an approach based on a multivariate analysis that minimizes the loss of information when going from the overall data to one value is optimal in this respect. With such an index, effluents exhibiting high values could be investigated more fully, using chemical analyses or more sophisticated biological methods in order to improve environment protection.

This index could be considered as a tool to calculate a fine for each effluent in application of a "polluter pays" principle. Such a fine certainly bears relevance to effluent chronic toxicity and would encourage the reduction of toxicant discharge to aquatic environments. Furthermore, such a fine could be considered fair and realistic since it is based on a statistical approach that minimizes the loss of information.

9. Conclusion

An index of toxicity is intended to be a simple tool that allows integrating and summarizing several variables into a single value. Realistically, this cannot be inferred without a judgement by environmental protection experts who consider all parameters available for their classification. PLS regression helped calculate an index fitted to expert judgement. The loss of information owing to the transformation of a multivariate situation to a univariate one was thus minimized since it is an inherent characteristic of multivariate analytical tools.

The index is easily calculated and depends mainly on chronic toxicity tests associated with a sensitive measurement endpoint chosen to be the EC10 (concentration incurring a 10% effect). It can be used as a regulatory criterion that is more protective of aquatic ecosystems than those based on a single acute toxicity test.

The need for an appropriate battery of genotoxicity tests is also important and cannot be compensated or replaced by toxicity tests since the variables are clearly independent. Further research, especially on genotoxicity tests with eukaryotes, applied to effluents, should be encouraged.

References

Agence de bassin Rhône-Mediterranée-Corse. (2002) Pollution toxique et écotoxicologie: notions de base, Guide technique n°7, Nov. 2002.

Ames, B.N., Lee, F.D. and Durston, W.E. (1973) An improved bacterial test system for the detection and classification of mutagens and carcinogens, *Proceedings of the National Academy of Sciences,* USA, **70**, 782-786.

Caux, P.Y. and Moore, D.R.J. (1997) A spreadsheet program for estimating low toxic effects, *Environmental Toxicology and Chemistry* **16**, 802-806.

Chapman, P.M., Caldwell, R.S. and Chapman, P.F. (1996) A warning: NOECs are inappropriate for regulatory use, *Environmental Toxicology and Chemistry* **15**, 77-79.

Chung, Y.S., Ichikawa, K. and Utsumi, H. (1977) Application of micro nucleus *in vitro* assay to micropollutants in river water, *Water Science and Technology* **35**, 9-13.

Efron, B. (1981) Non parametric estimates of standard error: the jack-knife, the bootstrap and other methods, *Biometrika* **68**, 589-599.

European Union (2001) Decision 2455/2001/EC of the European Parliament and of the Council of 20 November 2001 establishing the list of priority substances in the field of water policy and amending Directive 2000/60/EC (Text with EEA relevance), OJ L 331, 2001-12-15, pp. 0001-0005.

Garric, J., Migeon, B. and Vindimian, E. (1990) Lethal effects of draining on brown trout. A predictive model based on field and laboratory studies, *Water Research* **24**, 59-65.

Geladi, P. and Kowalski, B.R. (1986) Partial least squares regression: A tutorial, *Analytica Chimica Acta* **1**, 19-32.

Hill, A.V. (1910) The possible effects of aggregation of the molecules of haemoglobin on its dissociation curves, *Journal of Physiol*ogy - *London* **40**, IV-VII.

Isnard, P., Flammarion, P., Roman, G., Babut, M., Bastien, P., Bintein, S., Essermeant, L., Férard, J-F., Gallotti-Schmitt, S., Saouter, E., Saroli, M., Thiebaud, H., Tomassone, R. and Vindimian, E. (2001) Statistical analysis of regulatory ecotoxicity tests, *Chemosphere* **45**, 659-669.

ISO 20665 Determination of 7 days chronic toxicity to *Ceriodaphnia dubia* – Population growth Inhibition test.

Kusui, T. and Blaise, C. (1999) Ecotoxicological assessment of Japanese industrial effluents using a battery of small-scale toxicity tests, in S.S. Rao (ed.) *Impact assessment of hazardous aquatic contaminants,* Ann Arbor Press Inc., Chelsea, Michigan, USA, pp. 161-181.

Legault, R., Blaise, C., Trottier, S. and White, P.A. (1996) Detecting genotoxic activity in industrial effluents using the SOS chromotest microplate assay, *Environmental Toxicology and Water Quality* **11**, 151-165.

Latif, M., Persoone, G., Janssen, C., Coen, W. and de Svardal, K. (1995) Toxicity evaluations of waste waters in Austria with conventional and cost-effective bioassays, *Ecotoxicology and Environmental Safety* **32**, 139-146.

Oda, Y., Nakamura, S.I., Oki, I., Kato, T. and Shinagawa, H. (1985) Evaluation of the new system (UMU-test) for the detection of environmental mutagens and carcinogens, *Mutation Research* **147**, 219-229.

U.S. Environment Protection Agency (2002) Methods for Measuring the Acute Toxicity of Effluents and Receiving Waters to Freshwater and Marine Organisms, Fifth Edition, EPA-821-R-02-012.

Van Hummelen, P., Zoll, C., Paulussen, J., Kirsch-Volders, M. and Jaylet, A. (1989) The micronucleus test in *Xenopus*: a new and simple *in vivo* technique for detection of mutagens in fresh water, *Mutagenesis* **4**, 12-16.

Vasseur, P., Férard, J-F. and Babut, M. (1991) The biological aspects of the regulatory control of industrial effluents in France, *Chemosphere* **22**, 625-633.

Vindimian, E. (2001) The biological monitoring of toxic impacts in the environment, *Cellular and Molecular Biology* **47** (6), 1309-1318.

Vindimian, E., Garric, J., Flammarion, P., Thybaud, E. and Babut, M. (1999) An index of effluent aquatic toxicity designed by PLS regression, using acute and chronic tests and expert judgements, *Environmental Toxicology and Chemistry* **18**, 2386-2391.

Vindimian, E., Robaut, C. and Fillion, G. (1983) A method for co-operative and non co-operative binding studies using non-linear regression analysis on a microcomputer, *Journal of Applied Biochemistry* **5**, 261-268.

Wong, W-Z., Wen, Y-F., Steward, J. and Ong, T. (1986) Validation of the SOS/UMU test with mutagenic complex mixtures, *Mutation Research* **175**,139-144.

Abbreviations

ADE	Analyse des données écologiques
ECx	Effective concentration (for x% effect)
IF	Induction Factor
ISO	International Standardization Organisation
LOEC	Lowest Observed Effective Concentration
NOEC	No Observed Effect Concentration
NPDES	National Pollutant Discharge Elimination System
NT	Non Toxic
PCA	Principal Component Analysis
PEEP	Potential Ecotoxicity Effluent Probe
PLS	Partial Least Squares
PLS-R	PLS regression
TB	Test Battery.

3. THE PT-METHOD AS A HAZARD ASSESSMENT SCHEME FOR WASTEWATERS

FALK KREBS
German Federal Institute of Hydrology (BfG)
Am Mainzer Tor 1, 56068 Koblenz, Germany
krebs@bafg.de

1. Objective and scope of HAS

The pT-method is an evaluation strategy for investigating the toxic effects of wastewaters. With this Hazard Assessment Scheme (HAS), wastewater is tested with standardized bioassays, using dilution series in geometric sequence with a dilution factor of two. Its toxic status is then equated with the first dilution stage at which the effluent is no longer toxic. The numerical designation of toxicity for an effluent is described as the pT-value which is determined using the negative binary logarithm of the first non-toxic dilution factor. The pT-value indicates the number of times a sample must be diluted at a ratio of 1:2 with standardized dilution water before there are no longer any observed toxic effects on aquatic test organisms.

While individual toxicity tests measure specific endpoints, a single test cannot be used to adequately reflect the general hazard potential of a liquid sample. A hazard assessment can only be approximated using a multidisciplinary approach based on a large number of different toxicity tests within a test battery. An adequate strategy is the application of a multi-trophic testing scheme. The pT-value of the most sensitive organism within a test battery is known as the pT_{max}-value and it determines the toxicity class of an investigated sample. All bioassays in a test battery are considered equal in rank and Roman numerals are assigned to each toxicity class based on the magnitude of toxic effects observed in the most sensitive test organism. For instance, if the highest pT-value is 7, the tested material is then assigned to toxicity class VII which corresponds to a pT-index of VII (see Section 5.4). Hence, the pT-index derived from the most sensitive organism in a test battery constitutes a numerical classification of wastewater effluents based on ecotoxicological principles. With the aid of this simple index, the potential toxic hazard of any aqueous sample can be quantified in an easily understandable way.

The pT-method can be used to assess the toxic potential of both treated and untreated wastewater, surface water (receiving streams), groundwater, pore water, elutriates and chemical extracts of sediments. This paper deals with wastewater, the application of the pT-method to sediments is described in Chapter 9 of this volume. The method can be applied universally and is user-friendly. Any number of aquatic

C. Blaise and J.-F. Férard (eds.), Small-scale Freshwater Toxicity Investigations, Vol. 2, 115-137.

toxicity tests deemed necessary for the detection of general or special water pollutants can be integrated into this scheme. More specifically, pT-values derived from individual toxicity tests and the resulting pT-index can provide valuable information on the degree of ecotoxicity that wastewater discharges may have within an entire river basin.

2. Summary of the pT-method procedure

The toxicity of an effluent, classified by the pT-index, is based on the pT-value of the most sensitive organism within a given test battery. Thus, pT-values and pT-indices are numerical designations on an open scale to characterize the degree of hazard represented by effluent wastewaters. Sensitive aquatic ecosystems can be protected using the pT-index as a guide for ensuring sound management decisions and environmental protection. Table 1 provides synthetic information on the pT-method concept related to effluent assessment.

3. Historical overview and applications reported with the HAS procedure

The pT-method was designed to produce quantitative toxicity data to ensure protection of aquatic ecosystem biota. It offers a classification index for appraising effluent wastewaters and receiving streams, based on ecotoxicological principles, as described in Section 5. Application of the pT-method was further extended to solid media hazard assessment (sediments and dredged materials). In this case, the pT-method is used for monitoring, decision-making, and regulatory needs (see Chapter 9 of this volume). The pT-scale is a useful new scheme for the purposes intended, but there are other useful strategies as well that are already being employed, for example the US National Pollution Discharge Elimination System (U.S. EPA, 1991).

3.1 COUPLING OF BIOLOGY AND CHEMISTRY IN ECOTOXICOLOGICAL INVESTIGATIONS

Proper ecotoxicological investigations require both chemical and biological tools to determine the type, degree, and extent of pollution caused by toxic substances originating from industries, agriculture, and urban activities. As such, they are necessary complements to one another. The more evident cause(s) of pollution can normally be identified through chemical analysis. This information, however, is strictly restricted to measured parameters and chemical analysis can never hope to be exhaustive owing to the vast number of constituents present in complex wastewaters or surface waters. The toxicological risk linked to wastewaters can only be assessed if there are sufficient toxicity data for those toxic substances that have been detected. These data are frequently not available, thereby limiting the usefulness of chemical analysis and pointing out its shortcomings when used alone.

Table 1. Summary of the pT-index for wastewaters.

pT-Index **(pT = *potentia Toxicologiae* = toxicological exponent)**
Purpose The index was developed as a management tool to incorporate bioassay data in the decision making process for assessing and comparing the relative toxic hazards of industrial effluents. Stakeholders can make informed decisions using pT-values and pT-indices to ensure appropriate actions regarding effluent discharges that are identified as potentially harmful to a particular receiving aquatic system (*i.e.*, those labelled with high pT-values). Therefore, the pT-scale can contribute to cost-effective environmental assessment aimed at lowering toxic inputs from industrial origin into receiving waters.
Principle The pT-value of the most sensitive organism within a test battery determines the toxicity class of an effluent, and is designated as the pT-index. Thus, the pT-index can be used as a tool to act in the interest of environmental protection. When applied, it ensures that wastewater toxicity will be reduced, thereby protecting the most sensitive aquatic species. In fact, toxicity-based strategies of the pT-index are designed to protect aquatic biota in the receiving stream. Together, pT-values and pT-indices provide a valid system to describe different types of toxic pollution effects. This HAS procedure is a relevant discriminator of toxicity when hazard is being evaluated for a designated group of industrial effluents.
Bioassays employed The classic test battery used pursuant to the German Water Act (WHG, 2002; AbwV, 2002) is comprised of the following organisms: • *Desmodesmus subspicatus*, micro-algal growth inhibition assay (DEV L33, 1991). • *Vibrio fischeri*, bacterial luminescence inhibition test or Microtox® assay (DEV L34, 1998). • *Daphnia magna*, cladoceran acute immobilisation test (DEV L30, 1989). • *Leuciscus idus melanotus*, golden orfe, fish lethality test (DEV L31, 1989). Additional test systems newly added to the German Wastewater Charges Act (AbwV, 2002) include the following organisms: • *Danio rerio*, zebra fish, fish egg test (DEV T6, 2003). • *Salmonella typhimurium*, bacterial genotoxicity test or umu-test (DEV T3, 1996).

Table 1 (continued). Summary of the pT-index for wastewaters.

Determination of effluent hazard potential with the pT-value and the pT-index

1. pT-value for the numerical designation of aquatic toxicity measured in a single bioassay

Hazard potential is determined with standardised aquatic toxicity tests, using 2-fold serial dilutions. The toxicity endpoint corresponds to the first dilution stage that does not produce any toxic effects to the target organisms. The numerical designation of toxicity is the pT-value (*potentia toxicologiae* = toxicological exponent) which is the negative binary logarithm of the first non-toxic dilution factor in a dilution series in geometric sequence with a dilution factor of two. The pT-value indicates the number of times a sample must be diluted in a 1:2 ratio until test organisms no longer exhibit toxic effects. The toxic potential of any aqueous sample can be readily quantified in an easily understandable way. The pT-scale is unlimited, as values can theoretically range from 0 to ∞.

2. pT-index for the numerical classification of aquatic toxicity measured within a test battery

The pT-value of the most sensitive organism within a test battery determines the toxicity class of the sample. Different bioassays are considered equal in rank. Roman numerals are assigned to the toxicity classes. For example, if the highest pT-value is 7, then the tested material is assigned to toxicity class VII (*i.e.,* the pT-index is VII).

Additional remarks regarding the pT-index and pT value

In theory, the pT-value and the pT-index can range from 0 to infinity. In practice, the values usually range from 0 to about 12, similar to the PEEP scale (see Chapter 1 of this volume), also making the pT method a readily-understandable "environmental Richter scale". The pT method is designed for indicating point source industrial toxicity and can equally be used to describe the temporal changes of toxicity of industrial discharges. Again, pT-values can be determined for any aquatic bioassay.

Documented applications of the pT-method

The pT-values were generated for industrial effluents in the Rhine river basin (Krebs, 1988).

The pT-method is also applicable to identify ecotoxicological effects in surface waters. This was demonstrated in several rivers in Germany including the river Saar, a tributary in the Rhine river basin (Krebs, 1992b) and in rivers discharging into the Sepetiba Bay, Federal State of Rio de Janeiro, Brazil (Soares, 2000).

In contrast, bioassays reflect the integrated effects of water constituents on test systems (organisms or biological functions). They are not, however, suitable for qualitative or quantitative identification of contaminants, and cannot pretend to be substitutes for chemical analyses. Their inherent benefit is in being able to determinethe presence or absence of noxious compounds that are bio-available to living systems. The detection of toxic effects can clearly be test-specific and will not necessarily apply to all levels of biological organization. Generally, comprehensive assessment of toxicity will only be achieved by using several toxicity tests within a test battery.

3.2 IN SEARCH OF A SIMPLE METHOD TO MEASURE THE TOXIC POTENTIAL OF WASTEWATERS

In searching for ways to determine the toxic potential of contaminated environmental samples, one might choose to measure simply the "percent value effect" obtained with the undiluted sample material. While this approach is certainly simplistic, it may be prone to interferences. Algal toxicity tests, for example, conducted on wastewaters containing both auxinic and toxic chemical components, will often mask toxicity at high concentrations because of the presence of high nutrient levels in the undiluted test water. Only when these enhancing effects are eliminated through dilution will genuine inhibiting effects owing to sample toxic constituents become apparent. Hence, the toxic potential of such samples can be underestimated if they are only appraised in their undiluted state and they must also be evaluated at different dilutions to account for possible enhancement effects. A "dilution method", therefore, prevents such interferences and involves establishing a dilution series to find the sample dilution at which toxicological effects are no longer observed. Hence, the dilution factor can be used as a measure of toxicity, which is the basis of the pT-scale method. This method considers both the percent inhibition effect of the undiluted sample and that of the sample dilution factor indicative of absence of toxicity.

3.3 CONSIDERATIONS ON THE SELECTION OF TOXICITY TESTS FOR A REPRESENTATIVE TEST BATTERY

Selection of toxicity test methods should take into account the aquatic ecosystem (freshwater, brackish, or marine waters) that receives the wastewaters. Test organisms should be representative of three trophic levels: producers (*e.g.*, micro-algae), consumers (*e.g.*, crustaceans), and decomposers (*e.g.*, bacteria). Test battery bioassays are also selected according to the environmental protection objectives which have been set. If the focus of protection is on surface waters, tests representative of natural conditions are preferred with sensitivities comparable to those of endemic organisms. If several tests with different organisms at the same trophic level are available, the most sensitive one should be preferred, so that other sensitive organisms not represented in the test battery may benefit from the test response. All tests of the same battery should be ranked equal in importance in terms of their assessment capability. If information on both short-term and long-term effects is sought, then the battery should be composed of acute and chronic tests. Assessment endpoints can include any of several types of survival, growth and reproduction parameters.

3.4 PAST AND PRESENT APPLICATIONS OF THE pT-SCALE PROCEDURE

The first report describing the pT-scale procedure was presented at the annual meeting of the German speaking branch of the International Association of Theoretical and Applied Limnology (*Societas Internationalis Limnologiae* or SIL) in Hamburg in 1984 and a first publication followed thereafter (Krebs, 1987). At the time, German rivers such as the rivers Saar and Wupper were heavily polluted and toxic. As a result, application of the pT-method was recommended both for wastewater control purposes and for characterizing the toxic properties of surface waters in longitudinal sections of rivers (Krebs, 1992b).

Presently, the process of determining a pT-value follows directives outlined in the German Wastewater Charges Act which requires that each wastewater be investigated with standardized tests in a dilution series. Toxicological evaluation is reduced to finding the first dilution stage at which an effluent is no longer toxic. For determination of the pT-value for individual bioassays, a dilution series in geometric sequence with a dilution factor of two is sufficient. In the Wastewater Charges Act, these series are supplemented by intermediate dilution steps[1], *e.g.*, 1:2, 1:3, 1:4, 1:6, 1:8, 1:12, 1:16. The reason for these intermediate dilution steps is that the polluter has to pay for the toxic load and the non-toxic factor is used to calculate the amount of the wastewater charge. Industrial effluents are ordinarily monitored with only a few bioassays selected according to the type of industry being evaluated (AbwV, 2002), and a pT-index is not determined.

The pT-method has gained attention in other countries. In a GTZ project (German government-owned corporation for international technical co-operation or GTZ), the pT-method was applied successfully by the Environmental Agency (FEEMA) of the Federal State of Rio de Janeiro, Brazil, to identify the toxicity of industrial effluents discharging into the Rio Paraiba do Sul and to characterize the toxicity of surface water in water bodies discharging into the Sepetiba Bay (Soares, 2000). In both cases toxic effects could be detected. The pT-value functioned as a useful discriminator that could distinguish different types of hazard potentials.

4. Advantages of applying the pT-scale procedure

The advantages of the pT-method can be succinctly stated as follows:

- simple to use and to interpret (Sections 5.1, 5.2, 5.3, 8),

[1] "When testing wastewater by means of a graduated dilution (D), the most concentrated test batch at which no inhibition, or only minor effects not exceeding the test-specific variability, were observed is expressed as the "Lowest ineffective Dilution (LID)". This dilution is expressed as the reciprocal value of the volume fraction of wastewater in the test batch (*e.g.* if wastewater content is 1 in 4 (25% volume fraction) the dilution level is D = 4)", quoted from DEV L34, 1998 (draft 2004). (LID = 4 means pT = 2 and LID = 8 means pT = 3, cf. Table 2.)

- cost-effective battery of bioassays (Section 5.6),
- universal and flexible in application (Sections 5.1, 5.2, 5.3, 7),
- simple sample dilution method for toxicity appraisal (Section 3.2),
- good discriminatory potential for effluent toxicity (Section 5.9),
- no special software required for data reduction and pT calculations (Sections 5.3, 5.7, 6, 7),
- ease of technology transfer (Sections 3.4, 8),
- based on a sound scientific conceptual framework (Section 5.2),
- user-friendly management tool (Sections 5.7, 5.8, 5.10).

5. pT-method description

5.1 THE pT-VALUE AS A PARAMETER OF ECOTOXICITY

Based on the principle that the first non-toxic dilution level is used for numerical classification of liquid samples, toxicity can be expressed in the form of the negative binary logarithm of the dilution factor. In similar fashion to the pH value introduced by S.P.L. Sørensen in 1909 as a measure in chemistry, this measure of toxicity is called the pT-value (Krebs, 1987).

The pH (*potentia Hydrogenii* = hydrogen exponent) is the negative decadic logarithm (logarithm to the base 10) of the hydronium-ion concentration in a solution. The Latin term *potentia Hydrogenii* means “potency of hydrogen”. The word *potentia* was used in mathematics to name the exponent. Therefore, pH stands for hydrogen exponent. A pH unit of 4 indicates 1×10^{-4} moles of hydronium-ions per litre.

$$\mathrm{pH} = -\log_{10}\left(\mathrm{H_3O^+}\right) \tag{1}$$

$$\left(\mathrm{H_3O^+}\right) = 10^{-\mathrm{pH}} = \text{antilog}_{10}\left[-\mathrm{pH}\right] \ \text{moles/L}$$

$$\textit{e.g.},\ \left(\mathrm{H_3O^+}\right) = 1 \times 10^{-4} \ \text{moles/L}$$

$$\mathrm{pH} = -\log_{10}\left(\frac{1}{x}\right) = \log_{10}(x)$$

$$\mathrm{pH} = -\log_{10}\left(\frac{1}{10000}\right) = \log_{10}(10000) = 4$$

Or, pH = 4 stands for a hydronium-ion concentration of

$$\left(H_3O^+\right) = \text{antilog}_{10} - 4 = 10^{-4} \text{ moles/L}$$

Similarly, the exponent of toxicity is designated a *potentia Toxicologiae* (pT). The pT-value is equal to the toxicological exponent and is determined by the negative binary logarithm (logarithm to the base 2) of the first non-toxic dilution factor in a dilution series in geometric sequence with a dilution factor of 2. The pT-value indicates the extent to which a sample must be diluted at a ratio of 1:2 in order for it to no longer produce any toxic effects.

$$\text{pT} = -\log_2(\text{dilution factor}) \tag{2}$$

$$\text{dilution factor} = 2^{-\text{pT}} = \text{antilog}_2\,(-\,\text{pT})$$

$$\text{pT} = -\log_2\left(\frac{1}{x}\right) = \log_2(x)$$

The following example explains how a pT-value is determined. Within a dilution series, light inhibition percentages in the luminescent bacteria test at a dilution of 1:16 are below the threshold of 20 % (see Sections 5.2 and 5.4). In exponential form, 1:16 is written as $1{:}2^4 = 2^{-4}$. The negative logarithm on a base of 2 of the 1:16 dilution factor is 4, or explained differently, 2 raised to the negative 4th power corresponds to 1:16. Thus, the pT-value of 4 can be attributed to the tested material.

$$\text{pT} = -\log_2\left(\frac{1}{16}\right) = \log_2(16) = 4$$

In general, the threshold at which toxic effects are no longer expected can only be approximated (Section 5.3). Consequently, the pT-value method determines the dilution level where (for non-quantal tests involving micro-organisms or cells) the inhibition value is < 20% (this is analogous to the statutory test for wastewaters in Germany). Hence, the pT-value allows an ecotoxicological classification that is numerical and open-ended (*i.e.*, ranging from 0 to ∞ in theory). This open-ended scale is specific for each test organism and is a simple means of classifying the toxic potential of environmental samples.

5.2 GENERAL PRINCIPLE AND SCOPE OF APPLICATION OF THE PT-METHOD

The pT-method strategy describes the magnitude of toxic effects exerted by contaminants present in an environmental sample. The pT-value of the most sensitive organism within a test battery, the pT_{max}-value, determines the toxicity class of the sample, which is identified by the pT-index. Different bioassays are considered equal in rank. Roman numerals are assigned to identify toxicity classes. The pT-value, which relates to a single bioassay, and the pT-index, derived from the most sensitive organism in a test battery, permit a numerical classification of environmental samples based on ecotoxicological principles. With the aid of this scale, the toxicity status of any water sample can be quantified simply in an easily understandable way.

Although this chapter focuses on applications with effluent wastewaters, all types of aquatic environmental media (freshwater, brackish, marine) can be appraised with the pT-scale procedure. Testing of liquid samples is virtually unlimited and can include untreated and treated wastewater, surface water, ground water, porewater, elutriates and organic extracts of sediments. Applications could also be extended to assess toxicity of particle-bound substances in suspended matter and sediments. In this case, sample dilutions can be made with reference sediment material (Höss and Krebs, 2003). The pT-method can also capture the effects of both soluble and particulate toxicity in a sample, provided that appropriate bioassays are employed.

It is essential that proper sampling, storing and processing procedures of collected liquid media are followed to ensure the validity of subsequent bioassay results. In Germany, the DEV L1 - DIN EN ISO 5667-16 (1998) guideline is applicable to all pT-scale bioassay protocols (DEV L30, L31, L33, L34, T3, T6).

5.3 CONCENTRATION-EFFECT RELATIONSHIPS AND THRESHOLD VALUES

In toxicity examinations, relationships between sample concentration and effect are established with the aid of dilution series. A classical measurement endpoint in aquatic toxicology consists of determining the "50% effect concentration", where a quantitative relationship exists between a specific sample concentration and 50 % of the maximum attainable effect (*e.g.*, LC50, EC50). Today, the discipline of ecotoxicology also places emphasis on the threshold concentration at which harmful effects are no longer observed and endpoints such as LC0, EC0 or NOEC tend to be favored. Albeit useful, some of these values will be only approximations as their determination depends directly on the dilution series selected. Indeed, they are experimentally-derived values, rather than true values in terms of metabolic/physiologic parameters, dependent on selected test dilutions. To circumvent constraints linked to "threshold endpoints", it is often preferred to determine 10 or 20% effect concentrations (*e.g.*, EC10 or LC20).

Investigations of wastewater effluents made in the German Federal Institute of Hydrology have consistently shown that dilution series in geometric sequence with a dilution factor of 2 produce good evaluative concentration-effect curves. Dilution series of this type were thus selected in applications with the pT-scale where the finding of the hazard potential is reduced to the accurate measurement of the first sample dilution stage that is no longer toxic.

An example illustrating the pT-scale dilution method is that of mortality measured in the fish test according to DEV L31 (1989), where a dilution stage is identified ensuring 100% animal survival for a designated test exposure period. The requirement of 100% animal survival is connected with the low number of test animals. For this fish test the German Wastewater Charges Act demands only three fish in each dilution stage. In the protocol for *Daphnia* according to DEV L30 (1989) with 10 animals per dilution step, 9 of 10 animals must survive. This proportion is generally recommended by the author for other quantal tests with 10 or more test animals per dilution step. For non-quantal tests conducted with microorganisms, where a physiologic response is measured for large numbers of cells (*e.g.*, light inhibition of *Vibrio fischeri*), the first dilution stage indicative of the absence of effects is generally defined by the absence of statistical effects. This usually happens at values less than 20 %. If the test sample is diluted in sequence at a ratio of 1:2, the dilution series represented in Table 2 is obtained. The dilution factor described as a decimal fraction reflects the volume share of the original (undiluted) sample.

Quantal toxicity tests employing organisms such as daphnids or fish do not alter the concentration of contaminant(s) in a given volume of water because they are directly introduced into their respective experimental containers. In contrast, bioassays undertaken with algae and bacteria somewhat dilute the test material since they must be introduced into test containers (*i.e.*, flasks, tubes or microplate wells) via a certain volume of inoculum. In such tests, the volume share of the test culture can sometimes reach 20 % in the test preparation, which corresponds to a dilution of 1:1.25 (Tab. 2). This dilution stage is therefore the highest concentration that can be examined with such microbial tests.

5.4 CALCULATION OF THE pT-VALUE WHEN USING A PRE-DILUTION STEP

If a preliminary dilution step is deemed necessary before the geometric dilution series, as recommended in the case of heavily polluted wastewater, the preliminary dilution step, for example, should be 1:64 or 1:128. If it is necessary to obtain LC0 values for determining the pT-value, then the geometric dilution series with the factor of 1:2 must be followed through from the first dilution step. This is required in the fish test according to DEV L31 (1989) (Section 5.3). The same is required for other "quantal tests" like the *Daphnia* test, where 9 of 10 animals must survive.

It is possible to use a preliminary dilution step of 1:100 when the pT-scale relies, for instance, on a < 20% effect endpoint measurement (*e.g.*, determination of IC < 20 in "non quantal tests" with algae and bacteria). In this case, the pT-value should be calculated from the concentration-effect curve. The pT-value is determined by using the first sample concentration that generates an effect below 20%. If, for example, the concentration incurring a 19% effect is equated with a dilution factor of 1:3200, the pT-value is calculated with equation 5 as demonstrated below. The modulus for transforming decadic logarithms into binary ones, "$1 / \log_{10} 2$", yields equation 6. For the 1:100 diluted wastewater example above, the pT-value can then be calculated with the help of equation 8.

Table 2. Geometric dilution series, pT-values and pT-indices for wastewater effluents. The pT-values of receiving surface waters and of final effluents from wastewater treatment plants determined thus far are marked by the sign +.

Dilution factor as cardinal fraction	*Dilution factor as decimal fraction*	*Dilution factor as exponential fraction*	*pT-value*[a]	*Toxicity class*		*Measured ecotoxicity in environmental sample*	
				pT-index[b]	*Designation*	*Effluent*	*Receiving stream*
Original sample	1	2^{0}	0	0	non-detectable toxicity	+	+
1:1,25	0.8	$2^{-0,3}$	0	0	non-detectable toxicity	+	+
1:2	0.5	2^{-1}	1	I	very slightly toxic	+	+
1:4	0.25	2^{-2}	2	II	slightly toxic	+	+
1:8	0.125	2^{-3}	3	III	moderately toxic	+	+
1:16	0.0625	2^{-4}	4	IV	distinctly toxic	+	+
1:32	0.0313	2^{-5}	5	V	highly toxic	+	+
1:64	0.0156	2^{-6}	6	VI	extremely toxic "Mega toxic"	+	-
1:128	0.00781	2^{-7}	7	VII		+	-
1:256	0.00391	2^{-8}	8	VIII		+	-
1:512	0.00195	2^{-9}	9	IX	"Giga toxic"	+	-
1:1024	0.000977	2^{-10}	10	X		+	-
1:2048	0.000488	2^{-11}	11	XI		+	-
1:4096	0.000244	2^{-12}	12	XII	"Tera toxic"	+	-
1:8192	0.000122	2^{-13}	13	XIII		-	-
1:16384	0.0000610	2^{-14}	14	XIV		-	-

[a] **pT-value:** the highest dilution level devoid of adverse effects is used for the numerical designation of toxicity with regard to a single test organism. The pT-value (*potentia Toxicologiae* = toxicological exponent) is the negative binary logarithm of the first non-toxic dilution factor in a dilution series in geometric sequence with a dilution factor of two.

[b] **pT-index:** the numerical toxicological classification of an environmental sample attained with a test battery. The pT-value of the most sensitive organism within a test battery determines the toxicity class of the tested material. Roman numerals are assigned to each toxicity class. If the highest pT-value is 9, for instance, the tested material is then assigned as toxicity class IX (*i.e.*, the pT-index is IX).

$$y = a^x \tag{3}$$

$$x = \log_a(y) \tag{4}$$

$$\log_a(b) \cdot \log_b(x) = \log_a(x) \tag{5}$$

$$\log_2(x) = \frac{\log_{10}(x)}{\log_{10}(2)} \tag{6}$$

$$\mathrm{pT} = -\log_2\left(\frac{1}{x}\right) = \log_2(x) \tag{7}$$

$$\mathrm{pT} = -\frac{\log_{10}\left(\frac{1}{x}\right)}{\log_{10}(2)} = \frac{\log_{10}(x)}{\log_{10}(2)} \tag{8}$$

$$\mathrm{pT} = -\frac{\log_{10}\left(\frac{1}{3200}\right)}{\log_{10}(2)} = \frac{\log_{10}(3200)}{\log_{10}(2)} = 11.64386$$

$$\mathrm{pT} = 11.6$$

The pT-values of geometric dilution series are always integers, and the resulting pT-value of 11.6, as in this example, is ultimately reported as a pT-value of 12. Hence, by convention, calculated pT-values, determined from the relationship between concentration and effect, are always rounded-up to integers.

5.5 CALCULATION OF PT-VALUES WITH THE CONCENTRATION-RESPONSE APPROACH

Inhibition values measured at 11 different time periods in an effluent from a biological wastewater treatment plant with the luminescent bacteria test are shown in Figure 1. The pT-values of this industrial wastewater discharged into the Rhine River ranged between 4 and 12 (Tab. 3). It should be noted, however, that high values were measured prior to an internal processing change in the plant and that low values were obtained after the adjustment.

The results of this investigation generally showed that:

- the shapes of the concentration-effect curves confirm the suitability of the dilution factor of two as proposed for the pT-method;
- the luminescent bacteria test was found to be very useful for detecting effluent hazard potential and for describing changes over time.

Sigmoidal curves of the concentration-response data are presented in Figure 1, and their probit transformations in Figure 2, where a log-normal model is assumed. Here, the log-concentration transformation (pT-scale) is paired with the probit parameter, which is indicative of the proportion of "percentage inhibition".

While the median effective concentration (*e.g.*, EC50 or IC50[2]) is often the endpoint of choice (because medians are more consistent and tend to have small confidence intervals), the pT-value like other measurement endpoints based on thresholds favor the determination of lower effective concentration values. The rationale for this approach is based on the premise that it is more meaningful to estimate lower values for determining the hazard/risk posed by toxicant releases to the environment. These values are clearly more helpful in assessing adverse effects, and are indispensable whenever the measured response is less than 50 percent in an undiluted sample.

In the example provided (Figures 1 and 2), the slopes were relatively constant over the whole pT-scale for different time intervals. However, this may not always be the case and concentration-effect curves generated for other types of effluents could show much steeper slopes. Although IC50 values might be the same for two different effluents, a small change in concentration for the effluent with the steeper slope would have a more pronounced effect on test organisms. The inclusion of a "dilution step without toxicity" is thus crucial whenever a toxic threshold value is desired. Changes of slope for different toxicants and their impact on bioassay results, particularly with the luminescent bacteria test, have been reported previously (Krebs, 1992a).

Calculated values of IC50s and IC19s (from Figure 2 data) are listed in Table 3. The first concentration with an effect below 20% serves as the reference when the pT-value is calculated from the concentration-effect relationship.

The pT-values obtained without calculation from the standard pT-method are also listed in Table 3. As explained previously (Section 5.4), pT-values of each geometric dilution series are rounded up to whole numbers from IC 19 values.

---

[2] Microbiological tests such as the Microtox test report a non-quantal endpoint, which is an IC50. ECs are for quantal data (*e.g.*, # of immobilised daphnids).

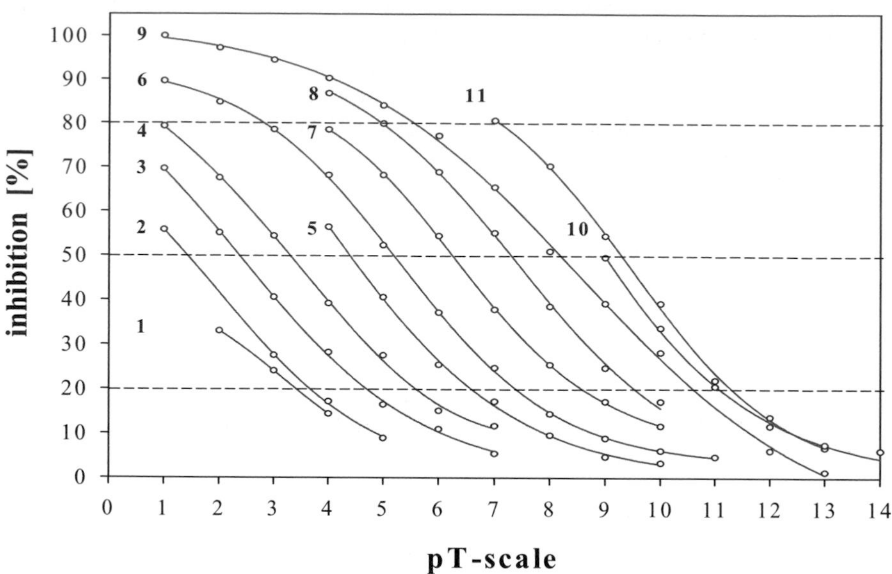

Figure 1. Inhibition values in percent measured at multiple intervals in effluents from an industrial wastewater treatment plant with the Microtox® luminescent bacteria test.

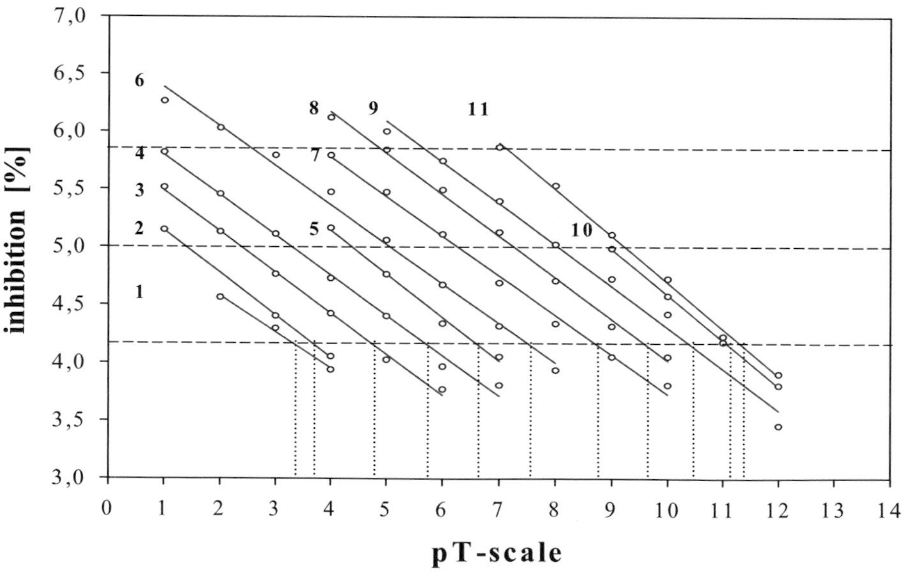

Figure 2. Probit transformation of the sigmoidal concentration-effect curves of Figure 1 for calculation of effective concentrations as IC50, IC20, or IC19.

Table 3. IC50 and IC19 values calculated from the concentration-effect lines of Figure 2. The pT-values are obtained without calculation from the first dilution stage at which the effluents were no longer toxic.

Curve No.	***IC 50***			***IC 19***			***pT-value***
	IC 50 pT-scale	***Dilution factor***	***Dilution factor***	***IC 19 pT-scale***	***Dilution factor***	***Dilution factor***	
1	(0.6)	($2^{-0.6}$)	(1:1.6)	3.5	$2^{-3.5}$	1:11	4
2	1.4	$2^{-1.4}$	1:2.6	3.8	$2^{-3.8}$	1:14	4
3	2.4	$2^{-2.4}$	1:5.2	4.9	$2^{-4.9}$	1:29	5
4	3.3	$2^{-3.3}$	1:9.8	5.8	$2^{-5.8}$	1:57	6
5	4.4	$2^{-4.4}$	1:21	6.7	$2^{-6.7}$	1:110	7
6	5.1	$2^{-5.1}$	1:34	7.7	$2^{-7.7}$	1:200	8
7	6.3	$2^{-6.3}$	1:78	8.9	$2^{-8.9}$	1:460	9
8	7.3	$2^{-7.3}$	1:150	9.7	$2^{-9.7}$	1:840	10
9	8.1	$2^{-8.1}$	1:270	10.5	$2^{-10.5}$	1:1470	11
10	9.0	$2^{-9.0}$	1:500	11.2	$2^{-11.2}$	1:2330	12
11	9.2	$2^{-9.2}$	1:610	11.4	$2^{-11.4}$	1:2750	12

5.6 TOXICITY TESTS RECOMMENDED FOR APPLYING THE pT-METHOD

These tests, described in Box 1, are performed according to the German Standard Methods for Examination of Water, Wastewater and Sludge (DEV) and are always based on DEV L1 - DIN EN ISO 5667-16 (1998).

Box 1. Test battery recommended for the examination of freshwater, groundwater, and wastewater.

Algal test *Desmodesmus subspicatus* HEGEWALD and SCHMIDT, 2000, formerly known as *Scenedesmus subspicatus* CHODAT, 1926. Taxonomy: Chlorophyta, Chlorophyceae, Chlorococcales. Test performed according to DEV L 33 - DIN 38 412 Part 33 (1991). Test duration: 72 h. Toxicity endpoint: cell growth inhibition. Number of test organisms per dilution step: 10^4 cells per mL. Threshold value for the determination of the pT-value: IC < 20%.
Luminescent bacteria test *Vibrio fischeri* BEIJERINCK, 1889; LEHMANN and NEUMANN, 1896, formerly known as *Photobacterium phosphoreum* COHN, 1878. Taxonomy: Bacteria; Proteobacteria; Gammaproteobacteria; Vibrionales; Vibrionaceae. Test performed according to DEV-L 34 - DIN EN ISO 11348-3 (1998), Microtox® bacteria. Test duration: 30 min. Toxicity endpoint: luminescence inhibition. Number of test organisms per dilution step: about 10^6 cells per mL. Threshold value for the determination of the pT-value: IC < 20%.
***Daphnia* test** *Daphnia magna* STRAUS, 1820, water-flea. Taxonomy: Crustacea, Branchiopoda, Cladocera, Daphniidae. Test performed according to DEV L30 - DIN 38 412 Part 30 (1989). Test duration: 24 h. Toxicity endpoint: microcrustacean acute immobilisation. Number of test animals per dilution step: 10. Threshold value for the determination of the pT-value: 90% survival.
Fish test *Leuciscus idus melanotus* L., 1758, golden orfe. Taxonomy: Osteichthyes, Teleostei, Ostariophysi, Cypriniformes, Cyprinidae. Test performed according to DEV L31 - DIN 38 412 Part 31 (1989). Test duration: 48 h. Toxicity endpoint: acute lethality. Number of test animals per dilution step: 3. Threshold value for the determination of the pT-value: 100% survival.

Box 1 (continued). Test battery recommended for the examination of freshwater, groundwater, and wastewater.

Fish egg test

Danio rerio HAMILTON, 1822, formerly known as *Brachydanio rerio* HAMILTON-BUCHANAN, 1822, zebra fish or zebra danio.
Taxonomy: Osteichthyes, Teleostei, Ostariophysi, Cypriniformes, Cyprinidae.
Test performed according to DEV T6 - DIN 38 415-6 (2003).
Test duration: 48 h.
Toxicity endpoint: embryo lethality.
Number of test eggs per dilution step: 10.
Threshold value for the determination of the pT-value: 90% survival.

umu-test

Determination of the genotoxicity of water and wastewater with the genetically engineered bacterium *Salmonella typhimurium* TA 1535/pSK 1002.
Taxonomy: Bacteria; Proteobacteria; Gammaproteobacteria; Enterobacteriales; Enterobacteriaceae.
Test performed according to DEV-T3 DIN 38 415-3 (1996), ISO 13828 (2000).
The test is based on the capability of genotoxic agents to induce the umuC-gene in the *Salmonella* strain in response to genotoxic lesions in the DNA. The induction of the umuC-gene is thus a measure for the genotoxic potential of the sample.
Incubation time: 2 h.
Threshold value for the determination of the pT-value is the lowest value of the dilution series at which the measured induction rate is < 1.5.

The threshold values are test-specific. In this recommended test battery all specific threshold values are part of the protocols standardized by DEV, DIN, or ISO. The choice of different threshold values can alter the pT-values measured in a single bioassay and finally the pT-index measured within a test battery.

In the regulations of the German Wastewater Charges Act (AbwV, 2002) the fish test was replaced in 2004 by the fish egg test.

5.7 NUMERICAL CLASSIFICATION OF WASTEWATER EFFLUENTS WITH THE pT-INDEX

The pT-index scale appraises the relative hazard of aquatic environmental samples by assigning them to a numerical class. As bioassays within a test battery are considered equal in rank, the most sensitive test with its pT_{max}-value defines the toxicity class of the test material. If, for example, an effluent yields pT-values of 7, 2, 8 and 0 for the bacterial, algal, daphnid and fish tests, respectively, its pT-index is therefore assigned to toxicity class # 8, based on the most sensitive response obtained with the *Daphnia* test. Since, by convention, sample toxicity classes are designated by Roman numerals, the test material is then assigned to toxicity class VIII. A toxicity class is not a strictly defined value, but rather the consequence of the respective test battery used. Assigning a toxicity class for a set of liquid samples can be derived from a single bioassay (when experimental logistics limit bioanalytical assessment) or, ideally, from a large number of different bioassays. Reporting the test or test battery employed to generate a pT-index is essential. Use of similar test organisms and procedures is also critical in order to properly compare the relative hazard potential of a set number of effluents discharging into a common aquatic system.

5.8 VERBAL CLASSIFICATION OF WASTEWATER EFFLUENTS

A given wastewater is assigned a specific pT-value only with respect to a given test organism, or a particular pT-index with respect to a given test battery. In order to provide consistency for classification purposes, the terminology shown in Table 2 is proposed for a relative ranking. This verbal designation was arbitrarily chosen and could still be amended. Currently, it is necessary to indicate the pT-index value in addition to the verbal designations. Above pT-index VI, verbal designations lack the power to further discriminate effluent hazard, as language does not facilitate reasonable differentiation at these values. The pT-values obtained thus far with different toxicity tests are indicated in Table 2 and suggest that effluents can have large differences in hazard potential. Once again, it can be seen that pT-values and pT-indices permit a wide-ranging toxicological classification of wastewater effluents. With the aid of this scale, any wastewater can be characterized in terms of its aquatic toxicity.

5.9 THE pT-VALUE AS AN ECOTOXICOLOGICAL DISCRIMINATOR

The sample dilution method employed in the pT-scale is a necessary step to provide improved differentiation of the toxic potential over simple effect percentages reported for undiluted test material. In cases where 100 % effects are manifest in an undiluted sample, the pT-value for effluents functions as an ecotoxicological discriminator that can distinguish between different environmental hazard potentials. For instance, 100% inhibiting effects produced by undiluted samples may be undetectable after a dilution of 1:2, or after much higher dilutions have been reached. Thus far, the highest pT-value measured in a wastewater effluent was 12, corresponding to a dilution level of 1:4.096 (Krebs, 1988). This finding demonstrates the importance of measuring the effects at several sample dilutions, which the pT-scale offers to properly differentiate toxicity hazard and classify wastewater effluents.

5.10 THE pT-METHOD AS A POSSIBLE HAZARD IDENTIFICATION SCHEME (HIS)

To be complete, toxicity hazard assessment should consider the specific effects caused by wastewaters to a receiving stream. As applied thus far, the pT-method only provides a quantitative measurement of wastewater toxicity.

Since the quality of wastewaters has been improved through biological treatment, the German Wastewater Charges Act requires that effluents be monitored with ecotoxicological tests. If an effluent remains toxic, those parties responsible for effluent quality are liable for its toxic loading to the receiving stream. In this matter, polluters are required to pay an amount that corresponds to toxic loading emissions (*e.g.*, toxicity units x effluent volume flow in m^3/h). This calculation is independent of the quality or volume of the receiving water. The Law focuses on "pollution pressure" as opposed to the actual effects impacting the water body.

As used currently, the pT-scale describes the "toxicity pressure potential" of wastewaters discharging to rivers. The pT-scale could therefore play a future role as a Hazard Identification Scheme (HIS) pursuant to the European Water Framework Directive (EU WFD). HIS provides the basis for assessing potential impacts to an aquatic system. Decisions regarding the necessity for improvements to wastewater treatment could be derived from pT-bioassay results. For example, an industry whose effluent discharge has been categorized in toxicity class IV with a pT_{max}-value of 4 (Tab. 2) could be requested to lower its toxicity status to class II with a pT_{max}-value of 2 following a designated period of time dedicated to improving wastewater treatment. Prospective effluent evaluations employing the pT-scale in this way will be governed, of course, by applicable national legislation and/or scientific criteria.

5.11 STATISTICS / CALCULATIONS / EXAMPLES OF DATA ASSOCIATED WITH THE pT-METHOD

Dilutions used in applying the pT-method are described in Table 2 (Section 5.3) and are easy to perform. In the event that a pre-dilution step is undertaken prior to undertaking the recommended dilution series, an example of a stepwise calculation yielding a binary pT-value is given in Section 5.4. No other calculations are needed or associated with the pT-scale.

5.12 CHEMICAL ANALYSES EMPLOYED IN PARALLEL TO THE pT-METHOD

Chemical analyses are an indispensable complement to ecotoxicological investigations, but they are not directly integrated in the application of the pT-scale for industrial effluent assessment. Performed in parallel, assessment of chemical substances is governed by a set of criteria, which are derived from ecotoxicological threshold (NOEC) values, in order to derive quality objectives or standards. When bioassay and chemistry results are both available for effluent assessment, they should be ranked equally. For determining the overall hazard of wastewaters, the most conservative information with respect to aquatic health will serve as the yardstick to ensure environmental protection.

6. Factors capable of influencing interpretation of the pT-scale

Selection of representative bioassays is crucial for successful application of the pT-scale. Sub-cellular tests such as enzyme assays may not be sufficiently relevant in their indicative power of environmental hazard instead of tests using cells or whole organisms. Future refinements to the pT-scale might be the incorporation of additional genotoxicity or immunotoxicity tests to a test battery to augment its diagnostic power in terms of hazard assessment.

7. Applications / case studies reported with the pT-method

When applied for characterizing the potential toxicity of wastewater effluents, the pT-method can draw attention to certain groups of chemicals in the effluent. In Section 5.7, for example, the extremely high pT-values measured with the luminescent bacteria test (pT value = 7) and the *Daphnia* test (pT value = 8) indicate that chemicals eliciting these toxic effects must subsequently be identified and removed in the wastewater treatment process, so as to protect decomposers and micro-crustaceans.

Determining pT-values can provide important information for identifying the toxic intensity of wastewater effluents (case study linked to Figure 1) which exhibited a 4-12 range in pT-values (Tab. 3). As toxicity of an industrial wastewater effluent can change with time, this trend can be measured and quantified with the pT-scale as shown in Section 5.5.

In Germany and Brazil, pT-values were shown to be suitable units of measurement to appraise surface water toxicity. These findings were the result of investigations carried out in the Saar River (Krebs, 1992) and in rivers discharging to the Bay of Sepetiba, Federal State of Rio de Janeiro, Brazil (Soares, 2000).

Again, the pT-scale concept has demonstrated applicability for toxicity evaluations of groundwater with samples collected in observation wells downstream of landfills or waste disposal sites. The German Federal Institute of Hydrology performs this type of monitoring during routine inspections of dumping areas filled with dredged material.

8. Accessory / miscellaneous information useful for applying the pT-method

Test batteries applicable for the pT-scale employ standardized test systems and procedures recognized by standardization organizations such as AFNOR, ASTM, DIN, EN, ISO, or OECD. The level of expertise required by personnel to ensure the proper application of pT-method bioassays must therefore comply with established norms. The necessary qualification will also depend on the test system being used.

All liquid media, in general, were found to be suitable for pT-scale assessment, provided that dilution series can be prepared from test samples. Investigations can be carried out with liquid samples such as:

- effluents from industrial and municipal wastewater plants;

- surface water, groundwater;
- porewaters, elutriates and organic extracts from sediments and dredged material. This area of application is described in Chapter 9 of this volume.

Toxic effects of all classes of contaminants including metals, pesticides, and organic substances can be captured with the pT-method, as long as the test battery employed reflects a sufficiently wide spectrum of sensitivity. Furthermore, test organisms in a battery should be representative of aquatic biota. The composition of test batteries can be varied according to different aquatic environments and country-specific issues. Clearly, application of the pT-method is suitable for several experimental designs which are linked to toxicity testing (methods, test species, endpoints).

The number of environmental samples that can be tested per day, per week, or per month will depend on the bioassays selected. Available equipment in the laboratories will also dictate sample throughput capabilities. In the German Federal Institute of Hydrology, one technical assistant can process six samples per week with the test battery and protocol described herein.

9. Conclusions / prospects

The pT-method is a hazard assessment scheme initially developed for identifying the presence of toxicity in wastewater treatment effluents and for quantifying its potential. The method can classify effluents into toxicity categories of increasing strength but does not directly assess impact on the receiving environment. Toxicity data generated with the pT-method are useful inputs for subsequent Environmental Impact Assessment (EIA), as they facilitate the decision-making process. Furthermore, the pT method can be employed to assess toxic effects in polluted surface waters. The pT-method is a valuable tool for undertaking hazard assessment aimed at protecting aquatic ecosystems.

Acknowledgements

I am thankful to Bernd Uebelmann, Dierk-Steffen Wahrendorf, Shannon McDowell and Derek McDowell for their support in writing this paper.

References

AbwV (2002) Verordnung über Anforderungen an das Einleiten von Abwasser in Gewässer (Abwasserverordnung – AbwV), *Bundesgesetzblatt* **I 2002**, 4048 and 4550.

DEV L1 - DIN EN ISO 5667-16 (1998) Water quality – Sampling – Part 16: Guidance on biotesting of samples (ISO 5667-16: 1998). Edited by Wasserchemische Gesellschaft - Fachgruppe in der Gesellschaft Deutscher Chemiker in Gemeinschaft mit dem Normenausschuß Wasserwesen (NAW) im DIN Deutsches Institut für Normung e.V., Beuth Verlag, Berlin, Germany.

DEV L30 - DIN 38 412 Part 30 (1989) German standard methods for the examination of water, wastewater and sludge – Determination of the non-poisonous effect of waste water to *Daphnia* by dilution limits, Beuth Verlag, Berlin, Germany.

DEV L31 - DIN 38 412 Part 31 (1989) German standard methods for the examination of water, wastewater and sludge – Determination of the non-acute-poisonous effect of wastewater to fish by dilution limits. Beuth Verlag, Berlin, Germany.

DEV L33 - DIN 38 412 Part 33 (1991) German standard methods for the examination of water, wastewater and sludge – Determination of the non-poisonous effect of wastewater to green algae (*Scenedesmus* chlorophyll fluorescence test) by dilution limits. Beuth Verlag, Berlin, Germany.

DEV L34 - DIN EN ISO 11348-3 (1998) Water quality – Determination of the inhibitory effect of water samples on the light emission of *Vibrio fischeri* (Luminescent bacteria test) - Part 3: Method using freeze-dried bacteria. Beuth Verlag, Berlin, Germany.

DEV T3 - DIN 38415-3 (1996) German standard methods for the examination of water, wastewater and sludge – Sub-animal testing – Part 3: Determination of the genotype potential of water with the umu-test. Beuth Verlag, Berlin, Germany.

DEV T6 - DIN 38415-6 (2003) German standard methods for the examination of water, wastewater and sludge – Sub-animal testing – Part 6: Toxicity to fish; Determination of the non-acute-poisonous effect of wastewater to fish eggs by dilution limits. Beuth Verlag, Berlin, Germany.

Höss, S. and Krebs, F. (2003) Dilution of toxic sediments with unpolluted artificial and natural sediment – Effects on *Caenorhabditis elegans* (Nematoda) in a whole sediment bioassay - Proceedings 13th Annual meeting, Society of Environmental Toxicology and Chemistry – Europe Branch (SETAC-Europe), 28.04 -01.05. 2003, Hamburg, pp. 148.

ISO 13829 (2000) Water quality – Determination of the genotoxicity of water and wastewater using the umu-test. Geneva, Switzerland.

Krebs, F. (1987) Der "pT-Wert" - ein gewässertoxikologischer Klassifizierungsmaßstab bei Abwasser- und Vorfluteruntersuchungen (The pT-value as a classification index for wastewater and receiving streams in aquatic toxicology), Proceedings, 26th IAD Conference International Association for Danube Research, 14-18. 09. 1987, Passau, pp. 543-546.

Krebs, F. (1988) Der pT-Wert: ein gewässertoxikologischer Klassifizierungsmaßstab (The pT-value as a classification index in aquatic toxicology), *GIT Fachzeitschrift für das Laboratorium* **32**, 293-296, simultaneously *GIT Edition Umweltanalytik-Umweltschutz* **1**, 57-63.

Krebs, F. (1992a) Der Leuchtbakterientest für die Wassergesetzgebung (The luminescent bacteria test for water legislation), *Schriftenreihe des Vereins für Wasser-, Boden- und Lufthygiene* **89**, 591-624.

Krebs, F. (1992b) Gewässeruntersuchung mit dem durch Alkali- und Erdalkaliionen-Zugabe optimierten DIN-Leuchtbakterientest, dargestellt am Beispiel der Saar (River water analysis with the luminescent bacteria test according to DIN, optimized by the addition of alkaline and alkaline-earth ions, with the River Saar as an example), *Schriftenreihe des Vereins für Wasser-, Boden- und Lufthygiene* **89**, 657-673.

Soares, F. de Freitas Lopes (2000) Avaliação da Qualidade de Água e Sedimentos da Baía de Sepetiba, Anexo 5: Ecotoxicologia, 1995 a 1999, Projeto de Cooperação Técnica Brasil/Alemanha - FEEMA/GTZ. FEEMA Fundação Estadual de Engenharia do Meio Ambiente, Rio de Janeiro, RJ, Brasil. (Water and Sediment Quality Assessment of Sepetiba Bay, Annex 5: Ecotoxicology, 1995 to 1999, Brazil/Germany Cooperation Project - FEEMA / GTZ. Report (CD-ROM), FEEMA Environmental Agency of the Federal State of Rio de Janeiro, Rio de Janeiro, Brazil).

U.S. EPA (1991) The technical support document for water quality-based toxics control, Office of Water, Washington, DC, EPA/505/2-90-001.

WHG (2002) Gesetz zur Ordnung des Wasserhaushalts (Wasserhaushaltsgesetz – WHG), *Bundesgesetzblatt* **I 2002,** 3246.

Abbreviations

AbwV	German Wastewater Charges Act
AFNOR	Association Française de Normalisation
antilog	antilogarithm
ASTM	American Society for Testing Materials

BfG	German Federal Institute of Hydrology
DEV	German Standard Methods for Examination of Water, Wastewater and Sludge
DIN	German Organization for Standardization
EC	Effective Concentration
EIA	Environmental Impact Assessment
EN	European Organization for Standardization
EU WFD	European Water Framework Directive
FEEMA	Environmental Agency of the Federal State of Rio de Janeiro, Brazil
GTZ	German government-owned corporation for international technical co-operation
HABAB-WSV	German Directive for the Management of Dredged Material in Inland Waters
HABAK-WSV	German Directive for the Management of Dredged Material in Coastal Waters
HAS	Hazard Assessment Scheme
HIS	Hazard Identification Scheme
IC	Inhibition Concentration
ISO	International Organization for Standardization
LC	Lethal Concentration
LID	Lowest Ineffective Dilution
NOEC	No Observed Effective Concentration
OECD	Organization for Economic Cooperation and Development
pH	potentia Hydrogenii
pT	potentia Toxicologiae
SIL	*Societas Internationalis Limnologiae*
U.S. EPA	United States Environmental Protection Agency
WHG	German Water Act.

4. STRATEGIES FOR MONITORING ENVIRONMENTAL EFFECTS OF INDUSTRIAL EFFLUENTS

RICK P. SCROGGINS
Biological Methods Division
Environmental Technology Centre
Environment Canada
Ottawa, Ontario, Canada
rick.scroggins@ec.gc.ca

ANNE I. BORGMANN
Environmental Conservation Branch
Environment Canada, Ontario Region
Burlington, Ontario, Canada
anne.borgmann@ec.gc.ca

JENNIFER A. MILLER
Miller Environmental Sciences Inc.
Innisfil, Ontario, Canada
miller.smith@sympatico.ca

MARY J. MOODY
Environment and Minerals Division
Saskatchewan Research Council
Saskatoon, Saskatchewan, Canada
moody@src.sk.ca

1. Objective and scope of Hazard Assessment Schemes for monitoring environmental effects

The use of laboratory toxicity tests to monitor industrial effluent discharges has become a common approach to estimating the potential for environmental effects in North America and Europe. Numerous schemes have been developed to characterize and assess potential toxic effects in aquatic receiving environments. The first regulatory application of Environmental Effects Monitoring (EEM) in Canada was within the 1992 Pulp and Paper Liquid Effluent Regulations, promulgated under the Fisheries Act. A second application of EEM in Canada was within the 2002 Metal

C. Blaise and J.-F. Férard (eds.), Small-scale Freshwater Toxicity Investigations, Vol. 2, 139-167.

Mining Liquid Effluent Regulations. Regulatory provisions that use EEM focus on assessing whether there is protection of fish populations, fish habitat, and use of the fisheries resource, in waterbodies that receive effluents. Results of EEM are interpreted by a "weight-of-evidence" approach using a suite of complementary field and laboratory monitoring tools, including sublethal toxicity tests on the main effluent(s). The program for pulp and paper mills requires three types of sublethal tests on the final effluent: 1) early-life-stage development of fish; 2) reproduction of an invertebrate; and 3) growth of an aquatic plant.

Sublethal toxicity tests complement other components of EEM: 1) chemical measurements in the effluent, the receiving water, and fish tissue and 2) biological surveys of organisms living in the waterbody (*e.g.*, benthic invertebrates and fish). Within the EEM program, sublethal toxicity testing on effluent or receiving water can be used to:

(1) measure changes in discharge quality over time, as a result of process changes or effluent treatment at the facility;
(2) estimate, in multiple discharge situations, the relative contributions from various point and/or non-point sources, to any observed effects in the receiving environment; and
(3) estimate the potential for effects in the receiving water environment.

This chapter presents two Hazard Assessment Schemes that have been recently used to assess the relationship between laboratory sublethal toxicity data and field measurements of the Canadian pulp and paper Environmental Effects Monitoring program. The two methods are 1) the estimation of Zone of Potential Effect (ZPE); and 2) the Lab-to-Field Rating Scheme (LTF). The application of these schemes illustrates how to estimate the potential for effects in the receiving water environment (third use above).

Both methods have been shown to be effective in illustrating the relationships between laboratory sublethal toxicity tests (using fish, invertebrates, and algae) and receiving environment measurements of fish and benthic invertebrates. The applications, strengths, and weaknesses of both the ZPE and LTF methods are discussed and compared.

1.1 ZONE OF POTENTIAL EFFECT HAZARD ASSESSMENT SCHEME

The Zone of Potential Effect (ZPE) Hazard Assessment Scheme can be used to estimate whether there is potential for effects of an effluent on organisms living in the local aquatic receiving environment. The ZPE scheme can also be used to qualitatively assess the relationship between the predicted zone of effect and the actual extent of effects observed in the receiving environment using field observations (benthic community structure and fish population measurements) in the near field zone. This Hazard Assessment Scheme is part of a weight-of-evidence approach for the monitoring of effluents from regulated industries (Moody, 2002).

1.2 LAB-TO-FIELD HAZARD ASSESSMENT SCHEME

The Lab-to-Field Rating Scheme (LTF) is a ranking system based on biology, which requires both laboratory sublethal toxicity data and ecological measures (*e.g.*, benthic community indices and fish population metrics). The objective for using this Hazard Assessment Scheme is to use a weight-of-evidence approach for the regulatory monitoring of effluents from pulp and paper mill and metal mining industries (Borgmann et al., 2004).

2. Summary of Hazard Assessment Schemes

2.1 ZPE SCHEME	**2.2 LTF SCHEME**
Purpose	
To estimate the extent of the toxic effects from effluent discharged to an aquatic receiving environment.	To examine the relationship between effluent sublethal toxicity results from laboratory testing and field biological measurements at a specific EEM study site.
Principle	
The potential effects based on results of sublethal toxicity tests are illustrated by zones superimposed on the industrial effluent plume and then compared to field survey components of a monitoring program.	The field survey components of a monitoring program are rated on a similar scale as the sublethal toxicity tests for weight-of-evidence comparison.
Toxicity tests employed	
*Pimephales promelas (*fathead minnow) growth inhibition, *Ceriodaphnia dubia* reproduction inhibition, and *Selenastrum capricornutum* growth inhibition.	*Pimephales promelas (*fathead minnow) growth inhibition, *Ceriodaphnia dubia* reproduction inhibition, and *Selenastrum capricornutum* growth inhibition.
Determination of effluent hazard potential	
Step 1. Determine effluent dilution in the receiving environment through a plume delineation study, and map the effluent plume.	Step 1. Assign a rating of 1 to 5 to each sublethal test based on the lowest IC25 subtracted from 100. See Table 1.
Step 2. Determine the lowest IC25 from a battery of sublethal toxicity tests.	Step 2. Assign an LTF rating of 1 to 5 to the fish survey based on the percentage of potentially effluent-related effects relative to all the endpoints measured.
Step 3. On the plume map, match the lowest IC25 for each test conducted with the same concentration of the effluent plume to estimate the extent of the effects zone.	Step 3. Assign an LTF rating of 1 to 5 to the benthic invertebrate community survey based on the percentage of effluent-related effects compared to the total number of descriptors measured.

2.1 ZPE SCHEME	**2.2 LTF SCHEME**
Determination of effluent hazard potential (continued)	
Step 4. If biological measurements from monitoring indigenous organisms in the near-field receiving environment are available, examine the level of agreement between laboratory and field results.	Step 4. Assess the strength of the relationship between toxicity tests and ecosystem indicators.
Notes of interest	
Once illustrated on a map of the effluent plume, the ZPE can be seen visually as larger or smaller than the area of the plume defined by the isopleth for the 1% concentration of effluent (Environment Canada, 1999).	The method is flexible in that any number of endpoints/descriptors can be used. However, redundancy in endpoints or descriptors should be scrutinized.
Documented applications of the hazard assessment schemes	
Scroggins et al., 2002 Moody, 2002	Moody, 2002 Borgmann et al., 2004

3. Historical overview and applications reported with the ZPE and LTF Hazard Assessment Schemes

U.S. EPA has shown that single species tests provide reliable qualitative predictions of biological community impacts or adverse effect concentrations (U.S. EPA, 1999). Sublethal toxicity predictions often correlate well with observations on sedentary organisms such as the benthic invertebrates (Sprague, 1997; U.S. EPA, 1999). A review of the literature found 63 studies that were suitable for comparing results of sublethal effluent tests and effects in the aquatic communities receiving the effluents. Of these, there were 53 cases of agreement and 10 cases of disagreement, for an overall 84% rate of agreement (Sprague, 1997).

Measurement of sublethal toxicity in each Canadian pulp and paper mill's final effluent is one of the monitoring requirements for the first cycle of EEM (1992 to 1996), as well as the second cycle (summer 1997 to winter 2000) and the third cycle (summer 2000 to winter 2004). During the first cycle, a battery of three sublethal toxicity tests was used during four consecutive quarters within the principal year of field work to assess effects on: (1) early-life-stage development of fish; (2) reproduction of an invertebrate; and (3) growth of an aquatic plant. During the second and third cycles, a similar battery of sublethal toxicity tests was conducted twice a year (summer and winter) for each of the three years in the cycle. The choice of tests and species depended on whether the mill discharged to a fresh or estuarine/marine waterbody. Data were reported to government within 90 days, and were interpreted at the end of each three-year cycle, along with results of field monitoring (EC, 1999).

Table 1: LTF rating scheme for scoring the results of laboratory testing and field surveys

LTF Rating Scheme for Laboratory Testing				*LTF Rating Scheme for Field Surveys*	
Observation	*Lowest IC25 (% v/v)*	*Lowest IC25 subtracted from 100 (z)[1] (%v/v)*	*LTF Rating*	*Observation*	*% PERE[2]*
Very Low Response or No Response	IC25 > 92	z < 8	**1**	Very Low Level of Effects or No Effects	% PERE < 8
Low toxicity	83 < IC25 ≤ 92	8 ≤ z < 17	**2**	Low Level of Effects	8 ≤ % PERE < 17
Moderate toxicity	66 < IC25 ≤ 83	17 ≤ z < 34	**3**	Moderate Level of Effects	17 ≤ % PERE < 34
High toxicity	33 < IC25 ≤ 66	34 ≤ z < 67	**4**	High Level of Effects	34 ≤ % PERE < 67
Severe toxicity	0 ≤ IC25 ≤ 33	67 ≤ z ≤ 100	**5**	Severe Level of Effects	67 ≤ % PERE ≤ 100

[1]z = (100 – lowest IC25); [2]PERE = Potential Effluent Related Effects
Modified from Borgmann *et al* 2004

Environment Canada has prepared written technical guidance for mill operators and their environmental consultants or government regulators on how sublethal data can be used to estimate the zone of potential effect (ZPE) in the receiving water (EC, 1999; 2000). Results from two cycles of sublethal toxicity tests conducted on effluent from a pulp and paper mill in Ontario showed that the predicted ZPE in receiving water, agreed with effects observed in biological surveys (Scroggins et al., 2002). The ZPE scheme was estimated at 16 pulp and paper EEM study sites in the Province of Ontario following the completion of Cycle 2. The relationship between the effluent sublethal toxicity results and field monitoring data from the same site was fairly strong, especially between the *Ceriodaphnia* and *Selenastrum* test results and benthic community indices (Moody, 2002).

Cycle 1 toxicity test results from pulp and paper EEM studies completed in the Atlantic region of Canada was characterized with the aid of a qualitative scale. For median IC25s equal or greater than 100% effluent, the sample was considered to be non-toxic; values between 50 and 99% effluent were considered to be slightly toxic, values between 15 and 49%, moderately toxic and values less than 15%, highly toxic (Parker and Smith, 1999). A qualitative scale was also developed for characterizing toxicity tests in the Ontario Region. In this case, grouped results of Cycle 2 sublethal tests utilized geometric means of IC25s (Borgmann et al., 1999; 2002). Development of the LTF rating scheme began as a modification of the latter scale. The LTF rating scheme was applied at 16 pulp and paper EEM study sites in the Province of Ontario following the completion of Cycle 2. The relationship between the effluent sublethal toxicity results and field monitoring data from the same site was fairly strong, especially between the *Ceriodaphnia* and *Selenastrum* test results and benthic community indices (Moody, 2002; Borgmann et al., 2004).

4. Advantages of applying the EEM-HAS procedures

Both the ZPE and LTF schemes can bring together laboratory sublethal toxicity data and statistically significant field observations for a more comprehensive or weight-of-evidence approach to hazard assessment. Once the relationship between laboratory and field results have been established at an effluent discharge location, sublethal toxicity data from continued monitoring can be used, not only to estimate whether effluent quality is improving or worsening, but also to estimate how the corresponding zones of potential effect are changing. In addition, both the ZPE and LTF schemes are flexible enough to include improvements in methodology used in future EEM cycles or in other applications.

Thus, the ZPE scheme can be used to estimate the potential for effects from industrial effluent discharges in the local aquatic receiving environment without the need for field survey data or in cases where field surveys cannot be easily conducted. With sublethal toxicity data and a thorough plume delineation study, a ZPE can be estimated and can help establish the priority sites where confirmatory field studies might be required or help an EEM study team to locate their near-field biological sampling locations.The LTF rating scheme includes all endpoints determined for fish and benthos, and rates them on the same scale, allowing the addition of further endpoints and indices, such as the Bray-Curtis Index. Rating of the sublethal test results, based on the lowest IC25, applies a similar scale to the fish and benthos field surveys, reducing the subjectivity of lab-to-field comparisons. Relating laboratory and field data, as done in the LTF scheme, increases the possibility of linking a specific effluent source to measured effect(s) in indigenous organisms, both biota and flora.

5. Factors capable of influencing the interpretation potential of the EEM-HAS procedures

Sublethal toxicity tests that use species of relatively low sensitivity (*i.e.*, fathead minnow) reduce the usefulness of both EEM Hazard Assessment Schemes to estimate potential effects observed in the field. Insensitive laboratory measurements can lead to an underestimation of potential field effects and reduce the strength of laboratory toxicity tests as good estimators of effects.

The presence of other discharge sources can influence the interpretation of the hazard estimate of both the ZPE and LTF schemes. These discharge sources could be either point (*e.g.*, treated municipal sewage discharge) or non-point (*e.g.*, bark or chip pile runoff), entering the watershed up-stream or within the immediate near-field zone that receives the industrial effluent.

The determination of a zone of potential effect depends on the availability of a thorough delineation of the effluent plume. Historically, few confirmation measurements of effluent concentration in the receiving water close to the outfall may be obtained. In this situation, the zone of potential effect might be only an approximation expressed as less than 100 m. This indicates a small ZPE in the near-field. In other cases, the plume may be well delineated with a more gradual dilution of the effluent making a ZPE estimate more accurate. The ZPE must be looked upon as an approximation, particularly as an effluent plume is not a static entity and is subject to change from a wide variety of influences. Application of the ZPE method can be strengthened by the collection of more data to better characterize the location of effluent plumes.

Endpoints should be checked for redundancies in the LTF Scheme (Borgmann et al., 2004). For example, if an index based on sensitive groups of invertebrates is used (*e.g.*, the EPT Index), then taxa included should not be identified as the toxicity indicator species. Of the seven Ontario mills where EPT indices were calculated, only one mill had significantly higher abundance of EPT taxa (*Ephemera*, *Hexagenia*, *Oxyethira*, *Mystacides*, *Lepidostoma*), and a corresponding high EPT Index in the reference area compared to the exposure area (Moody, 2002). However, there was significantly higher abundance of species other than Ephemeroptera, Plecoptera and Trichoptera, such as *Monoporeia affinis*, *Psidium*, *Pyrgulopsis*, *Amnicola*, *Fossaria* and *Lirceus* (Moody, 2002) observed in the reference area as well. These non-EPT species verified that there were mill effects in the receiving environment and the two endpoints could both be used because they were not redundant.

6. General description of EEM-HAS schemes

6.1 ZPE SCHEME

6.1.1 Determining the Zone of Potential Effect (ZPE)

The zone of potential effect (ZPE), based on the results of a sublethal test, is best illustrated by a zone superimposed on the area where effluent concentrations are above the lowest IC25 for a specific toxicity test on the plume delineation map for a given industrial site. The outer extent of this area is determined by the distance downstream of the industrial outfall at which the estimated effluent concentration in the receiving water is diluted to the effluent concentration of the lowest IC25 for a specific sublethal test. The ZPE should be estimated from the lowest IC25 in a series of the same sublethal toxicity test, which is the concentration most protective of the environment and is the worst-case toxic concentration to which near-field fish or benthos would be exposed. Environment Canada (1999) recommends using the lowest IC25 for comparison with concentrations observed in the effluent plume if there are 10 or more endpoints in a series of tests. Two additional methods for selecting the IC25 are described if there are fewer endpoints. These include: 1) the geometric mean of the IC25s; and 2) a statistical procedure which calculates the Predicted Minimum Toxic Concentration (PMTC), which is the concentration of effluent below which only 5% of samples would be expected to have a deleterious sublethal effect (at the 95% level of confidence). Further guidance on the use of these methods is provided by Environment Canada (EC, 1999).

Once illustrated on a map of the industry's effluent plume, the ZPE can be seen visually as larger or smaller than the area of the plume defined by the isopleth for the 1% concentration of effluent (EC, 1999). A ZPE should be estimated for each test species and then illustrated on a site map. As well, it is possible to compare the zones of potential effect for sublethal tests with the locations of exposure areas (generally the near-field) that have been or are to be sampled for fish and benthic invertebrates. This comparison illustrates the relationship between the sublethal tests and potential industry related effects observed in field measurements of fish and benthic invertebrates.

To describe a zone of potential effect, information about the initial concentration of the effluent, its dilution in the receiving water, and the extent of the 1% plume is required. Some pulp mills calculate a range of concentrations from the outfall to the 1% effluent plume boundary. Tracer studies and additional conductivity measurements taken during field work can support previous plume delineation studies to identify the areas in the receiving waters having effluent concentrations greater than 1%. The zone of potential effect for a sublethal test result can then be mapped onto the 1% effluent plume based on where the effluent concentrations are estimated to be ≥ the lowest IC25 that would be found under 'worst case' conditions (*e.g.*, low flow, specific wind conditions, etc.).

To aid in interpretation of the relationship between sublethal tests and field measurements, a rating of the relative quality of plume delineation at each site is made. The quality is determined based on the availability of information required in order to map the zones of potential effect. At sites where dilution of the effluent in the receiving water is well documented, the plume delineation can be considered 'strong'. In these cases, mapping of a ZPE is uncomplicated. In other cases, the significant variability due to inflow, water movement, currents, wind or a lack of ground truthing data makes estimation of the ZPE more problematic, basically because insufficient information about the plume's location has been collected during the field studies. In these cases, plume delineation is considered 'moderate' or 'weak'. A map illustrating a zone of potential effect for each sublethal test should accompany each site study.

6.1.2 Rating the relationship between ZPE and field measurements

The relationship between sublethal toxicity tests and field measurements can be rated on the basis of zones of potential effect (Environment Canada, 1999). The following points describe the criteria used for rating the relationship between zones of potential effect for each sublethal test (lowest IC25) and potential effluent-related effects on fish or the benthic invertebrate community (Moody, 1992).

A strong relationship between ZPE and field effects is described by the following two situations:

- If the ZPE includes at least part of a sampling area (generally the near-field) for the fish or benthic invertebrate survey and potential effluent-related effects are present in this area, then the relationship between the sublethal test and the field measurement is termed strong.
- Similarly, if there is a general low number or lack of effects reported from field measurements (rated 'low' or 'no response') and the ZPE is 0 (because the lowest IC25 was >100%), then the relationship between the two is strong.

A moderate (or moderately strong) relationship between ZPE and field effects is described by the following two situations:

- If the ZPE is measurable or can be estimated (*i.e.*, the lowest IC25 is <100%) and extends to an area reasonably near to the near-field, then the relationship is termed moderate if a number of potential effluent-related effects are reported.
- If the ZPE is very small, close to the outfall and quickly diluted and the field measurement shows a lack of effects or a low number of effects, the strength of the relationship is termed moderate.

A weak relationship between ZPE and field effects is described by the following three situations:

- If the ZPE is large and includes sampling areas in which no potential effluent-related effects were seen, the relationship is weak.
- If the ZPE is small relative to the distance at which potential effluent-related effects are seen and does not include sampling areas in which these effects were observed, the relationship is weak.

Given significant effects in fish or benthic organisms observed at relatively distant sampling areas, a ZPE that estimates a small area of potential effect would indicate a weak relationship between the sublethal test in question and the field survey.

6.2 LAB-TO-FIELD RATING SCHEME

The lab-to-field (LTF) rating scheme is a second approach for comparing sublethal test data with field measurements of fish and benthos. It is based on a four-step method which utilizes the effects data from fish and benthic surveys carried out in the receiving environment. The number of potential effluent-related effects in a field measurement (for example, fish) is compared with the severity of the effluent toxicity as determined by the sublethal laboratory tests. This allows a numerical approach to judging the relationship between field measurements, based on the relative number of observed effects to endpoints measured (expressed as a percentage), and upon lab tests based on degree of effluent toxicity (test endpoint being expressed as 100 minus lowest IC25 for comparison purposes). The same scale for assigning the LTF rating is applied to field measurements or sublethal tests (Tables 1, 2 and 4).

Table 2. Rating scheme for scoring the three sublethal tests.

Descriptor	*Lowest IC25 (% v/v)*	*Lowest IC25 subtracted from 100 (%v/v)*	*LTF [a] rating*
No response	> 92	< 8	1
Low toxicity	≥ 84 and < 92	≥ 8 and < 16	2
Moderate toxicity	≥ 66 and < 84	≥ 16 and < 34	3
High toxicity	≥ 33 and < 66	≥ 34 and < 67	4
Severe toxicity	< 33	≥ 67	5

[a] LTF = lab-to-field

Data from each study are assigned a rating in each of five categories. The categories include the three sublethal tests, the fish survey and the benthic invertebrate survey measurements. Ratings of 1 to 5 are assigned based on increasing toxicity or relative number of effects. For example, sublethal tests that have low IC25s indicate highly toxic effluent and thus the endpoint (100 minus lowest IC25) is high. The sublethal tests therefore receive a high LTF rating of 4 or 5 (Step 1).

Field surveys of fish and benthos are assigned a numerical rating and descriptor (Tab. 4) based on the number of potential effluent related effects (PERE) relative to the number of endpoints measured in each survey expressed as a percentage (Tables 3 and 5). The calculation is illustrated in Table 3.

The presence of confounding influences is typically addressed in an industry's sampling design and is reviewed as part of the detailed assessment of each study. A statistically significant difference between the reference area(s) and the exposure area that is potentially associated with an industry's effluent (potential effects of exposure to effluent, water quality, sediment deposits) is designated as a potential effluent-related effect, abbreviated PERE.

The four steps of the Lab-to-Field rating scheme are:

Step 1: assign a rating of 1 to 5 for each sublethal test (fathead minnow growth inhibition, *Ceriodaphnia dubia* reproduction inhibition, *Selenastrum capricornutum* growth inhibition) (Tab. 2).

Step 2: assign a rating of 1 to 5 to the fish survey based on the percentage of potential effluent-related effects (Tables 3 and 4) relative to endpoints measured.

Step 3: assign a rating of 1 to 5 to the invertebrate community survey (ICS) based on the percentage of potential effluent-related effects (Tables 4 and 5) relative to endpoints measured.

Step 4: assess the strength of the relationship between a sublethal test and a field measurement (fish survey or ICS) based on similarity between ratings.

Step 1: sublethal test results are assigned a numerical rating and descriptor based on the lowest IC25 subtracted from 100 for secondary-treated effluent (Tab. 2) so the more significant the response, the higher the number. The ranges chosen are suitable for test results of secondary-treated effluent. This system of grouping the results of sublethal tests is a modification of the qualitative scale developed for characterizing toxicity tests in the Ontario Region (Borgmann et al., 1999; 2002)

Table 3. Calculation of number of effects for fish survey.

	Endpoints	***PERE [a] Yes/No***
Age Structure	Age	Yes/No
Energy Expenditure	Total weight	Yes/No
	Weight and age	Yes/No
	Total (or fork) length	Yes/No
	Total (or fork) length and age	Yes/No
Energy Investment	Size and age (growth)	Yes/No
	Age and maturity	Yes/No
	Size and maturity	Yes/No
	Gonad weight and weight or length	Yes/No
	Fecundity	Yes/No
	Fecundity and weight, length or age	Yes/No
	Length, length and age	Yes/No
Energy Storage	Condition (weight and length)	Yes/No
	Liver weight and weight or length	Yes/No
	Total number of endpoints = y	Total number of effects = x
	% PERE [a]	$(x/y)\cdot 100$

[a] Potential Effluent Related Effects

Each range of values (100 minus lowest IC25) is assigned a rating from 1 to 5 with increasing toxicity (Tab. 2). This step yields a single rating of 1 to 5 for each sublethal test at each mill.

Step 2: rating the fish survey. Statistically significant differences between reference and exposed fish, including significant interactions from covariate analysis, in either sentinel species are considered to be potential effluent-related effects (PERE). Typical endpoints may be grouped into categories of age structure, energy expenditure, energy investment and energy storage (Tab. 3). Following the approach of Munkittrick (personal communication, 2001), the number of endpoints having effects is compared with the total number of endpoints reported by industry, and is expressed as a percentage (% PERE = total number of effects/total number of endpoints x 100). The endpoints should not used if 12 or fewer fish were caught at either the reference zone or the exposure zone.

The final LTF rating for the fish survey is assigned according to Table 4 using the percentage PERE value. A rating from 1 ('no effects') to 5 ('severe effects') is the LTF rating for the fish survey.

Table 4. Rating scheme for fish and benthic field measurements.

Descriptor	***% PERE***[a]	***LTF***[b] ***rating***
No effects	< 8%	1
Low effects	≥ 8 and < 17%	2
Moderate effects	≥ 17 and < 34%	3
High effects	≥ 34 and < 67%	4
Severe effects	≥ 67 and < 100%	5

[a] Potential Effluent Related Effects ; [b] LTF= lab-to-field.

Step 3: rating the benthic invertebrate community survey. Rating data from a benthic invertebrate survey is carried out by enumeration of potential effluent-related effects. These effects include significant differences between reference and exposure areas (or along a gradient) in abundance/density (total organisms/m^2) and richness (number of taxa). These indices and descriptors indicate exposure to the industry's effluent and are required components of a weight-of–evidence approach to interpretation of benthic survey results in the pulp and paper or mining EEM programs. Expected environmental effects relating to the presence of pulp mills are an increase in total abundance and a decrease in species richness in exposure areas relative to reference areas. Additional endpoints described in the EEM metal mining guidance document (EC, 2001) can also be included in the LTF scheme, such as the Bray-Curtis, Simpson's Diversity and Simpson's Evenness Indices.

A number of benthic species are highlighted in the Ontario EEM reports as indicators of pollution because they were considered either more or less sensitive to impacts that may be found in association with pulp and paper mills (Borgmann et al., 2002). Organisms that have shown significant differences in their abundance between reference and exposure areas in EEM Cycle 2 interpretive reports were classified on the basis of sensitivity to pollution. Toxicity indicator organisms are those having pollution tolerances of 1 to 5 and are expected to be significantly less abundant in exposure areas than in reference areas, possibly due to toxic chemicals in the mill effluent, than in reference areas. Enrichment indicator organisms have pollution tolerances of 6 to 10 and are expected to be significantly more abundant in exposure areas, possibly because of nutrient enrichment, than in reference areas. One or more such indicator organisms in either category is counted as a potential effluent-related effect (PERE, see Table 5) if a significant difference at $\alpha = 0.05$ in abundance was calculated. These effects are included as part of the LTF rating scheme for the benthic invertebrate survey. The number of endpoints having effects is compared with the total number of endpoints reported or calculated, and is expressed as a percentage (% PERE = total number of effects/total number of endpoints x 100).

Table 5. Calculation of number of effects for benthic invertebrates.

Parameter	***Expected PERE*** [a]	***Observed Effect Yes/No***
Total abundance	Exposure > Reference	Yes/No
Number of taxa/diversity	Reference > Exposure	Yes/No
Simpson's Diversity Index	Reference > Exposure	Yes/No
Simpson's Evenness	Reference > Exposure	Yes/No
Bray-Curtis Index	Exposure > Reference	Yes/No
EPT Index [b]	Reference > Exposure	Yes/No
Enrichment indicator organism	Exposure > Reference	Yes/No
Toxicity indicator organism	Reference > Exposure	Yes/No
	Total number of endpoints = y	Total number of effects = x
	% PERE [a]	(x/y)·100

[a] Potential Effluent Related Effect(s); [b] EPT Index = Ephemeroptera, Plecoptera and Trichoptera.

The final rating for the benthic survey is assigned according to Table 4. A rating from 1 ('no effects') to 5 ('severe effects') is the LTF rating for the benthic survey.

Step 4: the lab-to-field method of rating results of sublethal tests and field measurements yields a series of five ratings for each mill (*i.e.*, ratings for fathead minnow, *Ceriodaphnia* and *Selenastrum* tests, and for benthic invertebrate community and fish surveys). The strength of the relationship between a sublethal test and a field measurement can be made by observing the degree of similarity in ratings. LTF ratings at mills are used for this purpose as follows:

If ratings are equal, the strength of the relationship is termed 'strong'.

If ratings differ by 1 point, the strength of the relationship is termed 'moderate'.
If ratings differ by ≥ 2 points, the strength of the relationship is termed 'weak'.

7. Application of EEM-HAS schemes in case studies

7.1 ZPE CASE STUDY - KIMBERLY CLARK, TERRACE BAY

The Kimberly-Clark pulp mill at the town of Terrace Bay, Ontario employs a kraft process, producing market kraft pulp fibre from both softwood and hardwood. Before being discharged into Blackbird Creek, effluent undergoes secondary treatment consisting of a three-celled aerated stabilization basin. Effluent then passes through Blackbird Creek that extends for 15 km before discharging into Jackfish Bay on Lake Superior. Two lakes (Lake A and Lake C) along the length of the creek are used by the mill as settling and extended treatment basins (Fig. 1). Effluent concentrations in these lakes were typically always greater than 80% v/v, therefore the effluent concentration entering Jackfish Bay has been estimated in this case study to be 80%. The presence of historical deposits of fibre in Jackfish Bay contributes to the organic enrichment of the area and to the effects of organic material entering the bay from Blackbird Creek. Contamination of sediments by oil was observed in the near-field and far-field areas. A description of the EEM study is found in Second Cycle EEM Final Interpretive Report for Kimberly-Clark Inc. - Terrace Bay Mill (Beak International, 2000).

7.1.1 Determination and mapping of the effluent plume

The effluent plume in Jackfish Bay is subject to great variability due to water temperature and weather patterns. The plume boundaries on Figure 1 are a composite estimating a worst-case scenario, based on several plume delineation studies using Rhodamine WT dye dilution and conductivity completed in 1982, 1988 and 1991 and estimate the largest expected extent of the 1% effluent into Jackfish Bay. Under most conditions, the 1% plume occupies only the western part of the bay. Plume delineation information was sufficient to support mapping of zones of potential effect inside which the effluent concentration (% v/v) was greater than or equal to the lowest IC25 (% v/v) for each sublethal test. Plume delineation information at this site is abundant and was collected under a variety of conditions; it is therefore rated as 'good.'

7.1.2 Sublethal toxicity test results and determination of ZPE

Sublethal testing of the Kimberly-Clark final mill effluent indicated inhibition of growth in fathead minnow and *Selenastrum capricornutum* and inhibition of reproduction of *Ceriodaphnia dubia* during the period from 1994 to 2000. Zones of potential effects were estimated (Tab. 6) and plotted (Fig. 1) to illustrate the location of effluent concentrations ≥ the lowest IC25 (%v/v) for each test.

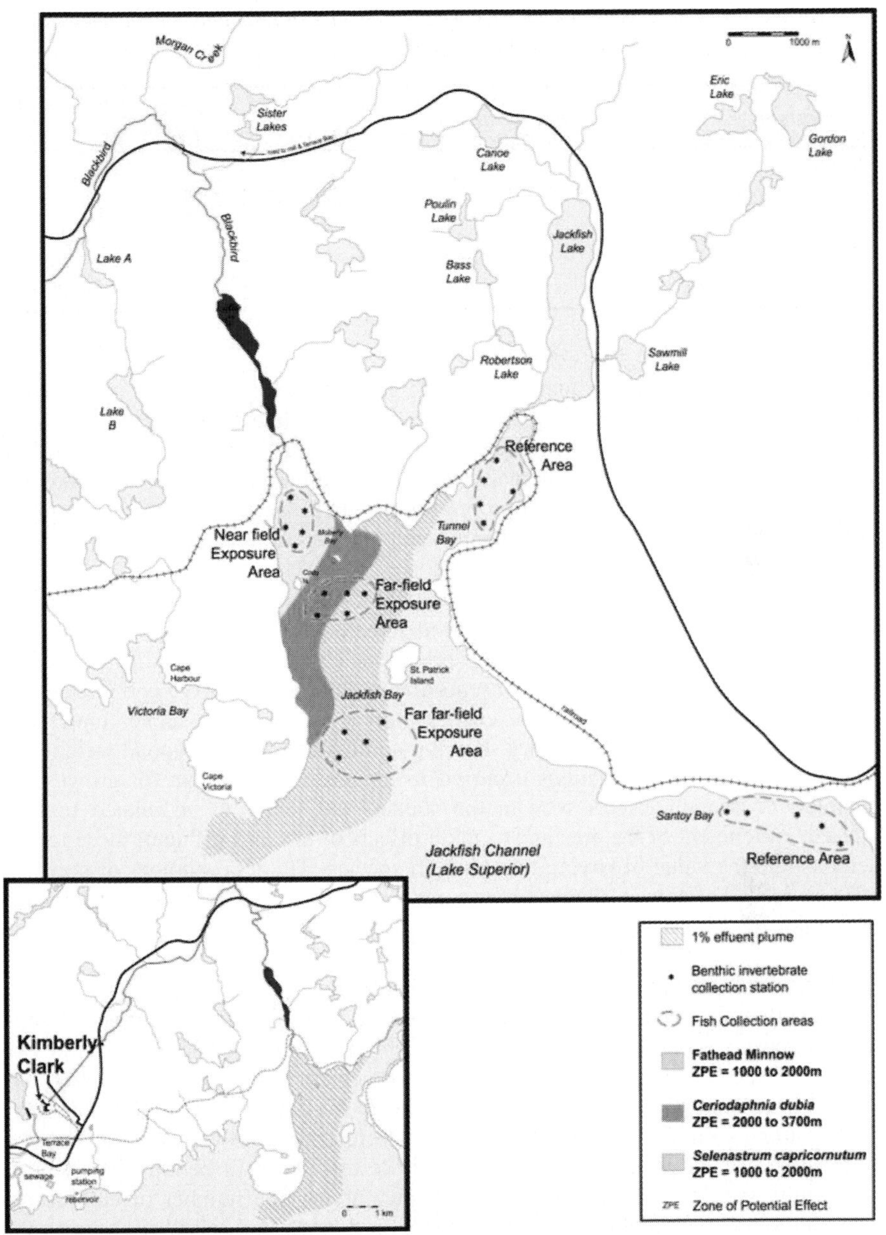

Figure 1. Effluent dispersion map for the Kimberly-Clark Mill, Terrace Bay.

Table 6. Sublethal toxicity tests for secondary treated ASB final effluent collected at Kimberly-Clark mill.

Organism	***Number of tests***	***Lowest IC25 (% v/v)***	***Zone of Potential Effect length (m)***
Fathead minnow (*Pimephales promelas*)	10	21.9	1000 – 2000
Ceriodaphnia dubia	10	5.9	2000 – 3700
S. capricornutum	6	17.7	1000 – 2000

7.1.3 Determination of effects on fish

Two sentinel species were caught in the reference and exposure zones in sufficient numbers (at least 12 fish) for statistical interpretation. The data for the white sucker, *Catastomus comersoni* and longnose sucker, *Catastomus catastomus*, are listed in Table 7. All endpoints (Beak International, 2000) and comparisons of exposure and reference area data involved appropriate statistical methods. Any significant difference or significant interaction in any endpoint was considered potentially mill-related.

Different patterns of effects relating to effluent exposure were observed in white sucker and longnose sucker. Both male and female white suckers were affected in age, reproduction and energy storage while females were additionally affected in growth. Relationships in the areas of reproduction and growth were complicated by statistical interactions. Growth and energy storage of longnose sucker males and females were affected, with females showing additional effects in gonad weight and fecundity. Statistical interactions occurred in longnose female data for growth and fecundity. Although effects seen in the suckers are likely to be related both to nutrient enrichment of the area and to toxic effects of the mill effluent, these results demonstrate the value of having two sentinel species. The total number of potential effluent-related effects is calculated for comparison with the LTF method case study in Section 7.2.

7.1.4 Determination of effects on benthic invertebrates

The invertebrate community survey followed a multiple-control/impact design. Three exposure areas (near-field (NF), far-field (FF), far-far-field (FFF)) and two reference areas (Tunnel Bay and Santoy Bay) were sampled (Fig. 1). Santoy Bay (Ref-1) is located approximately 10 km southeast of Moberly Bay and Tunnel Bay (Ref-2) about 4 km east of the mill outlet. Both reference areas were outside the area of the 1% effluent plume. There was very little difference in benthic invertebrate community structure between the two reference areas. The number of chironomid taxa and density of lumbriculids were higher in Santoy Bay, primarily due to the more sandy substrate (Beak International, 2000). Conductivity readings in exposure areas indicated a gradient in effluent exposure ranging from a high of 164 in the near-field to 98 μhos/cm in the far-far field area (EC, 2000). Conductivity in the reference areas was 100 μhos/cm.

Table 7. Fish survey summary statistics analyses[1] for white and longnose sucker, Kimberly-Clark.

Categories	*Endpoints (log)*	*Covariate (log)*	*PERE[2] observed for males?*	*PERE[2] observed for females?*
White sucker (males N: Ref = 25, Exp = 25; females N: Ref = 21, Exp = 21)				
Age Structure	Age	None	Yes (Ref > Exp)	Yes (Ref > Exp)
	Age distribution	None	Yes (Ref > Exp)	No
Energy Expenditure	Total length	Age	No	Yes (SI)
	Total wt	Age	No	Yes (SI)
Energy Investment	Gonad wt	Total length	Yes (SI)	Yes (Exp > Ref)
	Gonad wt	Total wt[3]	Yes (SI)	No
	Fecundity	Total length	N/A	No
	Fecundity	Total wt	N/A	No
	Fecundity	Age	N/A	Yes (SI)
	Egg weight	Total length	N/A	No
	Egg weight	Total wt	N/A	No
	Egg weight	Age	N/A	No
Energy Storage	Total wt	Total length	Yes (Exp > Ref)	Yes (SI)
	Liver wt	Total length	Yes (Exp > Ref)	Yes (Exp > Ref)
	Liver wt	Total wt[4]	No	No
Longnose sucker (males N: Ref = 30, Exp = 14; females N: Ref = 30, Exp = 19)				
Age Structure	Age	None	No	No
	Age distribution	None	No	No
Energy Expenditure	Total length	Age	Yes (Ref > Exp)	Yes (SI)
	Total wt	Age	Yes (Ref > Exp)	Yes (SI)
Energy Investment	Gonad wt	Total length	No	No
	Gonad wt	Total wt[3]	No	Yes (Ref > Exp)
	Fecundity	Total length	N/A	Yes (Exp > Ref)
	Fecundity	Total wt	N/A	Yes (Exp > Ref)
	Fecundity	Age	N/A	Yes (SI)
	Egg weight	None	N/A	Yes (Ref > Exp)
Energy Storage	Total wt	Total length	Yes (Exp > Ref)	Yes (Exp > Ref)
	Liver wt	Total length	No	No
	Liver wt	Total wt[4]	No	No
	Total number	**46**	**9**	**15**

[1]ANOVA, ANCOVA and Chi-square tests at $\alpha = 0.05$; [2] PERE = Potential Effluent Related Effects; [3]when used as a covariate of reproductive parameters, the total body wt is represented by total body wt minus gonad wt; [4]when used as a covariate for liver wt, total body wt represented by wt minus liver wt; SI = Significant Interaction; wt = weight.

Levels of total organic carbon (TOC) followed a gradient pattern, and ranged from 4.9% at the near-field to 0.7% in the far-far field (mean 3.1%), primarily due to historical fibre deposits. Oil contamination was present in the near-field and far-field sediments. Triplicate samples were collected using a standard Ponar grab and pooled. Invertebrates were preserved until they were counted and identified. Statistically significant differences between reference and exposure areas were observed for three of the five invertebrate community indices calculated (Tab. 8).

Benthic taxa exhibiting significant differences in abundance (at $\alpha = 0.05$) between reference and exposure areas were noted. If a taxon was considered to be relatively more tolerant of pollution (such as enrichment in pulp mill effluent), it was listed as an "Enrichment Indicator Organism". On a scale from 1 to 10, these taxa have pollution tolerances from 6 to 10 (Moody, 2000). If a taxon was considered to be relatively less tolerant of pollution, it was listed as a "Toxicity Indicator Organism". These taxa have pollution tolerances from 1 to 5. Species representing both types of indicator groups were found to show potential effluent-related effects at the Kimberly-Clark mill. The total number of potential effluent-related effects is calculated for comparison with the LTF method case study in Section 7.2.

Table 8. Significant differences of benthic invertebrate communities and selected benthic taxa between sites for Kimberly-Clark.

Descriptors	***Expected PERE***	***Significant differences[1] in densities of selected benthic taxa***	***Expected PERE[2] Observed?***
Total abundance (total organisms/m^2)	Exp > Ref		Yes (NF[3] > Ref)[4]
Richness (number of taxa = 54)	Ref > Exp		Yes (Ref > Exp)
Simpson's Diversity Index[5]	Ref > Exp		No
Simpson's Evenness[5]	Ref > Exp		No
Bray-Curtis Index[5]	Exp > Ref		Yes[6]
Enrichment indicator organism	Exp > Ref	Tubificidae, Isopoda, *Chironomus*, *Procladius*	Yes (Exp > Ref)[1]
Toxicity indicator organism	Ref > Exp	*Monoporeia* *Stylodrilus heringianus*	Yes (Ref > Exp)[1]
Total number of descriptors	7	Total number of effects	5

[1] Mann-Whitney Rank Sum Test, Sigmastat v 2.0; [2] PERE = Potential Effluent Related Effect; [3] NF (Nearfield) significantly greater than both Tunnel Bay and Santoy Bay reference areas, [4] t-tests, Mann-Whitney U-tests, Mann-Whitney rank sum tests, $\alpha = 0.05$,[5] National EEM database, unpublished data; [6] significantly less at Santoy Bay reference area, [6] significant difference at Tunnel Bay.

7.1.5 Rating the relationship between sublethal toxicity tests and ecosystem indicators using the Zones of Potential Effects

The relationship between sublethal toxicity tests and field measurements can be rated on the basis of zones of potential effects (EC, 1999). The criteria used for rating the relationship between zones of potential effect for each sublethal test (after determining the lowest IC25) and potential effluent-related effects on fish or the benthic community are described in Section 6.1.2.

The relationships between zones of potential effect for fathead minnow, *Ceriodaphnia dubia* and *Selenastrum capricornutum* tests and the locations of the exposure areas for the fish and benthic surveys are illustrated in Figure 1. Because the ZPE for the sublethal tests include part or all of the sampling areas for fish and benthos and a number potential effluent related effects are present, the relationships between the sublethal tests and the two ecosystem surveys have been rated "strong" in all cases (Tab. 9). If the fish and benthic surveys had been conducted in areas separate from one another and a PERE had been observed, the relationships between ZPE and the individual survey locations would have to be considered separately.

Table 9. Summary of qualitative relationship between sublethal toxicity and field measurements using the ZPE method, Kimberly-Clark, Terrace Bay.

Sublethal toxicity test	***Fish survey***	***Benthic survey***
Fathead minnow	Strong	Strong
Ceriodaphnia dubia	Strong	Strong
Selenastrum capricornutum	Strong	Strong

7.2 LTF CASE STUDY - PROVINCIAL PAPERS, THUNDER BAY

Provincial Papers Inc. is located on Lake Superior at Thunder Bay, Ontario. Sulphite/mechanical processes are used in the production of coated and uncoated fine papers from purchased kraft pulp and groundwood from softwood sources. Secondary treatment is carried out in an aerated lagoon system (a serpentine basin) and was installed at the mill in late 1995 (see Fig. 2). Effluent flow rates ranged from 25,000 to 35,000 m^3/day in the two years following installation of secondary treatment. Treated effluent is discharged into the inner basin of Lakehead Harbour, a sheltered embayment created by a series of breakwaters. The Current River is the other major discharge to the immediate area of the mill outfall. A description of the EEM study is found in ESG (2000).

7.2.1 Rating the sublethal toxicity tests

A rating of 1 to 5 was assigned to each sublethal test (fathead minnow growth inhibition, *Ceriodaphnia dubia* reproduction inhibition and *Selenastrum capricornutum* growth inhibition) conducted on secondary-treated effluent at

Provincial Papers. Toxicity test results were assigned a numerical LTF rating based on the lowest IC25 subtracted from 100 so that the greater the toxicity, the higher the LTF rating number, in order to make the scale similar to the field survey rating scale. Table 10 illustrates how the effluent toxicity tests were rated using data from the Provincial Papers mill (ESG, 2000).

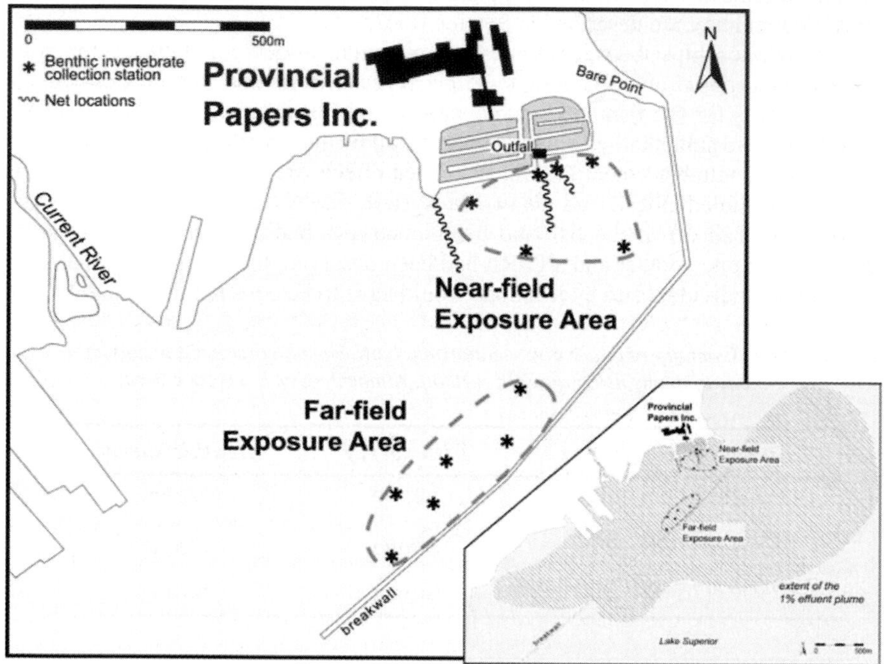

Figure 2. Effluent dispersion map for the Provincial Paper Mills, Thunder Bay.

Table 10. Sublethal toxicity tests, EEM Cycles 1 and 2, for secondary treated final effluent collected at Provincial Papers.

Organisms	***Number of tests***	***Lowest IC25 (% v/v)***	***Lowest IC25 subtracted from 100 (% v/v)***	***LTF [a] rating***
Fathead minnow	6	65.3	34.7	4 (high)
Ceriodaphnia dubia	6	36.0	64	4 (high)
Selenastrum capricornutum	6	30.7	69.3	5 (severe)

[a] *LTF* = lab-to-field.

7.2.2 Rating ecosystem indicators - fish population.

The two species of fish targeted at Provincial Papers were white sucker (*Catastomus commersoni*) and spottail shiner (*Notropis hudonsius*). Fishing for both species was carried out in the same reference or exposure areas. The exposure area was located inside the breakwall of the harbour, close to the effluent discharge. Substrate there was primarily wood fibre and mud. The reference area was located at Cloud Bay, about 45 km south of the exposure area on Lake Superior. Cloud Bay has a narrow entrance similar to the breakwater entrance at the Thunder Bay Harbour. The substrate is dominated by sand with some bedrock. Cloud Bay has some relatively dense beds of macrophytes along the shoreline. This type of habitat was not available in the exposure area. White suckers were collected using gill nets. Spottail shiners were collected using minnow traps and seine net (reference area only). Seining was not conducted in the exposure area because there was no suitable habitat within the effluent plume. Data analysis for spottail shiner was inconclusive due mainly to insufficient catch of mature fish.

More than 12 white suckers of both sexes were caught in both reference and exposure areas at Provincial Papers, so all of these endpoints were used. However, insufficient numbers of spottail shiner (*Notropis hudonsius*) were caught, and therefore the endpoints for this species were not included. Endpoints measured at Provincial Papers are reported in Table 11 (ESG, 2000).

Endpoints can be grouped into categories of age structure, energy expenditure, energy investment and energy storage (see Section 6.2, Step 2). Following the approach of Munkittrick (2001), the number of endpoints having effects was compared with the total number of endpoints reported by the mill, and was expressed as a percentage (*i.e.*, % PERE = total number of effects/total number of endpoints multiplied by 100).

A LTF rating of 1 ('very low level of effects' or 'no effects observed') to 5 ('severe level of effects') was assigned to the fish survey based on the percentage of potentially effluent-related effects relative to all the endpoints measured. Statistically significant differences at $\alpha = 0.05$ between reference and exposed white sucker (*Catastomus commersoni*) collected for the Provincial Papers study, including significant interactions from covariate analysis, in either sentinel species and in either sex, were considered to be possibly effluent-related effects (PERE).

7.2.3 Rating ecosystem indicators - benthic invertebrate community

The benthic invertebrate study design used at the Provincial Papers site was a Multiple Control/Impact study including near-field (NF) and far-field (FF) areas in both exposure and reference areas. The NF exposure area was located as close as possible to the outfall (Fig. 2) in the area showing elevated conductivity, while the FF exposure area was located approximately 1000 m from the NF area. A petite Ponar was used to collect duplicate samples at six stations in each exposure and reference area. Substrates in the reference and exposure areas were different; the exposure area was 100% organic, comprised of wood chips, while the reference area was dominated by silt and clay with small amounts of fine sand at two stations.

Table 11. Fish survey summary statistics analyses[1] for white sucker, Provincial Papers, Thunder Bay.

White sucker, *Catastomus commersoni*
(males N: Ref = 27, Exp = 30; females N: Ref = 28, Exp = 31)

Categories	***Endpoints (log)***	***Covariate (log)***	***PERE[2] observed for males?***	***PERE[2] observed for females?***
Age Structure	Age	None	Yes (Ref > Exp)[2]	Yes (Ref > Exp)
Energy Expenditure	Fork length	Age	Yes (Exp > Ref)	Yes (Exp > Ref)
	Total wt	Age	Yes (Exp > Ref)	Yes (Exp > Ref)
Energy Investment	Gonad wt	Fork length	Yes (SI)	Yes (Ref > Exp)
	Gonad wt	Total wt[3]	Yes (SI)	Yes (Ref > Exp)
	Fecundity	Fork length	N/A	No
	Fecundity	Total wt	N/A	No
	Egg weight		N/A	Yes (Ref > Exp)
Energy Storage	Total wt	Fork length	No	Yes (SI)
	Liver wt	Fork length	Yes (Ref > Exp)	No
	Liver wt	Total wt[4]	No	No
	Total number of endpoints measured y = 19		Total number of effects for males $x_1 = 6$	Total number of effects for females $x_2 = 7$
	% PERE		$(x_1 + x_2)/y \cdot 100$	$(6 + 7)/19 \cdot 100 =$ 68%
			LTF rating	5 (severe)

[1]ANOVA, ANCOVA tests at $\alpha = 0.05$; [2]PERE = Potential Effluent Related Effects; [3]when used as a covariate of reproductive parameters, the total body wt is represented by total body wt minus gonad wt; [4]when used as a covariate for liver wt, total body wt is represented by wt minus liver wt; SI = Significant Interaction; wt = weight.

Similar to the sublethal toxicity tests and the fish survey, an LTF rating of 1 ('no effects or very low effects observed') to 5 ('severe level of effects') was assigned to the benthic invertebrate community survey based on the percentage of PERE relative to the total number of descriptors measured. Effects included significant differences at $\alpha = 0.05$ between reference and exposure areas (or along a gradient) in abundance (total organisms/m^2) and richness (number of taxa). Expected environmental effects relating to the presence of pulp mills are an increase in total abundance and a decrease in species richness in exposure areas relative to reference areas.

Of the additional endpoints, Simpson's diversity, Simpson's Evenness, and Bray-Curtis indices, only the Bray-Curtis Index was calculated for Provincial Papers, but no PERE was observed for this index.

The number of endpoints having effects was compared with the total number of endpoints reported or calculated, and was expressed as a percentage (*i.e.*, % PERE =

total number of effects/total number of endpoints multiplied by 100). Table 12 illustrates the rating of benthic survey using Provincial Papers mill data (ESG, 2000).

Table 12. Significant differences of benthic invertebrate communities and selected benthic taxa between sites for Provincial Papers.

Descriptors	***Expected PERE***[2]	***Significant differences***[1] ***in densities of selected benthic taxa***	***Expected PERE***[2] ***observed?***
Total abundance (total organisms/m^2)	Exp > Ref		No (Ref > Exp)[3]
Richness (number of taxa = 65)	Ref > Exp		Yes (Ref > Exp)[3]
Simpson's Diversity Index[4]	Ref > Exp		N/A[5]
Simpson's Evenness[4]	Ref > Exp		N/A
Bray-Curtis Index[4]	Exp > Ref		No
Enrichment indicator organism	Exp > Ref	Oligochaetes (*Arcteonais lomondi, Limnodrilus hoffmeisteri, Quistadrilus multisetosus*)	Yes (Exp > Ref)[1]
Toxicity indicator organism	Ref > Exp	Amphipoda (*Monoporeia affinis*); Mollusca (*Gastropoda*)	Yes (Ref > Exp)[1]
Total number of descriptors	5	Total number of effects	3
		% PERE[6]	60
		LTF rating [7]	4 (high)

[1] Mann-Whitney Rank Sum Test, Sigmastat v 2.0; [2] PERE = Potential Effluent Related Effect; [3] ANOVA Analysis, $\alpha = 0.05$; [4] Canadian EEM database, unpublished data; [5] not available, not applicable or not calculated; [6] % PERE = Percent Potential Effluent Related Effect Observed = no. of expected effects divided by no. of endpoints calculated and multiplied by 100; [7] LTF = lab-to-field.

7.2.4 Assessing strength of relationship between toxicity tests and ecosystem indicators

The LTF hazard assessment scheme is based qualitatively on similarities between ratings. The strength of the relationship using LTF ratings at pulp and paper mills is indicated as strong, moderate or weak. A **strong** LTF relationship means that the ratings are equal. At Provincial Papers, *Ceriodaphnia* sublethal toxicity testing and benthic communities appeared to have a strong relationship because both rated an LTF score of 4. A **moderate** LTF relationship means that the ratings differ by one point. At Provincial Papers, the fathead minnow test was assigned a LTF score of 4, whereas the white sucker survey rated an LTF score of 5. Therefore these scores would indicate a moderate relationship between the fish in the lab and field. A **weak** LTF relationship means that the ratings differ by two or more points. Provincial Papers did not have any weak relationships between lab and field. Table 13 summarizes the assessment of the strength of relationships between toxicity tests and ecosystem indicators using data from the Provincial Papers mill EEM Interpretative Report (ESG, 2000).

Table 13. Summary of qualitative relationship between sublethal toxicity and field measurements using the LTF method, Provincial Papers.

Sublethal toxicity test	***Fish survey***	***Benthic survey***
Fathead minnow	Moderate	Moderate
Ceriodaphnia dubia	Moderate	Moderate
Selenastrum capricornutum	Strong	Strong

8. Potential improvements to the ZPE and LTF schemes

Potential improvements to the ZPE scheme are as follows:

- The design of the study should include enough measurements such as dye dilution or conductivity measurements that the gradient in effluent concentration can be mapped with confidence and be relevant to environmental conditions such as wind, currents, and changes in effluent flow rate.
- Measurements of effluent concentration must also be taken at the effluent outfall and at fish and benthic invertebrate field collection sites during the surveys.
- In some cases, aerial photography or remote sensing may be useful in the measurement of plume extent.
- Compilation of lists of benthic indicator species specific to regions and ecosystems would make the scheme useful to other regions.
- The use of a more sensitive species/life stage of fish in toxicity tests would improve the ZPE hazard assessment scheme.

Potential improvements to the LTF rating scheme are as follows:

- The use of a more sensitive species or life stage of fish in toxicity tests would improve the LTF hazard assessment scheme.
- Compilation of lists of benthic indicator species specific to regions and ecosystems would make the scheme useful to other regions.
- Further validation of the lab-to-field relationship through the collection of both effluent and receiving water samples for sublethal tests at the time of the field biological collections to generate synoptic data.

- Incorporating a lab test that quantifies the effects of organic enrichment in effluent would also improve this hazard assessment scheme.

9. Regional applications of the ZPE and LTF schemes

Using the ZPE scheme, the study of effluent discharge situations at 16 Ontario pulp and paper mills demonstrated a majority of strong or moderately strong relationships between sublethal toxicity tests and ecosystem indicators (fish populations and benthic invertebrate communities). The locations of effects in benthic organisms corresponded in 100% of cases with zones predicted by the *Ceriodaphnia* test and in 81% of cases with predictions from the *Selenastrum* test. The fathead minnow test did not perform as well, predicting effects on fish in only 53% of cases (Moody, 2000).

Using the LTF scheme, the study of effluent discharge situations at 16 Ontario pulp and paper mills has illustrated predominantly moderate to strong qualitative relationships between toxicity tests and ecosystem indicators (fish populations and benthic invertebrate communities). *Ceriodaphnia*-to-benthos, *Selenastrum*-to-benthos and fathead-to-fish survey relationships were qualitatively rated strong or moderate in 94%, 75% and 60% of the sixteen studies, respectively. Regression analysis of LTF scores has revealed that the relationship between the *Ceriodaphnia* reproduction test and benthic invertebrate field survey measurements was significant ($p < 0.001$, $r = 0.79$). However, there were not sufficient data to determine if this can be used as a predictive tool (Borgmann et al., 2004).

A comparison of ZPE and LTF results for the 16 Ontario pulp and paper mill studies is shown in Table 14. Similar results were obtained for both ZPE and LTF (Moody, 2002). The relationship between the fathead minnow test and the fish survey was described as strong or moderately strong at 40% of the mills on rivers using ZPE and 50% of the mills on rivers using LTF. For mills on lakes, both ZPE and LTF showed strong to moderately strong relationships between fathead minnow (FHM) and fish survey for 80% of the mills. For all mills, this relationship was 53% and 60% for ZPE and LTF, respectively (Tab. 14). Relationships between sublethal tests (*Ceriodapnia* and *Selenastrum*) and the benthic invertebrate field studies were, in general, strong with similar overall relationships determined using both ZPE and LTF. Relationships between the benthic invertebrate studies and *Ceriodaphnia* tests were strong or moderately strong in 100% and 94 % of the cases for ZPE and LTF, respectively. Correlations with *Selenastrum* were not quite as evident with strong or moderately strong correlations in 81% and 75% of the cases for ZPE and LTF, respectively. Generally, plume delineations were rated strong or moderately strong at 13 of 16 mills, and confounding influences were observed at 14 of 16 mills.

Table 14. Summary of relationship between sublethal toxicity tests and field studies using the ZPE and LTF methods at 16 Ontario Pulp and Paper Mills[1].
(**S**: Strong Relationship, **M**: Moderately Strong Relationship, **W**: Weak Relationship)

Mill ID #	*CF*[2]	*P*[3]	*Fish survey*		*Benthic survey*			
			FHM[4]		*Ceriodaphnia*		*Selenastrum*	
Mills on Rivers			**ZPE**	**LTF**	**ZPE**	**LTF**	**ZPE**	**LTF**
1	Y	W	W	W	M	M	M	S
2	Y	M	M	M	M	S	S	S
3	Y	S	M	S	M	S	S	S
4	N	S	W	W	M	W	M	W
5	Y	S	M	S	M	M	S	M
6	Y	S	M	M	S	M	M	M
7	Y	W	W	W	S	S	S	S
8	Y	S	W	W	S	M	S	M
9	Y	S	W	W	M	S	M	M
10	Y	S	W	M	M	M	W	W
Percent Strong & Moderate			**40%**	**50%**	**100%**	**90%**	**90%**	**80%**
Mills on Lakes								
11	Y	M	NS[5]	NS	S	M	S	M
12	Y	W	M	M	M	S	W	W
13	N	S	W	W	S	S	W	W
14	Y	M	S	S	S	S	S	S
15[6]	Y	S	S	M	S	S	S	S
16[7]	Y	S	S	M	S	M	S	S
Percent Strong & Moderate			**80%**	**80%**	**100%**	**100%**	**67%**	**67%**
Summary of all mills	Strong (S)		3	3	8	8	9	7
	Moderate (M)		5	6	8	7	4	5
	Weak (W)		7	6	0	1	3	4
	% Strong & Moderate		**53%**	**60%**	**100%**	**94%**	**81%**	**75%**

[1] Data from Moody (2002).
[2] FHM: Fathead Minnow.
[3] **C**onfounding **F**actors present (**Y**es or **N**o).
[4] **P**lume Delineation Rating (**S**trong, **M**edium, **W**eak).
[5] NS: No Survey.
[6] ZPE Case Study: Kimberly Clark, Terrace Bay (see Section 7.1).
[7] LTF Case Study: Provincial Papers, Thunder Bay (see Section 7.2).

10. Conclusion

The ZPE and LTF Hazard Assessment Schemes have been successfully used to assess the relationship between laboratory sublethal toxicity data and field measurements of the Canadian pulp and paper Environmental Effects Monitoring Program. Based on these results and the flexibility of the methods, both ZPE and LTF may be used for regulatory monitoring of other types of industrial effluents.

The application of these schemes illustrates how sublethal toxicity tests can be used to estimate the potential for effects in the receiving water environment. Both ZPE and LTF use laboratory sublethal toxicity data and statistically significant field observations for a more comprehensive or weight-of-evidence approach for regulatory monitoring of industrial effluents.

Although ZPE and LTF provide similar results in many cases, they differ in the way they are applied. ZPE uses sublethal toxicity data and effluent plume delineation data to estimate whether there is a potential for effects of an effluent on organisms living in the local receiving water environment, and how far the effects might extend. With just sublethal data and thorough plume delineation, the ZPE can guide the selection of near-field biological sampling sites and/or priority sites in an environmental monitoring program. The ZPE can also assess the relationship between the predicted zone of effect and the actual extent of effects observed in the field.

LTF uses laboratory sublethal toxicity data and biological field observations ranked on the same scale to examine the strength of their relationship. The LTF rating scheme can use all endpoints determined for biota (fish and invertebrates) in field observations, and allows for the addition of further endpoints and indices as they are developed. LTF has the potential of linking a specific effluent source to a measured field effect in indigenous organisms.

Sublethal toxicity tests that are relatively less sensitive can reduce the usefulness of both ZPE and LTF. In addition, confounding factors such as the presence of other discharge sources can complicate the interpretation of both hazard assessment schemes.

The investigation of Ontario mills demonstrated that the two schemes for determining the relationship between sublethal tests and field measurements were equally effective. Relationships between lab and field measurements were, in general, strong for both *Ceriodaphnia* and *Selenastrum* tests but weaker for fathead minnow. The fathead-to-fish relationship was weak because, in general, fathead minnow testing significantly underestimated potential mill related effects observed by the fish survey. The underestimations of potential field effects resulting from less sensitive laboratory toxicity tests (7-day fathead minnow growth and reproduction test) can be overcome by replacing the test with one that is more sensitive and provides a better estimation of field-effects.

A more sensitive fathead minnow (*Pimephales promelas*) toxicity test was recently evaluated using the effluent from Kimberly-Clark (see case study # 1). The short-term reproductive bioassay developed by Ankley et al. (2001) has a number of advantages. Various biological and biochemical endpoints can be assessed over only

21 days resulting in reduced time and cost compared to a full life-cycle toxicity test. Each breeding pair of fathead minnows was placed in a 16 L aquarium under control conditions (16:8 light:dark photoperiod, temperature 20°C) and fed frozen brine shrimp twice daily. Each aquarium contained one spawning tile that was examined daily for eggs. A breeding trial, consisting of 24 breeding pairs, was conducted over 21 days to determine activity of each pair and to acquire pair-based baseline data for the following endpoints: survival, egg production, fertilization success, hatching success, larval deformities, and secondary sex characteristics of breeding adults. Secondary sex characteristics of fathead minnows consist of banding, nuptial tubercles, dorsal pad and fin dot in males and ovipositor size in females. Three breeding pairs were then randomly assigned to each of four treatments (control, 100% and 50% treated pulp mill effluent, and a positive control - ethynylestradiol 10 ng/L). The fish were exposed for 21 days and the biological endpoints re-measured. Significant reductions in number of spawning events, reduction in survival over time and number of normal larvae hatched were observed in the two treatments compared to controls. These effects relate well with the effects seen in the fish survey (Rickwood et al., 2003).

Acknowledgements

The assistance of Charlene Hudym and Leanne Crone of the Saskatchewan Research Council for document preparation and editing is appreciated.

References

Ankley, G.T., Jensen, K.M., Kahl, M.D., Korte, J.J. and Makynen E.A. (2001) Description and Evaluation of a Short-Term Reproduction Test with the Fathead Minnow (*Pimephales promelas*), *Environmental Toxicology and Chemistry* **20** (6), 1276–1290.

Beak International (2000) Second cycle EEM Final Interpretative Report for Kimberley-Clark Incorporated Terrace Bay Mill, Beak International Incorporated, Brampton, Ontario.

Borgmann, A., Michajluk, S. and Humphrey, S. (1999) Environmental effects monitoring at 22 pulp and paper mills in Ontario: Cycle 2 interim report – Summary of summer 1997 and winter 1998 EEM effluent toxicity results, Environmental Protection Branch – Ontario Region, Environment Canada, Downsview, Ontario, 7 pp. Available at http://www.on.ec.gc.ca/eem/intro-e.html

Borgmann A., Tuininga, K., Ali, N. and Audet, D. (2002) Environmental Effects Monitoring at twenty-three pulp and paper mills in Ontario – Synthesis of Cycle 2 study results and recommendations for Cycle 3, Environmental Protection Branch – Ontario Region, Environment Canada, Downsview, Ontario, 67 pp. Available at http://www.on.ec.gc.ca/eem/intro-e.html

Borgmann, A., Moody, M. and Scroggins, R. (2004) The Lab-to-Field (LTF) Rating Scheme: A New Method of Investigating the Relationships between Laboratory Sublethal Toxicity Tests and Field Measurements in Environmental Effects Monitoring Studies, *Journal of Human and Environmental Risk Assessment*, August 2004.

EC (Environment Canada) (1999) Guidance document for implementing and interpreting single-species tests in environmental toxicology, EPS 1/RM/34, Environment Canada, Environmental Technology Centre, Ottawa, ON.

EC (Environment Canada) (2000) Pulp and paper technical guidance for aquatic environmental effects monitoring, EEM/2000/2, Environment Canada, Ottawa, ON.

EC (Environment Canada) (2001) Metal Mining Guidance Document for Aquatic Environmental Effects Monitoring, Environment Canada Report EEM/2001/1.

ESG (Ecological Services Group) (2000) Environmental Effects Monitoring (EEM) Cycle 2 Final Report for Provincial Papers Inc., ESG International Inc., Guelph, Ontario.

Moody, M. (2002) Assessment of relationship between laboratory sublethal toxicity and field measurements through the review of Ontario Region Environmental Effects Monitoring (EEM) studies, Environment Branch, Saskatchewan Research Council, SRC Publication No. 11415-1E01/172.

Munkittrick, K. (2001) *Personal Communication*, Dept. of Biology, University of New Brunswick.

Parker, W.R. and Smith, N. (1999) A Synopsis of the first cycle of the pulp and paper mill Environmental Effects Monitoring program in the Atlantic Region, Environment Canada, Dartmouth, NS, Surveillance Report EPS-5-AR-99-3/51. Available at http://www.ec.gc.ca/eem/English/default.cfm

Rickwood, C.J., Dubé, M., Hewitt, M., MacLatchy, D.L. and Parrott, J.L. (2003) Assessing effects of pulp and paper mill effluent using a fathead minnow *(Pimephales promelas*) bioassay. Poster presentation at the 30th Aquatic Toxicity Workshop, Ottawa, Ontario.

Scroggins, R.P., Miller, J.A., Borgmann, A.I. and Sprague, J.B. (2002) Sublethal toxicity findings by the pulp and paper industry for cycles 1 and 2 of the Environmental Effects Monitoring program, *Water Quality Research Journal of Canada* **37**, 21-48.

Sprague, J.B. (1997) Review of methods for sublethal aquatic toxicity tests relevant to the Canadian metal-mining industry, Natural Resources Canada, Canada Centre for Mineral and Energy Technology, Aquatic Effects Technol. Eval. Progr. (AETE), Ottawa, ON, AETE Project 1.2.1, 102 pp.

U.S. EPA (U.S. Environmental Protection Agency) (1999) A review of single species toxicity tests: Are the tests reliable predictors of aquatic ecosystem community responses? EPA/600/R-97/114, U.S. EPA, Office of Research and Development, Washington D.C., 58 pp.

Abbreviations/acronyms

CF	Confounding Factors
EEM	Environmental Effects Monitoring
EPT	Ephemeroptera, Plecoptera and Trichoptera
Exp	Near-field exposure area or zone is within 1% effluent plume, close to the mill diffuser
FF	far -field
FFF	far-far-field
FHM	Fathead Minnow
HAS	Hazard Assessment Scheme
ICS	Invertebrate Community Survey
LTF	Lab-To-Field
N	number (*i.e.*, counted number of fish)
NF	near-field
P	plume delineation rating
PERE	Potential effluent related effects; probable mill-related effects
PMTC	Predicted Minimum Toxic Concentration
Ref	Reference area, usually upstream of mill diffuser
SI	Significant Interaction
TOC	Total Organic Carbon
ZPE	Zone of Potential Effect.

5. OVERVIEW OF TOXICITY REDUCTION AND IDENTIFICATION EVALUATIONS FOR USE WITH SMALL-SCALE TESTS

LESLEY J. NOVAK
& KEITH E. HOLTZE
Stantec Consulting Ltd.
11B Nicholas Beaver Road, Guelph
Ontario N1H 6H9, Canada
lnovak@stantec.com
kholtze@stantec.com

1. Objective and scope

The Toxicity Reduction Evaluation (TRE) approach developed by the United States (U.S.) Environmental Protection Agency (U.S. EPA) (1989; 1991a and b; 1993a and b) is a "site-specific study designed to identify the substances responsible for toxicity, isolate the source, evaluate the effectiveness of control options, and confirm toxicity reduction". Although the approach to any TRE may have similar components, the sequence of events or steps will be site-specific and depend on the nature of the toxicants, the test species of interest, regulatory requirements, as well as the results and findings from each phase of work. The information provided in this chapter is not intended to replace the existing U.S. EPA methods, but rather to provide an overview of existing approaches, and supplementary guidance specific for application with small-scale tests. Although the focus of this chapter applies to those investigations conducted using industrial effluent samples (*e.g.*, metal mining, pulp and paper, organic chemical), the approaches are flexible and can be used with different types of aquatic media, including municipal effluent, receiving water (surface water), groundwater, leachates, and sediment porewater. Methods discussed are relevant for both acute and chronic freshwater testing.

C. Blaise and J.-F. Férard (eds.), Small-scale Freshwater Toxicity Investigations, Vol. 2, 169-213.

2. Summary of TRE procedures

Table 1. Summary of Toxicity Reduction Evaluation (TRE) procedures for effluents.

Purpose
• A Toxicity Reduction Evaluation (TRE) is a "site-specific study designed to identify the substance(s) responsible for toxicity, isolate the source, evaluate the effectiveness of control options, and confirm the reduction in toxicity of the effluent" (U.S. EPA, 1989). It is an approach that combines laboratory testing, chemical analysis and on-site investigations to achieve compliance with toxicity based effluent limits.
TRE components
• Three fundamental TRE components are: Toxicity Identification Evaluations (TIEs), Source Investigations (SIs), and Toxicity Treatability Evaluations (TTEs). • The objective of the TIE is to characterize and identify the specific substances responsible for toxicity. The TIE process is divided into three phases. Usually each phase is completed sequentially, but they may be conducted simultaneously when patterns of toxicity begin to emerge during Phase I. Phase I involves characterization of the toxicants through a variety of effluent treatments (U.S. EPA, 1991a and b). After completion of the Phase I characterization of an effluent, the TRE can proceed to: i) TTE to evaluate various treatment methods for removal of the toxicant, ii) SI to identify the source of the toxicant, or iii) Phase II and III TIE to identify, and confirm the specific substance responsible for toxicity prior to conducting a TTE or SI. • TTEs and SIs can be conducted with or without identification of the specific toxicant(s), but will be more effective if a specific substance can be targeted for treatment. In the case that the TTE or SI approach is selected, confirmation testing will still be required to ensure that the method selected consistently removes toxicity. • Establishing the frequency (*i.e.*, toxicity is consistent or transient between samples), degree (*i.e.*, magnitude) and persistency (*i.e.*, how toxicity changes over time) of toxicity will be important, since these factors can provide insight into the type of substance responsible for toxicity, and can also influence subsequent TRE activities. • A successful TRE requires teams of individuals with a variety of expertise including, aquatic toxicologists, chemists, treatment and process engineers, and industry personnel. Effective communication among all TRE participants will increase TRE success and lead to complete transfer of information.

Table 1 (continued). Summary of Toxicity Reduction Evaluation (TRE) procedures for effluents.

Test organisms and methods
• Various test organisms and methods can be used. Examples of small-scale freshwater tests include: *Daphnia magna* acute lethality test (Environment Canada, 2000a); *Ceriodaphnia dubia* survival and reproduction test (Environment Canada, 1992a); *Vibrio fischeri (*Microtox® light inhibition test) (Environment Canada, 1992b); *Daphnia* IQ ®, Thamnotoxkit F®, Rotoxkit F®, and early-life stage rainbow trout (embryo, swim-up fry) (Environment Canada, 1998; Pollutech, 1996).
• In the case where a small-scale test is used in place of the species of interest (or regulatory test method), sufficient testing prior to, and during the TRE should be conducted to determine and confirm that the small-scale test (or surrogate species) responds to the untreated effluent in a similar manner as the species of interest under a variety of conditions.
Recommended reading for conducting TREs with effluent samples
• U.S. EPA 1989. Generalized methodology for conducting industrial toxicity reduction evaluations. EPA-600/2-88/070.
• U.S. EPA 1991a. Methods for aquatic toxicity identification evaluations: Phase I toxicity characterization procedures. EPA-600/6-91/003.
• U.S. EPA 1991b. Toxicity identification evaluation: characterization of chronically toxic effluents, Phase I. EPA-600/6-91/005.
• U.S. EPA 1993a. Methods for aquatic toxicity identification evaluations: Phase II toxicity identification procedures for samples exhibiting acute and chronic toxicity. EPA-600/R-92/080.
• U.S. EPA 1993b. Methods for aquatic toxicity identification evaluations: Phase III toxicity confirmation procedures for samples exhibiting acute and chronic toxicity. EPA-600/R-92/081.
• U.S. EPA 1999. Toxicity Reduction Evaluation Guidance for Municipal Wastewater Treatment Plants. EPA/833B-99/002.
• Norberg-King T, Ausley L, Burton D, Goodfellow W, Miller J. and Waller WT. 2005. Toxicity Identification Evaluations (TIEs) for effluents, ambient waters, and other aqueous media. Workshop on Toxicity Identification Evaluation (TIE): what works, what doesn't, and developments for effluents, ambient waters, and other aqueous media; 2001 Jun 23-28; Pensacola Beach FL. Pensacola FL, USA: Society of Environmental Toxicology and Chemistry (Norberg-King et al., 2005).

A Toxicity Reduction Evaluation (TRE) is a site-specific and systematic approach that combines laboratory testing, chemical analysis and on-site investigations to achieve compliance with toxicity based effluent limits. Three fundamental TRE components include: 1) Toxicity Identification Evaluations (TIEs), 2) Source Investigations (SIs), and 3) Toxicity Treatability Evaluations

(TTEs). Toxicity Identification Evaluations (TIEs) incorporate the responses of organisms into the assessment of complex effluent mixtures to characterize and identify the substance(s) responsible for toxicity. Source Investigations (SI) and Toxicity Treatability Evaluations (TTEs) may be used in combination with, or as alternatives to a TIE. Source Investigations determine whether the toxicants may be isolated in one or more waste streams that comprise the final effluent discharge. A TTE involves the systematic evaluation of various treatment technologies or management options (*i.e.*, process or operational changes) to assess the ability of these technologies (or operational/process changes) to reduce levels of contaminants that are causing toxicity (Novak et al., 2005). Although the initial approach to any TRE may have similar components, the sequence of events or steps taken will be site-specific and depend on the nature of the toxicant, as well as the results and findings from each phase of work.

Establishing the frequency (*i.e.*, is toxicity consistent or transient between sample), degree (*i.e.*, magnitude) and persistency (*i.e.*, how toxicity changes over time) of toxicity will be important, since these factors can provide insight into the type of substance responsible for toxicity, and can also influence subsequent TRE activities. The actual number of samples required to assess these factors will be site-specific and depend predominantly on effluent variability.

The U.S. EPA TRE methods were developed for use with small-scale freshwater tests using fathead minnows (*Pimephales promelas*) and *Ceriodaphnia dubia*. Examples of other small-scale freshwater tests which have been used in TRE studies include: *Daphnia magna* acute lethality test (Environment Canada, 2000a); *Ceriodaphnia dubia* survival and reproduction test (Environment Canada, 1992a); *Vibrio fischeri* (Microtox® light inhibition test) (Environment Canada, 1992b); *Daphnia* IQ ®, Thamnotoxkit F®, and early-life stage rainbow trout (embryo, swim-up fry) (Environment Canada, 1998; Pollutech, 1996). In the case where a small-scale test is used in place of the species of interest (or regulatory test method), sufficient testing prior to, and during the TRE should be conducted to determine and confirm that the small-scale test (or surrogate species) respond to the untreated effluent in a similar manner as the species of interest under a variety of conditions. Failure to adequately compare the surrogate and regulatory species could lead to incorrect conclusions regarding the substance that is responsible for toxicity.

3. Historical overview

One of the major benefits of aquatic toxicity tests is that they provide an integrative indicator of potential biological impact. Test organisms respond to all chemicals present in a sample, thereby providing a measure of the bioavailability and the true toxicity potential of its constituents (Environment Canada, 1999). In recognition of this, many governments and international agencies (*e.g.*, Environment Canada, American Society for Testing and Materials (ASTM), International Standards Organization (ISO), Organization for Economic Cooperation and Development (OECD), and the U.S. EPA) have developed and adopted standardized aquatic toxicity test methods for assessing individual chemicals and complex mixtures (*i.e.*, industrial and municipal effluent).

The availability of standardized tests has lead to their incorporation into regulations that set specific compliance limits for toxicity. For example, Canadian provincial and federal effluent discharge regulations exist for a variety of industrial sectors (*e.g.*, pulp and paper, metal mining, petrochemical, iron and steel, metal casting, inorganic and organic chemicals). These regulations often include, among other chemical parameters (*e.g.*, metals, pH, biochemical oxygen demand, total suspended solids, ammonia), indicators of aquatic toxicity, such as acute lethality to rainbow trout or *Daphnia magna*. Beginning in 2003, operating Canadian metal mines were required to conduct monthly rainbow trout and *Daphnia magna* acute lethality tests using full strength (100%) effluent. Once 12 consecutive passes (≤ 50% mortality in 100% effluent) with rainbow trout were obtained, testing frequency could be reduced to quarterly assessments of acute lethality (using both species). However, if a rainbow trout test produced mortality of more than 50% of the test organisms in 100% effluent, the sample is considered to "fail" the acute lethality test and investigations into the cause of toxicity are required.

With increasing use of aquatic toxicity tests in a regulatory framework, methods were required to determine the cause of toxicity in the event that an effect was observed or a toxicity limit was exceeded. Acknowledging the complexity of effluent matrices and limitations associated with using chemical analysis alone to determine the cause of toxicity (see Section 4), the U.S. EPA developed the toxicity based TRE methods, which were first published in 1989, and later revised in 1999. The methods provide a general structure to: i) evaluate the operation and performance of existing effluent treatment, ii) identify and correct treatment deficiencies contributing to effluent toxicity (*e.g.*, operation or process problems, chemical additives), iii) identify the substance(s) responsible for toxicity, iv) identify the source of the toxicants, and v) evaluate and implement toxicity reduction methods or technologies to control effluent toxicity (U.S. EPA, 1989; 1999). Other publications focused on laboratory methods for freshwater effluent with both acute and chronic tests (U.S. EPA, 1991a, b; 1993 a, b). Additional methods have been developed for marine species (U.S. EPA, 1996), and draft methods are available for sediments (U.S. EPA, 2003).

Since publication of the U.S. EPA TRE and TIE methods, numerous studies have been published on their application and use; however, many more remain unpublished as "grey literature" because they are often conducted in reaction to regulatory toxicity failures or exceedances of a toxicity limit. Significant advances have also been made in TRE and TIE methods and approaches. In recognition of this, the Society of Environmental Toxicology and Chemistry (SETAC) sponsored a workshop in June 2001 to advance the science of TREs. The proceedings from this workshop (Norberg-King et al., 2005) included development of a document that identifies various advances in TREs, provides TRE examples through more than 30 case studies, identifies research needs, and includes a comprehensive summary of available TRE literature. In recognition that the SETAC proceedings represent a thorough summation of recent advances in TREs, every effort was made to include key findings and recommendations in this chapter.

TRE methods have been applied and adapted throughout North America, Australia, Asia and Europe using various types of aquatic media, including municipal effluent, receiving water (surface water), groundwater, leachates, and sediment porewater (Norberg-King et al., 2005). However, by far the largest application has focused on effluent associated with industrial operations. TRE procedures have been successfully used to identify numerous classes of substances as contributors to toxicity, including total dissolved solids (Chapman et al., 2000; Goodfellow et al., 2000; Tietge et al., 1997), metals (Kszos et al., 2004; Van Sprang and Jassen, 2001; Bailey et al., 1999; Hockett and Mount, 1996; Wells et al., 1994; Schubauer-Berigan et al., 1993), pesticides (Bailey et al., 1996; Amato et al., 1992; Norberg-King et al., 1991), chlorine (Maltby et al., 2000; Burkhard and Jenson, 1993), ammonia (Bailey et al., 2001; EVS, 2001; Jin et al., 1999; Wenholz and Crunkilton, 1995) and a variety of non polar organics (Gustavson et al., 2000; Yang et al., 1999; Ankley and Burkhard, 1992; Lukasewycz and Durhan,1992; Jop et al., 1991).

4. Advantages of Toxicity Reduction Evaluations

Chemical analysis of complex mixtures (such as an industrial effluent) can provide some useful information on substances that might be responsible for toxicity. Using this process, chemical concentrations can be compared to available toxicity data in the literature. If the effluent concentration exceeds the toxicity value for the test species of interest, then it is possible that the substance could have caused the observed response. However, a chemical based approach alone is insufficient to determine the cause of toxicity, and development of the toxicity-based TRE approach is necessary for several reasons. First, it is not possible to analytically identify and quantify all possible contaminants in a complex effluent. Second, even if the chemicals of concern were measured, toxicity data may be unavailable for many of the substances, or if available, not applicable to the species of interest. Third, the potential synergistic, antagonistic or additive interactions that occur among chemicals are difficult to predict based on chemical concentrations alone. Finally, other properties of an effluent (*e.g.*, TOC, hardness, pH) can have an effect on the manner in which others chemicals exert their toxicity (often referred to as a matrix effect). The TRE process incorporates the responses of organisms into the assessment of complex effluent mixtures to determine the identity of the substance(s) responsible for toxicity, allows matrix effects and toxicant bioavailability to be quantified and allows for increased analytical precision and sensitivity by providing characteristics of the suspected toxicants. Without some knowledge of the toxicant characteristics, broad-spectrum analyses (*e.g.*, GC/MS, HPLC) alone are less effective and more expensive (Ankley et al., 1992; U.S. EPA, 1991a).

TREs provide a systematic approach to reduce toxicity in a variety of aqueous samples. The approach incorporates the responses of organisms into the assessment of complex effluent mixtures and allows for focused investigations on a particular class or group of substances responsible for toxicity, without which it would be near impossible to determine the specific cause(s) of toxicity. In cases where an effluent is comprised of multiple streams, TREs include procedures for identifying upstream sources of the toxicant, followed by implementation of controls (*i.e.*, treatment technologies or alteration of upstream management systems) that can translate into elimination of final effluent toxicity. A key advantage is that treatment of smaller, more concentrated streams can often be performed more efficiently and economically than treatment of larger, more dilute streams (*e.g.*, the final effluent) (U.S. EPA, 1989). Source investigations can also be a viable alternative to eliminate toxicity in cases where the identity of a specific toxicant is unknown, or when toxicity is transient or not persistent. Identification of the source of toxicity can also lead to opportunities to recycle concentrated or heavily contaminated streams back to the process for substance recovery, reagent savings, and reduced effluent toxicity. TRE guidance is also available for evaluation of various treatment technologies, combinations of technologies, or management options (*i.e.*, process or operational changes) to assess the ability of these technologies (or operational/process changes) to reduce elevated levels of contaminants that are causing toxicity.

A TRE study is not limited to existing operating facilities, but can be undertaken during the design and bench-scale stage of an Effluent Treatment Plant (ETP). ETPs are traditionally designed by engineers to meet specific chemical limits. Achieving compliance with toxicity limits has only recently been a consideration in the design stages. Assessments of effluent toxicity during bench, pilot and full-scale evaluations should be included in the selection of appropriate treatment technologies (Norberg-King et al., 2005; Novak et al., 2005). In cases where the proposed effluent treatment does not eliminate toxicity, the TIE process could be used to identify the substance(s) responsible. Once the toxicant is identified, the proposed treatment system could then be modified to ensure removal of the toxicant. Toxicity results should be assessed throughout the process since performance changes often occur when implementing any system from the bench or pilot to full-scale operation.

5. Overview of the TRE approach

The objective of the TRE is to determine the actions necessary to reduce effluent toxicity to acceptable levels (U.S. EPA, 1989). The U.S. EPA approach includes six tiers, which are outlined in Figure 1. Tier I, Data Acquisition and Facility-Specific Information, involves the collection and analysis of available information and data that might be useful in designing and directing the most cost-effective study. Tier II, Evaluation of Remedial Actions to Optimize Facility Operations, includes an evaluation of: i) housekeeping practices, ii) treatment plant optimization, and iii) chemical optimization. Tier III involves application of the Toxicity Identification Evaluation (TIE) procedure. The objective of the TIE is to identify the specific substances responsible for acute lethality. Tier IV, Sources Investigations (SI), and Tier V, Toxicity Treatability Evaluations (TTEs), may be used in combination with, or as alternatives to, a TIE. Source Investigations determine whether the toxicants may be isolated in one or more waste streams and can include chemical or toxicity based tracking. If the TTE approach is selected, then the performance of different effluent treatment technologies (to reduce or eliminate toxicity) is evaluated either at the bench-scale, or directly at the ETP. Tier VI, Follow-Up and Confirmation, involves implementing a monitoring program to confirm that the required regulatory or compliance toxicity limit has been achieved.

5.1 INITIATION OF A TRE

The first step in the TRE process should begin prior to experiencing the first toxic event, and involves the development of a Toxicity Prevention/Response Plan (Norberg-King et al., 2005; Novak et al., 2005). This plan is designed to increase the speed and efficiency with which toxicity failures can be addressed, by facilitating the data acquisition phase, and assisting with decision-making processes. The prevention/response plan should include (but is not limited to):

- description of processes/operations and effluent treatment facilities,
- line diagrams showing the major areas of operation and the main inputs to the ETP (noting that effluent retention times need to be considered and understood when attempting to identify cause and effect),
- documentation of facility operations during collection of samples for routine toxicity testing,
- characterization of effluent chemistry and toxicity over time to provide baseline data to be used for comparisons to samples collected during a toxicity episode,
- results from toxicity tests and chemical analysis for routinely monitored parameters (summarized in an electronic format for ease of retrieval and statistical analysis of data),
- history of compliance with other regulated chemical parameters that may influence toxicity,
- Material Safety Data Sheets (MSDS) for chemicals used in the process and effluent treatment (with available toxicity data for species of interest),
- development of a notification protocol (*i.e.*, procedures for identifying personnel who should be notified of the toxicity event),
- selection of a toxicity response team that is prepared to assist with TRE studies. This may include consultants (*i.e.*, aquatic toxicologists, chemists, engineers experienced in the TRE process, if not already available within the facility) and facility personnel (*i.e.*, management, operations, support personnel for sampling).

5.2 PRELIMINARY ASSESSMENT OF TOXICITY

Effective tools for a preliminary assessment of toxicity are summarized in this section (corresponding to Tiers I and II in Figure 1). The information is derived directly from guidance provided by U.S. EPA (1989 and 1991a, b) and Novak et al. (2002). Readers are directed to these documents for additional details.

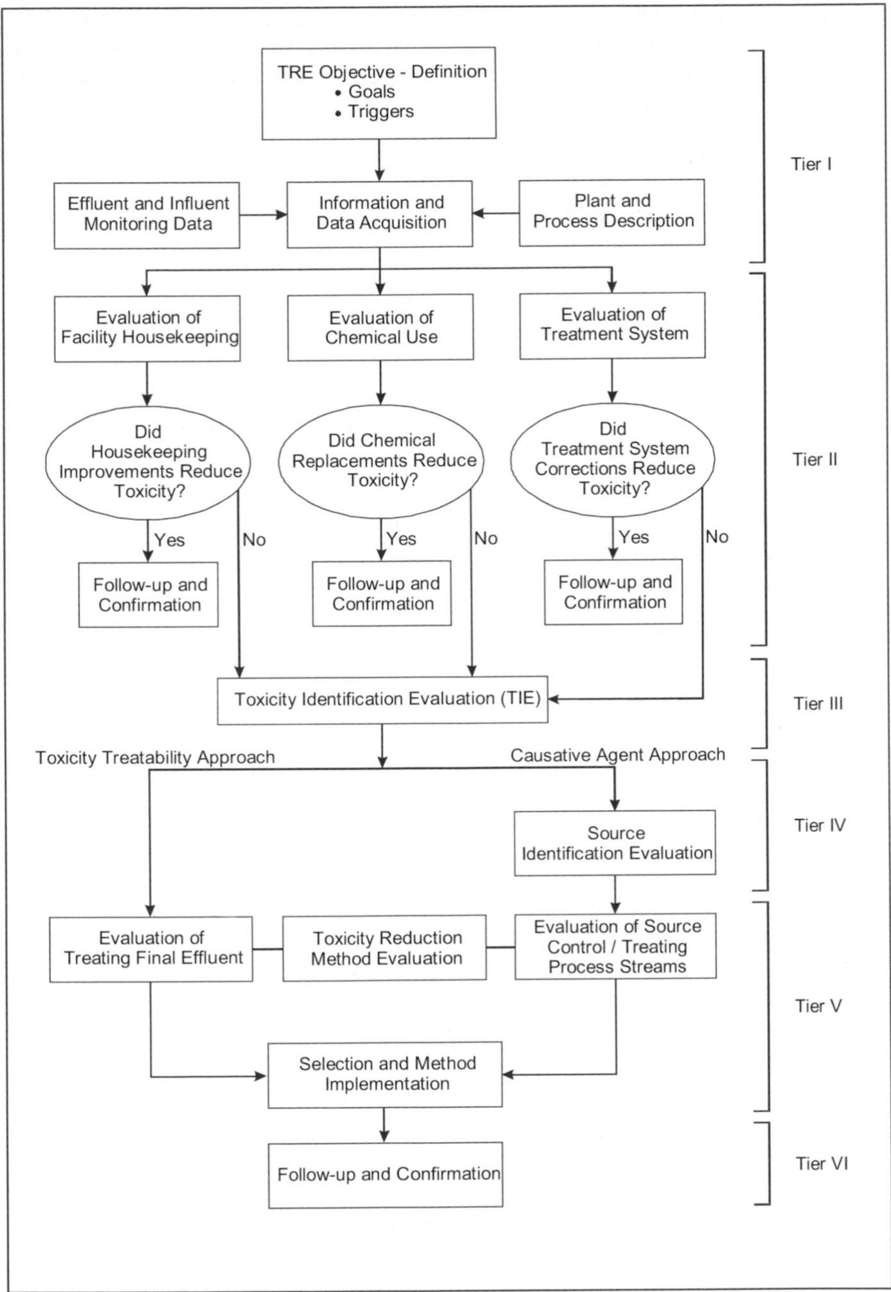

Figure 1. U.S. EPA (1989) Toxicity Reduction Evaluation (TRE) flow chart.

One of the initial TRE procedures includes the collection and review of available data and facility specific information, as well as an evaluation of remedial actions to optimize facility operation. The key components of this preliminary assessment include an evaluation of:

- Historical toxicity and chemistry data
- Facility and process information
- Effluent treatment plant operation
- Housekeeping procedures
- Chemical usage

Any actions taken as a result of the preliminary assessment phase can result in a reduction or elimination of toxicity, negating the need for further investigation. In addition, at this early stage of the TRE, management attention will often lead to subtle operational changes, which in turn, may result in a reduction or elimination of acute lethality without a clearly identified cause (Ausley et al., 1998).

5.2.1 Historical toxicity and chemistry data

A review of historical toxicity and chemistry data is conducted to obtain information on the potential cause(s) toxicity and to assist the investigator in determining an effective approach for the TRE (U.S. EPA, 1989; 1999). All data and reports should be critically reviewed to ensure their reliability (*i.e.*, all tests were conducted in accordance with the required test methods and associated QA/QC practices). This review ensures that data which triggered the investigation are valid. Water quality parameters measured during the toxicity tests can provide useful clues as to the cause(s) of toxicity. For example, conditions of low dissolved oxygen and extreme pH values can be lethal to test organisms either alone, or in combination with other substances. Changes in pH during a test can also alter chemical forms (*e.g.*, ammonia, metals), which can in turn alter their toxicity. If multiple species were evaluated, the test results should be compared, since species sensitivity comparisons can provide useful information about the possible cause(s) of toxicity (see Section 5.3.3). The data review may also suggest that accelerated testing (beyond that required by specific regulations) is necessary, so as to: i) characterize effluent variability, ii) evaluate the magnitude of toxicity in order to determine if potential future TIEs will involve an evaluation of the full strength (100%) sample, or require the use multiple concentrations, and iii) determine if toxicity changes over time within a single sample (*i.e.*, is/are the toxicant(s) persistent?).

5.2.2 Facility and process information

A review of facility and process information should be conducted in order to identify what is already known, provide information on the potential cause(s) and source(s) of toxicity, and help design a comprehensive TRE study (U.S. EPA, 1989; 1999). All TRE team members must have a clear understanding of how the facility/process is designed and operated. However, much of this information should already be available as part of the Toxicity Prevention/Response Plan (see Section 4.1). It is also useful to document facility and ETP operations during collection of toxicity samples. If a toxic sample is observed, this information could lead to a quick solution (*i.e.*, toxicity was a result of an upset, spill or atypical operation which can be identified and corrected), or to the establishment of a relationship between a particular process/operation and toxicity (Novak et al., 2002).

5.2.3 Effluent Treatment Plant operations

Information about effluent treatment plant operations must be reviewed (prior to conducting toxicant identification) to determine if the facility is functioning in an optimal manner with respect to its design parameters (U.S. EPA, 1989). Process or operational changes often occur over time and can alter the chemical composition of the effluent, resulting in a wastewater containing contaminants that were not present in the original effluent at the time of treatment plant design. A review of routinely monitored effluent treatment parameters (*i.e.*, pH, TSS, BOD, COD) should be conducted to determine if they were within predicted/expected operating ranges. These parameters may not be the cause of toxicity, but are used to identify upset or abnormal conditions within the treatment system or operation. Furthermore, most treatment systems are not designed to achieve a specific toxicity limit. In fact, we have observed instances where the final treated effluent was actually more toxic than the raw influent prior to treatment. Comparisons of influent and effluent toxicity may be useful in assessing the effectiveness of the treatment plant to reduce toxicity, while ensuring that toxicity is not being added during treatment (*i.e.*, from a treatment chemical). The frequency and duration of treatment plant by-passes (*e.g.*, during or after a rainfall) and shock-loads (*e.g.*, from spills, cleaning/maintenance) should be assessed and a thorough effort made to correlate these occurrences with toxicity. Changes in wastewater residence times or short-circuiting may also affect contaminant degradation or precipitation, with a resultant change in toxicity. Understanding the retention time of the system will help in selecting the frequency of testing required to detect effluent variability (U.S. EPA, 1989; 1999).

5.2.4 Housekeeping practices

A review of housekeeping practices is conducted in an attempt to optimize chemical usage, and reduce chemical losses that could contribute to the contaminant load and toxicity. Factors to be evaluated include; i) general facility cleanliness, ii) documentation of cleaning or maintenance activities, iii) spill prevention, control and containment, iv) waste handling, storage and disposal, and v) disposal strategies for domestic (laundry and shower) and sewage waste (U.S. EPA, 1989).

5.2.5 Chemical usage

A review of the use of process and treatment plant chemicals is required to identify those that have the potential to contribute to toxicity. For each chemical the following should be determined: i) availability of current MSDS and toxicity test data for species of interest, ii) purpose and volume used (volumes used are typically available from the supplier), iii) whether the amount can be reduced or reused, iv) whether less toxic alternatives are available, and v) if it is possible to avoid discharge of the chemical (U.S. EPA, 1989; 1999). Even a slight overdosing of effluent treatment chemicals (*e.g.*, polymers, chlorine) could result in potentially toxic concentrations in the final effluent, since these chemicals do not have the opportunity for degradation during normal plant operations. In cases where process or treatment chemicals (*e.g.*, polymers, water treatment agents, corrosion inhibitors, etc.) are suspected as possible causes of effluent toxicity, the use of chemical specific toxicity "finger-prints" may be of use when combined with TIE results. For example, the Phase I TIE treatments could be conducted on the process or treatment chemical to characterize its toxicity. Once the chemical "finger-print" is identified (*i.e.*, C18 at pH 3, and Amberlite XAD™ resin are effective at eliminating toxicity), these results can be compared to the "finger-print" of the effluent. If results were similar, this would be suggestive of the chemical as a possible contributor to toxicity. This approach may require detailed exchange of information with the chemical supplier and can often necessitate confidentiality or non-disclosure agreements, since the composition of many process and treatment chemicals are considered proprietary (Bailey et al., 2000). Where possible, it may also be useful to predict the concentration of each process or treatment chemical in the final discharge. The predicted environmental concentration (PEC) could be compared to the available toxicity data for each chemical. However, this approach will not take into consideration synergistic effects, binding of the chemicals to particulates or fibres in the effluent, or by-products resulting from transformation or breakdown of the chemicals during use and in the treatment system (Norberg-King et al, 2005; Novak et al., 2002).

5.3 TOXICITY IDENTIFICATION EVALUATIONS

If the aforementioned activities (Tiers I and II in Figure 1) were ineffective at resolving toxicity, the investigator will need to proceed to the next TRE tier – a Phase I Toxicity Identification Evaluation (TIE). The TIE process is divided into three phases, which usually occur sequentially, but may be conducted simultaneously when patterns of toxicity begin to emerge. Phase I involves characterization of the toxicants through a variety of effluent treatments. Phase II involves identification of the suspected toxicant(s), while confirmation of the suspected toxicants occurs in Phase III. This section provides an overview of a Phase I TIE for acutely toxic samples. Although approaches for chronic toxicity are similar, the tests methods are usually modified due to the additional level of effort required for longer-term chronic tests, and readers are directed to U.S. EPA (1991b; 1993a, b) methods for specific details.

5.3.1 Sample collection

TREs require a high degree of planning and co-ordination. Sample collection schedules must be coordinated with the test laboratory to ensure the availability of staff (*e.g.*, technical staff performing tests) and resources (*e.g.*, test organisms, bench-space). Prior to collection of a sample for a TIE study, the approximate volume required to complete the tests must be determined. The volumes required will vary depending on the test species, test type (*i.e.*, acute versus chronic), test conditions (*i.e.*, exposure volumes), degree or magnitude of toxicity (*i.e.*, single or multiple concentration tests) and the number of treatments to be conducted on a single sample (Novak et al., 2002). Particularly, the magnitude of toxicity will have an effect on the sample volume and level of effort required to conduct a TIE. If toxicity is only observed in the full strength (100%) effluent, the TIE will involve an evaluation of the undiluted effluent (single concentration tests). Multi-concentrations tests will be required if toxicity is observed in the diluted effluent concentrations. The single concentration tests require less effluent volume, are easier to complete (with respect to effluent manipulations), and are therefore less costly than the multi-concentration LC50 tests (Novak et al., 2002).

The selection of grab or composite samples will depend on the discharge situation, (*i.e.*, access to sampling locations), variability in effluent quality, questions to be answered by the TIE, and stage of the TIE. Initially, the type of sample collected should be similar to that used for the test(s) that "triggered" the TIE investigation. Composite samples may be more representative when effluent quality is variable and results are difficult to interpret. Grab samples may be preferable if toxicity is marginal, not persistent or intermittent.

To minimize changes in chemical composition of the effluent, samples should be delivered as quickly as possible to the laboratory and be kept cool (but not allowed to freeze) during transport. Special attention to sample delivery will be essential for facilities located in remote or isolated areas where delivery services may be limited. Unlike testing of samples for routine testing purposes, TIEs conducted on a single sample can take several weeks to complete. Therefore, sample toxicity (and chemical composition) must be measured periodically during storage to document any changes that may occur. The holding time and extent of analysis on any given sample must be weighed against the cost of additional sampling, persistence of toxicity, representativeness of the sample, and the need to test samples that represent the range of toxicity and toxicants occurring at the site (U.S. EPA, 1991a,b).

5.3.2 TIE Quality Assurance and Quality Control

Quality assurance (QA) and quality control (QC) measures are necessary in any toxicity test if reliable and accurate data are expected. Phase I TIE tests are not required to follow all aspects of a standardized test method, or necessarily require exacting quality control, because the data are only preliminary. Phase I, and to a lesser extent Phase II, are more tentative in nature compared to the confirmation tests conducted in Phase III (U.S. EPA, 1991a). Due to the large number of effluent manipulations and the time required to conduct the treatments and tests, the level of QA/QC effort is generally reduced during Phase I. This does not imply that a QA/QC program should not exist when conducting a Phase I TIE, but rather that the level of QA effort increases as the results become more definitive. Factors that will help ensure the generation of quality toxicity test data include careful documentation of all observations during testing, use of similar test conditions (*i.e.*, temperature, exposure volume, dilution water), adherence to exposure times and monitoring routines, use of organisms approximately the same age or size and use of reference toxicant tests.

In addition, all Phase I TIE tests should include system blanks or controls to detect toxic artifacts added during the effluent characterization manipulations. Common sources of toxicity artifacts include: i) excessive ionic strength resulting from addition of acid/base during pH adjustments, ii) contaminated reagents, acid/base solutions or air/nitrogen sources, iii) formation of toxic by-products by acids/bases, iv) inadequate mixing of test solutions and v) contaminants leached from filters, pH probes, solid phase extraction columns. The U.S. EPA (1991a) recommends the use of two types of controls to detect these potential sources of artifactual toxicity: "Toxicity Controls" and "Toxicity Blanks". Toxicity Controls involve comparison of the untreated and treated effluent sample, and are used to determine if the effluent manipulation was effective in reducing toxicity, and that it did not cause an unintended increase in mortality. Toxicity Blanks involve the performance of a Phase I test on dilution water to determine if toxicity is added by the effluent manipulation itself.

5.3.3 Test organsims

In theory any test species could be used for a TRE, however, the most common approach involves the use of the regulatory test organism that initially triggered the toxicity investigation. In the United States, the small-scale *Ceriodaphnia dubia* and fathead minnows are the most commonly used species because they are the main regulatory test organisms (Norberg-King et al., 2005). Examples of other small-scale freshwater tests include: *Daphnia magna* acute lethality test (Environment Canada, 2000a), *Ceriodaphnia dubia* survival and reproduction test (Environment Canada, 1992a); *Vibrio fischeri* (Microtox® light inhibition test) (Environment Canada, 1992b); *Daphnia* IQ ®, Thamnotoxkit F®, Rotoxkit F®, and early-life stage rainbow trout (embryo, swim-up fry) (Environment Canada, 1998; Pollutech, 1996).

In Canada, the main regulatory test organisms are rainbow trout and *Daphnia magna*. The existing TRE methods can easily be adapted for use with the small-scale *Daphnia magna* test. However, adaptation of methods for use with rainbow trout (or any other larger scale test) is generally not as straightforward and can require greater effort and expense since this test requires larger sample volumes. In cases where the effluent is toxic to a larger scale test, two options are available: 1) the use of a surrogate "small-scale" test species or 2) modification of the larger scale test method (*e.g.*, reduced exposure volumes). Both options may be acceptable in a Phase I TIE (and to a lesser extent Phase II). However, tests conducted in Phase III must avoid modifications to the test methods. Specifically, the conditions of the test that triggered the investigation should be followed, with particular attention to test conditions, replication, test organism quality, representativess of effluent sample tested, and analytical procedures (U.S. EPA, 1993b).

It is generally recommended that all tests be conducted with the species that originally triggered the TIE. Even if more expensive, the benefits of using the regulatory species often outweigh the use of surrogate tests because; i) correlations may not be strong with the surrogate species, ii) the surrogate test organism may not be sensitive to the same toxicants affecting the regulatory species, and iii) results can led to erroneous conclusions if results are not confirmed by testing with the target species. However, the use of a surrogate test species may be necessary for those TIE treatments limited by the ability to treat only small effluent volumes (*e.g.*, solid phase extraction with C18). A surrogate species requiring smaller test volumes may also be more practical for facilities located in remote areas where collection and shipment of large volumes of effluent could be difficult (Novak et al., 2002).

Sufficient testing prior to, and during the TIE should be conducted to determine and confirm that the target and surrogate species respond similarly to the untreated effluent under a variety of conditions. Considerable time and resources will be wasted if the surrogate and regulatory species are responding to different toxicants, or if the effluent is not toxic to the surrogate species to begin with. The U.S. EPA (1991a) Phase I guidance document provides additional details on approaches to demonstrate that the toxicant is the same for both species, including comparison of LC50s, comparison of Phase I results, and symptom comparisons for similar organisms.

The use of multiple test organisms that exhibit different sensitivities to known contaminants (*e.g.*, ammonia, copper) can provide important clues on the cause of toxicity. For example, ammonia is more toxic to rainbow trout than to either *Daphnia magna* or fathead minnows. Based on this information (combined with measured chemical concentrations and calculations of un-ionized values), ammonia could generally be eliminated as a possible cause of toxicity in cases where the effluent is acutely lethal to *Daphnia magna*, but non-lethal to fish. In comparison, *Daphnia magna* are more sensitive than rainbow trout to total dissolved solids (TDS). For example, sodium LC50 values for rainbow trout and *Daphnia magna* have been estimated at 6.4 g/L and 2.3 g/L, respectively (unpublished data from internal laboratory reference toxicant tests). In our experience with effluents where acute lethality was attributed to elevated TDS, conductivities greater than 6000 μS/cm were shown to have the potential to cause *Daphnia magna* mortality, yet were relatively harmless to rainbow trout. This information on its own cannot be used to conclude elevated TDS as a cause of acute lethality, particularly since the acute lethality of freshwater with high TDS is dependent on the specific ion composition (Mount et al., 1997). However, when combined with supporting TIE investigations, the differences in trout and *Daphnia magna* sensitivity can be a powerful component of the "weight of evidence" implicating TDS as the cause of toxicity (Novak et al., 2005).

5.3.4 Phase I TIE

A Phase I TIE involves a series of physical and chemical manipulations (treatments) of an effluent sample, which are designed to classify or characterize the type of substance responsible for toxicity (*e.g.*, metal, non-polar organic, volatile substance). Toxicity of the untreated effluent (as received by the testing laboratory) is compared to the treated effluent following each chemical or physical manipulation. The relative degree to which the manipulations result in an improvement in toxicity provides an indication of the types of contaminants that may be involved. An example Phase I TIE strategy is present in Figure 2, and is based on U.S. EPA (1991a) acute methods, but also includes some additional treatments proposed by Novak et al. (2002) and Norberg-King et al. (2005). A description of each Phase I TIE treatment is provided in Table 2.

At the start of the Phase I TIE (Day 1), the untreated sample is initially tested for routine water quality parameters required by the test method (*e.g.*, pH, dissolved oxygen, conductivity) and toxicity (Initial Test). Additional "baseline" tests using the untreated effluent are conducted at the start of the TIE sample manipulations and on each day that tests on any treated sample are initiated. These "baseline" tests are conducted in order to monitor changes in toxicity of the untreated effluent over time. Depending on what is already known about the toxicant(s), limited chemical analysis (*e.g.*, metals scan) can also be conducted when the sample arrives at the laboratory.

On Day 2, sub-samples of the effluent are adjusted to pH 3 and 11 (pH 9 for the C18 treatment), and then filtered, aerated or passed through a C18 column. After the manipulations are done, the samples are adjusted back to the initial pH of the effluent (pH *i*) and tested for toxicity. The Ethylenediaminetetraacetate (EDTA), sodium thiosulfate and graduated pH adjustments are also conducted on Day 2, but testing is delayed until Day 3. The delay allows EDTA to complex with metals, thiosulfate to react with oxidative substances and metals, and pH to stabilize in the graduated pH test.

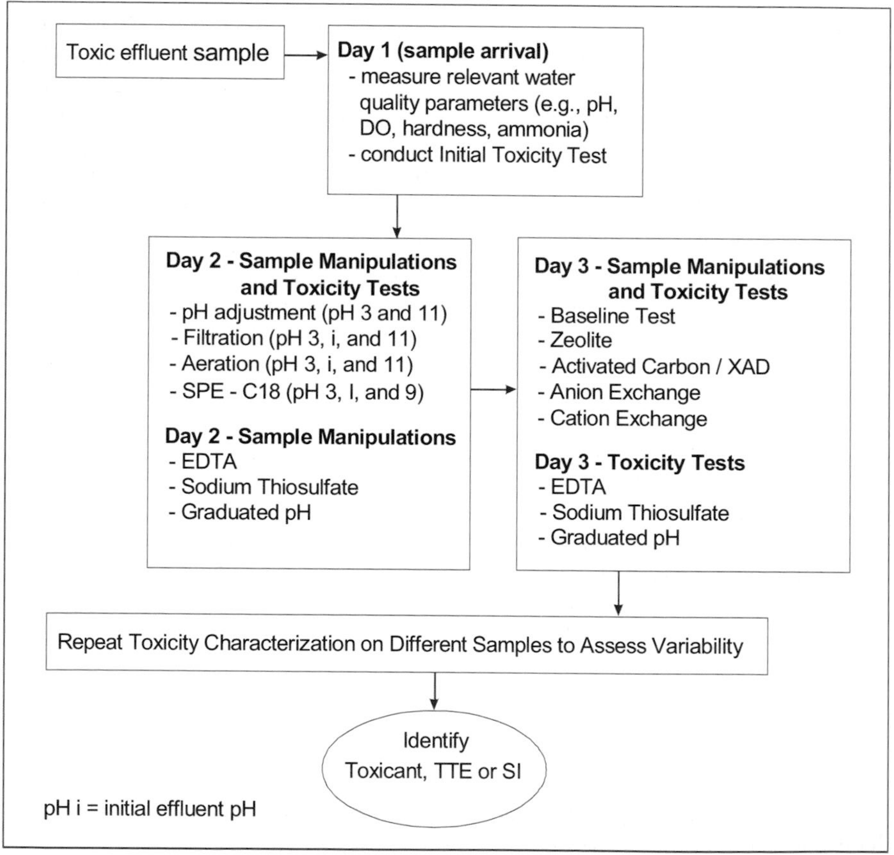

Figure 2. Phase 1 Toxicity Identification Evaluation (TIE) strategy (based on U.S. EPA, 1991a).

On Day 3, the manipulations conducted on Day 2 are tested. Treatments with zeolite, activated carbon, Amberlite XAD™, anion and cation exchange resins are conducted and the toxicity tests initiated.

Additional sample manipulations and tools that may be considered for inclusion in the Phase I TIE include:

- air-stripping and head-space CO_2 to characterize ammonia toxicity,
- addition of SO_2 to characterize oxidants,
- addition of piperonyl butoxide (PBO) to characterize organophosphate (OP) pesticide toxicity,
- use of enzyme linked immunosorbent assays (ELISA) and antibody techniques to characterize OP pesticide toxicity,
- use of semi-permeable membrane devices (SPMDs) to characterize surfactant toxicity,
- use of the Freshwater Salinity Toxicity Relationship (FSTR) model (Tietge et al., 1994) to predict the acute lethality of seven common ions (Na, K, Ca, Mg, Cl, SO_4 and HCO_3) using three freshwater organisms, *Ceriodaphnia dubia, Daphnia magna* and fathead minnows,
- use of Biotic Ligand Model (BLM) (DiToro et al., 2000) and chemical equilibrium modelling to predict speciation and bioavailability of metals.

Although it is beyond the scope of this chapter to describe these new methods in detail, we considered it important to demonstrate that significant advances in TIE procedures have occurred since the publication of the original methods. Readers are directed to Norberg-King et al. (2005) and Novak et al., (2002) for additional information on these new approaches.

The timing and sequence of treatments can be altered based on test requirements and persistency of the toxicant(s). However, it is recommended that the full battery of treatments be used. There is often a tendency to alter or modify the Phase I TIE treatment based on preconceived theories about the specific substance(s) responsible for toxicity. However, it is also possible that classes of compounds contributing to toxicity may be missed if the Phase I process is modified. Abbreviated Phase I TIEs may result in the loss of valuable and necessary information about the characteristics of the substances responsible for toxicity, which may lead to inconclusive results or erroneous conclusions, and can in fact delay resolution of the toxicity event (Norberg-King et al., 2005).

Following completion of the Phase I TIE the toxicant will have been thoroughly characterized. However, it is ill advised to assume that results from a single sample provide sufficient evidence on the cause of toxicity. To assess effluent variability and ensure that all toxicants have been accounted for, the Phase I characterization should be repeated on different samples (*e.g.,* samples collected on different days, during different operating cycles, etc.) (U.S. EPA, 1991a).

Table 2. Summary of Phase I TIE procedures (U.S. EPA, 1991a; Novak et al., 2002).

Test or treatment	***Description***
Initial test	Untreated effluent test started when sample is received by test laboratory.
Baseline test	Each day a sample manipulation is performed, an untreated effluent test (or baseline test) is set. The results from each manipulation are compared to the untreated effluent to assess the effectiveness of the manipulation on reducing toxicity.
Adjustment of pH	Adjustment of pH (to 3 and 11) provides additional information on the nature of the toxicants, and provides blanks for subsequent pH adjustment tests performed in combination with other treatments (*i.e.*, filtration, aeration). Samples are adjusted to pH 3 and 11, and then subjected to filtration, aeration, or solid phase extraction with a C18 column. The treated samples (including the pH adjusted samples without additional treatment) are re-adjusted to the initial pH of the effluent (pH *I*) prior to testing.
PH adjustment / filtration	The pH adjustment/filtration test evaluates the effect of pH change and filtration on the toxicity of substances associated with filterable material, focusing on irreversible chemical reactions. Effluent samples, at pH 3, 11 and *I* are filtered (using positive pressure) through a glass fibre filter (1.0 μ pore size). The pH of each filtered sample is re-adjusted to pH *I* prior to testing.
PH adjustment / aeration	The pH adjustment/aeration test evaluates the effect of pH change and aeration on the toxicity of the sample that may be due to volatile, sublatable or oxidizable substances. Effluent samples, at pH 3, 11 and *I* are placed in graduated cylinders and vigorously aerated for a standard time interval. The pH of each aerated sample is re-adjusted to pH *I* prior to testing.
PH adjustment/C18 Solid Phase Extraction	The pH adjustment/C18 Solid Phase Extraction (SPE) test evaluates the extent to which toxicity may be due to relatively non-polar organics and certain metals. Filtered effluent samples, at pH 3, 9 and *I* are passed through prepared C18 columns. The pH of each sample is re-adjusted to pH *I* prior to testing.
Oxidant Reduction test	The Oxidant Reduction test evaluates the extent to which oxidative substances (*e.g.*, chlorine, iodine, bromine) and some cationic metals (*e.g.*, Cd, Cu, Ag, Hg) can be made less toxic or non-toxic by the addition of sodium thiosulfate. Sodium thiosulfate is typically added as a gradient of concentrations (based on its toxicity to the species of interest) to a single effluent concentration.

Table 2 (continued). Summary of Phase I TIE procedures (U.S. EPA, 1991a; Novak et al., 2002).

Test or treatment	***Description***
EDTA chelation	The EDTA chelation test evaluates the extent to which cationic metals (*e.g.*, Al, Ba, Cd, Co, Cu) can be made less toxic or non-toxic by the addition of EDTA (Ethylenediaminetetraacetate). A cationic metal may be suspected as the cause of toxicity if both EDTA and sodium thiosulfate reduce toxicity. EDTA is typically added as a gradient of concentrations (based on its toxicity to the species of interest) to a single effluent concentration.
Graduated pH test	The graduated pH test evaluates the effect of pH on the toxicity of a variety of contaminants. Effluent samples are adjusted to three different pH values (*e.g.*, pH 6, 7 and 8), without readjustment to pH *i*. The pH values selected will be based on the specific characteristics of the effluent under investigation.
Activated carbon	Activated carbon will remove a broad-spectrum of chemicals (*e.g.*, organics, metal) from solution. Although non-selectivity of activated carbon is a limitation (and toxicants cannot typically be recovered) it can be useful in cases where toxicity is not removed by any of the other Phase I treatments, or when combined with chemical analysis before and after treatment. Samples are passed through a column (or mixed as a slurry) containing carbon and then tested for toxicity.
Amberlite XAD™ resin	Amberlite XAD™ resin will remove a broad range of relatively lower molecular weight organic contaminants. Unlike carbon, toxicants can often be recovered from XAD™ resin using methanol or other solvents. Samples are passed through a column (or mixed as a slurry) containing the resin, and the pH re-adjusted to pH *i* prior to testing.
Anion and cation exchange resins	Ion exchange resins can be classified as cation exchange resins, which have positively charged mobile ions available for exchange, and anion exchange resins whose ions are negatively charged. Samples are passed through a column (or mixed as a slurry) containing the resin, and the pH re-adjusted to pH *i* prior to testing. Particular attention to the use of blanks is required due to the potential for extreme changes in pH and osmotic strength following treatment using these resins. A variety of resin types used in TREs are reported in Norberg-King et al. (2005).

Table 2 (continued). Summary of Phase I TIE procedures (U.S. EPA, 1991a; Novak et al., 2002).

Test or treatment	***Description***
Zeolite	Zeolites are crystalline aluminosilicates, which exhibit high selectivity for ammonia, but can also remove some heavy metals (or add small amounts of calcium to a sample). Effluent samples are passed through a column containing zeolite and then tested for toxicity.

5.3.5 Interpretation of Phase I TIE results

Interpretation of Phase I results is a critical part of TIE and should be conducted by experienced investigators. Statistical comparison between the untreated and manipulated samples should not be the only tool used to evaluate TIE results, "judgement and experience in toxicology must be allowed to guide the interpretation of the results" (Norberg-King et al., 2005) and direct subsequent investigations. The U.S. EPA (1991a) suggests the following general approach for interpreting Phase I results. If multiple toxicants are present, focus on identification of one toxicant (once this toxicant is identified, it should be easier to identify the other). Also concentrate on those manipulations observed to have the most dramatic effect on toxicity (*e.g.*, those treatments that eliminated mortality or growth effects). Lastly, focus on those treatments that remove the toxicant from other effluent constituents (*e.g.*, solid phase extraction with C18 where the toxicant could be recovered from the resin).

Even with an experienced TIE researcher, a number of factors can complicate interpretation of results including lack of good pH control, increased TDS during pH adjustment (caused by the addition of HCl or NaOH), the presence of multiple toxicants, variable effluent quality, marginal toxicity and seasonal effects. The potential for matrix effects must also be considered when interpreting Phase I TIE results. Matrix effects occur when toxicants interact with other effluent constituents in ways that change their toxicity. As described by the U.S. EPA (1993b), matrix effects can fit into one of two categories. The first involves toxicants that undergo a change in form, such that they exhibit a different toxicity. For example, as pH increases the toxic fraction of ammonia (un-ionized ammonia) increases. Other examples include cyanide and hydrogen sulphide, which increase in toxicity as pH decreases. The second involves substances that undergo a physical change (*i.e.*, binding to particulates) making them biologically unavailable to the organism. For example, a particulate bound toxicant may be unavailable to fish, but readily available to cladocerans as the particulates are ingested via filter feeding. Again, experience of the investigator as well as detailed knowledge of sample chemistry will be vital in recognizing matrix effects.

Two examples of the interpretation of Phase I TIE results are provided. They represent only the very simplest of tools, and are in no way intended as definitive diagnostic characterizations (U.S. EPA, 1991). Examples for other toxicants can be found in the U.S. EPA Phase I manual. A cationic metal may be suspected as the cause of toxicity if:

i) toxicity is removed or reduced by EDTA addition,
ii) toxicity is removed or reduced by sodium thiosulfate addition (note that some metals may not be removed by thiosulfate, but by EDTA, and vice versa),
iii) toxicity is removed or reduced by solid phase extraction with a C18 column,
iv) toxicity is removed or reduced by filtration (under alkaline conditions),
v) erratic (non-linear) dose response is observed.

Ammonia may be implicated as the substance responsible for fish (*e.g.*, fathead minnow or rainbow trout) mortality if:

i) the un-ionized concentration at the start or end of the test is > 0.2 mg/L (in tests with rainbow trout),
ii) the sample is more toxic to fish than cladocerans (*e.g.*, *Daphnia magna* or *Ceriodaphnia dubia*),
iii) toxicity is removed after treatment with zeolite, with a corresponding decrease in total ammonia concentrations,
iv) toxicity increases as pH increases (or conversely, toxicity decreases as pH decreases),
v) toxicity is removed after extended air-stripping at high pH (*e.g.*, > pH 11).

Multiple substances may be responsible for toxicity if: i) no single Phase I manipulation eliminates toxicity, but several cause a reduction, or ii) different treatments reduce or eliminate toxicity to different species. If multiple toxicants are suspected, combinations of the effective manipulations could be conducted in sequence on a single sample. If toxicity is eliminated in the combined manipulations (compared to an individual manipulation), then it is likely that multiple substances are responsible for toxicity. If the results are similar, then it is probable that all of the manipulations were successful in reducing the same substance. For example, if multiple toxicants are present and aeration and EDTA both removed some toxicity, the addition of EDTA to the post-aerated sample may suggest that metals contributed to toxicity (U.S. EPA, 1991a).

Detection of hidden toxicants (those that do not express their toxicity because of the presence of a second toxicant) can be one of the most difficult aspects of TIE testing and can be difficult to identify when ammonia is the main toxicant (U.S. EPA, 1993b). Ammonia toxicity is attributable to the free or un-ionized (NH_3, N) form as opposed to the ionized (NH_4^+, N) species (Thurston et al., 1981). The relative concentration of un-ionized ammonia increases proportionately with pH and water temperature. Although toxicity due to ammonia can be observed in a variety of effluents, it is commonly observed in effluent associated with metal mining and municipal discharges (Novak et al., 2002; U.S. EPA, 1999). Because of its ability to mask the presence of other toxicants, it may be more effective to address toxicity due to ammonia before proceeding with a full Phase I TIE. The approach would include the use of multiple species with differing sensitivities to ammonia (*e.g.*, fish versus cladocerans), and include the following effluent treatments: lowered ambient pH, air-stripping at pH 11 and treatment with zeolite. Measured pH and ammonia concentrations (before and after treatment) are also required in order to track changes in un-ionized ammonia concentrations. If all tests (combined with measured ammonia concentrations and calculated un-ionized ammonia) consistently indicated toxicity due to ammonia, then a sufficient "weight of evidence" has been provided attributing toxicity to ammonia (Norberg-King et al., 2005).

Linking toxicity and chemical data can be useful in the early stages of a TRE, but may be most valuable when the TIE treatments implicate a particular substance(s) as the cause of toxicity. A common approach is the use of Toxic Units (TUs). Lethal TUs express the degree of effluent toxicity, or the concentration of substance, as a fraction of the LC50. Similarly, sublethal TUs are expressed as a fraction of the IC25 or IC50 (Environment Canada, 1999). TUs are dimensionless, and allow for normalization of LC50 data. The TU can be used to predict toxicity of an effluent based on the measured toxicant concentrations.

The lethal TU for an effluent is obtained by dividing 100% by the LC50 (or to obtain a sublethal TU, 100% is divided by the IC25 or IC50). A TU equal to 1 indicates a "marginally" lethal sample (*i.e.*, LC50 = 100%). If the total TUs > 1, then the effluent is expected to be lethal to the test organism. The suspect toxicant concentration is converted to a TU by dividing the measured toxicant concentration by the LC50 (or IC25or IC50) for that toxicant. If more than one toxicant is present, the concentration of each one is divided by the respective LC50, and the TUs can then be summed (Environment Canada, 1999). The total TUs for the individual contaminants are compared to the actual TUs for the effluent sample. If the TUs are equal then it is likely that all toxicity has been accounted for.

There are several limitations associated with the TU approach. First, the prediction of mixture toxicity is based on the assumption that the effects of the individual contaminants are additive (Environment Canada, 1999). The potential for synergistic or antagonistic effects is not taken into consideration. Second, toxicity data for the species of interest may not be available for those substances suspected to be responsible for toxicity. Databases, such as the US EPA Ecotox Database System (www.epa.gov/ecotox), are useful sources for obtaining toxicity data for a variety of substances. However, it is often not possible to find relevant toxicity data for single chemical tests conducted using water quality conditions that mimic the sample being investigated. In this case, attempts should be made to use relatively conservative values. Alternatively, the better choice may be to conduct additional tests to generate toxicity data for the substance of interest under water quality conditions (*i.e.*, pH, hardness, TOC, TSS) that mimic the effluent.

5.4 OPTIONS FOLLOWING COMPLETION OF PHASE I TIE

TRE options available following completion of a Phase I TIE are presented in Figure 3. After completion of the Phase I characterization of an sample, the TRE can proceed to: i) a TTE to evaluate various treatment methods for removal of the toxicant, ii) an SI to identify the source of the toxicant, or iii) a Phase II and III TIE to identify and confirm the specific substance responsible for acute lethality prior to conducting a TTE or SI. There will be more uncertainty associated with TTE or SI studies based on toxicant characteristics alone, rather than the known identity of the substance(s) responsible for acute lethality (U.S. EPA, 1989). However, TTEs or SIs are often valuable when toxicant identification is not possible. Regardless of the approach, confirmation testing will be required to ensure that effluent variability (*e.g.*, resulting from production/process schedules, weather conditions) is taken into consideration, and verify that the toxicity is consistently removed.

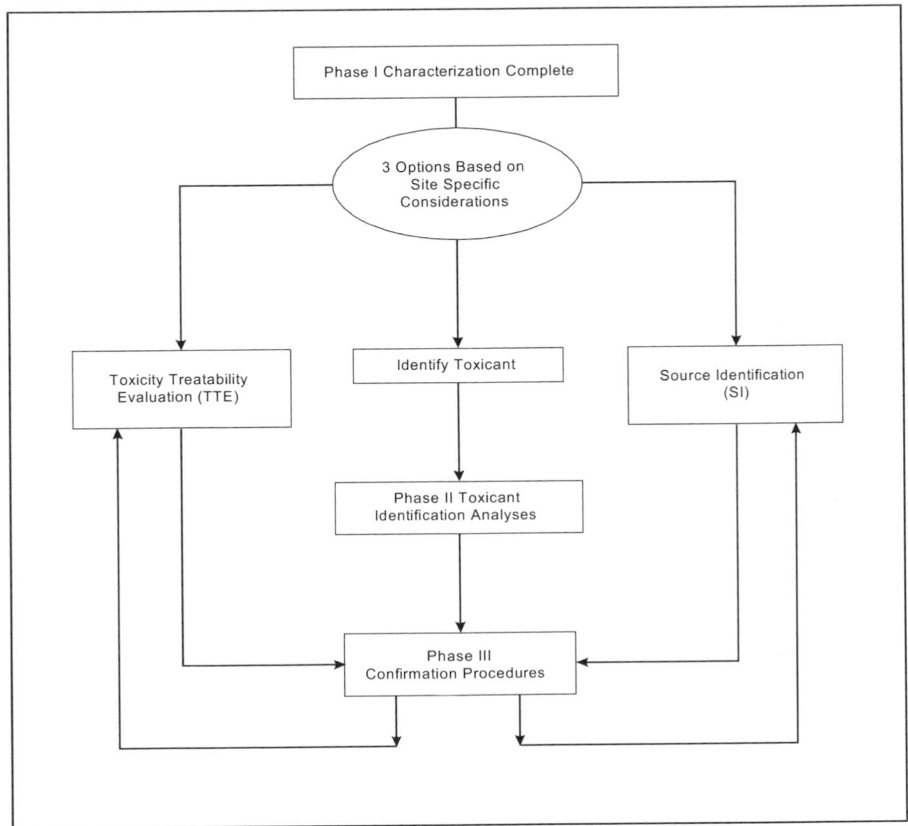

Figure 3. Options following completion of Phase 1 TIE (U.S. EPA, 1999; 1991a).

5.4.1 Phase II and III Toxicity Identification Evaluations

An overview of Phase II and III TIE procedures outlined by the U.S. EPA (1993a, b) are provided in Tables 3 and 4, and were previously presented in Novak et al. (2002). The treatments, procedures and analytical methods selected for a Phase II and III TIE are directly related to those treatments observed to effectively eliminate or reduce toxicity during Phase I. Consequently, the specific approach can only be determined after Phase I is complete. As Phases II and III proceed, QA requirements must be revisited and modified as required. If modified test methods or surrogate test species were used during Phase I, increased standardization should be used during Phase II, and predominantly in Phase III, to confirm that the suspected substance is responsible for the toxicity observed in the test that originally triggered the TIE (U.S. EPA, 1993a, b).

In Phase II further effluent treatments are conducted to identify the specific substance(s) responsible for toxicity. Toxicity tests are combined with chemical analysis to obtain a quantitative measurement of the suspected toxicants. The objectives in Phase III are to confirm that the substances responsible for toxicity have been correctly identified, and ensure that all of the toxicity has been accounted for. A "weight-of-evidence" approach is used to confirm that the substances responsible for toxicity have been identified. In both Phase II and III there are many possible approaches to identifying and confirming the substance(s) responsible for toxicity.

Table 3. Summary of U.S. EPA (1993a) Phase II TIE approach.

Objective	• Identify the specific substance(s) responsible for toxicity.
Preliminary testing	• Methods are directly related to specific treatments that eliminated or reduced toxicity during Phase I characterization.
Quality Assurance and Quality Control	• Increased QA/QC required as Phase II proceeds. • Standardized test methods should be applied to confirm that the suspected substance is responsible for the toxicity observed in the test that originally triggered the TIE.
Test species	• Recommend that all tests be conducted with the species that originally triggered the TIE. • If an alternative species is selected, tests must demonstrate that the original and alternative species are responding to the same toxicant.
Toxicity test procedures	• Sample toxicity must be tracked to assess if the substance responsible has degraded over time. • Tighter concentration intervals may be required in order to detect smaller incremental changes in toxicity.
Effluent sampling and handling	• Type of sample (*i.e.*, grab versus composite) used during Phase II should be the same as that used during Phase I. • One composite or one grab sample should be subjected to both Phases II and III. • Once the substance responsible is identified, multiple samples may be analysed for the presence of the toxicant.

Table 3 (continued). Summary of U.S. EPA (1993a) Phase II TIE approach.

Treatments	• Treatments used are specific for each suspected toxicant. • Non-polar organics – separation of toxic and non-toxic fractions; subsequent analyses using analytical methods, *e.g.*, High Pressure Liquid Chromatography (HPLC), Gas Chromatography (GC), Gel Permeation Chromatography (GPC). • Ammonia – measurement of ammonia and pH in effluent (combined with calculation of un-ionized ammonia concentrations); graduated pH testing; zeolite resin to remove ammonia; air-stripping of ammonia from the effluent at high pH. • Metals – measurement of metals in effluent; treatment with EDTA and sodium thiosulfate; graduated pH and ion exchange tests; use of BLM and metal speciation models. • Chlorine – measurement of total residual chlorine (TRC) in effluent; treatment with sodium thiosulfate. • Filterable toxicants – use of other filter types (*i.e.*, nylon, Teflon) or pore sizes; centrifugation; extraction and concentration of filtered material.
Interpretation of test results	• Specific for each suspected toxicant. • Interpretation is often different from the "standard" acceptance or rejection of a hypothesis; judgement and experience of investigator will be critical to interpretation of results. • The presence of multiple toxicants complicates interpretation. • Focus on the toxicant that appears easiest to identify. • Effects of effluent toxicants are not always additives; at least one toxicant must be identified before additivity can be established.

Table 4. Overview of U.S. EPA (1993b) Phase III confirmation approach.

Correlation approach	• Objective is to determine if there is a consistent relationship between the concentration of the suspected toxicant(s) and effluent toxicity. • A wide range of toxicity responses with several samples must be obtained in order to provide an adequate range of effect concentrations for analysis. • Two key problems associated with the correlation approach are; 1) lack of additivity requires careful data analysis, and 2) analysis is difficult when matrix effects are present.
Symptom approach	• Approach involves the use of organism behaviour and time to death in comparing the responses of organisms to the effluent and then to the suspected toxicant(s). • Different toxicants could produce similar or different symptoms in a test species; if symptoms are different, the toxicants are unquestionably different, but similar symptoms could indicate the toxicants are the same or different. • If organisms exposed to the effluent and the suspected toxicant display different symptoms, the substance thought to be responsible for toxicity is either not the actual toxicant, or other toxicants are present.
Species sensitivity approach	• If the suspected toxicant(s) has been correctly identified, effluent samples with different LC50, IC25 or IC50s for one species should have the same ratio for a second species with different sensitivity. • When two or more species exhibit different sensitivities to the suspected toxicant during single chemical testing, and the same pattern is observed in the whole effluent, this provides supporting evidence that the chemical tested is the cause of effluent toxicity.
Spiking approach	• In spiking tests, the concentration of the suspected toxicant(s) can be increased in the sample to determine if toxicity increases proportionally to an increase in chemical concentration. • The suspected toxicant could also be added to a non-toxic sample, to dilution water or to a sample of effluent where the suspected toxicant(s) has been removed. • Matrix effects, toxicant solubility, and equilibrium time could all impact upon the outcome of these experiments.

Table 4 (continued). Overview of U.S. EPA (1993b) Phase III confirmation approach.

Mass balance approach	• The mass balance approach is used when the toxicant(s) can be effectively removed from the effluent and subsequently recovered (*e.g.*, C18 Solid Phase Extraction).
Deletion approach	• The deletion approach involves removal of the suspected toxicant(s) from a waste stream. The waste stream containing the suspected toxicant is removed for a short period and the effluent is tested; this approach offers the strongest evidence that the suspected toxicants identified are the correct ones.
Additional approaches	• Manipulation of water quality conditions (*e.g.*, pH, hardness); measurements of body uptake, to assess bioavailability; combined Phase I characterizations; effluent simulations.

5.4.2 Toxicity Treatability Evaluations

A Toxicity Treatability Evaluation (TTE) involves the systematic evaluation of various treatment technologies or management options (*e.g.*, process or operational changes) to reduce or eliminate the substance responsible for toxicity. TTEs can be conducted without identification of the specific toxicant, but will be more effective if a specific substance can be targeted for treatment. Treatment technology selection based on the Phase I TIE results can also help focus the TTE and increase the likelihood of success. However, prevention or management strategies to eliminate toxicity (*e.g.*, source reduction by waste minimization/control or chemical optimization) will generally be more cost-effective than technology based treatment (Novak et al., 2002).

Criteria for the selection of the preferred treatment technology or management option should be defined at the beginning of the TTE, and may include; performance (*i.e.*, ability to consistently reduce toxicity), cost, complexity (*i.e.*, ease of installation/implementation and routine operation), service life and flexibility. Once a viable option has been selected and implemented, a follow-up monitoring program must be established to confirm its effectiveness, which could include more frequent monitoring (but will be site specific) (U.S. EPA, 1999).

Toxicity testing will be used throughout the TTE to gauge the effectiveness of the treatment options in reducing toxicity. Often, it is an engineering group that performs the bench- or pilot-scale treatments, with sub-samples of the treated effluent provided to an analytical and toxicology laboratory for testing. Therefore, key to the success of any study will be co-ordination among the TTE participants (*e.g.*, toxicologist, chemists and engineering groups). For example, it will be important to ensure sufficient sample volumes are treated (particularly at the bench-scale) to allow for both toxicity testing and chemical analysis of the treated effluent. Modified toxicity test methods may be used during an initial screening assessment of a large number of treatment options. However, as the number of treatment options is narrowed, all tests should follow the methods that initially triggered the TRE to verify that the treatment selected will be able to meet the regulatory limits (Novak et al., 2002).

Once the toxicant(s) is characterized or identified, the TTE will require the selection of technologies that will remove the specific toxicant(s) or class of substances. The treatment technologies identified may be commercially available, or experimental. However, experimental technologies may be more risky in implementation. The treatment options should be developed and evaluated first at the bench-scale, to assess and verify process constraints, efficiency, operating costs, robustness of the technology to process upsets, and to develop information for engineering scale-up and capital cost. Once this information has been evaluated, decisions can be made to pilot the technologies that have been most successful (Novak et al., 2002).

Although it is beyond the scope of this chapter to provide detailed approaches for bench-, pilot- and full-scale testing, some key considerations for TTEs are presented. Readers are directed to U.S. EPA methods (1989; 1999), Novak et al. (2002) and to the case study in Section 7 for detailed approaches and examples. Key considerations for bench-scale testing include:

- Designing an appropriate sampling methodology and toxicity test procedures.
- Ensuring the existing ETP (if one exists) can be adequately simulated at the bench-scale.
- Developing treatment flow sheets, calculation of chemical addition ratios (and methods of addition), temperatures and retention times.
- Coordinating technology evaluation with toxicity testing laboratory.
- Repeating testing on different samples to consider effluent variability, process/operation changes, and seasonal changes in order to have a high degree of certainty as to the effectiveness of the treatment.
- Evaluating and ranking of the technologies, taking into consideration toxicity removal and costs (Novak et al., 2002).

Pilot testing is conducted after the most effective bench treatment technology has been selected. Pilot-scale testing involves experimental treatment of a portion of the facility effluent flow in actual climate conditions (or in a laboratory equipped to simulate these conditions). Key considerations for pilot-scale testing include:

- Determining location of pilot-scale testing (on site or at a laboratory remote from the facility).
- Determining if batch or continuous pilot-scale tests are to be conducted.
- Applying technologies selected, ensuring that: tests are representative of field conditions (*e.g.* temperature), steady-state conditions are achieved (and samples for toxicity testing are taken during steady-state conditions), conditions are reproducible, operating and process data are obtained for full-scale tests or full-scale application.
- Repeating testing on different samples to consider effluent variability and confirm effectiveness of the treatment (Novak et al., 2002).

Full scale testing involves experimental treatment of all of the facility effluent in actual climate conditions. For facilities with existing ETPs, full-scale testing is undertaken to provide a more reliable assessment of the impact on final effluent toxicity of the treatment. For facilities without an ETP, full-scale testing is a reliable way to test effluent treatment or modifications to effluent generation. Such testing can also be the final stage in the TTE. Key considerations for full-scale testing include:

- Assessing current effluent management or treatment system and its operation to ensure steady-state plant or optimisation prior to initiation of the TTE.
- Designing the approach for adjustment or change of existing facility.
- Selecting the most effective location for a full-scale test on site.
- Using bench- and pilot-scale test data to engineer and install the temporary treatment technology.
- Commissioning and operating treatment plant, and confirming removal of toxicity (Novak et al., 2002).

5.4.3 Source Investigations

The objective of a Source Investigation (SI) is to identify the upstream source of the toxicant, followed by implementation of controls (*i.e.*, treatment technologies or alteration of upstream management systems) that will translate into elimination of final effluent toxicity. A key advantage is that treatment of smaller, more concentrated streams can often be performed more efficiently and economically than treatment of larger, more dilute streams (i.e., the final effluent) (U.S. EPA, 1989; 1999). Source investigations can also be viable alternatives to eliminate final effluent toxicity in cases where a specific toxicant(s) cannot be identified, or if toxicity is transient or non-persistent. The selection of a SI or TTE approach will be facility-specific. However, an SI conducted prior to a TTE can be more beneficial and more cost-effective, since the source of toxicity can be identified and then treated separately from the final effluent (Novak et al., 2002).

The U.S. EPA provides both a generic approach for conducting a SI (U.S. EPA, 1989), as well as methods specific to municipal effluent (U.S. EPA, 1999). Steps involved in a SI (as described by Novak et al., 2002) could include:

(1) **Identification of sewers, discharge locations and inputs to the final effluent or ETP (if one exists).** The main streams and inputs to the ETP (if present) should be identified, along with sub-component streams (information should already be available as part of a Toxicity Prevention/Response Plan described in Section 5.1). Inputs should include process, site-runoff, stormwater runoff and groundwater sources. A description of the process and operation at each location should also be provided (including operation frequency). Locations with existing monitoring equipment and flow control devices should also be identified.

(2) **Selection of sampling locations, protocols, methods for flow monitoring and implement sampling program.** Based on the results from task #1, sampling locations, methods for sample collection and flow monitoring options should be identified. Sampling locations should be readily accessible. The choice of grab or composite samples will depend on the variability of the individual stream. Flow measurements will be required for each sampling location. Simple methods (*e.g.*, using a bucket and stopwatch) may be sufficient, while other locations may require more complicated approaches (*e.g.*, installation area-velocity meters). For certain locations, where it is not possible to measure flow directly, the sum of individual upstream flows could be used as an estimate.

(3) **If the toxicant has been identified, i) use chemical specific analysis for tracking the sources, and ii) evaluate the effects of the treatment plant on altering the toxicant.** This approach involves testing the source streams for the toxicant using chemical-specific analysis. Chemical specific analysis should also be conducted on the combined influent and effluent streams to assess the effects of the ETP on the substance responsible for toxicity. Effluent residence times must be considered to ensure the same batch of water from the influent and effluent is analyzed (U.S. EPA, 1989). Once the source of the toxicant(s) has been identified, the SI could go further into the process to identify sub-component streams, or the TRE could proceed to a TTE for the source stream(s).

(4) **If toxicant has not been identified, i) use bench-scale model to simulate treatment plant and track toxicity, and ii) characterize the bench-scale treated samples using Phase I treatments**. This approach, also referred to as Refractory Toxicity Assessment (U.S. EPA, 1999), involves systematic sampling and a process of elimination to track toxicity to the source. It can be useful in cases where toxicity is transient, non-persistent, or when the substance responsible for toxicity has not been identified. Methods involve collection of samples from each source, followed by simulation of the ETP and testing for toxicity. A key component of every SI (and TTE) is the ability to simulate existing effluent treatment. The amount of toxicity that could potentially pass through the treatment system must be estimated by treating each source stream in a simulation of the ETP prior to toxicity testing. Because toxicity is not degraded by the existing effluent treatment, the primary source of toxicity can be identified. To provide additional certainty that the source streams contain similar classes of toxicants, the suspect source streams could also be characterized using the Phase I TIE methods used for the final effluent (U.S. EPA, 1999).

(5) Analysis of results and selection of the number of streams to be treated and the treatment technology (or management option) to be used for each stream (*e.g.*, TTE). If multiple toxicant sources have been identified, or if the toxicity tracking approach was used, the results from the process stream characterization can be analyzed to identify the streams representing the largest contributors (in terms of toxicity and chemical load) to the ETP (referred to as mass balance). The objective of the mass balance approach is to identify those streams that represent the largest contributors (in terms of toxicity and chemical load) to the final effluent or ETP (if one exists). This approach could be used if the identified toxicant is found in multiple streams, or if the substance(s) responsible for toxicity is only suspected, but has not been conclusively identified. In the latter case, the risk associated with source stream misidentification is increased. Key steps in mass balance approach include:

- Calculating mass loading of toxicity; TUs are calculated for each stream, and then multiplied by the proportion of the total flow for that stream to arrive at a value representing "relative" contribution to final effluent toxicity.
- Comparing and ranking TUs to determine which stream contributed the greatest toxicity to the final effluent.
- Calculating mass loading for measured chemical parameters; individual chemical concentrations for each stream are multiplied by the proportion of total flow for that stream, to arrive at a total loading for each parameter.
- Comparing and ranking loadings to determine which stream contributed the greatest chemical load for each parameter.
- Selecting the number of streams to be treated and the treatment technology or management option to be used for each stream.

The majority of a specific chemical may originate from a single source, in which case further investigations may proceed directly to a TTE designed to remove this substance. However, it is possible that the SI indicates multiple sources of the toxicant, which can occur if the substances are widely dispersed throughout the system. The inability to locate the toxicant(s) may also suggest the sampling points did not include all possible sources. In this case, it may be necessary to evaluate additional input lines in the collection system (U.S. EPA, 1999).

In cases where the toxicant has not been conclusively identified in the final effluent, the TTE approach must be used cautiously. Streams selected for TTE should be based on some knowledge of the type of process stream, effect of the effluent treatment plant on source stream toxicity, or Phase I TIE results. For example, a Phase I TIE conducted on the suspect source stream(s) could be used to ensure that it contains the sample class of toxicant(s) as the final effluent. The Phase I TIE may also suggest a possible treatment option for the TTE (*i.e.*, addition of EDTA: Novak et al., 2002).

6. Factors influencing success or failure of Toxicity Reduction Evaluations

The approach to any TRE will be site specific and success will depend to great extent on the nature of the toxicant(s), effluent variability, education and experience of the investigator, as well as the level of communication among all stakeholders (*i.e.*, the discharger requesting the investigation, the toxicologists conducting the studies, and regulatory authority).

The efficiency with which the toxicity event can be addressed will be greatly increased through the use of the Toxicity Prevention/Response Plan as described in Section 5.1. In response to a toxic event, dischargers (at times in response to regulatory / public pressures) will often immediately collect and submit samples to a testing laboratory for a Phase I TIE. From a regulatory / public perspective, this approach could be interpreted as the most proactive response. However, based on experience the preferred approach is to develop a logical strategy for each study, which take into consideration episode- and site-specific conditions and allows sufficient time for evaluation of process and operational data, review of available chemistry and toxicology data, and flexibility in the study design (including the use of alternative tools and techniques).

In general, TREs will be most successful when an effluent is consistently toxic, if the loss of toxicity is minimal over time and factors contributing to toxicity do not vary between samples. Conversely, the process can be more difficult if toxicity is transient, if the samples quickly lose toxicity over time, or if the factors contributing to toxicity are variable (*i.e.*, different causative agents). Data interpretations can also be complicated by low contaminant concentrations and marginal toxicity. For example, it can be difficult to discern differences in toxicity between a toxic final effluent and TIE treatments when the mortality in the full strength (100%) effluent is close to 50% (Novak et al., 2002).

Toxicant identification will be most effective when used in conjunction with the knowledge of the process (*e.g.*, mining, pulp and paper, textile, municipal discharge) and operation (*e.g.*, changes, upsets, reagent usage etc.). This information can be valuable in preliminary design of a limited and well-defined investigation for a specific chemical or cause of toxicity. Yet significant modification or “short-cutting” of the TRE process could result in certain contaminants or causes of toxicity being disregarded and in fact increase the duration and cost of the investigation (Norberg-King et al., 2005).

Beyond Phase I, the TIE approach is not standardized and subsequent studies to identify the specific causes of toxicity require experienced personnel. Phase II (identification) and Phase III (confirmation) studies must be well planned and scientifically defensible if they are to be successful. However, limitations still exist for the characterization and identification of polar organic substances. These compounds (generally with a log $K_{ow} < 3$) are water-soluble, pass through a C18 column and are not easily characterized using existing TIE manipulations (Norberg-King et al., 2005). Further studies are needed to ensure characterization methods include polar organics. Similarly, certain complex polar organic compounds (*e.g.*, polymers, surfactants, breakdown products) can be difficult to identify. Often the class of substance can be characterized during the Phase I TIE (*i.e.*, using C18 solid phase extraction and recovery methods), but the specific chemical not identified using traditional organic analytical techniques (*i.e.*, GC or HPLC). Unique analytical approaches (*e.g.*, reverse phase HPLC, GPC, MS or UV detectors, or different columns such as C8 and silica) are often required, but can require extensive method development, since procedures are generally not standardized.

Identification of the specific chemical(s) responsible for toxicity is not always necessary in order to develop sufficient control options for achieving and maintaining a consistently non-toxic effluent. For example, source investigations may identify an opportunity to recycle a heavily contaminated stream back to the process, for additional reagent recovery/savings, and reduced effluent toxicity (see case study example in Section 7). Treatment of individual concentrated sources prior to discharge could prevent the contamination of larger volumes that are more dilute and potentially more expensive to treat (Novak et al., 2002).

While statistical analyses are used throughout the process, TREs do not "prove" the cause of toxicity, but rather use a weight of evidence approach. Knowledge and experience of the investigator will be vital to interpretation of TIE results and in the design of subsequent studies. Moreover, conclusions as to the cause of toxicity cannot be based on a single sample. In all TRE components, repeated testing on different effluent samples must be conducted in order to account for effluent and process variability and confirm that the cause of toxicity is the same under all conditions.

Effective communication between all team members (*i.e.* toxicologist, chemists, engineers, site personnel) during all stages of the TRE will increase the likelihood of success. Lack of understanding and communication among TRE participants (*i.e.*, toxicologists, industry and regulators) can lead to the incomplete transfer of information. It is particularly important that the TRE team have a clear understanding of the information gathered from all components. The better understanding all TRE team members have of the site/facility, the greater the chance of achieving and maintaining a non-toxic effluent.

7. Toxicity Reduction Evaluation case study

In 2001, SETAC sponsored a workshop aimed at advancing the science of TREs, the results of which were released as a SETAC publication (Norberg-King et al., 2005). Included in this book are over 30 case studies provided by the workshop participants, as well as a literature search identifying relevant TRE journal articles published between 1991 and 2003. The following case study was a TRE conducted by Novak et al. (1998, 2002) and was also included in the SETAC publication.

The TRE study was conducted (between 1998 and 2000) for metal refinery that was intermittently non-compliant with toxicity limits (> 50% mortality in 100% final effluent) for *Daphnia magna* and rainbow trout. Although not classified as a "small-scale" test, the results from the rainbow trout tests were also included in this case study. Key elements of the study included Phase I TIEs, Source Investigations and Toxicity Treatability Evaluations. The study team included site operations and process representative, toxicologists, chemists and engineers.

The main refinery operation/process includes cobalt/nickel and precious metal refining. The ETP treated all process streams, surface run-off, and water from groundwater recovery projects. The ETP also received run-off from some of the adjacent residential areas and wetland drainage. Prior to entering the ETP, the wastewater was automatically pH-adjusted to between 7.5 and 8.0 with sulphuric acid to control influent pH. A coagulant was added to the raw water as it was pumped to a reactor clarifier. To precipitate metals, the untreated wastewater was adjusted with slaked lime slurry to an approximate pH of 10.6, and a polymer was added to assist with solids settling. The clarified water was automatically adjusted with carbon dioxide to achieve a maximum pH of 9.5 at the regulated control point. The treated effluent entered a polishing pond before being discharged to the receiving environment.

The effluent complied with all regulated chemical limits, but still experienced periodic non-compliant events (> 50% mortality in 100% effluent) with rainbow trout and *Daphnia magna*. Historical data indicated that the effluent was more frequently lethal to *Daphnia magna* than to rainbow trout. This data also indicated that effluent quality was variable and that mortality often coincided with high TDS. For example, conductivity (as a surrogate measure of TDS) was generally greater than 7,000 μS/cm for samples lethal to *Daphnia magna* and greater than 12,000 μS/cm for samples lethal to rainbow trout.

7.1 TOXICITY IDENTIFICATION EVALUATIONS

The Phase I TIEs were conducted over a 3-month period. All tests were conducted using rainbow trout and *Daphnia magna* following Environment Canada (2000a, b) toxicity test methods and U.S. EPA (1991) TIE procedures. Separate TIEs were conducted for each species. Phase I treatments included zeolite, activated carbon and Amberlite XAD™ resin (for removal of relatively low molecular weight organics). Chemical analyses (total and dissolved metals, chloride, sulphate, nitrite, nitrate, TDS, total organic carbon (TOC) and dissolved organic carbon - DOC) were conducted on all untreated effluent samples and on any treated sample observed to reduce or eliminate toxicity. For comparison purposes, selected treatments that did not eliminate toxicity (activated carbon) were also submitted for analysis. The decision to include fairly extensive chemical analysis during the Phase I TIE was based on knowledge of the operation and historical data reviews which suggested metals and certain TDS components could have contributed to mortality. Care was taken to ensure that analytical method detection limits (MDLs) were sufficiently low so as not to impede data interpretation and the detection of subtle changes in chemistry. Furthermore, dissolved metal concentrations were used for most data interpretations. While particulate bound metals can be a source of toxicity (*e.g.*, through ingestion), dissolved metals are more readily available to the organism of interest and are consequently the most relevant in terms of toxicity.

7.1.1 Phase I TIE - Daphnia magna

Zeolite was only the treatment to effectively reduce *Daphnia magna* mortality. Comparisons of effluent constituents in the untreated effluent to available toxicity data reported in the literature indicated that sodium (Na) concentrations (2,410 mg/L) were sufficiently high to account for approximately 50% of the mortality (Na LC50 ~ 2,340 mg/L). Similarly, dissolved copper (Cu) concentrations (0.041 mg/L) were at the LC50 for *Daphnia magna* (Cu LC50 ~ 0.04 mg/L). Changes in several chemical parameters were observed following zeolite and activated carbon treatment. For example, calcium (Ca), potassium (K) and magnesium (Mg) increased following zeolite treatment, but remained unchanged after treatment with activated carbon. The latter was more effective than zeolite at reducing metal (*e.g.*, coblat (Co), copper (Cu), nickel (Ni)) concentrations. However, activated carbon had little effect on *Daphnia magna* survival, while zeolite reduced mortality. TDS was relatively unchanged following either treatment, and it was hypothesized that the cause of toxicity could have been due to a change in the concentration of individual components of TDS, rather than elevated TDS alone.

7.1.2 Phase I TIE - Rainbow trout

Rainbow trout mortality was eliminated following treatment with zeolite and activated carbon. None of the other Phase I TIE treatments eliminated or reduced trout mortality. The concentration of Cu in the untreated effluent (0.107 mg/L) was slightly above the LC50 for rainbow trout (~ 0.09 mg/L). Copper concentrations were reduced following treatment with both zeolite (0.057 mg/L) and carbon (< 0.005 mg/L). The observed increase in Ca, K and Mg concentrations following treatment with zeolite supported results obtained during testing with *Daphnia magna*. Ca, K and Mg were unchanged after treatment with activated carbon.

7.1.3 Ion balance experiments with D. magna

A review of historical toxicity test data, suggested that rainbow trout and *Daphnia magna* mortality could have resulted from imbalances of major ions (specifically Na, Ca and K). This hypothesis was supported by Ingersoll et al. (1992) and Dwyer et al. (1992), who demonstrated that addition of Ca and Mg to sodium enriched waters reduced toxicity to a variety of organisms. In other words, the TDS content actually increased while toxicity decreased; Ca and Mg were thought to have a protective effect against Na toxicity. Based on these studies, ion balance experiments were conducted, where varying amounts of Ca (as calcium chloride) and K (as potassium chloride) was added to lethal refinery effluent samples. The highest additions of Ca and K, at a Na/(Ca+K) ratio of 15:1, resulted in a reduction in *Daphnia magna* mortality. Na/(Ca+K) ratios of 40:1 and 88:1 had little effect on overall survival. Na/(Ca+K) ratios were also calculated for all toxicity test data for which chemical data were available (> 60 samples). Based on this analysis, it appeared that a Na/(Ca+K) threshold existed for *Daphnia magna* at a ratio of approximately 75:1. A threshold could not be determined for rainbow trout due to an insufficient number of toxic samples.

7.1.4 Data analysis

The relationships between *Daphnia magna* mortality and the concentrations of the components in the exposure solutions were explored by regression analysis.

Emphasis was placed on analysis of the *Daphnia magna* toxicity data due to the larger number of lethal samples and consequently an expected greater chance of producing a significant model. It was hypothesized that Na levels were sufficient to account for at least 50% of the *Daphnia magna* mortality. However, the concentration of Na did not vary sufficiently to be a significant parameter in the regression model. Additional potential sources of toxicity identified through multiple regressions, included Cu, K and bicarbonate (H_2CO_3). The final regression equation ($y = -0.708 - 1.49(K) + 574(Cu) + 0.363(bicarbonate)$, $r^2 = 0.96$) assumed Na could explain 50% of the observed mortality. Based on the limited available data, regression models could not be developed for rainbow trout, however, it was suspected that periodic peaks in Na (*e.g.*, > 6,000 mg/L) or Cu (*e.g.*, > 0.1 mg/L) concentrations contributed to the sporadic toxicity.

The FSTR ("Freshwater Salinity Toxicity Relationship") model (Mount et al., 1997) was also applied with selected *Daphnia magna* samples. The results indicated that for most samples, the model accurately predicted toxicity. Percent differences between measured and predicted TUs were generally less than or equal to 20%. In several cases the model either over- or under-estimated toxicity. Its failure to consistently predict mortality was supported by Mount et al. (1997), who observed that in its preliminary application in field-collected samples, the *Daphnia magna* model tended to over-predict toxicity. Alternatively, the presence of Cu could also explain why measured toxicity was occasionally higher than the predicted toxicity.

7.2 SOURCE INVESTIGATIONS

Because the substances responsible for final effluent toxicity were not adequately characterized using the Phase I TIE procedures (*e.g.*, zeolite and activated carbon were the only treatments that reduced *Daphnia magna* and rainbow trout mortality), an alternative approach was undertaken to focus on characterization of upstream sources of toxicity. The specific objectives were to: i) characterize upstream sources in terms of toxicity and chemical composition, ii) investigate possible treatment options for these streams and iii) develop a conceptual treatment approach such that the selected treatment would result in a compliant final effluent.

A mass balance approach was used to identify those streams that represent the largest contributors, in terms of toxicity and chemical loading to the treatment plant. Twenty-two sampling locations, representing the main inputs and sub-components to the treatment plant, were identified and selected for characterization. For each selected location, flow measurements were taken, and samples for chemical analysis (as in Section 7.1) and toxicity testing were collected.

For many locations, flow could be easily measured (*i.e.*, using a bucket and stopwatch). For other locations where it was not possible to measure flow directly the sum of individual upstream flows was used as an estimate. For several of the main pipes entering the treatment plan, a velocity meter was used to estimate flows. In these cases the velocity, pipe diameter and depth of water within the pipe were used to calculate flow. There were a total of four sampling events. Initially, all samples were to be collected under dry conditions to maximize contaminant concentrations. However, samples were also collected during precipitation events to assess impact from runoff sources (*e.g.*, landfill and wetlands drainage).

To determine the chemical loading of each main and sub-component stream, measured effluent parameter concentrations were matched with observed flows within each stream for each sampling event. These values were multiplied to arrive at a total chemical loading for each parameter within each stream. Loadings were then compared to determine which stream contributed the greatest load for each water quality parameter measured. To determine which stream contributed the greatest toxicity to the treatment plant, TUs were calculated and then multiplied by the proportion of total flow for that stream in order to arrive at a value representing the "relative contribution" to toxicity. These values were ranked (according to toxic contribution) and grouped (according to stream component).

Variability in water quality parameter concentrations, the retention time experienced by effluent within the discharge network, as well as inaccuracies in flow rate measurements within the effluent streams impacted the loading calculations. However, the results did allow for identification of major chemical and toxicity sources within the wastewater network. Four process streams were identified as the main contributors to toxicity. One of the four streams (Stream #1), associated with Co/Ni refining, contributed the greatest portion of toxic levels of Ni, Co and Cu. The greatest proportion of TDS contaminants (Na, SO_4, HCO_3, Cl) originated from Streams #2 or #3 (both associated with Co/Ni refining). Stream #4 (associated with precious metal refining) also contributed a considerable amount of Cu during one of the four sampling events. The remaining 18 streams were either non-lethal or only caused marginal toxicity.

Because the specific toxicant had not been identified, it was recognized the substances selected for possible treatment had to be based on the TIE Phase I characterization results, knowledge of the type of process stream, and the ETP design parameters (*i.e.*, ETP not designed to remove TDS). Using this approach, a reduction in the toxicity of the source streams by removal of specific contaminants did not guarantee an elimination of toxicity of the final effluent. However, the weight-of-evidence strongly supported TDS components and copper as the main contributors to final effluent toxicity. Therefore, a decision was made to proceed with bench-scale simulations to determine if treatment or adjustment of these streams would reduce final effluent toxicity.

7.3 TOXICITY TREATABILITY EVALUATIONS

Based on the results from the source identification, further studies were conducted to investigate possible treatment options for Streams #1, 2 and 3. Two approaches were taken: 1) investigation of full-scale process adjustment options (*e.g.*, removal of Stream #1 from the process), and 2) bench-scale testing of potential treatment technologies focusing on the TDS components in Streams #2 and #3.

Stream #1 was temporarily shut down for a period of 10-days. To ensure this was a sufficient amount of time for its clearance from the wastewater distribution system, a dye test was conducted to determine the retention time of waste from this stream in the sewer system, including the length of time to pass from the influent, through the treatment plant and to the final discharge. Removal of this stream reduced concentrations of Ni, Cu and Co in the final effluent. Prior to removal of this stream, final effluent LC50s for rainbow trout ranged from 71 to 80%. All samples met the rainbow trout toxicity limit (< 50% mortality in the 100% effluent) after removal of Stream #1. Elevated Cu was the most likely cause of rainbow trout mortality. Prior to removal of Stream #1, Cu concentrations in the final effluent ranged from 0.118-0.124 mg/L. These values exceeded the rainbow trout LC50 for Cu of 0.09 mg/L. After removal of Stream #1, Cu concentrations remained below the LC50 and the final effluent was non-lethal to trout.

In the case of *Daphnia magna*, it was suspected that high TDS from Streams #2 and #3 contributed to toxicity even in the absence of elevated Cu concentrations from Stream #1 (suggesting that Cu was not the only contributing factor to *Daphnia magna* mortality). Although not all of the observed toxicity could be easily explained, the data suggested that removal of Stream #1 would be beneficial to *Daphnia magna* survival, when accompanied by a reduction in TDS.

The bench-scale testing of potential treatment technologies focused on the TDS components in Streams #2 and #3, and included an evaluation of i) Ion Exchange (IX) with K and Ca, ii) Zeolite, iii) Evaporation, and iv) Selective Precipitation. Stream #1 (primary source of Cu) was not included in any of the bench-scale trials.

The first step in the bench-scale trials involved treatment of individual streams using the proposed technologies. Each stream was analyzed before and after treatment to properly assess the treatment efficiency. After treatment of Streams #2 and #3, all process streams were combined in the correct proportions (according to flow). The combined sample was then subjected to effluent treatment (which simulated the full-scale effluent treatment plant - ETP). This sample was then tested for toxicity to determine the technologies that demonstrated the most promise. The ETP was also simulated on the combined effluent prior to treatment with the proposed technologies. The results from this "baseline" test were used for comparison purposes to the ETP simulated effluent after treatment with the proposed technology. Results indicated that only evaporation of Stream #2 eliminated final effluent toxicity to *Daphnia magna*. Elimination of final effluent toxicity following removal of both Streams #1 and #2, provided strong evidence to support the premise that both Cu (from Stream #1) and elevated TDS (from Stream #2) were the main causes of *Daphnia magna* mortality.

Although cost-effective treatment options for TDS are scarce at best (Goodfellow et al., 2000), a feasibility study for the evaporation treatment of Stream #2 was conducted to assess actual costs and potential disadvantages of this treatment method. The technology selected was mechanical vapour compression evaporation, a technology that adds energy for evaporation by compressing and recycling the vapour produced by the evaporation process. The total construction cost was estimated at $11 million (Canadian $), with operating costs estimated at $1.5 million per year.

Key environmental disadvantages regarding the use of evaporation technology in this application were identified, included disposal of the salt cake and substantially increased electrical consumption. With respect to disposal of the evaporated salts, the crystals formed would be readily soluble in water, and as a result, dissolve in any water with which they come into contact, including rainwater and leachate from other sources.

Based on costs and environmental disadvantages, evaporation was not a viable option for effluent treatment. As an alternative, it was hypothesized that toxicity could be controlled by management of TDS (or conductivity used as a surrogate measurement). Although elevated TDS could not account for all of the observed toxicity (due to the presence of Cu), the data indicated a significant relationship between *Daphnia magna* mortality and TDS. Furthermore, the hypothesis that both Cu (from Stream #1) and elevated TDS (from Stream #2) were the main causes of *Daphnia magna* mortality was supported by results generated during the bench-scale treatability testing (*i.e.*, removal of both streams eliminated toxicity). Analysis of conductivity and toxicity data suggested that final effluent samples should achieve compliance with the toxicity requirement for both species (*e.g.*, < 50% mortality in 100% effluent) if the conductivity remained below approximately 7,000 μS/cm.

7.4 SUMMARY

Stream #1 was successfully removed from the process with the added benefit of metals recovery back into the refining process. Further testing was conducted to confirm the conductivity threshold for *Daphnia magna* in the absence of toxicity due to Cu. Following permanent removal of Stream #1 and control of TDS (using conductivity as a surrogate measurement) the effluent has been non-lethal to trout and *Daphnia magna*.

8. Accessory TRE Information

8.1 COST AND TIME FRAMES FOR TREs

TREs can take several weeks, months, or years to complete. Norberg-King et al. (2005) provided some general timeframes for completion of a TRE:

(1) Preparation of TRE Plan – 1 to 3 months
(2) Data/Process/Housekeeping Review – 1 to 3 months
(3) Phase I TIE – 1 to 6 months
(4) Phase II TIE – 1 to 6 months
(5) Phase III TIE – 1 to 3 months
(6) Source Investigations – 1 to 3 months
(7) Toxicity Treatability Evaluations– 1 to 9 months

However, the above can only be used as a guide, since a number of effluent- and site-specific factors will influence the cost and time frames for TREs, including; complexity of the effluent (*i.e.*, multiple processes/waste streams), number of substances responsible for toxicity, magnitude or degree of toxicity (*i.e.*, single versus multiple concentration tests), variability between samples (*i.e.*, consistent or transient toxicity), persistency (*i.e.*, does toxicity in a single sample degrade overtime), regulatory issues (*e.g.*, meetings, planning, negotiations, approvals from regulatory bodies), and financial constraints (*e.g.*, corporate funds allocated to toxicity events).

Costs can often only be provided or estimated for the initial stages of a TRE (*e.g.*, bullets 1 through 3), since subsequent treatments, analyses and methods selected (*e.g.*, Phase II or III TIE, TTE or SI) are directly related to the initial TRE findings combined with those treatments observed to effectively eliminate or reduce toxicity during the Phase I TIE. Therefore, the specific approaches (and associated costs) cannot be provided until the Phase I TIE is completed. However, allocating TRE funding for each fiscal year (as part of the Toxicity Prevention/Response Plan) can reduce regulatory and internal corporate pressures when a toxicity event is observed since funds to support the initial investigation would be readily available. In certain cases, the costs of the TRE and subsequent process adjustments to achieve toxicity compliance can be recovered during the process. For example, in the case study described in Section 7, the TRE results lead to removal of the most toxic stream from the process. The cost for removal of Stream #1 was estimated at $400,000 CDN. However, the corresponding recovery of metals back into the refining process would recover most of this initial capital investment within approximately 10 years.

8.2 COMMON SUBSTANCES ASSOCIATED WITH DIFFERENT INDUSTRIAL OPERATIONS

Industrial and municipal effluents are complex wastewaters comprised of many different constituents which can vary in terms of their concentration and form in response to factors such as process changes, quality of the feed material, waste treatment practices or environmental conditions (*e.g.*, temperature) which can affect their relative toxicity. Although preconceived assumptions regarding the cause of toxicity must be avoided, a review of available chemistry data during the initial phases of a TRE can be useful, particularly for those substances that have been extensively studied (Norberg-King et al., 2005; Novak et al., 2005). Examples of contaminants commonly associated with pulp and paper effluent include resin and fatty acids, sulphide and ammonia (Kovacs and O'Connor, 1996). Contaminants commonly associated with base metal mining effluent include, free acidity, depressed or high pH, dissolved metals, ammonia, thiosalts and xanthates. Common toxicants at gold mines include cyanide and cyanide related compounds, arsenic, dissolved metals, ammonia and total suspended solids (TSS). Common toxicants at uranium mines include solvent extraction organics, arsenic, uranium, TSS and dissolved metals. TSS is a commonly encountered toxicant at iron ore mines (Novak et al., 2005). Examples of contaminants commonly associated with municipal effluent include ammonia, chlorine, surfactants, pesticides, metals and TDS (U.S. EPA, 1999). The aforementioned information must not be used in place of conducting a TRE, and is only intended as a guide since the specific substances responsible for toxicity will be site specific and unique to each operation.

9. Conclusions

The Toxicity Reduction Evaluation process is a logical site-specific set of procedures which can be used for preventing and resolving toxicity with a variety of aqueous media, including industrial and municipal effluent, receiving water (surface water), groundwater, leachates and sediment porewater. Factors influencing success of a TRE include establishing a Toxicity Prevention/Response Plan prior to the initial toxicity episode, experience and knowledge of the investigators, inclusion of a multi-disciplinary team, effective mechanisms for communication and co-ordination among all stakeholders, and flexibility in site-specific studies. The methods for the prevention and reduction of toxicity are not limited to the guidance provided in the preceding sections. Significant new and useful advances in toxicant identification are made on a regular basis; as these methods or approaches are developed, they should be included for consideration as part of the TRE.

Acknowledgements

The authors wish to extend our gratitude to Kent Burnison and Larry W. Ausley for the manuscript review and comments. We acknowledge the United States Environmental Protection Agency as the source for Figures 1, 2 and 3. We thank the copyright holder for permission to present the case study provided in Chapter 7, which was reprinted with permission from Norberg-King TJ, L Ausley, D Burton, W Goodfellow, J Mille and WT Waller, (eds.), 2005. Toxicity identification evaluations (TIEs) for effluents, ambient waters, and other aqueous media. Workshop on Toxicity Identification Evaluation (TIE): what works, what doesn't, and developments for effluents, ambient waters, and other aqueous media; 2001 June 23-28; Pensacola Beach FL. Pensacola FL, USA: Society of Environmental Toxicology and Chemistry (SETAC). Copyright SETAC, Pensacola, Florida, USA (in press).

References

Amato, J.R., Mount D.I., Durhan, E.J., Lukasewycz, M.T., Ankley, G.T. and Robert, E.D. (1992) An example of the identification of diazinon as a primary toxicant in an effluent, *Environmental Toxicology and Chemistry* **11**, 209-216.

Ankley, G.T. and Burkhard, L.P. (1992) Identification of surfactants as toxicants in a primary effluent, *Environmental Toxicology and Chemistry* **11**, 1235-1248.

Ankley, G.T., Schubauer-Berigan, M.K. and Holke, R.A. (1992) Use of toxicity identification evaluation techniques to identify dredged material disposal options: a proposed approach, *Environmental Management* **16**, 1-6.

Ausley, L.W., Arnold, R.W., Denton, D.L., Goodfellow, W.L., Heber, M., Hockett, R., Klaine, S., Mount, D., Norberg-King, T., Ruffler, R. and Waller, W.T. (1998) Application of TIEs/TREs to whole effluent toxicity: principles and guidance. A report by the Whole Effluent Toxicity TIE/TRE Expert Advisory Panel. Pensacola, FL: Society of Environmental Toxicology and Chemistry (SETAC).

Bailey, H.C., DiGiorgio, C., Kroll, K., Miller, J.L., Hinton, D.E. and Starrett, G. (1996) Development of procedures for identifying pesticide toxicity in ambient waters: carbofuran, diazinon, chlorpyrifos, *Environmental Toxicology and Chemistry* **15**, 837-845.

Bailey, H.C., Elphick, J.R., Potter, A., Chao, E., Konasewich, D. and Zak, J.B. (1999) Causes of toxicity in stormwater runoff from sawmills, *Environmental Toxicology and Chemistry* **18**, 1485-1491.

Bailey, H.C., Krassoi, R., Elphick, J.R., Mulhall, A., Hunt, P., Tedmanson, L. and Lovell, A. (2000) Whole effluent toxicity of sewage treatment plants in the Hawksbury-Nepean watershed, New South Wales, Australia, to *Ceriodaphnia dubia* and *Selenastrum capricornutum*, *Environmental Toxicology and Chemistry* **19**, 72-81.

Bailey, H.C., Elphick, J.R., Krassoi, R. and Lovell, A. (2001) Joint acute toxicity of diazinon and ammonia to *Ceriodaphnia dubia, Environmental Toxicology and Chemistry* **20**, 2877-2882.

Burkhard, L.P. and Jenson, J.J. (1993) Identification of ammonia, chlorine and diazinon as toxicants in a municipal effluent, *Archives of Environmental Toxicology and Chemistry* **25**, 506-515.

Chapman, P.M., Bailey, H. and Canaria, E. (2000) Toxicity of total dissolved solids associated with two mine effluents to chironomid larvae and early lifestages of rainbow trout, *Environmental Toxicology and Chemistry* **19**, 210-214.

DiToro, D.M., Allen, H.E., Bergman, H.L., Meyer, J.S., Paquin, P.R. and Santore, R.C. (2000) A biotic ligand model of the acute toxicity of metals. 1. Technical basis, *Environmental Toxicology and Chemistry* **20**, 2383-2396.

Dwyer, R.J., Burch, S.A., Ingersoll, C.G. and Hunn, J.B. (1992) Toxicity of trace element and salinity mixtures to striped bass (*Morone saxatilis*) and *Daphnia magna*, *Environmental Toxicology and Chemistry* **11**, 513-520.

Environment Canada. (1992a) (including November 1997 amendments) Biological test method: test of reproduction and survival using the cladoceran *Ceriodaphnia dubia*. Environmental Protection, Conservation and Protection, Environment Canada. Ottawa, Ontario, Report EPS 1/RM/21.

Environment Canada. (1992b) Biological test method: toxicity test using luminescent bacteria (*Photobacterium phosphoreum*). Environmental Protection, Conservation and Protection, Environment Canada. Ottawa, Ontario, Reference Method EPS 1/RM/24.

Environment Canada. (1998) Biological Test Method: Toxicity Tests Using Early Life Stages of Salmonid Fish (Rainbow Trout), Environmental Protection, Conservation and Protection, Environment Canada, Ottawa, Ontario, Second edition, Report EPS 1/RM/28.

Environment Canada. (1999) Guidance Document on Application and Interpretation of Single-species Tests in Environmental Toxicology. Environmental Protection, Conservation and Protection, Environment Canada. Ottawa, Ontario, Reference Method EPS 1/RM/34.

Environment Canada. (2000a) Biological test method: Reference method for determining acute lethality of effluents to *Daphnia magna*. Environmental Protection, Conservation and Protection, Environment Canada. Ottawa, Ontario, Reference Method EPS 1/RM/14.

Environment Canada. (2000b) Biological test method: Reference method for determining acute lethality of effluents to rainbow trout. Environmental Protection, Conservation and Protection, Environment Canada. Ottawa, Ontario, Reference Method EPS 1/RM/13.

EVS Environment Consultants. (2001) Acute toxicity identification evaluations of Greater Vancouver Sewerage and Drainage District wastewater treatment plant effluents. Prepared for Greater Vancouver Regional District, 89 pp.

Goodfellow, W.L., Ausley, L.W., Burton, D.T., Denton, D.L., Dorn, P.B., Grothe, D.R., Heber, M.A., Norberg-King, T.J. and Rodgers, J.H.Jr. (2000) Major ion toxicity in effluents: a review with permitting recommendations, *Environmental Toxicology and Chemistry* **19**, 175-182.

Gustavson, K.E., Sonsthagen, S.A., Crunkilton, R.A. and Harkin, J.M. (2000) Groundwater toxicity assessment using bioassay, chemical and toxicity identification evaluation analyses, *Toxicology* **177**, 131-142.

Hockett, J.R. and Mount, D.R. (1996) Use of metal chelating agents to differentiate among sources of acute aquatic toxicity, *Environmental Toxicology and Chemistry* **15**, 1687-1693.

Ingersoll, C.G., Dwyer, F.J., Burch, S.A., Nelson, M.K., Buckler, D.R., and Hunn, J.B. (1992) The use of freshwater and saltwater organisms to distinguish between the toxic effects of salinity and contaminants in irrigation drain water, *Environmental Toxicology and Chemistry* **11**, 503-511.

Jin, H., Yang, X., Yu, H. and Yin, D. (1999) Identification of ammonia and volatile phenols as primary toxicants in a coal gasification effluent, *Bulletin of Environmental Contamination and Toxicology* **63**, 399-406.

Jop, K.M., Kendall, T.Z., Askew, A.M. and Foster, R.B. (1991) Use of fractionation procedures and extensive chemical analysis for toxicity identification of a chemical plant effluent, *Environmental Toxicology and Chemistry* **10**, 981-990.

Kszos, L.A., Morris, G.W. and Konetsky, B.K. (2004) Source of toxicity in storm water: zinc from commonly used paint, *Environmental Toxicology and Chemistry* **23**, 12-16.

Kovacs, T. and O'Connor, B. (1996) Insights for toxicity-free pulp and paper mill effluents. Pulp and Paper Research Institute of Canada (PAPRICAN) Report No. MR331 (June 1996), 23 pp.

Lukasewycz, M.T. and Durhan, E.J. (1992) Strategies for the identification of non-polar toxicants in aqueous environmental samples using toxicity-based fractionation and gas chromatography-mass spectrometry, *Journal of Chromatography* **580**, 215-28.

Maltby, L., Clayton, S.A., Yu, H., McLoughlin, N., Wood R.M. and Yin, D. (2000) Using single species toxicity tests, community level responses and toxicity identification evaluations to investigate effluent impacts, *Environmental Toxicology and Chemistry* **19**, 151-157.

Mount, D.R., Gulley, D.D., Hockett, J.R., Garrison, T.D. and Evans, J.M. (1997) Statistical models to predict the toxicity of major ions to *Ceriodaphnia dubia*, *Daphnia magna* and *Pimephales promelas* (fathead minnows), *Environmental Toxicology and Chemistry* **16**, 2009-2019.

Norberg-King, T.J., Durhan, E.J. and Ankley, G.T. (1991) Application of toxicity identification evaluation procedures to the ambient waters of the Colusa Basin drain, California, *Environmental Toxicology and Chemistry* **10**, 891-900.

Norberg-King, T.J., Ausley, L., Burton, D., Goodfellow, W., Mille, J. and Waller, W.T. (eds.) (2005) Toxicity identification evaluations (TIEs) for effluents, ambient waters, and other aqueous media, Workshop on Toxicity Identification Evaluation (TIE): what works, what doesn't, and developments for effluents, ambient waters, and other aqueous media, 2001 June 23-28, Pensacola Beach FL., Pensacola FL., USA: Society of Environmental Toxicology and Chemistry (SETAC) (in press).

Novak, L.J., Holtze, K.E. and Roy, R. (1998) Evaluation of Toxicity Reduction Evaluation (TRE) and Toxicity Identification Evaluation (TIE) application to the Canadian mining industry, AETE Report No. 1.2.5. Prepared by ESG International Inc.

Novak, L.J., Holtze, K.E., Wagner, R., Feasby, G. and Liu, L. (2002) Guidance document for conducting Toxicity Reduction Evaluation (TRE) investigations of Canadian metal mining effluents. Prepared ESG International Inc. and SGS Lakefield for TIME (Toxicological Investigations of Mining Effluents) Network, 85 pp.

Novak, L.J., Holtze, K.E., Gilron, G., Wagner, R., Zajdlik, B., Scroggins R. and Schroeder, J. (2005) Guidance for conducting and evaluating rainbow trout and *Daphnia magna* acute lethality toxicity tests using Canadian metal mining effluent, *Environmental Toxicology and Chemistry*, submitted.

Pollutech. (1996) Environmental comparison of results from alternative acute toxicity tests with rainbow trout for selected mine effluents, AETE Report No. 1.1.4.

Schubauer-Berigan, M.K., Amato, J.R., Ankley, G.T., Baker, S.E., Burkhard, L.P., Dierkes, J.R., Jenson, J.J., Lukasewycz, M.T. and Norberg-King, T.J. (1993) The behaviour and identification of toxic metals in complex mixtures: examples from effluent and sediment pore water toxicity identification evaluations, *Archives of Environmental Contamination and Toxicology* **24**, 298-306.

Tietge, J.E., Mount, D.R. and Gulley, D.D. (1994) The Gas Research Institute freshwater salinity toxicity relationship model and computer program: overview, validation and application, Topical Report, Gas Research Institute, Chicago, IL, USA.

Tietge, J.E., Hockett, R., Russell, J. and Evans, J.M. (1997) Major ion toxicity of six produced waters to three freshwater species: application of ion toxicity models and TIE procedures, *Environmental Toxicology and Chemistry* **16**, 2002-2008.

Thurston, R.V., Phillips, G.R., Russo, R.C. and Hinkins, S.M. (1981) Increased toxicity of ammonia to rainbow trout (*Salmo gairdneri*) resulting from reduced concentrations of dissolved oxygen, *Canadian Journal of Fisheries and Aquatic Sciences* **38**, 983-998.

U.S. EPA (U.S. Environmental Protection Agency) (1989) Generalized methodology for conducting industrial toxicity reduction evaluations, EPA-600/2-88/070.

U.S. EPA (U.S. Environmental Protection Agency) (1991a) Methods for aquatic toxicity identification evaluations: Phase I toxicity characterization procedures, EPA-600/6-91/003.

U.S. EPA (U.S. Environmental Protection Agency) (1991b) Toxicity identification evaluation: characterization of chronically toxic effluents: Phase I, EPA-600/6-91/005.

U.S. EPA (U.S. Environmental Protection Agency) (1993a) Methods for aquatic toxicity identification evaluations: Phase II toxicity identification procedures for samples exhibiting acute and chronic toxicity, EPA-600/R-92/080.

U.S. EPA (U.S. Environmental Protection Agency) (1993b) Methods for aquatic toxicity identification evaluations: Phase III toxicity confirmation procedures for samples exhibiting acute and chronic toxicity, EPA-600/R-92/081.

U.S. EPA (US Environmental Protection Agency) (1996) Marine toxicity identification evaluation, EPA/600/R-96/054.

U.S. EPA (US Environmental Protection Agency) (1999) Toxicity reduction evaluation guidance for municipal wastewater treatment plants, EPA/833B-99/002.

U.S. EPA (US Environmental Protection Agency) (2003) DRAFT. Porewater and whole sediment toxicity identification evaluation for freshwater and marine sediments: Phase I (Characterization), Phase II (Identification) and Phase III (Confirmation) modifications of effluent procedures. Office of Research and Development, Narragansett, RI and Duluth, MN.

Van Sprang, P.A. and Janssen, C.R. (2001) Toxicity Identification of metals: development of toxicity identification fingerprints, *Environmental Toxicology and Chemistry* **20**, 2604-2610.

Wells, M.J.M., Rossano, A.J.J. and Roberts, C. (1994) Textile wastewater effluent toxicity identification evaluation, *Archives of Environmental Contamination and Toxicology* **27**, 555-560.

Wenholz, M. and Crunkilton, R. (1995) Use of toxicity identification evaluation procedures in the assessment of sediment pore water toxicity from an urban stormwater retention pond in Madison, Wisconsin, *Bulletin of Environmental Contamination and Toxicology* **54**, 676-682.

Yang, L., Yu, H., Yin, D. and Jin, H. (1999) Application of the simplified toxicity identification evaluation procedures to a chemical works effluent, *Chemosphere* **93**, 3571-3577.

Abbreviations

Al	Aluminum
ASTM	American Society for Testing and Materials
Ba	Barium
BLM	Biotic Ligand Model
BOD	Biochemical Oxygen Demand
Ca	Calcium
Cd	Cadmium
Cl	Chloride
Co	Cobalt
CO_2	Carbon dioxide
COD	Chemical Oxygen Demand
Cu	Copper
DO	Dissolved Oxygen
DOC	Dissolved Organic Carbon
EDTA	Ethylenediaminetetraacetate
ELISA	Enzyme Linked Immunosorbent Assays
ETP	Effluent Treatment Plant
FSTR	Freshwater Salinity Toxicity Relationship
GC	Gas Chromatography
GC/MS	Gas Chromatography/Mass Spectrophotometry
GPC	Gel Permeation Chromatography
HCl	Hydrochloric acid
HCO_3	Bicarbonate

HPLC	High Pressure Liquid Chromatography
ISO	International Standards Organization
IX	Ion Exchange
K	Potassium
Mg	Magnesium
MMER	Metal Mining Effluent Regulation
MSDS	Material Safety Data Sheet
Na	Sodium
NaOH	Sodium Hydroxide
NH_3, N	Un-ionized ammonia
NH_4^+, N	Ionized ammonia
OECD	Organization for Economic Cooperation and Development
OP	OrganoPhosphate
PBO	Piperonyl ButOxide
PEC	Predicted Environmental Concentration
QA	Quality Assurance
QC	Quality Control
SETAC	Society of Environmental Toxicology and Chemistry
SI	Source Investigation
SO_4	Sulphate
SPE	Solid Phase Extraction
SPMD	Semi-Permeable Membrane Device
TDS	Total Dissolved Solids
TIE	Toxicity Identification Evaluation
TOC	Total Organic Carbon
TRC	Total Residual Chlorine
TRE	Toxicity Reduction Evaluation
TSS	Total Suspended Solids
TTE	Toxicity Treatability Evaluation
TU	Toxic Unit
U.S. EPA	United States Environmental Protection Agency
UV	Ultraviolet
Zn	Zinc

6. DETERMINATION OF THE HEAVY METAL BINDING CAPACITY (HMBC) OF ENVIRONMENTAL SAMPLES

GABRIEL BITTON
MARNIE WARD
& ROI DAGAN
Laboratory of Environmental Microbiology and Toxicology
Department of Environmental Engineering Sciences
University of Florida
Gainesville, FL 32611, USA
gbitton@ufl.edu
wcward@infionline.net
rdagan@ufl.edu

1. Objective and scope

The Heavy Metal Binding Capacity (HMBC) test is a bioassay that helps to quickly determine metal bioavailability in aquatic environments. HMBC can also be applied to soils and to root exudates from aquatic and terrestrial plants. The HMBC test is based on MetPLATE, a bacterial toxicity test that selectively detects metal toxicity.

2. Summary of HMBC procedure

HMBC test is based on the U.S. EPA concept of Water Effect Ratio (WER), except that a bacterial response (MetPLATE) is used to determine metal bioavailability. Briefly, the methodology consists of spiking samples of both laboratory water (moderately hard water) and site water with a given metal and the mixtures are shaken for 60 min at 25°C. Afterwards, both mixtures are assayed for metal toxicity using MetPLATE. HMBC is determined as the ratio of IC50 of the metal in site water over IC50 of the metal in laboratory water.

3. Historical overview and reported applications

The discharge of metal-laden effluents is regulated by the U.S. Environmental Protection Agency (U.S. EPA) which sets water quality criteria and standards for

C. Blaise and J.-F. Férard (eds.), Small-scale Freshwater Toxicity Investigations, Vol. 2, 215-231.

metals in receiving waters. However, these regulations were traditionally based on total recoverable metals. Dissolved metals were later considered because they are more available to biota and represent more closely the bioavailable metals which are toxic to aquatic organisms (Allen, 1993; Allen and Hansen, 1996). We are now addressing the concept of metal bioavailability as an essential factor to be incorporated in any regulation of metal discharges into receiving waters. It was estimated that basing water quality criteria on bioavailable metals might reduce the cost of removing metals by 85% to 90%. It is therefore useful to consider site-specific criteria and standards for heavy metals (Hall et al., 1992; Hall and Raider, 1993).

3.1 FACTORS AFFECTING METAL BIOAVAILABILITY IN THE ENVIRONMENT

It is generally accepted that free ionic forms of heavy metals are generally more toxic to biota than chelated or precipitated forms. Several factors control metal bioavailability and, thus, toxicity in environmental samples. These factors include pH, redox potential, alkalinity, hardness, adsorption to suspended solids, cations and anions, as well as interaction with organic compounds (Kong et al., 1995).

Metals generally are in the free ionic form at acidic pHs while they form precipitates at higher pHs. Thus, pH affects the solubility and bioavailability of metals and has an impact on their toxicity. The oxido-reduction potential of aquatic environments also affects metal bioavailability. Hydrogen sulfide is produced in reduced environments such as sediments and helps precipitate metals to form non toxic metal sulfides. Acid volatile sulfide (AVS) in sediments binds metals and generally contributes to lowering their toxicity to biota (Di Toro et al., 1990; 1992). AVS was proposed for predicting metal toxicity in sediments although some have reported its failure for this purpose (Ankley et al., 1996; Berry et al., 1996). In addition to sulfide, metals form complexes with anions such as chloride, phosphate, carbonate or bicarbonate. Water hardness contributes significantly to the reduction of metal toxicity. Finally, organic compounds such as fulvic, humic and organic acids are known to act as chelating agents for metals, thus reducing their toxicity in aquatic and terrestrial environments (Allison and Perdue, 1994; Benedetti et al., 2002; Christl and Kretzschmar, 2001; Ge et al., 2002; Koukal et al., 2003; Parat et al., 2002; Voelker and Kogut, 2001). Municipal solid wastes (MSW) landfill leachates contain high concentrations of dissolved organic matter that is involved in metal complexation (Calace et al., 2001). These organic complexes help in the transport of trace metals through soils (Kaschl et al., 2002). In some landfills, organic complexes account for up to 98% of the total metals (Kang et al., 2002).

3.2 U.S. EPA'S WATER EFFECT RATIO

As shown in the preceding section, toxic metals may be present in a wide variety of physicochemical forms in surface waters, wastewater, landfill leachates, soils, or sediments. Early on, metal speciation in surface waters was determined, using a chemical approach (Giesy et al., 1978). We now know that metal speciation affects their bioavailability and potential toxicity to aquatic organisms (Tessier and Turner,

1995). The U.S. EPA has proposed the concept of Water Effect Ratio (WER) to take into account the effect of the above mentioned factors on heavy metal bioavailability/toxicity in environmental samples (U.S. EPA, 1982, 1984, 1994). EPA water quality criteria for metals have often ignored the local water quality conditions. Thus, WER is the ratio of the LC50 derived from testing the toxicity of a metal to fish or invertebrates in site water to the LC50 derived from testing the toxicity of the same metal to the same test organism (fish or invertebrate) in laboratory water.

$$WER = LC_{50}\ (site\ water) / LC_{50}\ (lab\ water) \tag{1}$$

The site-specific water quality criterion for a given metal is obtained by multiplying the national ambient water quality criterion (AWQC) for that metal by the WER which should be significantly different from 1:

$$Site\ specific\ criterion = WER \times AWQC \tag{2}$$

The site-specific criterion gives a more accurate picture of metal bioavailability in aquatic environments. A modified version of this procedure was used to obtain the site-specific criterion for copper in the Duck River in Tennessee (Sinclair, 1989). The site-specific criterion for copper was found to be 43 μg/L, as compared to the 18 μg/L criterion proposed by EPA. The discrepancy between these two values suggests that water quality parameters of the Duck River were responsible for reducing copper bioavailability and toxicity. The MINTEQA1 metal speciation model of Brown and Allison (1987) confirmed these results. WERs were also determined for the Lehigh River in Pennsylvania, using *Ceriodaphnia dubia* and *Pimephales promelas* (fathead minnow) as the test organisms. Fathead minnow generally displayed higher WERs (for Pb, Zn, Cu, and Cd) than *Ceriodaphnia* (Diamond et al., 1997a). It was reported that WER for copper increased as the wastewater effluent contribution to the Lehigh River increased (Diamond et al., 1997b).

A streamlined WER procedure was published by the U.S. EPA to determine WERs for copper in environmental samples (U.S. EPA, 2001). This procedure uses *Ceriodaphnia dubia* or *Daphnia magna* as test organisms. This alternative method is simpler than the preceding one and requires a minimum number of two WER measurements spaced, at least one month apart, instead of four measurements required by the interim procedure (U.S. EPA, 1994). A comparison between the interim and the streamlined procedure is shown in Table 1 (U.S. EPA, 2001).

4. HMBC concept and procedure description

Prior to discussing the heavy metal binding capacity (HMBC) concept, we will introduce the MetPLATE toxicity assay, which is the sole test used to determine HMBC.

Table 1. A comparison between the 1994 interim procedure and the streamlined procedure for the determination of the water effect ratio (WER) in aquatic environments[1].

Characteristic	*1994 Interim procedure*	*Streamlined procedure*
Applicability	Universal	Copper from continuous discharges
Minimum number of sampling events	3	2
Minimum number of WER measurements	4	2
Minimum number of WER measurements considered in obtaining final site WER	3	2
Preparation of downstream water	Mix effluent and upstream samples at the dilution ratio occurring at the time of sampling	Mix effluent and upstream sample at the design low-flow dilution ratio
Calculation of sample WER	Site water LC ÷ lab water LC	Site water LC ÷ the greater of (a) lab water LC, or (b) SMAV[2]
Calculation of final site WER	Complicated scheme with six "if...then...else" clauses and 12 possible paths	Geometric mean of the two measurements

[1] From U.S. EPA (2001).
[2] SMAV = Species Mean Acute Value (the mean 50% effect concentration from a large number of published toxicity tests with laboratory water).

4.1 METPLATE: AN ASSAY FOR HEAVY METAL TOXICITY

MetPLATE™ is a bacterial/enzymatic test that is selective for heavy metal toxicity in water (Bitton and Morel, 1998). As compared to the MetPAD assay (Bitton et al., 1992a,b), it is a quantitative test that uses a 96-well microplate format, thus allowing the computation of IC50s for environmental samples. The MetPLATE™ toxicity test kit is based on the specific inhibition of the activity of the enzyme β-galactosidase by heavy metals in a freeze-dried *E. coli* strain. The MetPLATE kit includes a freeze-dried test bacteria, a buffer, a diluent (moderately hard water), a freeze-dried enzyme substrate (chlorophenol red-β-D-galactopyranoside CPRG), and a 96-well microplate. The test is run according to the flowchart displayed in Figure 1 (Bitton et al., 1994). Table 2 shows the sensitivity of MetPLATE to heavy metals and its comparison with other tests such as the Microtox, *Daphnia magna*, and Rainbow trout assays.

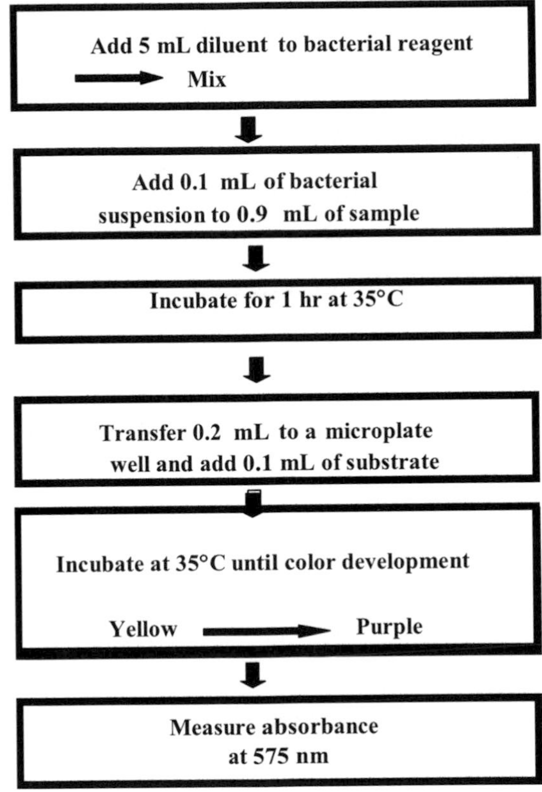

Figure 1. Flowchart of the MetPLATE toxicity test.

FluoroMetPLATE™, the fluorogenic version of MetPLATE™, uses a fluorogenic substrate, 4-methyl umbelliferyl galactopyranoside (MUGA) instead of the chromogenic substrate CPRG used in MetPLATE. FluoroMetPLATE™ displays a higher sensitivity to heavy metals (Tab. 3) but, as its chromogenic counterpart, is relatively insensitive to organic toxicants. Its sensitivity to metals is similar to that of the *Ceriodaphnia dubia* acute assay when testing pure metal solutions or industrial effluents (Jung et al., 1996).

Other applications of the MetPLATE test include the assessment of heavy metal toxicity in aquatic environments (Gupta and Karuppiah, 1996), soils (Bitton et al., 1996; Brohon and Gourdon, 2000; Kong et al., 2003), sediments (Bitton et al., 1992a; de Vevey et al., 1993; Kong et al., 1998), leachates from wood treated with CCA and other wood preservatives (Stook et al., 2001), municipal solid waste (MSW) landfill leachates (Ward et al., 2002), teapots (Boularbah et al., 1999) and metal accumulation in plants (Boularbah et al., 2000).

Table 2. Sensitivity of MetPLATE™ to heavy metals in comparison with Microtox, Daphnia, and fish bioassays[1].

	IC50 (mg/L)	***Range of IC50s, EC50s or LC50s (mg/L)***		
Metal	***MetPLATE™ 1 hr***	***Microtox[2] 15 min***	***Daphnia magna 48-hr[2]***	***Rainbow trout 96-hr[2]***
Cd	0.029 ± 0.001	19 – 220	0.041-1.9	0.15 - 2.5
Cr (III)	6.9± 0.31	13	0.10 - 1.8	11
Cu	0.22 ± 0.04	0.076 – 3.8	0.020 - 0.093	0.25
Pb	10 ± 0.3	1.7 – 30	3.6	8.0
Hg	0.038 ± 0.001	0.029 – 0.05	0.0052 - 0.21	0.033 - 0.2
Ni	0.97 ± 0.02	23	7.6	36
Zn	0.11 ± 0.001	0.27 – 29	0.54 - 5.1	0.55 - 2.2

[1] Adapted from Bitton et al., (1994)

[2] Data drawn from the literature (original references in Bitton et al., 1994).

4.2 HEAVY METAL BINDING CAPACITY (HMBC)

The Heavy Metal Binding Capacity (HMBC) concept initiated in our laboratory (Huang et al., 1999) is similar to that of WER, except that the ratio is obtained by using MetPLATE which, as mentioned above, is specific for heavy metal toxicity. This ratio assesses the binding and complexing ability of a given environmental sample toward added heavy metals.

Table 3. Comparison of the sensitivity of FluoroMetPLATE™ with the 48-h acute Ceriodaphnia bioassay[1].

Toxicant	***FluoroMetPLATE™ IC50 (mg/L)***	***C. dubia 48-hr EC50 (mg/L)***
Cadmium (II)	0.0029 ± 0.0003	0.054 ± 0.0026
Copper (II)	0.0124 ± 0.0007	0.011 ± 0.0010
Lead (II)	1.8675 ± 0.1998	0.118 ± 0.0045
Mercury (II)	0.0037 ± 0.0004	0.013 ± 0.0005
Zinc (II)	0.0521 ± 0.0037	0.060 ± 0.0122
SDS	> 2500	10 ± 2.9
Phenol	> 1250	14 ± 7.1
Pentachlorophenol	> 500	0.33 ± 0.058
2,4,6-Trichlorophenol	> 625	4.0 ± 0.53

[1]Jung et al. (1996).

The methodology for the determination of HMBC, using MetPLATE™ as a toxicity test, is shown in Figure 2 (Huang et al., 1999). This methodology was adapted from that proposed by the U.S. EPA for the determination of the Water Effect Ratio (U.S. EPA, 1982; 1984; 1994). Both the laboratory water (MHW) and the site water are spiked with a given metal from a stock solution and the mixtures are shaken for 60 min at 25°C. Both mixtures are serially diluted and assayed for metal toxicity, using the MetPLATE test. The IC50s for the laboratory and site waters were determined and HMBC was calculated as follows:

$$HMBC = \frac{IC50\ of\ metal\ in\ site\ water}{IC50\ of\ metal\ in\ MHW} \tag{3}$$

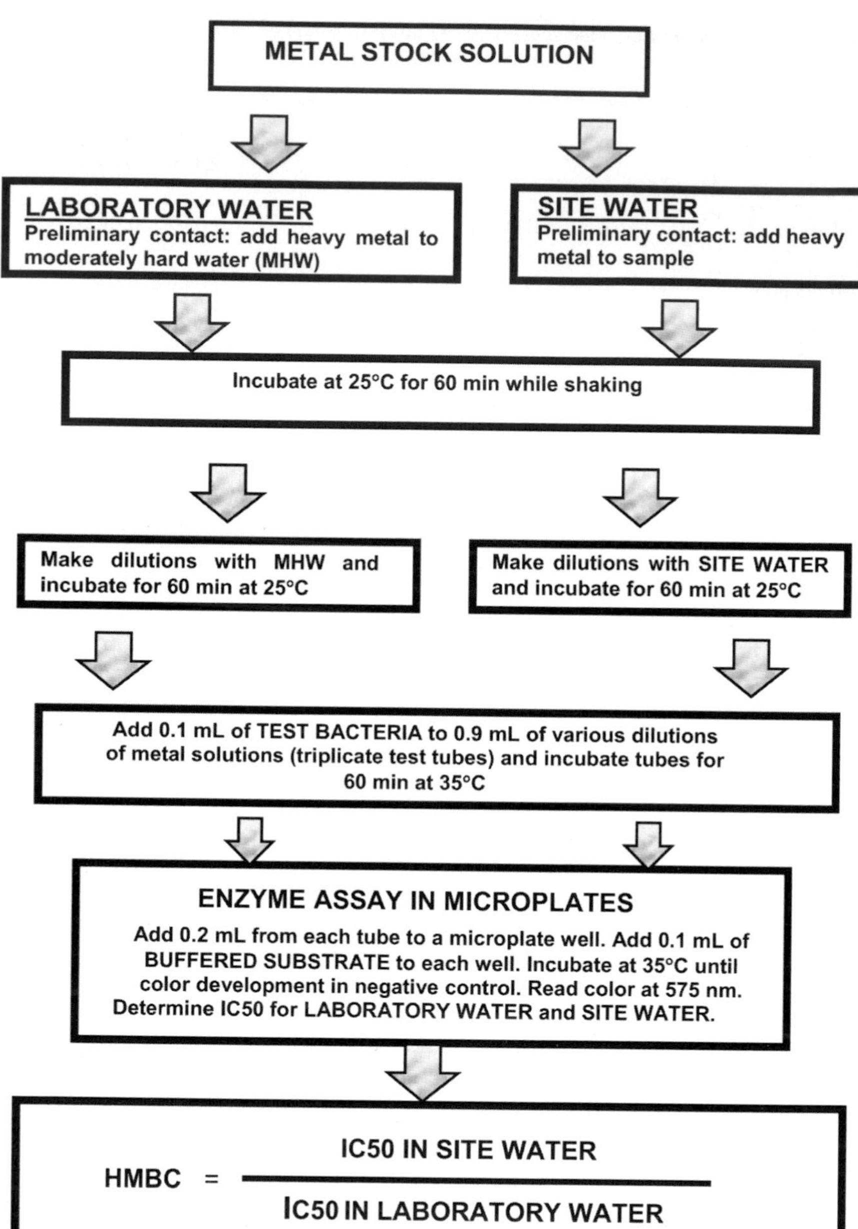

Figure 2. Methodology for HMBC determination.

5. Applications of HMBC test

5.1 HMBC OF SURFACE WATERS

HMBCs of surface waters (lakes, springs, creeks, estuarine water, seawater, and wetlands) were determined, using the MetPLATE toxicity test (Tab. 4; Huang et al., 1999). HMBC for Cd and Cu varied from 1.7 to 39.2 and < 1 to 11.9, respectively. Surprisingly, HMBC for Cd was generally higher than for Cu. Wetland water exerted the highest HMBC for Cd whereas seawater had the highest HMBC for Cu.

Table 4. Heavy Metal Binding Capacity (HMBC) for cadmium and copper, as determined by MetPLATE ™ (adapted from Huang et al., 1999).

Sample/Location	***HMBC-Cd***	***HMBC-Cu***
Hogtown Creek (Gainesville, FL)	5.4 ± 2.0*	2.8 ± 0.9*
Creek E (Gainesville, FL)	2.3 ± 0.4*	< 1
Glen Spring (Gainesville, FL)	1.7 ± 0.6	2.8 ± 0.3**
Newnan's Lake (Gainesville, FL)	3.5 ± 0.4**	< 1
Lake Alice (Gainesville, FL)	3.0 ± 0.2*	3.3 ± 2.4
Orange Lake (FL)	NT[a]	2.4 ± 0.8*
Okefenokee Swamp (GA)	39.2 ± 1.0**	2.4 ± 0.0**
Estuarine water (St John's River, Jacksonville, FL)	9.5 ± 0.0**	4.2 ± 0.1**
Seawater (St. Augustine, FL)	5.7 ± 2.2*	11.9 ± 2.3**

[a] NT = not tested.
* The difference between IC50 of sample and moderately hard water is significant at 95% confidence level.
** The difference between IC50 of sample and moderately hard water is significant at 99% confidence level.

5.2 SEASONAL VARIATION OF HMBC IN THE ST. JOHN'S RIVER, JACKSONVILLE, FL.

Water quality parameters (*e.g.*, dissolved organic carbon, suspended solids, hardness) and climate-related factors (*e.g.*, temperature, precipitation) vary seasonally. This, in turn, affects the HMBC of aquatic environments. As an example, Figure 3 shows the seasonal variation of HMBC-Cu in the St John's River, Jacksonville, FL (Huang et al., 1999). HMBC-Cu varied between 1.5 and 4.2 and displayed a peak in the fall season. Using *Ceriodaphnia dubia* as the test organism, Diamond et al. (1997a) also found that WER-Cu varied between 3.5 and 6.1 in the Lehigh River.

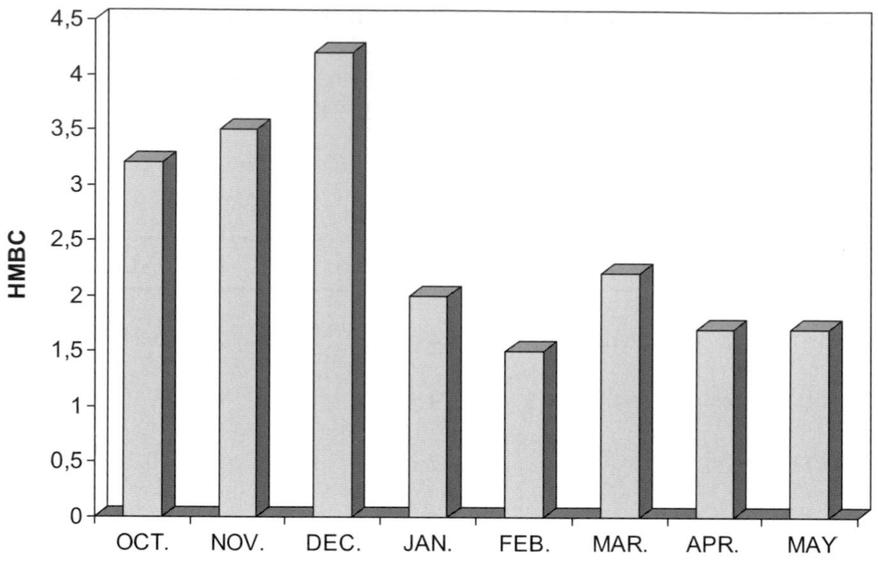

Figure 3. Seasonal variation of HMBC-Cu in the St. John's River, Jacksonville, FL.

5.3 HMBC OF MUNICIPAL SOLID WASTES (MSW) LEACHATES

The HMBC approach and methodology were used to assess the metal bioavailability/toxicity of MSW leachates and to compare them to wastewater effluents and lake water as regards their metal binding capacity.

Table 5 shows the HMBC of 7 MSW landfill leachates for copper, zinc and mercury. These data indicate that the MSW landfill leachate metal binding capacity was relatively high and was site-specific. HMBCs for the MSW leachate samples ranged from of 2.9 to 114.9, 4.9 to 45.2, and 3.6 to 100.8 for HMBC-Cu^{+2}, HMBC-Zn^{+2}, and HMBC-Hg^{+2}, respectively. Comparatively, much lower HMBC values were obtained for other environmental samples, such as lake water (Lake Alice and Lake Beverly) and a wastewater treatment plant effluent (data not shown).

For sites sampled on several occasions, a wide range of HMBC values was found in the landfill leachates. As an example, HMBCs for leachates sampled at site 3, varied from 7.9 to 327.9 while HMBCs for Zn varied from 8.8 to 101.1.

High strength leachates generally displayed the highest HMBCs. This is due to the elevated concentrations of complexing agents (*e.g.* organic and inorganic ligands) that are present in high-strength leachates (Sletten et al., 1995). A partial HMBC fractionation methodology was developed for MSW leachates in Florida (data not shown). This preliminary fractionation scheme essentially demonstrated that solids, organics and hardness were responsible for the metal binding capacity of landfill leachates.

Table 5. HMBC for Municipal Solid Waste leachates from Florida Landfills.

Landfill #	***HMBC***[a]		
	$Cu^{+2}(CuSO_4)$	***Zn^{+2} ($ZnCl_2$)***	***Hg^{+2} ($HgCl_2$)***
	Mean	***Mean***	***Mean***
1	54.5 (1.6-162.9)	25.7 (8.8-42.7)	30.5 ± 2.4
2	34.9 (7.2-56.2)	48.1 (6.2-115.7)	21.7 (16.7-26.7)
3	114.9 (7.9-327.9)	32.3 (8.8-101.1)	86.8 (85.2-88.3)
4	2.9 (2.7-3.1)	10.9 (10.7-11.1)	3.6 ± 0.2
5	27.5 (1.1-79.7)	6.1 (1.4-10.7)	12.9 (6.8-19.1)
6	59.7 ± 3.6	45.2 ± 9.9	100.8 ± 8.4
7	26.7 (9.9-43.5)	4.9 (3.8-5.9)	13.3 ± 1.1
Lake Alice	1.1 ± 0.1	1.8 ± 0.3	2.8 ± 0.1
Lake Beverly	< 1	< 1	< 1
WWTP Effluent	2.4 ± 0.1	2.6 ± 0.3	2.3 ± 0.1

[a] HMBC is unitless. Results are presented as mean and range (in brackets). Mean ± one standard deviation is shown for sites sampled only once. Ranges are shown for sites sampled on multiple occasions.

5.4 SOIL HMBC (SHMBC)

Soils also display a wide range of physico-chemical characteristics that influence the bioavailability of toxic metals to microorganisms, fauna, and plant roots. These characteristics include soil texture, organic matter, cation and anion type, pH, or alkalinity. The presence of colloidal clay minerals in soils greatly increases the adsorption and immobilization of metals in soils via adsorption and ion exchange, thus reducing their toxic effect. Similarly, humic substances in soils contribute to metal complexation and reduction in bioavailability. Thus, metal bioavailability in soils varies widely according to the soil type. Researchers studying metal uptake by plants often use clean sand spiked with metals and measure the metal content of the plants following a given growth period. This approach does not give a true picture of metal binding in other soils with varying physico-chemical characteristics.

It seemed therefore appropriate to develop a relatively rapid test to determine the soil potential for "binding" (*i.e.*, via adsorption, chelation, precipitation ...) metals. This has been traditionally accomplished by chemical analyses. However, this approach does not give complete information about metal bioavailability in soils. A preliminary investigation (unpublished data) was carried out on the use of a biotest (*e.g.*, MetPLATE) to gain knowledge on the *soil metal binding capacity* (SHMBC). The SHMBC methodology again incorporates the use of MetPLATE, a test specific for heavy metals, to determine metal bioavailability.

SHMBC methodology briefly consists of adding a given metal solution, at various concentrations, to the soil under investigation. The metal solution is also added to clean Ottawa sand, which serves as the control soil with no or little metal binding capacity. The soil-metal mixtures are shaken at room temperature for 4 hours. The mixtures are centrifuged, the supernatants are assayed by MetPLATE and an IC50 is determined for the metal.

The soil HMBC (SHMBC) is the ratio of the IC50 of a metal in a soil sample divided by the IC50 of a metal in a reference soil.

$$SHMBC = \frac{IC50\ of\ field\ soil\ spiked\ with\ a\ given\ metal}{IC50\ of\ reference\ soil\ spiked\ with\ the\ same\ metal} \quad (4)$$

To illustrate the SMBC concept, Table 6 shows the SHMBC of three different soils and a clay mineral (bentonite) for Cu, Zn, and Hg. As expected, the binding capacity of soils is quite high. Of the three soils tested, the sandy soil displayed the lowest SHMBC. Their SHMBC decreased in the following order for the three metals: organic soil > red clay soil > sandy soil. Pure bentonite clay mineral displayed the highest SHMBC with nearly 6,000 for Hg. For this metal, its binding was much higher than that of Cu and Zn in the red clay soil and organic soil.

In summary, the new methodology for soil HMBC demonstrated that the metal toxicity attenuating capacity of solids such as soils and minerals can be rapidly determined, with the entire procedure taking only a few hours, as compared to weeks required for soil column studies.

Table 6. SHMBC of soils and a clay mineral for Cu, Zn, and Hg.

Soil Type	***HMBC-Cu***	***HMBC-Zn***	***HMBC-Hg***
Sandy soil	91.2	33.9	72.2
Red clay soil	202.2	178.3	1,488.3
Organic soil	621.6	399.0	1,697.9
Bentonite clay	4,710.8	2,871.1	5,898.6

5.5 HMBC OF ROOT EXUDATES FROM CULTIVATED PLANTS

Root exudates from terrestrial plants are composed of a number of organic compounds that play a role as metal chelators for enhancing the uptake of certain trace metals, and may also protect plants from toxicity caused by high metal concentrations (Hall, 2002). Hydroponic cultures of *Triticum aestivum, Brassica napus*, *Matricaria inodora* and *Centaurea cyanus* were undertaken under sterile condition in the laboratory. Following 14 to 25 days of cultivation, the root exudates were collected and tested for their organic carbon content and their heavy metal binding capacity (HMBC) towards Cu, using MetPLATE. Table 7 shows that, as compared to the blank (Hoagland nutrient solution), the root exudates from *Triticum aestivum* displayed the highest HMBC (HMBC = 3.00) of the four plants tested (Dousset et al., 2001). Cu complexation appears to be related to the carbon content of the plant exudates. However, despite the similar organic content of root exudates from *Triticum aestivum,* and *Brassica napus,* the former exerted a higher HMBC than the latter (Tab. 7). This suggests that the nature of organic compounds in the root exudates might play a role in copper complexation and subsequent decrease in toxicity.

5.6 HMBC OF ROOT EXUDATES OF AQUATIC PLANTS FROM WETLANDS

Wetland waters, owing to their high levels of humic substances and other organic materials, have the ability to bind metals and render them less toxic. Root exudates from wetland aquatic plants are also involved in binding and detoxifying metals. Wetland plants, *Orontium aquaticum*, *Pontederia cordata* and *Sagittaria lancifolia*, were grown axenically in a chelator-free liquid medium (Hoagland solution minus EDTA). Root exudates from these plants were collected and their HMBC was determined. It was found that root exudates from the three plants detoxified water contaminated with metals. Their HMBC for Cd was 1.3, 1.8 and 28 for *Pontederia cordata, Orontium aquaticum* and *Sagittaria lancifolia*, respectively. The growth medium for the plants did not exert any HMBC and behaved as the moderately hard water, which served as the negative control (Neori et al., 1993). Therefore, some wetland plants have the capacity to release some metal-binding organic compounds (*e.g.*, humic substances) that probably allow them to withstand the adverse effects of metal contamination. Comparatively, some wetland waters (Okefenokee Swamp, GA) can exert HMBCs as high as 39 (see Tab. 4) (Huang et al., 1999). These preliminary results show the potential use of HMBC testing to determine the role of root exudates in metal bioavailabiliy although it is realized that the rhizosphere of aquatic and terrestrial plants is a much more complex environment (St-Cyr et al., 1993; St-Cyr and Campbell, 1996).

Table 7. Cu binding capacity of root exudates[1].

Plants	***Organic content of exudate solutions (mg/L)***	***HMBC***
Triticum aestivum	15.3	3.00
Brassica napus	15.7	1.73
Centaurea cyanus	10.8	1.40
Matricaria inodora	1.2	1.07
Blank (Hoagland nutrient solution)	ND[2]	1.00

[1]Adapted from Dousset et al., 2001.
[2]ND = none detected.

6. Conclusions

We have addressed the topic of metal bioavailability and metal toxicity in environmental samples. Traditionally, metal availability is investigated using a chemical approach. Afterwards, the concept of Water Effect Ratio (WER) was proposed by the U.S. EPA and employed bioassays (*e.g.*, fish and invertebrate tests) to assess metal bioavailability and toxicity. In the HMBC approach discussed in this review, we have made use of a bacterial assay that is specific for metal toxicity to achieve this goal. This is only a preliminary survey of the potential applications of the HMBC concept. Some preliminary results on the use of MetPLATE for the fractionation of HMBC to obtain information on the factor(s) that control metal bioavailability in environmental samples were also presented. Using MetPLATE eliminates or diminishes the confounding factor represented by the presence of organic toxicants in a given sample. Further work is needed to refine the fractionation scheme.

Microbial communities play an essential role in nutrient cycling and in the fate of trace metals in aquatic environments and in engineered systems (*e.g.*, wastewater treatment plants). However, some of the microorganisms (*e.g.*, nitrifying bacteria) are quite sensitive to metals in aquatic environments (Twiss et al., 1996). The HMBC test may thus also prove useful to assess the potential adverse effects of metals on microbial communities in aquatic environments.

Acknowledgements

This work was in part funded by National Science Foundation grant # BES-9906060.

References

Allen, H.E. (1993) Water quality criteria for metals, *Water Environment Research* **65**, 195- 198.

Allen, H.E. and Hansen, D.J. (1996) The importance of trace metal speciation to water quality criteria, *Water Environment Research* **68**, 42-53.

Allison, J.D. and Perdue, E.M. (1994) Modeling metal–humic interactions with MINTEQA, in N. Senesi and T.M. Miano (eds.), *Humic Substances in the Global Environment and Implications on Human Health*. Elsevier Science, Amsterdam, pp. 927-942.

Ankley, G.T., Di Toro, D.M., Hansen, D.J. and Berry, W.J. (1996) Technical basis and proposal for deriving sediment quality criteria for metals, *Environmental Toxicology and Chemistry* **15**, 2056-2066.

Benedetti, M., Ranville, J.F., Ponthieu, M. and Pinheiro, J.P. (2002) Field-flow fractionation characterization and binding properties of particulate and colloidal organic matter from the Rio Amazon and Rio Negro, *Organic Chemisrty* **33**, 269-279.

Berry, W.J., Hansen, D.J., Mahony, J.D., Robson, D.L., Di Toro, D.M., Shipley, B.P., Rogers, B., Corbin, J.M. and Boothman, W.S. (1996) Predicting the toxicity of metal-spiked laboratory sediments using acid-volatile sulfide and interstitial water normalizations, *Environmental Toxicology and Chemistry* **15**, 2067-2079.

Bitton, G. and Morel, J.L. (1998) Enzyme assays for the detection of heavy metal toxicity, in P.G. Wells, K. Lee, and C. Blaise (eds), *Microscale Aquatic Toxicology: Advances, Techniques and Practice*, CRC Press, Boca Raton, FL.

Bitton, G., Campbell, M. and Koopman, B. (1992a) MetPAD: A bioassay kit for the specific determination of heavy metal toxicity in sediments from hazardous waste sites, *Environmental Toxicology and Water Quality* **7**, 323-328.

Bitton, G., Koopman, B. and Agami, O. (1992b) MetPAD^{TM}: a bioassay for rapid assessment of heavy metal toxicity in wastewater, *Water Environment Research* **64**, 834-836.

Bitton, G., Jung, K. and Koopman, B. (1994) Evaluation of a microplate assay specific for heavy metal toxicity, *Archives of Environmetal Contamination and Toxicology* **27**, 25-28.

Bitton, G., Garland, E., Kong, I.-C., Morel, J.L. and Koopman, B. (1996) A Direct solid phase assay for heavy metal toxicity. I. Methodology, *Journal of Soil Contamination* **5**, 385-394.

Boularbah, A., Bitton, G. and Morel, J.L. (1999) Assessment of metal content and toxicity of leachates from teapots, *Science of the Total Environment* **227**, 69-72.

Boularbah, A., Bitton, G., Morel, J.L. and Schwartz, C. (2000) Assessment of metal accumulation in plants using MetPAD, a toxicity test specific for heavy metal toxicity, *Environmental Toxicology* **15**, 449-455.

Brohon, B., and Gourdon, R. (2000) Influence of soil microbial activity level on the determination of contaminated soil toxicity using Lumistox and MetPlate bioassays, *Soil Biology and Biochemistry* **32**, 853-857.

Brown, D.S. and Allison, J.D. (1987) MINTEQA1, an Equilibrium Metal Speciation Model: User's Manual., *EPA/600/3-87/012, U.S.* Environmental Protection Agency, Athens, Georgia.

Calace, N., Liberatori, A., Petronio, B.M. and Pietroletti, M. (2001) Characteristics of different molecular weight fractions of organic matter in landfill leachates and their role in soil sorption of heavy metals, *Environmental Pollution* **113**, 331-339.

Christl, I. and Kretzschmar, R. (2001) Relating ion binding by fulvic and humic acids to chemical composition and molecular size: I. Proton binding, *Environmental Science and Technology* **35**, 2505-2511.

Diamond, J.M., Koplish, D.E., McMahon III, J. and Rost, R. (1997a) Evaluation of the water-effect ratio procedure for metals in a riverine system, *Environmental Toxicology and Chemistry* **16**, 509-520.

Diamond, J.M., Gerardi, C., Leppo, E. and Miorelli, I. (1997b) Using a water-effect ratio approach to establish effects of an effluent-influenced stream on copper toxicity to the fathead minnow, *Environmental Toxicology and Chemistry* **16**, 1480-1487.

Di Toro, D.M., Mahony, J.D., Hansen, D.J., Scott, K.J., Hicks, M.B., Mayr, S.M. and Redmont, M.S. (1990) Toxicity of cadmium in sediments: The role of acid volatile sulfide, *Environmental Toxicology and Chemistry* **9**, 1487-1502.

Di Toro, D.M., Mahony, J.D., Hansen, D.J., Scott, K.J., Carison, A.R. and Ankley, G.T. (1992) Acid volatile sulfide predicts the acute toxicity of cadmium and nickel in sediments, *Environmental Toxicology and Chemistry* **26**, 96-101.

Dousset, S., Morel, J.L., Jacobson, A. and Bitton, G. (2001) Copper binding capacity of root exudates of cultivated plants and associated weeds, *Biology and Fertility of Soils* **34**, 230-234.

Ge, Y., Murray, P., Sauve, S. and Hendershot, W. (2002) Low metal bioavailability in a contaminated urban site, *Environmental Toxicology and Chemistry* **21**, 954-961.

Giesy, J.P., Briese, L. and Leversee, G.J. (1978) Metal binding capacity of selected Maine surface waters, *Environmental Geology* **2**, 257-268.

Gupta, G. and Karuppiah, M. (1996) Toxicity study of a Chesapeake Bay tributary - Wicomico River, *Chemosphere* **32**, 1193-1215.

Hall, J.C. and Raider, R.L. (1993) Interim guidance for metals criteria has CERCLA implications, *Water Enviroment. and Technology* **5**, 96-99.

Hall, J.C., Raider, R.L. and Grafton, J.A. (1992) EPA's heavy metal criteria strategies for obtaining reasonable limitations, *Water Environment and Technology.* **4**, 60-62.

Hall, J.L. (2002) Cellular mechanisms for heavy metal detoxification and tolerance, *Journal of Experimental Botany* **53**, 1-11.

Huang, F., Bitton, G. and Kong, I.-C. (1999) Determination of the heavy metal binding capacity of aquatic samples using MetPLATE: a preliminary study, *Science of The Total Environment* **234**, 139-145.

Jung, K., Bitton, G. and Koopman, B. (1996) Selective assay for heavy metal toxicity using a fluorogenic substrate, *Environmental Toxicology and Chemistry* **15**, 711-714.

Kang, K.-H., Shin, H.S. and Park, H. (2002) Characterization of humic substances present in landfill leachates with different landfill ages and its implications, *Water Research* **36**, 4023-4032.

Kaschl, A., Romheld,V. and Chen, Y. (2002) The influence of soluble organic matter from municipal solid waste compost on trace metal leaching in calcareous soils, *Science of The Total Environment* **291**, 45-57.

Kong, I.-C, and Bitton, G. (2003) Correlation between toxicity and metal fractions of contaminated soils in Korea, *Bulletin of Environmental Contamination and Toxicology* **70**, 557-565.

Kong, I.-C, Bitton, G., Koopman, B. and Jung, K-H. (1995) Heavy metal toxicity testing in environmental samples, *Reviews of Environmental Contamination and Toxicology* **142**, 119-147.

Kong, I.-C., Lee, C.W. and Kwon, Y.-T. (1998) Heavy metal toxicity monitoring in sediments in Jinhae Bay, Korea, *Bulletin of Environmental Contamination and Toxicology* **61**, 505-511.

Koukal, B., Gueguen, C., Pardos, M, and Dominik, J. (2003) Influence of humic substances on the toxic effects of cadmium and zinc to the green alga *Pseudokirchneriella subcapitata, Chemosphere* **53**, 953-961.

Neori, A., Clark, M., Beck, J., Huang, F., Kane, M.A. and Bitton, G. (1993) Heavy metal binding by natural waters and soluble exudates from micropropagated axenic wetland plants, measured with MetPLATE, a bioassay for rapid assessment of heavy metal toxicity, *Amer. Soc. Limnol. Oceanog. (ASLO)-Soc. Wetland Scientists Ann. Joint Meeting*, Univ. Alberta, Edmonton, May 31-June 3, 1993.

Parat, C., Chaussod, R., Leveque, J. Dousset, S. and Andreux, F. (2002) The relationship between copper accumulated in vineyard calcareous soils and soil organic matter and iron, *European Journal of Soil Science* **53**, 663-669.

Sinclair, R.M. (1989) *The development of a Site Specific Water Quality Standard for Copper*, Ph.D. Dissertation, Vanderbilt University, Nashville, Tennessee.

Sletten, R.S., Benjamin, M.M., Horng, J.J. and Ferguson, J.F. (1995) Physical-chemical treatment of landfill leachates for metals removal, *Water Research* **29**, 2376-2386.

St-Cyr, L. and Campbell, P.G.C. (1996) Metals (Fe, Mn, Zn) in the root plaque of submerged aquatic plants collected *in situ*: relations with metal concentrations in the adjacent sediments and in the root tissue, *Biogeochemistry* **33**, 45-76.

St-Cyr, L., D. Fortin, and Campbell, P.G.C. (1993) Microscopic examination of the iron plaque of a submerged aquatic plant (*Vallisneria Americana* Michx), *Aquatic Botany* **46**, 155-167.

Stook, K., Ward, M., Townsend, T., Bitton, G. and Booth, M. (2001) Toxicity evaluation of leachates from preserved wood (Abstract) *10th International Symposium on Toxicity Testing*, August 2001, Quebec City, Canada.

Tessier, A. and Turner, D.R. (1995) *Metal Speciation and Bioavailability in Aquatic Systems*, John Wiley and Sons, Hoboken, N.J.

Twiss, M.R., Campbell, P.G.C. and Auclair, J.-C. (1996) Regeneration, recycling, and trophic transfer of trace metals by microbial food-web organisms in the pelagic surface waters of Lake Erie, *Limnology and Oceanography* **41**, 1425-1437.

U.S. EPA (U.S. Environmental Protection Agency) (1982) *Water Quality Standards Handbook* (Draft), Office of Water Regulations and Standards, Washington, D.C., 20460.

U.S. EPA (U.S. Environmental Protection Agency) (1984) *Guidelines for deriving numerical aquatic site-specific water quality criteria by modifying national criteria*, EPA/600/384009, Environ. Res. Lab., Duluth, Minn.

U.S. EPA (U.S. Environmental Protection Agency) (1994) *Interim Guidance on determination and use of water-effect ratios for metals*. EPA-823-B-94-001, Washington, DC.

U.S. EPA (U.S. Environmental Protection Agency) (2001) *Streamlined Water-Effect Ratio Procedure for Copper Discharges*, Report 822-R-01-005, Office of Science and Technology, Washington, D.C., 41 pp.

de Vevey, E., Bitton, G., Rossel, D., Ramos, L.D., Guerrero, L.M. and Tarradellas, J. (1993) Concentration and bioavailability of heavy metals in sediments in Lake Yojoa (Honduras), *Bulletin of Environmental Contamination and Toxicology* **50**, 253-259.

Voelker, B.M. and Kogut, B. (2001) Interpretation of metal speciation data in coastal waters: the effects of humic substances on copper binding as a test case, *Marine Chemistry* **74**, 303-318.

Ward, M.L., Bitton, G., Townsend, T. and Booth, M. (2002) Determining toxicity of leachates from Florida municipal solid waste landfills using a battery-of-tests approach, *Environmental Toxicology* **17**, 258-266.

Abbreviations

AVS	Acid Volatile Sulfide
AWQC	Ambient Water Quality Criterion
CCA	Chromated copper arsenate
CPRG	chlorophenol red-β-D-galactopyranoside
HMBC	Heavy Metal Binding Capacity
MSW	Municipal Solid Wastes
MUGA	4-methyl umbelliferyl galactopyranoside
SDS	Sodium dodecyl sulfide
SHMBC	Soil Heavy Metal Binding Capacity
SMAV	Species Mean Acute Value
U.S. EPA	U.S. Environmental Protection Agency
WER	Water Effect Ratio.

7. THE APPLICATION OF HAZARD ASSESSMENT SCHEMES USING THE WATERTOX TOXICITY TESTING BATTERY

ALICIA RONCO[1], GUSTAVO BULUS ROSSINI[2], CECILIA SOBRERO[3], CARINA APARTIN[4]
CIMA, Facultad de Ciencias Exactas
Universidad Nacional de la Plata
La Plata, Argentina. 47 y 115, (1900)
[1]*cima@quimica.unlp.edu.ar*
[2]*gbulus@quimica.unlp.edu.ar*
[3]*csobrero@quimica.unlp.edu.ar*
[4]*apartin@quimica.unlp.edu.ar*

GABRIELA CASTILLO
Departamento de Ingeniería Civil
Facultad de Ciencias Físicas y Matemáticas
Universidad de Chile, Casilla 228-3 Santiago, Chile
gcastilo@ing.uchile.cl

M. CONSUELO DÍAZ-BAEZ[1], ADRIANA ESPINOSA RAMÍREZ[2]
Facultad de Ingeniería
Universidad Nacional de Colombia. Bogotá, Colombia
[1]*mcdiazb@unal.edu.co*
[2]*ajespinosar@unal.edu.co*

INÉS AHUMADA[1], JORGE MENDOZA[2]
Facultad de Ciencias Químicas y Farmacéuticas
Universidad de Chile, Santiago, Chile
[1]*iahumada@ciq.uchile.cl*
[2]*jmendoza@ciq.uchile.cl*

1. Objectives

The development or application of new or existing toxicity ranking systems, based on the use of a battery of tests inspired by the WaterTox Program are presented.

C. Blaise and J.-F. Férard (eds.), Small-scale Freshwater Toxicity Investigations, Vol. 2, 233-255.

These systems allow for the aquatic toxicity assessment of water-soluble contaminants from different type of matrices.

Application examples are provided for surface waters, groundwater, and pore waters from sediments, as well as complex environmental samples including industrial wastewaters, biosolids from municipal treatment plants, hazardous wastes and waste leachates.

The WaterTox battery, comprising standardized toxicity tests which have undergone intercalibration exercises, is simple to use, scientifically robust, cost-effective and user-friendly. Applications of different Hazard Assessment Schemes using the battery of test approach are discussed.

2. Summary

The application of a core battery of WaterTox Program toxicity tests were applied to different types of samples by three South American laboratories. The core battery included the following tests:

- *Lactuca sativa,* 120 h inhibition of germination and root elongation test (Dutka, 1989a);
- *Daphnia magna*, 24-48 h acute lethality test (Dutka, 1989b);
- *Hydra attenuata*, 48-96h acute lethality and sublethality test (Blaise and Kusui, 1997; Trottier et al., 1997); and
- *Pseudokirchneriella subcapitata* -formerly *Selenastrum capricornutum*- 72-h growth inhibition test (Blaise et al., 2000).

Different Hazard Assessment Schemes (HAS) were developed based on existing indexes. In Colombia, the PEEP index (Costan et al., 1993) was used to compare the toxic potential of industrial wastewaters. In Chile, a ranking scheme based on approaches proposed by the National Water Research Institute of Environment Canada (Dutka, 1988; Dukta and Kwan, 1988) and Gent University, Belgium (Persoone et al., 2003) was employed to assess the toxicity of soluble contaminants associated with biosolids from municipal wastewater treatment plants. In Argentina, a classification system for water samples, the Effect-Dilution Average Ratio Index (EDAR), was utilized to assess hazardous waste, leachates, water and sediment pore waters. In this chapter, applications of these Hazard Assessment Schemes are discussed based on the ranking scale of each HAS and toxicity test responses to pure compounds subsequent to an initial round-robin exercise. Some limitations are encountered in applying the test battery especially owing to the lack of sensitivity related to insoluble toxicants. Testing samples concentrated by pre-treatment with solvents proved unhelpful, since high dilutions were then required to avoid carrier toxic effects. Overall, the application of specific HAS schemes with the WaterTox battery of toxicity tests contributed ecotoxicological information that identified the more problematic water samples and wastes in three South American countries. Such information is crucial for subsequent decision-making that will lead to improved protection and conservation of aquatic ecosystems.

3. Historical overview and applications

The International Development Research Centre (IDRC, Ottawa, Canada) created an international network of laboratories (WaterTox) whose goal was to identify and use a battery of toxicity tests which were simple and easy to use, affordable, yet sensitive and reliable, for water toxicity testing (Forget et al., 2000). In the original WaterTox battery (Phase I) the following toxicity tests were used:

- onion root bundle growth assay (Fiskesjö, 1993);
- the lettuce seed germination (Dutka, 1989a) 120-h exposure assay (root and seedling length);
- the *Daphnia magna* 48-h lethality test (Dutka, 1989b);
- the *Hydra* 96-h lethality (tulip stage and disintegration of organisms) and sub-lethality (morphological changes: clubbing and shortening of tentacles) assay (Blaise and Kusui, 1997; Trottier et al., 1997);
- the Muta-Chromoplate mutagenicity test (conducted according to instructions provided with this commercial kit); and
- the nematode maturation bioassay (Samoiloff et al., 1980).

Based on criteria evaluating test performance, reproducibility, and user-friendliness, a standardization and calibration exercise was carried out by eight participating laboratories in different countries, involving the testing of 30 blind samples (Phase I). As a result of this exercise, a simplified battery was recommended (Phase II) that called for the use of the lettuce seed germination, *Daphnia* and *Hydra* tests. In addition, an algal test (72-h exposure *S. capricornutum* growth inhibition chronic toxicity), developed within the scope and framework of the WaterTox inter-calibration exercise, was also recommended (Blaise et al., 2000).

Phase II of the exercise involved toxicity screening of environmental (Diaz-Baez et al., 2002) and blind samples (Ronco et al., 2002) with the simplified battery. Critical analysis of each toxicity test was undertaken with the latter samples to evaluate their reliability. This involved looking at such factors as: 1) variability of responses among laboratories to negative controls; 2) conformance with test quality control criteria; 3) false positive responses induced by sample concentration; and 4) variability within and among laboratories of responses to toxic samples. Results indicated that the battery was generally reliable in detecting the presence of toxicity. However, some false positives were identified with a concentrated soft water sample and with the *Lactuca* and *Hydra* (sub-lethal end-point) tests. Probabilities of detecting false positives for individual and combined toxic responses of the four toxicity tests are presented. Overall, inter-laboratory comparisons confirmed good reliability for the battery.

After completion of the WaterTox program, the test battery continued to be applied by laboratories from Argentina, Chile and Colombia to assess different types of environmental matrices. These initiatives facilitated the development or application of new or existing ranking systems that enabled evaluation of the effectiveness of biological treatment for the toxicity reduction of wastes and combined effluents. These studies are described herein.

4. Procedures

4.1 TOXICITY TESTS AND QUALITY CONTROL

Tests employed to describe the studies outlined below on are indicated in Table 1. Each laboratory ran internal quality control charts (U.S. EPA, 1991; Environment Canada, 1999) with known reference toxicants using the following chemicals: Cr(VI) as $K_2Cr_2O_7$ for *D. magna* and *H. attenuata*; Zn(II) as $ZnSO_4 \cdot 7H_2O$ for *L. sativa*; and Cu(II) as $CuSO_4 \cdot 5H_2O$ for *S. capricornutm*. Probit analysis (for *Hydra* and *Daphnia* tests) and non-parametric linear interpolation (for seed and algae tests) were used for the LC/EC/IC50 estimation.

Table 1. Characteristics of small-scale toxicity tests used in the WaterTox battery of tests.

Trophic level	***Toxicity test with test species***	***Assessment and statistical endpoint***	***Reference***
Primary producer	Algal test *Selenastrum capricornutum*	Chronic sublethal growth inhibition (after a 72-h exposure), IC50	Blaise et al., 2000
Primary producer	Vascular plant *Lactuca sativa*	Inhibition of germination, root and shoot elongation (after 120-h exposure), IC50	Dutka, 1989a
Primary consumer	Cladoceran test *Daphnia magna*	Acute lethality (after a 48-h exposure), LC50	Dutka, 1989b
Secondary consumer	Cnidarian test *Hydra attenuata*	Acute lethality (after a 96-h exposure), LC50	Blaise and Kusui, 1997; Trottier et al., 1997
		Acute sublethal indicated by morphological changes (after a 96-h exposure), EC50	

4.2 TEST BATTERY APPROACH

The test battery approach used in toxicity testing is now widely advocated internationally for assessing complex mixtures such as municipal and industrial effluents, or hazardous wastes from different sources, as different trophic levels of aquatic biota can be impacted by specific groups of toxicants. However, ranking samples is complex because different tests in the battery will respond to toxicity to varying degrees. One way to resolve this problem is to integrate test responses into a toxicity index that expresses the relative hazard of different samples by a single numerical value.

5. HAS case studies

The following sections highlight case studies undertaken independently in three South American countries facing different types of environmental problems related to toxic discharges to aquatic environments.

5.1 ARGENTINA

The case study presented here was conducted by the Environmental Research Centre, CIMA, Faculty of Sciences, of the University of La Plata. The more frequently applied tests were those conducted with the *Lactuca sativa* seed germination assay, followed by the *Hydra*, *Daphnia* and *Selenastrum* tests. They were used for the assessment of toxicity from hazardous wastes and waste leachates, sediment pore water and sediment leachates, surface waters and groundwater. An additional test based on β-galactosidase (*in vitro*-free enzyme test) inhibition, known for its sensitivity to metals (Apartin and Ronco, 2001), was also incorporated in the battery.

5.1.1 Effect-Dilution Average Ratio Index (EDAR)

This index (Bulus Rossini et al., 2005), which integrates five tests (four toxicity tests from the WaterTox intercalibration exercise and an enzymatic test sensitive to metals) and six endpoints, was developed as a tool to assess and compare the hazard of water soluble contaminants in surface water bodies and ground water of the coastal region of the Río de la Plata estuary, Argentina (Ronco et al., 1995; 1996; Camilion et al., 2003). The index values were established in such a way that the interval limits for each level were associated with a 20% effect for all tests for a given sample dilution, except for the upper and lower interval of the first and second rank values (Tab. 2). The 0.15 upper interval of the first rank value was based on the consideration that one of the WaterTox toxicity tests produced a negative response and the other four a toxic effect of 20% with the undiluted sample. The responses to three pure compounds from the Phase II WaterTox intercalibration exercise (including the β-galactosidase test) were used to assess the behaviour of the index (Ronco et al., 2002). Application of the index to other types of environmental samples with a reduced battery of three toxicity tests was also conducted using the same principle, but by adapting the toxicity ranking scale.

Different approaches were considered for enhancing the index to rank the ecotoxicity hazard of aqueous samples according to results obtained with the test battery. The selected EDAR index makes use of the sample concentration for each test producing an effect of 20% in line with the following principles:

- The concentration producing a 20% toxic effect (LC/IC/EC20) estimated from the concentration-response curve
- When it is not possible to determine an LC/IC/EC20, the following data can be used:
    - The highest dilution (*i.e.*, lowest concentration) showing a toxic effect of 15% or higher,

Table 2. EDAR index scale for the hazard assessment of aqueous samples using a battery of five toxicity tests.

EDAR Index Interval	***Qualitative hazard description***	***Hazard rank***	***Response for each lower limit value of the interval***
[0-0.15]	Not hazardous	I	Absence of effect with the undiluted sample (100%)
[0.15-0.19]	Possibly hazardous	II	20% effect with the undiluted sample for three tests and no effect in the fourth
[0.19-0.38]	Slightly hazardous	III	20% effect with the undiluted sample with all tests
[0.38-1.9]	Hazardous	IV-1	20% effect in the 50% dilution with all tests
[1.9-3.8]		IV-2	20% effect in the 10% dilution with all tests
[3.8-19]		IV-3	20% effect in the 5% dilution with all tests
[19-38]	Very hazardous	V-1	20% effect in the 1% dilution with all tests
[38-189]		V-2	20% effect in the 0.5% dilution with all tests
> 189	Extremely hazardous	VI	20% effect in the 0.1% dilution with all tests

- The undiluted sample (100%), when toxic responses at this concentration are below 15%; and

➢ Dilutions producing a 100% toxic effect are not used in the index calculation.

To calculate an EDAR value for the given battery, each estimated effect is divided by the corresponding dilution. Since there is no evidence to support a difference in importance between each toxicity test, the same weight was assigned to all the tests, except for the β-galactosidase assay. The highest sample concentration compatible with this test is a 50% v/v dilution, and it was assigned a weight of 0.5. For the tests with more than one end point, the weight is equally divided between all the end points assessed (*i.e.*, for *Hydra*, with two end points, the total weight is 1 and each end point has a weight of 0.5). Since a 0% effect value could occur at a high dilution (a very intense toxic effect), the value of 1 unit has been added to the measured effect before the quotient is calculated in the formula below.

$$EDAR = \frac{\sum_{i=1}^{n} p_i \cdot ((e_i + 1)/d_i)}{\sum_{i=1}^{n} p_i} \tag{1}$$

Where:

p_i is the weight assigned to the endpoint assessed,
e_i is the measured effect corresponding to a d_i dilution,
n is the number of tests/end points in the battery.

A ranking scale of nine levels ranging from ´non-hazardous´ to ´extremely hazardous´ was developed (Tab. 2). The rationale behind the EDAR index is based on averaging out the ecotoxic effects of a given aqueous sample.

Since toxicity assessment of environmental samples do not always yield data conductive for the plotting of a concentration-response curve, quantitative response measurements are sometimes impossible to calculate. To compensate for these shortcomings, the EDAR index averages the % effect with the dilution producing this measured effect, hence normalizing the data from the different tests. Whenever sufficient data were available to obtain a concentration-response plot, we selected the sample concentration used in the index calculation as the one producing a 20% effect on the exposed test organisms. This 20% effect generally corresponds to the lowest concentration indicative of significant differences between negative controls and sample effects, based on the results produced with the WaterTox intercalibration exercise (Ronco et al., 2002).

The ranking scale limit values of the index were set considering the results that would be obtained if all tests yielded a response of 20% to the same concentration or dilution. Each interval of the reference scale was arbitrarily fixed according to valued judgment taking into account the authors' experience.

5.1.2 Application of the EDAR Index to a case study of surface water and groundwater pollution

Samples investigated with the five toxicity tests from the battery were surface water (*i.e.*, El Gato -S1 and 2-, Martin -S6- and Carnaval -S10- streams, Oeste Canal -S3-, water intake for the treatment plant -S7- and near the sewers discharge -S4- both from the Río de la Plata) and groundwater (S8 and 9), all the sites corresponding to the south eastern sector of coastal area of the Río de la Plata (Ronco et al., 1996, 2001; Camilion et al., 2003), and tap water (S5) with conventional treatment (see the location of sampling points in relation to possible contaminant sources in Figure 1). Physico-chemical parameters from all samples were within the following ranges: conductivity 0.3-1.8 mS/cm; hardness 50-450 mg $CaCO_3$/L; dissolved oxygen from non detectable to 8.8 mg/L; DOC mg/L < 20-82 mg/L; alkalinity 100-470 mg $CaCO_3$/L (low dissolved oxygen concentrations and higher DOC and conductivity was detected in surface waters close to contamination sources). Blind positive (*i.e.*, Hg(II) and 4-Nitroquinoline-*N*-oxide) and negative (soft water) samples from the Phase II WaterTox intercalibration exercise (Ronco et al., 2002) and methanol 2% were also tested and ranked using the EDAR index. Scoring results are summarized in Table 3. Samples from surface water bodies considered hazardous according to the scoring system correspond to sectors associated with direct industrial or urban contaminant discharges. Also, potential health hazard from groundwater samples was found to be related to chemical contamination from intensive agriculture. Tap water was sampled from an old lead water pipe. The pure compound index values were clearly higher and in a class apart from the rest of the samples. As expected, no positive toxic responses to the blind negative samples (results not shown in Table 3) were observed in any of the tests in the test battery.

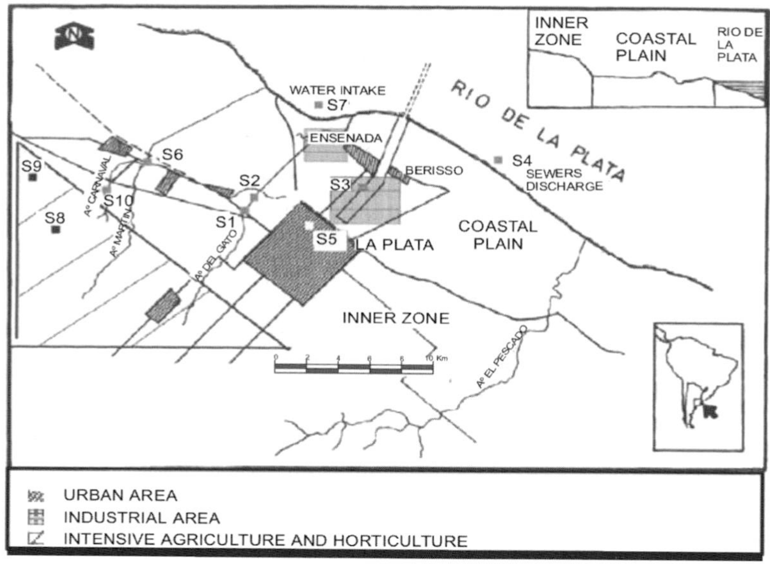

Figure 1. Study area with indication of surface water and groundwater sampling points and type of activity in each sector.

5.1.3 Ranking the toxicity of soluble toxicants in several types of matrix leachates and wastes with a reduced battery of tests.

Owing to possible future restrictions that could preclude the testing of samples with all five toxicity tests, we evaluated the index response with a reduced battery of toxicity tests. The basic rule for the selection of tests in a reduced battery was to maintain one primary producer, one primary consumer and a secondary consumer. The test combinations of two selected reduced batteries were: *Hydra, Daphnia* and *Lactuca* tests (H-D-L) and *Hydra, Daphnia* and *S. capricornutum* (H-D-S) tests

(Tab. 4). These batteries were applied for ranking hazardous waste, pore water and leachates from sediments and wastes. The scale was prepared using the same criteria as previously described (see Section 5.1.1).

Table 3. EDAR index application for the hazard assessment of water samples and pure compounds using a battery of five toxicity tests. Effect:dilution ratio values are indicated for each test.

Sample identification	Algal assay[a]	H-L[b]	H-SL[c]	DM[d]	LS[e]	E[f]	*EDAR Index*	*Sample Hazard Rank*[g]
Environmental samples*								
S1	0.36	0.65	1.0	0.59	0.88	0.01	0.59	IV-1
S2	0.41	0.56	2.9	0.67	0.52	0.01	0.84	IV-1
S3	0.42	0.54	10.10	0.04	0.14	0.28	1.92	IV-2
S4	0.01	0.31	1.3	0.82	0.48	0.01	0.48	IV-1
S5	0.68	0.01	0.01	0.44	0.26	0.35	0.29	III
S6	0.01	0.01	0.17	0.70	0.73	0.01	0.27	III
S7	0.01	0.01	0.28	0.01	0.16	0.01	0.08	I
S8	2.2	0.01	0.06	0.01	1.0	0.01	0.70	IV-1
S9	0.32	1.5	1.5	0.01	0.80	0.01	0.54	IV-1
S10	0.01	0.01	0.01	0.01	0.30	0.01	0.06	I
Pure compounds								
Hg(II) 5 mg/L	90	78	631	228	4	155	198	VI
4-NQO 2 mg/L	300	1.2	8.4	0.43	0.42	0.28	52	V-2
Methanol 2%	2.5	0.01	-	0.01	0.32	0.99	0.76	IV-1

* Sites correspond to the south eastern sector of coastal area of the Río de la Plata (Ronco *et al.*, 1996, 2001, Camilion et al., 2003): El Gato -S1 and 2-, Martin -S6- and Carnaval -S10- streams, Oeste Canal -S3-, water intake for the treatment plant -S7- and near the sewers discharge -S4- both from the Río de la Plata), and groundwater (S8 and 9), see Figure 1.
a) *Selenastrum capricornutum* assay (Blaise et al., 2000).
b) *Hydra attenuata* lethality assay (Blaise and Kusui, 1997; Trottier et al., 1997).
c) *Hydra attenuata* sublethal assay (Blaise and Kusui, 1997; Trottier et al., 1997).
d) *Daphnia magna* assay (Dutka, 1989b).
e) *Lactuca sativa* assay (Dutka, 1989a).
f) Enzyme assay with *β-galactosidase* (Apartin and Ronco, 2001).
g) See Table 2 for details.

Results of the EDAR index application to these types of matrices are provided in Table 5. Samples were selected for toxicity screening to ensure the presence of different types of contaminants commonly present in complex wastes (*e.g.*, hydrocarbons, DOC, nutrients, ammonia, inorganic anions and cations, pesticides) and other matrices (*e.g.*, sediments and sludges, solid materials, liquid phases). Water samples and pure compounds studied with the complete battery (Tab. 3) were also scored with the reduced batteries (H-D-L and H-D-S) for comparison (Tab. 5).

The comparison of the EDAR index values obtained for the same samples with the complete and reduced batteries (H-D-L and H-D-S) indicated a good agreement thus supporting the use of a reduced battery of tests when necessary.

Table 4. EDAR index scale for hazard assessment of aqueous samples using two reduced test batteries, H-D-L[a] and H-D-S[b], each with three toxicity tests.

EDAR Index Interval	***Qualitative Hazard description***	***Hazard rank***	***Response for each lower limit value of the interval***
[0-0.14]	Not hazardous	I	Absence of effect with the undiluted sample (100%)
[0.14-0.21]	Possibly hazardous	II	20% effect with the undiluted sample for two tests and no effect in the third
[0.21-0.42]	Slightly hazardous	III	20% effect with the undiluted sample with all tests
[0.42-2.1]	Hazardous	IV-1	20% effect in the 50% dilution with all tests
[2.1-4.2]		IV-2	20% effect in the 10% dilution with all tests
[4.2-21]		IV-3	20% effect in the 5% dilution with all tests
[21-42]	Very hazardous	V-1	20% effect in the 1% dilution with all tests
[42-210]		V-2	20% effect in the 0.5% dilution with all tests
> 210	Extremely hazardous	VI	20% effect in the 0.1% dilution with all tests

[a] *Hydra*, *Daphnia* and *Lactuca* tests.
[b] *Hydra*, *Daphnia* and *S. capricornutum* tests.

It was observed that EDAR index values and intervals (Tables 2 and 4) did not change markedly with the deletion of two toxicity tests. Results of applying the EDAR index to waste samples indicate that values and ranks relate to the solubility of toxicants in aqueous phases. Based on this evaluation, wastes from photographic and X-Ray laboratories were observed to be extremely hazardous in contrast to hydrocarbon-containing waste leachates, described as either slightly hazardous or hazardous. The existence of sub-levels for an equivalent hazard description allows for better sample discrimination (*e.g.*, Pharmaceutical solid waste leachate *versus* liquid waste with pesticides in Table 5).

Table 5. Results of applying the EDAR index for hazard assessment of waste samples, sediment extracts or pore water and pure compounds using the reduced battery of tests.

Sample identification	***EDAR Index***	***Sample rank***	***Hazard description***	***Battery used***
Industrial waste samples				
Soil with hydrocarbons from land farming (leachate)	1.23	IV-1	Hazardous	H-D-L
Sludge with oil, grease and hydrocarbons (leachate)	0.33	III	Slightly hazardous	H-D-L
Food industry sludge (pore water)	0.36	III	Slightly hazardous	H-D-L
Food industry solid waste (leachate)	0.29	III	Slightly hazardous	H-D-L
Food industry solid waste (pore water)	0.72	IV-1	Hazardous	H-D-L
Food industry liquid waste	0.02	I	Not hazardous	H-D-L
Photoshop liquid waste	508	VI	Extremely hazardous	H-D-L
Photoshop liquid waste II	117	V-2	Very hazardous	H-D-L
Pharmaceutical solid waste (leachate)	0.70	IV-1	Hazardous	H-D-L
Liquid waste with pesticides	10.2	IV-3	Hazardous	H-D-L
Pure compounds				
Hg(II) 5 mg/L	235	VI	Extremely hazardous	H-D-L
Hg(II) 5 mg/L	257	VI	Extremely hazardous	H-D-S
4-NQO 2 mg/L	2.7	IV-2	Hazardous	H-D-L
4-NQO 2 mg/L	33	V-1	Very hazardous	H-D-S
4-NQO 2 mg/L	2.6	IV-2	Hazardous	H-D-L
4-NQO 2 mg/L	78	V-2	Very hazardous	H-D-S
Methanol 2%	0.11	I	Not hazardous	H-D-L
Methanol 2%	0.83	IV-1	Hazardous	H-D-S
Environmental samples*				
S1	0.79	IV-1	Hazardous	H-D-L
S1	0.66	IV-1	Hazardous	H-D-S
S2	1.2	IV-1	Hazardous	H-D-L
S2	1.1	IV-1	Hazardous	H-D-S
S3	2.7	IV-2	Hazardous	H-D-L
S3	2.8	IV-2	Hazardous	H-D-S
S4	0.72	IV-1	Hazardous	H-D-L
S4	0.60	IV-1	Hazardous	H-D-S
S5	0.18	II	Possibly hazardous	H-D-L
S5	0.28	III	Slightly hazardous	H-D-S
S6	0.4	III	Slightly hazardous	H-D-L
S6	0.22	III	Slightly hazardous	H-D-S
S7	0.11	I	Not hazardous	H-D-L
S7	0.076	I	Not hazardous	H-D-S
S8	0.28	III	Slightly hazardous	H-D-L
S8	0.56	IV-1	Hazardous	H-D-S
S9	0.97	IV-1	Hazardous	H-D-L
S9	0.85	IV-1	Hazardous	H-D-S
S10	0.08	I	Not hazardous	H-D-L
S10	0.01	I	Not hazardous	H-D-S

* Sites correspond to the south eastern sector of coastal area of the Río de la Plata (see Fig. 1).

When comparing the scores of the water samples and reference toxicants in both scales, the EDAR index description was similar when using the complete battery (Tab. 3) and the reduced H-D-S battery (Tab. 5). Some differences were observed, however, and these were mainly attributable to the pure organic compounds, when the complete and H-D-L reduced batteries are compared. These differences could be associated with a lower sensitivity of the seed test to toxicants and the weight assigned to this assay in the EDAR equation. One notable example is that for methanol (full battery EDAR index = 0.76, Table 3; reduced battery H-D-S EDAR index = 0.83, Table 5; reduced battery H-D-L EDAR index = 0.11, Table 5). Reduction of the number of toxicity tests within a battery certainly favours cost-effectiveness, but selection of those maintained in a reduced battery should be given careful consideration in order to avoid lowering the EDAR index toxicity detection potential. Future applications with different classes of chemical compounds will further explore those factors capable of influencing EDAR index values (*i.e.*, use of full and reduced batteries and toxicity test weight factors) in order to optimize this simple and user-friendly toxicity scale.

5.2 CHILE

One goal of the Chilean Government is the treatment of all domestic wastewaters by the year 2010. The generation of 220 tons year^{-1} of sludge is expected as a treatment by-product (SISS, 2003). At the University of Chile a team of investigators from different centres is carrying out studies on land application of sewage sludge and biosolids, considering their sanitary quality, heavy metal content and bioavailability, as well as their ecotoxicity. Their main objectives are to generate information for the environmental administration officials that are setting specific regulations for agricultural use.

5.2.1 HAS description

To assess soluble contaminants associated with sewage sludge and biosolids from different municipal treatment plants in Chile, a core battery of toxicity tests including *D. magna*, *H. attenuata* and *L. sativa* was used. Two Hazard Assessment Schemes (HAS) toxicity ranking systems were applied to categorize sample toxicity. The first scheme [HAS1] is based on a point ranking system that integrates toxicity data obtained for different tests (Dutka, 1988; Dutka and Kwan, 1988; Dutka, 1993; Castillo et al., 2000). This ranking depends on the number of tests and the weight assigned to each one. The scale comprises five degrees of hazard and ranges from “non toxic”, to “extremely toxic”. The range scheme used in this study was adapted to the three toxicity tests applied here (Tab. 6). Because of the generally lower sensitivity responses elicited with the *L. sativa* toxicity test in response to chemical contaminants, a higher score was allocated to it as compared to the *D. magna* and *H. attenuata* tests. Essentially, higher scores corresponded to more toxic samples with this ranking system.

Table 6. Point allocation scheme for sample ranking and hazard classification based on a toxicity test core battery [HAS 1].

Ranking interval		***Test score***			***Total battery score***	***Hazard description***
L(I)C50%	***TU****	***D. magna***	***H. attenuata***	***L. sativa***		
> 90	< 1.1	0	0	0	0	Non toxic
90 – 75	1.1–1.33	1	1	3	1 – 5	Slightly toxic
74.9 – 50	1.34 – 2	2	2	5	6 – 9	Toxic
49.9 – 25	2.01 – 4	4	4	9	10 – 17	Highly toxic
< 25	> 4	6	6	13	18—25	Extremely toxic

*TU (Toxic Units) = [1/(L(I)C50] x 100.

Table 7. Hazard classification scheme for wastes discharged into the environment [HAS 2].

Class	***Hazard description***	***Characteristics***
I	No toxicity	- none of the tests show a toxic effect (< 0.4 TU)
II	Slight toxicity	- LOEC is reached at least for one test - the effect level is below 50% (0.4 - < 1TU)
III	Toxicity	- the L(I)C50 is reached in at least one test - in the 10-fold dilution of sample, the effect is lower than 50% (1-10 TU)
IV	High toxicity	- the L(I)C50 is reached in the 10-fold dilution for at least one test - in the 100-fold dilution of sample, the effect is lower than 50% (> 10-100 TU)
V	Very high toxicity	- the L(I)C50 is reached in the 100-fold dilution for at least one test (> 100 TU)

* TU (Toxic Units) = [1/(L(I)C50] x 100.

The second scheme [HAS2], proposed by Persoone et al. (2003), is based on toxicity responses of one or more tests applied to wastes, and involves two steps: (i) an acute ranking in five classes (Tab. 7) and, (ii) a weight score for each toxicity class. The class describes hazard from "no toxicity", if no toxic effects are detected in a sample, to "very high toxicity" when toxic effects for a 100-fold dilution of sample are observed. The class weight quantifies the degree of toxicity in that class. The weight score is expressed in percentage (%), and ranges from 25% - if only one test of the battery reaches the toxicity level of the class - to 93% - if all tests but one reach it. For calculating the class weight, an allocation of a test score is applied for each toxicity test of the battery (Tab. 8). Then, the total score is divided by the total number of tests. This result is then divided by the maximum particular score obtained, and expressed as a percentage. The higher the weight score obtained, the more toxic hazard the class represents (Persoone et al., 2003). For example, sample AS1-b (HAS2 classification results given in Table 10) yielded the following classification based on its bioanalytical data:

- Toxic units of 28.3 (*D. magna*), 556 (*H. attenuata*) and 30.6 (*L. sativa*) giving individual scores of 3, 4 and 3, respectively (see Tab. 8), thereby placing this sample in class V (on the basis of class criteria outlined in Table 7).
- Sample score = [3 + 4 + 3] ÷ 3 bioassays = 3.33.
- Class weight % = [3.33 x 100] ÷ 4 (the highest score reached by the *H. attenuata* result of 556 TU, as per Table 8 criteria) = 83.3.

Table 8. Score allocation based on the toxic effect of each core battery bioassay for class weight calculation [HAS 2].

Toxic effect	***Score***
No significant toxic effect (< LOEC)	0
LOEC < % effect < L(I)C50 (=< 1 TU)	1
1 – 10 TU	2
10 – 100 TU	3
> 100 TU	4

5.2.2 Application of the HAS schemes to biosolids toxicity

This study included sludge samples from five different wastewater treatment facilities: (i) one stabilization pond (SP), (ii) two conventional activated sludges (AS), (iii) one compact activated sludge (CAS), and (iv) one trickling filter (TF). The conventional AS plants treat sewage produced by close to two and a half million people; the sludge obtained is anaerobically digested, mechanically dewatered, and dehydrated in sand drying beds. The other plants are located in small towns (~ 25,000 inhabitants). Sludge from the SP is auto-digested in the bottom of the pond, remaining there for approximately one year prior to being extracted and air dried; the TF sludge is anaerobically digested in tanks, and dried in conventional sand drying beds; the CAS sludge is not treated.

A total of eight sludge samples and two soils to be amended with sludge were tested with the core testing battery. In addition, two amended soils with AS1 sludge applied in rates 0, and 30 tons per hectare (ton ha^{-1}), incubated during 60 days for agricultural use, were also analyzed. Sludges and soils were air-dried and sieved through a 2 mm mesh-size polyethylene sieve. Portions of the < 2 mm fractions from sludges and soils were ground in an agate mortar and stored in polyethylene sealing bags. Forty g of sludges, soils and amended soils were extracted with the respective culture media from each toxicity test, using a ratio of 1:4. The mixture was shaken at 180 rpm for one hour and centrifuged under refrigeration at 3000 rpm for 20 min. The supernatant was then kept for toxicity testing.

Results of global acute toxicity of sludge and soils (dry-weight basis) are presented in Table 9. According to HAS1, independent of source and moisture, all sludges were classified as "extremely toxic", reaching the maximum battery score (25 points). Neither of the soils exerted toxic effects on the bioassay battery (0 points). In contrast, "high toxicity" was found in soils (12 points), after two months of application of the final sludge (biosolid) from one of the activated sludge treatment plants, at the rate 30 tons ha^{-1} (dry-weight basis).

Table 9. Sewage sludge and agricultural soil toxicity[1] [HAS 1].

Sample	*Moisture %*	*D. magna LC50-48h (%)*	*UT*	*H. attenuata LC50-96h (%)*	*UT*	*L. sativa IC50-5d (%)*	*UT*	*Total score*	*Hazard description*
SP	28.3	9.46	10.5	2.48	40.3	2.6	38.5	25	*Extremely toxic*
CAS	5.9	9.3	11.6	0.11	909	7.37	13.6	25	*Extremely toxic*
AS_{1a}	7.0	3.5	28.3	0.32	316	3.6	27.5	25	*Extremely toxic*
AS_{1b}	8.4	3.5	28.3	0.18	556	3.3	30.6	25	*Extremely toxic*
AS_{1c}	78.0	1.84	54.5	0.15	667	1.02	98.0	25	*Extremely toxic*
AS_{1d}	65.4	1.48	67.6	0.14	714	1.17	85.5	25	*Extremely toxic*
AS_{2a}	34.7	1.26	79.4	0.1	1000	2.17	46.1	25	*Extremely toxic*
TF	98	3.8	26.3	0.13	769	5.1	19.6	25	*Extremely toxic*
$Soil_1$	2.3	>100	<1.1	>100	<1.1	>100	<1.1	0	*Non toxic*
$Soil_2$	1.3	>100	<1.1	>100	<1.1	>100	<1.1	0	*Non toxic*
$Soil_1^2$	2.5	17.4	5.8	16.5	6.1	>100	<1.1	12	*Highly toxic*
$Soil_2^2$	1.2	17.3	5.8	15.1	6.6	>100	<1.1	12	*Highly toxic*

[1]Dry-weight basis.
[2] Amended soil with final sludge (AS_{1-a}) of conventional activated treatment sludge (rate 30 tons ha^{-1} x 60 days).

Although HAS 1 scheme cannot discriminate into different sub-categories the tested sludge samples (and therefore their relative toxicity), the results are of interest to set acceptable toxicity levels in specific regulations for sludge land application and agriculture reuse. The tested sludge comes from different types of environments (*i.e.*, small towns with mining and agriculture as their main productive activities, and a large city with a great diversity of economical activities), and also different types of sewage treatment, showing similarly high toxicity profiles, posing a potential risk of contamination to surface water and groundwater.

Similarly, the HAS2 classification system confirmed the high toxicity of sludges and the negative responses of both soils (Tab. 10). Most sludges fell into class V, with a weight of 83.3 %. SP sludge proved to be somewhat less toxic with a weight of 100% into class IV. In contrast, the hazard toxicity of amended soils decreased by two levels, falling into class III, with a weight of 66.7%.

Table 10. Toxicity of sewage sludge and amended soils using [HAS 2].

Sample	***Class***	***Hazard description***	***Class weight (%)***
SP	IV	High toxicity	100
CAS	V	Very high toxicity	83.3
AS1-a	V	Very high toxicity	83.3
AS1-b	V	Very high toxicity	83.3
AS1-c	V	Very high toxicity	83.3
AS1-d	V	Very high toxicity	83.3
AS2-a	V	Very high toxicity	83.3
TF	V	Very high toxicity	83.3
Soil[1]	I	No toxicity	- -
Soil[2]	I	No toxicity	- -
Soil[1]+AS1-a[2]	III	Toxicity	66.7
Soil[2]+AS1-a[2]	III	Toxicity	66.7

[1]Dry-weight basis.
[2]Amended soil with digested sludge of conventional activated sludge treatment (rate 30 tons ha^{-1} x 60 days).

Using the HAS1 framework, the *H. attenuata* test yielded the most sensitive toxic responses for all types of sludges. However, although the sensitivity of *D. magna* and *L. sativa* was of the same order of magnitude, classification of sludges as being "extremely toxic" (HAS1, Tab. 9) was in part attributable to the latter test, because of its high test score attribution (Tab. 6). In amended soils, both *D. magna* and *H. attenuata* assays generated maximum test scores (Tab. 6). In this instance, the negative response of *L. sativa* tended to reduce the hazard level of this matrix. The HAS2 classification scheme was similar in its ratings of samples and no major differences with respect to HAS1 were observed (Tab. 10).

In general, both hazard schemes were found to be simple and easy to apply and they can be considered complementary. When toxicity is present, both can discriminate between high, medium, low and absence of hazardous effects on tested organisms. HAS1 takes into account the response of each toxicity test included in the battery, assigning a particular score related to their respective response to toxicants. In contrast, HAS2 classifies hazard level based on the response of each test, but also includes a weight factor within a toxic class. Again, HAS1 attributes a toxic hazard based on all test scores while HAS2 gauges the hazard level. Based on the HAS1 scheme, all sewage sludge samples reached the highest classification, because their score was > 4 TU in all tests (Tab. 6). Because of the class and weight criteria imposed by the HAS2 scheme (Tab. 7), it appears to offer better possibilities to discriminate sludges on the basis of their toxic properties (Tab. 10).

Future studies should strive to improve upon these HAS schemes so as to better discriminate between highly toxic samples by separating them into sub-classes. This, in turn, will allow for the development of more precise criteria for the disposal of such hazardous wastes. Presently, some of the samples investigated, whose toxicity demonstrates effects at 1:10 and 1:1000 dilutions, are all grouped in the same class rank as "highly toxic". There is room for improvement in future optimization of HAS schemes to refine their judgement in terms of toxicity classification.

5.3 COLOMBIA

5.3.1 Principle of HAS and toxicity tests employed

The proposed hazard assessment scheme (HAS) used in Colombia is a ranking system where toxicity data obtained from the application of a test battery enables one to determine the degree of toxicity of liquid samples on a relative basis. Test battery results are then integrated into the Potential Ecotoxic Effects Probe (PEEP) index formula developed by Environment Canada for the comparison of wastewaters (Costan et al., 1993). This index can be applied to evaluate the potential toxicity of industrial and municipal wastewaters, and to assess the effectiveness of toxicity abatement measures for effluents. This procedure is easy to apply and can be used with different batteries of tests (see Chapter 1 of this volume).

As its principle, the PEEP index integrates the responses of a test battery of toxicity tests and determines the relative toxic loading contribution of a series of effluents to the toxic loading of the same receiving environment on a comparative basis using organisms from different trophic levels and taxonomic groups. In Colombia, the previously described toxicity tests were complemented with the agar plate method for rapid toxicity assessment of water-soluble and water-insoluble chemicals (Liu et al., 1991). In the agar plate method, pre-dried agar plates are thinly coated with a quantitative amount of fresh *Bacillus cereus* culture and the seeded plates are spotted with test chemicals at known concentrations. The plates are incubated at the optimal growth temperature for four hours and the diameter of the inhibition zone can be measured.

5.3.2 Determination of Effluent Hazard Potential

Hazard potential for each effluent was calculated using a mathematical formula (the PEEP index) proposed by Costan et al. (1993). This formula integrates the ecotoxic responses of the battery of tests before and after a biodegradation step. Toxicity test endpoint responses are first transformed to toxic units. The product of effluent toxicity and effluent flow (m^3/h) gives the toxic loading value. The log 10 value of an effluent's toxic loading corresponds to its PEEP index. In order to rank the effluents a toxicity classification scale is generated (Tab. 11).

$$P = \log_{10}\left[1 + n\left(\frac{\sum_{i=1}^{N} T_i}{N}\right) Q \right] \tag{2}$$

Where:

P = PEEP value,

n = number of endpoints exhibiting toxic responses,

N = maximum number of obtainable toxic endpoints,

T_i = Toxic Units from each test, before and after biodegradation,
Q = effluent flow in m^3/h.

Table 11. PEEP index scale for the hazard assessment of wastewater samples using a reduced test battery.

PEEP index values	***Toxic classification***
< 1.99	Practically non-toxic
2- 2.99	Slightly toxic
3-3.99	Moderately toxic
4- 4.99	Highly toxic
> 5	Very highly toxic

5.3.3 Application of the PEEP Index to a case study of industrial wastewaters

The Bogotá River basin is 375 km long and drains an area of about 6107 km^2. The river receives wastewater from a wide variety of industries, such as tanneries, organic and inorganic chemical production, metal plating, textile production, mining, agrochemical production, as well as sewage from the City of Bogotá and many other smaller municipalities. In 1995, the Colombian Ministry of the Environment, through the Regional Corporation for the control of the river, undertook a program to improve the water quality in the Bogotá River basin. The goal of this program was to reduce by 80% the load of organic compounds and toxicant concentrations by 50% that were being discharged into the Bogotá river.

To reduce the inflow of toxic substances into basin waters efficiently, the Corporation required information on which effluents posed the greatest hazard to the river. While chemical data on many of the effluents were available, their diverse composition was difficult to interpret in terms of hazard potential. Moreover, hazard is not only linked to chemical composition of an effluent, but also on the toxic effects it can have on a variety of freshwater organisms. Adverse effects on biota are also influenced by the volume of wastewater discharged at different times of the year. Hence, the Corporation searched for a cost-effective approach, based on ecotoxicological principles, to rank the various effluents (there are several hundred sources), in terms of their toxic loading, so that subsequent efforts (such as clean up actions) could be prioritized.

In light of these concerns, the National University of Colombia initiated a pilot study, through the application of the PEEP toxicity index to the Bogotá River. This study began by collecting wastewater samples from three effluent sources typical of industries with highest toxic load to the river; tanneries, a thermal power plant and a chlorine production. The assessment was conducted using a battery of three toxicity tests (*i.e.*, the *D. magna* 48-h motility inhibition assay, the Agar plate bacterial growth inhibition test and the *S. capricornutum* 72-h growth inhibition test); the results are presented below in Table 12. The PEEP index clearly identified the chlorine plant effluent as the most toxic for the receiving environment as it contributes, on its own, close to 98% of the toxic loading generated by this series of five effluents.

Another important industrial sector investigated was the textile industry. Ten different effluent samples were collected and each wastewater was characterized by standard chemical analyses as well as by the toxicity test battery. In this case, the test species included *Daphnia*, *Hydra* and *Lactuca*. Toxicity endpoint values were first transformed into toxic units (TU), a quantitative expression reflecting the resulting toxic potential of all chemical contaminants present in an effluent sample. Subsequently, their PEEP values were determined (Tab. 13).

Table 12. PEEP index characteristics for five effluent samples, and percentage contribution (%) of each effluent to total toxic charge.

Industrial effluent	*Effluent flow (m3/h)*	*Toxic print*[1]	*Toxic charge*[2]	*%*	*PEEP value*
Tannery	0.001	352	0.3	0.002	0.11
Tannery	0.001	443	0.4	0.003	0.14
Thermal Power Plant	25.2	1.2	29	0.23	1.48
Thermal Power Plant	248.4	1.1	279	2.17	2.45
Chlorine Plant	0.429	29172	12520	97.6	4.10
	Total toxic charge		12829		

[1] n (ΣT_i /N) in the PEEP formula
[2] Effluent flow x Toxic Print in the PEEP formula

Table 13. PEEP index characteristics for ten textile effluent samples, and percent contribution (%) of each effluent to the total toxic charge.

Plant	*Effluent flow (m^3/h)*	*Toxic print*[1]	*Toxic load*[2]	*%*	*PEEP value*
1	0.67	29.2	19.5	0.2	1.31
2	0.83	33.9	28.2	0.2	1.47
3	5.2	11.6	61.0	0.5	1.79
4	4.8	22.1	105.7	0.9	2.03
5	14.4	10.6	153.4	1.4	2.19
6	4.8	79.7	385.0	3.4	2.59
7	22.3	31.5	702.2	6.2	2.85
8	30.6	25.3	775.6	6.9	2.89
9	98	31.2	3080.2	27.2	3.49
10	360	16.7	5997.4	53	3.78
	Total toxic charge		113308		

[1] n (ΣT_i /N) in the PEEP formula
[2] Effluent flow x Toxic Print in the PEEP formula

The results demonstrated a wide range of toxic effects and loadings for textile effluents (Tab. 13). Inter-effluent toxicity differences could be attributed to factors such as: type of industrial process, degree of waste treatment, variability of effluent composition and dilution by process waters. Toxic loads from industries 9 and 10 indicate they contribute the greatest toxic load to the receiving waters; therefore based on PEEP results the first priority would be to reduce the toxic loading from industries 9 and 10.

The PEEP index also allows for monitoring the treatment efficiency of industrial effluents. This is illustrated in Figure 2 following the collection of seven composite effluent samples that were taken before and after treatment from a cosmetic industry. The objective of the waste treatment was to maximize the removal of the toxicants causing the toxic loading. In all cases, application of biological and chemical treatments proved to be beneficial in producing a reduction in toxic loading.

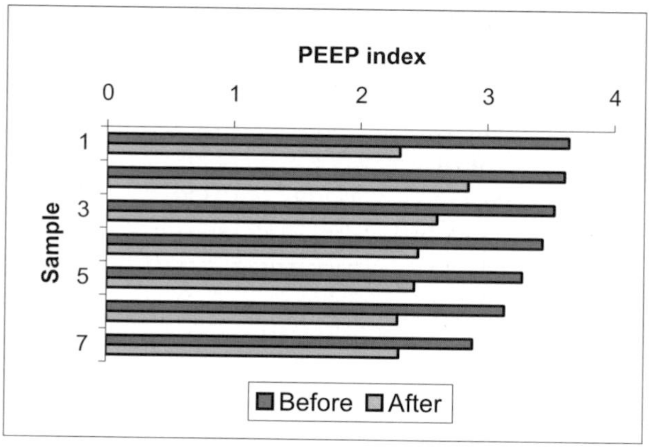

Figure 2. Effect of biological treatment on toxicity reduction of cosmetic industry effluents as indicated by PEEP index values.

In brief, the PEEP index is a useful HAS to apply in comparative studies of wastewater effluents to assess their ecotoxicity and toxic loading. Some of its advantages include the fact that it considers results from different toxicity tests and endpoints, while integrating all possible antagonistic, additive or synergistic interactions that can occur between toxicants in a complex liquid sample. Furthermore, the use of a single PEEP value becomes very useful for decision-makers who are then able to take science-based decisions to prioritize corrective actions on industries whose effluents are the most toxic for the aquatic environment. It is also noteworthy to point out that the PEEP index can be applied anywhere with any number or type of tests and endpoints to suit the needs and expertise of laboratories internationally.

6. Conclusions

The battery of test approach for toxicity testing is now a universally-accepted concept. It has recently been applied in Latin American countries and is presently recognized as a critical tool for the assessment of complex mixtures. Interpretation of hazard by reducing complex ecotoxicological data into a single numerical value (*e.g.*, PEEP index) is generally favoured by decision-makers involved in various facets of environmental regulation.

Applying the WaterTox battery of tests in Argentina, Chile and Colombia for toxicity assessment of chemical contaminants present in different types of complex matrices by means of existing, modified or developed HAS approaches has proven to be environmentally beneficial. Water and wastewater samples, sewage sludge and biosolids from municipal treatment plants and effluent toxic loads, as well as pure compounds, were effectively scored as toxic or non-toxic with the ranking systems employed, thereby allowing them to be differentiated in terms of their adverse potential. In all cases, similar bioanalytical tools were employed to conduct these evaluations.

Environmental programs in Chile and Colombia have set clear goals regarding the treatment of effluents and have already initiated research projects in cooperation with academic groups. Argentinean regulatory agencies have also been incorporating tools for the interpretation of results from toxicity testing and categorization of hazardous wastes. There are diverse applications for bioanalytical tools, particularly when they are integrated into testing batteries, as documented in this chapter. While the simple HAS systems described are unquestionably useful to assess and rank toxicity, future refinement will necessitate additional appraisals on types/numbers of tests and endpoints required to maximize detection of toxicity potential and to sharpen their power to discriminate samples based on more subtle levels of toxicity. Cost-efficiency, reliability of testing and ease in interpreting hazard will also be critical in future initiatives designed to integrate HAS in environmental risk assessment processes by local environmental protection agencies.

Acknowledgements

The authors wish to acknowledge the International Development Research Centre IDRC from Canada, the National Water Research Institute (NWRI, Burlington, Ontario) and Centre Saint-Laurent (Montreal, Quebec) of Environment Canada for their support at different stages of the research. Thanks are also extended to Drs. B. Dutka, G. Forget, A. Sanchez and C. Blaise for their support.

Part of this study was financed by Project Fondecyt/Conicyt Chile 1020129-02, and Project DID TNAC 18-02/01, U. of Chile; the National University of La Plata and National and Buenos Aires Province Research Councils from Argentina; and the National University of Colombia, Project DIN-1037, UNAL.

References

Apartin, C. and Ronco, A. (2001) Evaluation of a β-Galactosidase *in vitro* enzymatic test specific for heavy metal toxicity, *Environmental Toxicology* **16,** 117-120.

Blaise, C. and Kusui, T. (1997) Acute toxicity assessment of industrial effluents with a microplate-based *Hydra attenuata* assay, *Environmental Toxicology and Water Quality* **12**, 53-60.

Blaise, C., Forget, G. and Trottier, S. (2000) Toxicity Screening of Aqueous Samples using a cost-effective 72-hour exposure *Selenastrum capricornutum* Assay, *Environmental Toxicology* **15,** 352-359.

Bulus Rossini, G., Sobrero, C., Apartin, C., Grassi, V., Mugni, H. and Ronco, A. (2005) Bases conceptuales y metodología para la utilización del Indice PRED en la categorización de la toxicidad de muestras ambientales, Fundación Ciencias Exactas, Universidad Nacional de La Plata, La Plata, Argentina (in press).

Camilión, C., Manassero, M., Hurtado, M. and Ronco, A. (2003) Copper, Lead and Zinc distribution in soils and sediments of the South Western coast of the Río de la Plata estuary, *Journal of Soils and Sediments* **3**, 213-220.

Castillo, G., Vila, I. and Neild, E. (2000) Ecotoxicity assessment of metals and wastewater using multitrophic assays, *Environmental Toxicology* **15,** 370-375.

Costan, G., Bermingham, N., Blaise, C. and Férard, J. (1993) Potential Ecotoxic Effect Probe (PEEP): a novel index to assess and compare the toxic potential of industrial effluents, *Environmental Toxicology and Water Quality* **8**, 115-140.

Diaz-Baez, C., Sánchez, A., Dutka, B., Castillo, G., Ronco, A, Pica-Granados, Y., Castillo, L., Ridal, J., Arkhipchuk, V. and Srivastava, R. (2002) Overview of Results from the WaterTox intercalibration and environmental testing Phase II Program: Part 2, Environmental sample testing, *Environmental Toxicology* **17**, 241-249.

Dutka, B. (1988) Priority setting of hazards in waters and sediments by proposed ranking scheme and battery of test approach, *Zeit Angewandte Zool.* **75**, 303-316.

Dutka, B. and Kwan, K. (1988) Battery for screening tests approach applied to sediment extracts, *Toxicity Assess*ment **3**, 303-314.

Dutka, B. (1989a) Short term root elongation toxicity assay, in B. Dutka (ed.), *Methods for Toxicological Analysis of Waters, Wastewaters and Sediments*, National Water Research Institute, Environment Canada, Burlington, Ontario, pp. 120-122.

Dutka, B. (1989b) *Daphnia magna* 48 hours static bioassay method for acute toxicity in environmental samples, in B. Dutka (ed.), *Methods for Toxicological Analysis of Waters, Wastewaters and Sediments*, National Water Research Institute, Environment Canada, Burlington, Ontario, pp. 55-59.

Dutka, B. (1993) Ecotoxicological assessment of water, effluent and sediment quality using a battery of tests approach. Rivers research Branch, National Water Research Institute, Canada Center for Inland Waters, Burlington, Ontario, pp. 37.

Environment Canada (1999) Guidance document on application and interpretation of single-species tests in environmental toxicology, Method Development and Application Section, Environmental Technology Centre, EPS 1/RM/34, Ottawa.

Forget, G., Gagnon, P., Sanchez, A. and Dutka, B. (2000) Overview of methods and results of the eight country International Development Research Centre (IDRC) WaterTox project, *Environmental Toxicology* **15**, 264-276.

Fiskesjö, G. (1993) *Allium*Test I: A 2-3 day plant test for toxicity assessment by measuring the mean root growth of onions (*Allium cepa* L.), *Environmental Toxicology and Water Quality* **8**, 461-470.

Liu, D., Kwasniewska, K., Chau, Y. and Dutka, B. (1991) A four-hour agar plate method for rapid toxicity assessment of water-soluble and water-insoluble chemicals, *Environmental Toxicology and Water Quality* **6**, 437-444.

Persoone, G., Marsalek, B., Blinova, I., Törökne, A., Zarina, D., Manusadzianas, L., Nalecz-Jawecki, G.,Tofan, L., Stepanova, N., Tothova, L. and Kolar, B. (2003) A practical and user friendly toxicity classification system with microbiotests for natural and wastewater, *Environmental Toxicology* **18**, 395-402.

Ronco, A., Sobrero, C., Alzuet, P, Bulus Rossini, G. and Dutka, B. (1995) Screening for sediment toxicity in the Río Santiago basin: A base line study, *Environmental Toxicology and Water Quality* **10**, 35-39.

Ronco, A., Alzuet, P., Sobrero, C. and Bulus Rossini, G. (1996) Ecotoxicological effects assessment of pollutants in the coastal region of the Gran La Plata, Province of Buenos Aires. *Proceedings of International Conference on Pollution Processes in Coastal Environments*, Mar del Plata, Argentina, pp. 116-119.

Ronco, A., Camilion, C. and Manassero, M. (2001) Geochemistry of heavy metals in bottom sediments from streams of the western coast of the Rio de la Plata Estuary, Argentina, *Environmental Geochemistry and Health,* **23**, 89-103.

Ronco A., Gagnon, P., Diaz-Baez, C., Arkhipchuk, V., Castillo, G., Castillo, L., Dutka, B., Pica-Granados, Y., Ridal, J., Srivastava, R. and Sánchez, A. (2002) Overview of results from the WaterTox intercalibration and environmental testing Phase II Program: Part 1, Statistical analysis of blind sample testing, *Environmental Toxicology* **17**, 232-240.

Samoiloff, M.R., Schulz, S., Denich, K., Jordan, Y. and Arnott, E. (1980) A rapid simple long term toxicity assay for aquatic contaminants using the nematode *Panagrellus redivivus*, *Canadian Journal of Fisheries and Aquatic Sciences* **37**, 1167-1174.

Superintendencia de Servicios Sanitarios SISS Chile (2003) Informe anual de coberturas de servicios sanitarios, http://www.siss.cl

Trottier, S., Blaise, C., Kusui, T. and Johnson, M. (1997) Acute toxicity assessment of aqueous samples using a microplate based *Hydra attenuata* assay, *Environmental Toxicology and Water Quality* **12**, 265-272.

U.S. EPA (United States Environmental Protection Agency) (1991) Methods for measuring the acute toxicity of effluents and receiving waters to freshwater and marine organisms, Fourth Edition, Report EPA 600/4-90/027, Washington, DC.

Abbreviations

AS	Activated Sludge
CIMA	Environmental Research Centre
EC20	20 % effect inhibitory concentration
EC50	50 % effect inhibitory concentration
EDAR	effect:dilution average ratio index
HAS	Hazard Assessment Scheme
IC50	50% effect inhibitory concentration
IDRC	International Research Development Centre
LC50	50% effect on survival
LOEC	Lowest Observed Effect Concentration
PEEP	Potential Ecotoxic Effects Probe
SP	Stabilization Pond
TF	Trickling Filter
TU	Toxic Units.

8. THE SED-TOX INDEX FOR TOXICITY ASSESSMENT OF CONTAMINATED SOLID MATRICES

MANON BOMBARDIER
Environmental Technology Centre
Environment Canada
335 River Road, Ottawa
Ontario K1A 0H3, Canada
manon.bombardier@ec.gc.ca.

1. Objectives and scope

SED-TOX is an index that aggregates multiple toxicity data into an easily understandable and single value, the SED-TOX score. Although designed for the assessment of marine, estuarine and freshwater sediments, it could also easily be applied to soil samples, sludges and biosolids. This index can assist the environmental manager in the:

(a) prioritization of remediation action;
(b) ranking of environmentally-degraded sites;
(c) determination of temporal changes of environmental conditions;
(d) public communication of environmental decisions; and
(e) evaluation of remediation or monitoring activities.

The purpose of this paper is to describe the SED-TOX index for the toxicity assessment of contaminated solid matrices, particularly sediments, demonstrate its discriminative potential, and show its correlation with chemical analyses derived data and indices of benthic community structure.

2. Summary of the SED-TOX index

Environmental assessment studies usually involve the generation of a large set of data that may be difficult to analyze, interpret, and translate into simple terms that can easily be grasped. To this end, the use of environmental indices plays an important role in the translation of multiple complex technical data into a single number. Several indices have been developed for assessing water quality (most of which are based on chemical parameters or indicators of community structure), but very few exist for the assessment of sediment quality through the use of toxicity assays.

C. Blaise and J.-F. Férard (eds.), Small-scale Freshwater Toxicity Investigations, Vol. 2, 257-280.

The SED-TOX index was inspired by the Potential Ecotoxic Effects Probe (PEEP) approach (Costan et al., 1993), developed and used under the St. Lawrence Action Plan for ranking industrial effluents based on their toxic potential to a battery of aquatic organisms. The precursor to the development of the PEEP and SED-TOX indices at the St. Lawrence Centre was the common use of toxicity tests to rank environmental samples according to the severity of toxicity in laboratory assays. Ranking is useful for determining priority for action on the most toxic effluents or contaminated sites. However, ranking can become difficult if toxicity is measured with several tests that generate variations in rank for a given sample. Collapsing the toxic responses into a single toxicity index was viewed as a convenient way to express hazard in a single number. In both the SED-TOX and the PEEP indices, a high score implies that exposed organisms are more likely to be in danger at that site than in an area with a low score.

The SED-TOX index has been applied to a variety of sediment types (marine, estuarine, or freshwater), toxicity assays, and test phases (Bombardier and Bermingham, 1999; Bombardier and Blaise, 2000). It presently incorporates four exposure phases: pore water (PW), organic extract (OE), wet sediment (WT), and whole sediment (WS), but could also easily include tests on elutriates. Test organisms might include bacteria, algae, macrophytes, benthic as well as epibenthic and pelagic invertebrates. However, for comparability considerations, tests conducted on each sediment sample ideally should be the same although it is recognized that this is not always possible.

Table 1. The SED-TOX index at a glance.

Purpose
▪ As for the PEEP index, the SED-TOX index evolved from the need to make better interpretative use of bioassay-derived data. It was developed as a tool for environmental managers to evaluate the relative hazard at various sediment sites, based on the results of a battery of toxicity texts. The Index merges multiple toxicity data into a single index that can be used as a convenient measurement of overall sediment toxic potential. The SED-TOX index can be helpful in: (a) assisting environmental managers in prioritizing remediation action; (b) ranking environmentally-degraded sites; (c) determining temporal changes of environmental conditions; (d) enhancing public awareness; and (e) evaluating remediation or monitoring activities.
Principle
▪ The SED-TOX conceptual framework is composed of two stages: data conversion and data integration. In summary, toxicity data are organized into the four test phases, converted to a single scale of measurement (*i.e.*, Toxic Units) and combined in a single Index (SED-TOX score) which represents the aggregate sediment toxic hazard potential. Scores can then be compared to set priorities for remedial action or to target further investigation.

Table 1 (continued). The SED-TOX index at a glance.

Bioassays employed

- A critical component in the application of the SED-TOX index is the execution of a series of sediment laboratory bioassays. At the time of its conception (Bombardier and Bermingham, 1999), the index integrated the results of a battery of seven bioassays conducted with four test species (*Vibrio fischeri, Lytechinus pictus, Escherichia coli, and Amphiporeia virginiana*) and four exposure phases (*i.e.*, pore water, organic extract, wet sediment and whole sediment). In a subsequent application (Bombardier and Blaise, 2000), the index integrated toxicity data from tests conducted on four trophic groups, namely bacteria (*Vibrio fischeri*), cnidarians (*Hydra attenuata*), micro-crustaceans (*Thamnocephalus platyurus*), and benthic macro-invertebrates (*Hyalella azteca* and *Chironomus riparius*), and involved two phases of exposure (*i.e.*, pore water and whole sediment). Although time-consuming, the mathematical formulas necessary to generate the SED-TOX score (refer to the Appendix) are relatively simple and can accommodate any number and type of toxicity tests.

Determination of sediment toxicity hazard index

- The calculation steps are fully detailed in the Appendix to this chapter. The following lines attempt to provide a summary.
- To standardize the multivariate bioassay-derived data and facilitate their integration in the index, toxicity data are first converted into Toxic Units (TU), which are dimensionless ratios originally defined as the actual concentration of a particular toxic substance, divided by the incipient lethal concentration for that substance (Sprague and Ramsay, 1965). A similar term can be used to define the toxicity potential of a sediment sample, whereby:

$$TU = C_{max} \cdot C_{tox}^{-1} \qquad (1)$$

where: C_{max} is the highest tested concentration (*e.g.*, 100 % w/v), and C_{tox} is the concentration associated with the assessment endpoint of interest (*e.g.*, IC25, LC50).

- For each bioassay, toxic units are then adjusted with the test sensitivity (or detection limit, DL), to generate toxicity incremental factor (TIF) values, as follows:

$$TIF = TU_{dw} \bullet DL^{-1} \qquad (2)$$

where: TU_{dw} is expressed on a dry weight basis.

Table 1 (continued). The SED-TOX index at a glance.

Determination of sediment toxicity hazard index
▪ Test sensitivity is defined as the lowest *TU* at which a toxic response can be observed; for instance, a test run at 10% of the full concentration (*e.g.*, 10% v/v) would have a detection limit of 10. For quantal data, this normalization step requires the determination of a minimum non toxic percent response (R_{min}), *i.e.*, the minimum percent response which designates a sediment sample as non-toxic. The formula for quantal tests is as follows: $$TIF = (R_{min} - R_{tox}) \bullet \log (R_{max} - R_{tox}) \quad (3)$$ where: R_{max} is the maximum attainable value (*e.g.*, 100% survival) and R_{tox} is the response observed in the test sediment). Negative values are assigned a TIF value of zero.
▪ For each exposure phase, the TIFs are averaged (*i.e.* arithmetic mean) over all measured endpoints, to generate a weighted average of phase toxicity (WAPT).
▪ The WAPT values are summed to obtain the cumulative average of phase toxicity (CAPT), which is finally expressed on a logarithmic scale, to generate the SED-TOX score: $$SED\text{-}TOX = \log 10\ [1 + n(CAPT)] \quad (4)$$ where n is the number of test phases eliciting toxic effects.
▪ The detailed mathematical formula generating the SED-TOX score are provided in the Appendix.
Notes of interest
▪ An increase in SED-TOX scores is indicative of higher toxic potential. To facilitate the comparison of toxicity scores among sediment stations or sites, and make the index-derived values more meaningful to managers, an arbitrary ranking scheme has been proposed. SED-TOX scores are assigned to four classes of increasing toxicity hazard potential: – scores of zero indicate no hazard; – scores varying between 0.1 and 0.9 represent a marginal hazard; – scores between 1.0 and 1.9 indicate a moderate hazard; and – scores greater or equal to 2.0 represent a high hazard potential.

3. Historical overview and applications of the SED-TOX index

Over the last three decades considerable resources have been devoted to the assessment, management, and remediation of contaminated sediments. Their quality has become a serious and complex issue for dredging and disposal projects to maintain navigational channels, recreational and commercial fisheries management, water-quality protection, and natural resource restoration. Complexities arise from the great variability in sediment physical, chemical, geochemical and biological characteristics, as well as in the social and economic values associated with different freshwater, estuarine, and marine environments. Complexities also arise from the use of inconsistent approaches for assessing and managing contaminated sediments. As a result of these complexities, progress in addressing contaminated sediments in ports and harbours, rivers, lakes and at hazardous waste sites has been relatively slow.

To address the issue of contaminated sediments, environmental scientists throughout the world have developed a variety of approaches for evaluating the degree to which sediment-associated chemicals might adversely affect aquatic organisms. The best known approach is the use of Sediment Quality Guidelines (SQGs), which have been described as numerical chemical concentrations intended to be either protective of biological resources, or predictive of adverse effects to those resources, or both. Such values are being developed and used by a number of jurisdictions around the world, using many different approaches. In Canada, sediment quality guidelines for the protection of aquatic life are being developed under the auspices of the Canadian Council of Ministers of the Environment (CCME) on the basis of both the NTSP (National Status and Trends Program - Long and Morgan, 1990) and the SSTT[1] (Spiked-Sediment Toxicity Test) approaches. Together, these two approaches provide complementary information to support the development of national Sediment Quality Guidelines (Macdonald et al., 1992; CCME, 1995). The underlying assumption in the derivation of effects-based SQGs is that these guidelines can be used as a substitute for direct measures of potential adverse effects of contaminants in sediments on benthic organisms. Several concerns have been expressed regarding the use of SQGs in sediment quality assessments. Such concerns relate, for instance, to the ability of SQGs to: (a) adequately predict effects on sediment-dwelling organisms in the field, and (b) establish cause and effect relationships.

Although it is acknowledged that data derived from chemical analyses may be sufficient for decision-making in extreme conditions of contamination or non-contamination, especially if bioaccumulation is taken into account (Adams et al., 1985; Adams, 1987; Di Toro et al., 1991), there is growing recognition that sound decision-making on contaminated sediment can be best achieved with a comprehensive approach for assessing the quality of contaminated sediments. At IFREMER (France), a research program has been created that involves the application of a comprehensive approach for studying the physical, chemical and biological impacts of sediment dredging activities. Results from these studies can be found, at least partly, in publications from Ifremer (Alzieu, 1999; Alzieu et al.,

[1] A thorough review of this approach has been provided by Lamberson and Swartz (1992).

2003). Quiniou and Alzieu (1999) suggested a risk analysis methodology that is based on levels of chemical contamination and the global toxicity of the sediments as determined with a selected battery of laboratory bioassays. The "Geodrisk" software constitutes the practical application of this methodology (Alzieu, 2001). Another comprehensive approach in this regard is the Sediment Quality Triad (SQT) which was first introduced in Puget Sound (WA, USA) by Long and Chapman (1985), and which simultaneously investigates sediment chemistry, sediment toxicity, as well as alterations in the field, for example, modifications of benthic community structure. The SQT has repeatedly been used in North America in both marine and freshwater environments, as recently reviewed by Chapman (2000) and in this book (see Chapter 10 of this volume).

Laboratory bioassays examine potential toxicity of the contaminated matrix through acute or chronic exposures periods, after which measures of mortality, growth inhibition, reproductive impairment, genotoxicity and other lethal and/or sublethal effects on individuals are made. Major technical advancements in ecotoxicological tests have been made over the past decades for the aquatic environment (for a review on aquatic microbiotests, consult Blaise et al., 1988, and Wells et al., 1998) and more recently, for terrestrial systems (Novak and Scroggins, 2003). Multiple measures of toxicity are needed to provide an accurate estimate of toxicity, as there is no single measure of response to contaminated sediment that can clearly discriminate between contaminated and non-contaminated areas. However, difficulties arise when attempting to merge such multivariate data into binary decisions (pass/fail) or ordinal ranks (*e.g.*, not/possibly/likely/very different from reference sites) in order to assess the overall toxic potential of a contaminated matrix.

The search for an index that both integrates the diverse parameters of effects on a variety of species representing several trophic levels and distinguishes between degraded and non degraded areas has been a focus of research at the St. Lawrence Centre of Environment Canada during the past decade (Costan et al., 1993; Bombardier and Bermingham, 1999). The development of the PEEP and the SED-TOX indices was part of this research area.

So far, the SED-TOX index has been the subject of two scientific publications:

(1) In the first publication (Bombardier and Bermingham, 1999), the SED-TOX index was applied to toxicity data obtained from marine sediments collected at two sites in the Gulf of St. Lawrence: Anse-à-Beaufils (Gaspé, Quebec, Canada) and Cap-aux-Meules (Magdalen Islands, Quebec, Canada). Three areas (harbour, disposal, and reference) were evaluated for each site. The following questions were addressed in this particular study:
    - (a) What is the hazard potential of Anse-à-Beaufils and Cap-aux-Meules harbour, disposal, and reference sediments as determined by the SED-TOX approach?
    - (b) Is the index sensitive enough to discriminate sediment toxicity among sampling sites?

(c) Are SED-TOX scores related to sediment chemical concentrations?
(d) Are there any relationships between test phase toxicity and sediment physico-chemical characteristics?
(e) What is the influence of exposure phase sensitivity weighting factors on site ranking?

(2) In the second publication (Bombardier and Blaise, 2000), laboratory toxicity data derived from two larger projects conducted on freshwater sediments were integrated in the SED-TOX index and it was field validated using four benthic community metrics (species richness, number of taxa in the orders Ephemeroptera, Plecoptera, and Trichoptera, the Shannon-Wiener diversity index, and the ICI-SL which is a version of the Invertebrate Community Index modified for the St. Lawrence River).

4. Advantages and limitations of applying the SED-TOX index

The SED-TOX approach entails several advantages:

- It provides a single value of overall sediment toxicity that is easily grasped. However, this advantage may also be viewed as a limitation (see below).
- It facilitates the decision-making process when attempting to rank or prioritize for action sediment sites/samples based on their toxic potential to a wide variety of organisms.
- It offers the possibility of incorporating any bioassay that is currently available. It thus provides the current best estimate of relative hazard for the sites being investigated. There is however a continuing need to develop, both for freshwater and marine or estuarine ecosystems, a battery of validated toxicity tests with sensitive species for whole sediment, wet sediment, organic extract, and pore water.
- The approach is founded on generally accepted concepts and principles. It is instructive in that it examines toxicity responses associated with multiple routes of exposure.
- It allows the user to assign a different treatment to acute lethal data to account for low discrimination capability and low uncertainty compared to chronic responses, as demonstrated by Ingersoll et al. (1997).
- As demonstrated previously (Bombardier and Bermingham, 1999; Bombardier and Blaise, 2000), the SED-TOX index has a good discriminatory potential.
- Although originally designed for the assessment of contaminated sediments, it can also be used to evaluate the toxic potential of contaminated soil or other environmental solid matrices (*e.g.*, sludges and biosolids).

The index also has certain limitations:

- Although it can accommodate any number and type of toxicity tests, and different types of solid matrices (such as contaminated soils and sediments), calculating the SED-TOX index can be time consuming and professional judgment is required to interpret the final scores.

- Comparisons of SED-TOX scores are only meaningful when made between sediment samples collected at the same time and evaluated with the same toxicity tests; this is however not always possible. Therefore, the effects of inconsistency in test selection and in the temporal scale of sediment sampling on the SED-TOX scores has to be established.
- The index does not take into account test variability (*i.e.*, coefficient of variation of response for a set of laboratory replicates for each sampling station).
- Another noteworthy limitation of the SED-TOX index, which is actually a limitation of toxicity bioassays, relates to the fact that results measured in toxicity tests only provide a measure of what is occurring under very specific laboratory test conditions, and an indication of what is or could occur in the field. To maximize the ecological relevance of the laboratory toxicity tests, there is a need to relate the SED-TOX scores with a series of benthic community matrices.
- Finally, as for all single and dimensionless indices that integrate a number of different measurements, there is a loss of information in the final score. This again argues for the need to consider, in the decision-making process, other parameters of sediment quality such as benthic community structure and sediment chemistry.

5. Procedure description

The following details are not specific to the use of the SED-TOX index but to sediment toxicity assessment studies in general. The first step involved in sediment toxicity assessment is the study plan, which consists of defining the goals and objectives, the methods for sediment collection, the tests to be conducted and the analysis of the results. The next step is to define the study area and the actual study site, to select the location of the sampling stations by evaluating any historical data relevant to this site and to determine the sample size (*i.e.*, sample volume) and number of samples and replicates. Preferably, sediment samples should be collected synoptically from both exposed and reference (*i.e.*, relatively uncontaminated) sites. Once collected, the sediment samples need to be stored or prepared for toxicity testing. Storage times should be minimized with a preferred maximum time of two weeks, and a maximum permissible storage time of six weeks prior to testing (Environment Canada, 1994). Sample preparation may involve physical operations such as sieving to remove macro-fauna and large debris, freezing to kill indigenous organisms and inhibit microbial activity, chemical treatment for extraction or fractionation of some organic contaminants, homogenizing to achieve homogeneity of colour, texture and moisture, and centrifugation to remove pore water from the collected sediment. Appropriate quality assurance/quality control (QA/QC) procedures should be followed in all aspects of a sediment toxicity assessment study, from field collections through to data analyses to assure that steps to maintain and improve data quality are in place, and that the limits of uncertainty associated with the data are known (Sergy, 1987).

Several sediment phases can be used for assessing the toxicity potential of sediments. Pore water is frequently considered to be the major exposure route of aquatic organisms in overlying waters to sediment-associated contaminants. Studies have demonstrated that bioavailability of chemicals to benthic organisms is strongly correlated with pore-water concentrations (Adams et al., 1985; Di Toro et al, 1990, 1991; Ankley et al., 1993) and toxicity has often been shown to be more pronounced in pore-water exposures than in solid-phase sediment exposures (*e.g.*, Winger and Lasier, 1995). However, this increased sensitivity in pore-water tests could be, at least in some cases, an artefact of sediment manipulations during collection, storage and experimentation, and/or to other natural contaminants such as sulphur and ammonia. For instance, in anoxic sediments, metals are generally found in the form of sulphide and polysulfide complexes. During sampling and handling of pore water, sulfides are unintentionally oxidized, potentially leading to an increase in metal bioavailability and hence in toxicity. In some instances however, pore-water may not be the most important route of exposure (Warren et al., 1998; Hare et al., 2001; Chapman et al., 2002). Because many organisms, such as deposit feeding invertebrates, ingest sediment particles, tests conducted with whole sediments, where ingestion of particles is taken into account, would be more appropriate for such organisms. Microbes, in contrast, although they do not ingest sediment particles, are in intimate contact with sediments and indeed physical exposure to sediments is an important factor in microbial bioassays, such as the solid-phase Microtox®, the direct Chromotest®, and Toxi Chromopad®. However, these tests have been shown to suffer from high variability. For instance, Cook and Wells (1996) reported coefficients of variation from 18 to 36% for reference and control sediments and 30 to 135% for contaminated sediments using this solid-phase test. Another limitation of these tests is the occurrence of false positives due to adhesion of bacteria to the sediment particles (Bulich et al., 1992). Albeit less frequently, solvent extracts of sediment organics are sometimes assessed with the Microtox® test. Solvent extract tests are often used to assess carcinogenicity or genotoxicity and are considered to represent the worst-case scenario in terms of bioavailability of sediment associated organic contaminants. For this reason, they are generally believed to lead to overestimations. Since each test phase has its own specific limitations, the use of a suite of toxicity tests with different test phases is desirable. However, the selection of the appropriate test phases requires some understanding of the factors affecting bioavailability in the aquatic environment and in the laboratory. Furthermore, detailed measurements of physico-chemical factors (pore-water pH, total organic carbon, ammonia, and acid volatile sulfides), and contaminants of concern in each test phase used in the test battery, are necessary to correctly interpret results of toxicity tests.

The choice of test organisms and life stages also has a major influence on the relevance and interpretation of a test. Furthermore, because different species and life stages within the same species exhibit different sensitivities to toxicants, it is generally advised to conduct a series of tests using a range of organisms representing different trophic groups and life cycles (*i.e.*, short versus long) and using different endpoints. Several batteries have been recommended in the scientific literature for the assessment of sediment and dredged material. For instance, Côté et al. (1998a, b) recommended a two-battery tiered approach for the assessment of freshwater sediment in the St. Lawrence River (Quebec, Canada), consisting of seven micro-scale assays: whole sediment esterase inhibition test with *Selenastrum capricornutum* (now renamed *Pseudokirchneriella subcapitata*), ATP test for microbial biomass on whole sediment, survival test (pore-water) with *Thamnocephalus platyurus*, ingestion inhibition test (pore water) with *Daphnia magna*, survival test (pore water) with *Hydra attenuata*, light inhibition test (pore water) with *Vibrio fischeri*, and genotoxicity test (pore water) with *Escherichia coli*, as well as two benthic invertebrate (*Chironomus riparius* and *Hyallela azteca*) toxicity tests (whole sediment). In the United States, the "Inland Testing Manual" (U.S. EPA and U.S. ACE, 1998) specifies a tiered testing approach whereby early tier toxicity tests focus on acute responses, whereas later tier testing (when required) can reflect longer exposures and evaluate sublethal enpoints. In a nutshell, species representing five phyla and including 21 species of crustacean, 13 species of fish, and 7 species of bivalve are listed. In practice, however, the number of species used is generally much smaller, with arthropods, annelids and molluscs being the most commonly used. In the case of marine sediments, Nendza (2002) compiled an inventory of bioassays for the evaluation of dredged material and sediments on behalf of the Federal Environmental Agency of Germany. The selected bioassays are applicable to whole sediment, sediment suspension, sediment elutriate, pore water and/or sediment organic extract and the endpoints cover acute and chronic toxicity, bioaccumulation reproductive effects as well as carcinogenicity and mutagenicity. In another comparative study, Alzieu et al. (2003) assessed 9 marine bioassays for their sensitivity and discrimination potential when applied to a series of 10 sediment samples collected at different locations along the French coastline. Based on the test results, the authors recommended a battery of four bioassays: modified bivalve acute toxicity test (based on the protocol from U.S. EPA, 1996b), solid-phase Microtox® (test conducted according to ISO, 1998), a modified version of the bioassay using the amphipod *Corophium* sp. (Oslo and Paris Commissions, 1995), and the marine copepod acute toxicity test with *Acartia tonsa*, *Tisbe battagliai* and *Nitocra spinipes* (ISO, 1999).

Testing of solid environmental matrices such as soils and sediments is often done with single-concentration tests using 100% of the sample, and the observed effects are compared with the performance in the controls. The endpoint is simply the presence or absence of toxicity, and there is no quantitative estimate of the strength of the toxicity. Tests could also be performed with organisms exposed to a series of concentrations of the test material, normally in a geometric dilution sequence (*i.e.*, a logarithmic series). Dilution series are typically conducted on sediment pore water and solvent extracts, but can also be performed with whole or wet sediment by diluting with a "clean" (*i.e.*, relatively uncontaminated) sediment carefully chosen to be as chemically and physically similar to the test sediment as possible. Care must be taken in interpreting toxicity results of sediment dilution experiments because of the possible influence of the sediment used for dilution.

Multi-concentration tests measure the effect of test concentrations on a quantal (all-or-none) or graded/quantitative (*e.g.*, weight of the organism, number of offspring produced) variable. An example of a quantal endpoint would be a lethality/survival, where the measurement endpoint would be an LC50, or the median lethal concentration. The preferred measurement endpoint for graded tests is the ICp, or the inhibitory concentration for a given percent effect. A value is selected for "p", usually 25% (IC25) or 50% (IC50), and occasionally 10% (IC10). Although there is some reticence in employing the null hypothesis approach, it is sometimes used to analyze quantitative sublethal data. The lowest-observed-effect-concentration (LOEC) is the lowest concentration in a test series, at which a biological effect is observed. The no-observed-effect-concentration (NOEC) is the highest concentration at which an adverse effect is not observed (*e.g.*, size of exposed organisms is the same as that in controls); it is always the next lowest concentration after the LOEC in the dilution series. The threshold-effect-concentration (TEC) is the geometric mean of the NOEC and LOEC. Refer to the Appendix section for the detailed steps involved in the calculation of the SED-TOX index.

6. Factors capable of influencing the SED-TOX procedure interpretation potential

Indices such as the SED-TOX and the PEEP have great appeal because their single value is easily grasped. However, they could sometimes be deceptive. An index could encompass a large number of toxicity measurements that provides an assessment of water or sediment quality, but indices need to be used cautiously because there is a loss of information in the calculation of one summary value. Results of individual toxicity tests should be examined for questionable results and endpoints that might skew the index.

Caution is also required in the selection of test organisms, endpoints and protocols that are included in the battery to ensure their environmental relevance when integrated in the calculation of the index. As stated earlier, it would be preferable to include tests that cover a wide range of organisms representing different trophic groups, life cycles (*i.e.*, short versus long) exposure route, sensitivities to contaminants, and local relevance. Ideally, a battery of at least three different tests should be conducted. Obviously, the SED-TOX index is strongly dependent on the selection of toxicity tests. It is assumed that the tests selected and integrated in the battery are appropriate to the particularities of a given study.

Toxicity tests, which are used to generate the SED-TOX index, cannot substitute for chemical measurements or for surveys of benthic communities. On the contrary, the strength of toxicity tests, and hence the SED-TOX index, are best realized in conjunction with chemical and biological field measurements. These three approaches form a natural triad in which each component enhances the power of the others (Sergy, 1987).

7. Application of the SED-TOX index in a case-study

This section illustrates how the SED-TOX index has been effective in assessing the relative toxic potential of freshwater sediments to aquatic organisms. The index was applied to laboratory toxicity data derived from two larger projects conducted on freshwater sediments in the St. Lawrence river ecosystem (Bombardier and Blaise, 2000).

7.1 SEDIMENT COLLECTION SITES AND ANALYSES

Sediments were collected during the fall of 1995 and 1996 as part of two separate studies:

(1) The first study was designed to assess the suitability of various micro-scale bioassays and recommend an appropriate testing strategy for sediment toxicity assessment (Côté et al., 1998a,b). The recommended test batteries included seven micro-scale laboratory assays conducted on bacteria (*Vibrio fischeri*), cnidarians (*Hydra attenuata*), micro-crustaceans (*Thamnocephalus platyurus*), and benthic macro-invertebrates (*Hyalella azteca* and *Chironomus riparius*), and involved two phases of exposure (pore water and whole sediment). A total of 16 stations were included in the toxicity assessment scheme.

(2) The second study applied the weight-of-evidence approach to assess the quality of 17 sediment stations located in a highly industrialized sector along the St. Lawrence River. Five toxicity assays were conducted and encompassed four taxonomic groups, namely bacteria (*V. fischeri* and *Escherichia coli)*, microphytes (*Pseudokirchneriella subcapitata*), amphipods (*H. azteca*) and chironomids (*C. riparius*), and considered three exposure phases (*i.e.*, wet sediment, organic extract, and whole sediment).

The reader should refer to Table 2 for more details on each toxicity assay used in the two above-described studies. In both studies, superficial (top 15 cm) sediment samples were collected using a 0.23-m^2 Ponar dredge. Each sample was a composite of the 15 to 30 L of sediment taken per site. All sediment samples were subjected to chemical characterization using appropriate methodologies (U.S. EPA, 1983, 1986; Allen et al., 1993).

Two to three individual grab samples per station were also taken for benthic measures at a limited number of stations (total of 9 for both studies). Taxa of Nematoda, Anellida, Hydracarina, Harpacticoida, Ostracoda, Gammaridae, Amphipoda, Isopoda, Trichoptera, Ephemeroptera, Odonata, Lepidoptera, Ceratopogonidae, Chironomidae, Psychodidae, Gastropoda and Pelecypoda were determined according to descriptions found in standard literature. Data were used to compute the following four benthic measures: species richness (total number of species), the EPT index (number of taxa in the orders Ephemeroptera, Plecoptera, and Trichoptera), the Shannon-Wiener (H') diversity index, and a modified version of the integrative Invertebrate Community Index (Ohio-EPA, 1988), labeled "ICI-SL" after the Invertebrate Community Index-St. Lawrence (Willsie, 1993a,b).

Table 2. Bioassay testing conditions and reference methods for the selected species.

Toxicity test/ Species	***Study***[c]	***Assessment endpoint***	***Test phase***[d]	***Test duration***	***Measurement endpoint***[e]	***Number of replicates***[f]	***Test method reference***
BACTERIA							
Microtox® *V. fischeri*	BA	Light inhibition	PW	15 min	IC50	4	Microbics, 1992
Microtox® SPT[a] *V. fischeri*	ERA	Light inhibition	WT	20 min	IC50	2	Blaise et al., 1994
SOS Chromotest *E. coli* PQ37	ERA	Genotoxicity	OE	2 h	LOEC	4	Env. Canada, 1993
MICRO-ALGAE							
S. capricornutum ASPA[b]	ERA	Esterase inhibition	WT	24 h	IC50	3	Blaise & Ménard, 1998
MICRO-CRUSTACEAN							
ThamnotoxKit® *T. platyurus*	BA	Survival	PW	24 h	LC50	3	Creative Selling, 1992
CNIDARIAN							
H. attenuata	BA	Survival	PW	96 h	TEC	3	Trottier et al., 1997
MACRO-INVERTEBRATES							
H. azteca	BA, ERA	Survival	WS	14 d	% survival	5	Env. Canada, 1997a
C. riparius	BA, ERA	Survival	WS	10 d	% survival	5	Env. Canada, 1997b

[a] SPT: Solid-Phase Test.
[b] ASPA: Algal Solid-Phase Assay.
[c] BA : Battery Approach study; ERA : Ecotoxicological Risk Assessment study.
[d] PW: Pore-Water; OE: Organic Extract; WS: Whole sediment; WT: Wet Sediment (*i.e.*, that remaining after removal of pore water).
[e] IC50: concentration estimated to cause 50% reduction in the biological function of interest; LC50: concentration estimated to cause 50% reduction in survival; TEC: Threshold Effect Concentration, calculated as the geometric mean of the No-Observable Effect Concentration (NOEC) and the Lowest-Observable Effect Concentration (LOEC).
[f] Number of test replicates for a single sediment sample, as indicated in the testing methods.

7.2 SED-TOX INDEX CALCULATION AND INTERPRETATION

SED-TOX scores were determined for each sediment sample investigated in the two studies and Pearson's correlations were estimated with benthic community metrics and levels of contamination (SAS, 1988). Contamination levels were expressed as the mean ratios of individual contaminant concentrations in a sample relative to their respective SQG values. Logarithmic transformations were applied to mean SQG quotient values to respect the assumption of normality.

Figure 1 illustrates the relationship between sediment contamination and SED-TOX scores. Sites yielding high SED-TOX scores (≥ 2.0) had elevated levels of chemical contaminants (mean SQG quotients > 1) relative to sites which revealed marginal hazard potential (*i.e.* SED-TOX varying between 0.1 and 0.9). Moreover, respectively 70% of the sites showing a high hazard potential had mean SQG quotients > 5, while 86% of those with a marginal SED-TOX score had mean SQG quotients < 1. Hence the proportion of sediments showing a high toxicity hazard to exposed organisms increased with increasing contaminant concentrations.

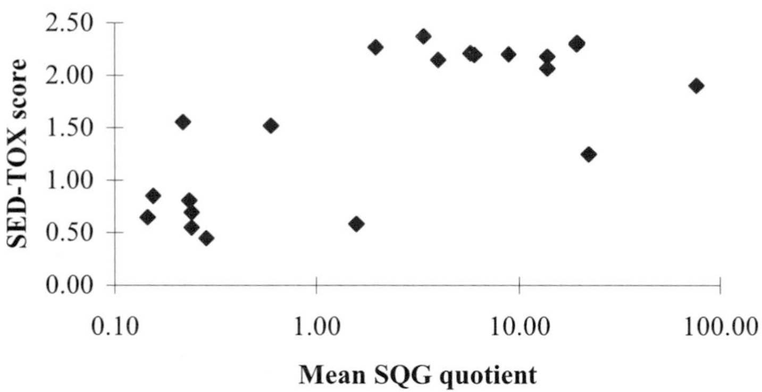

Figure 1. Relationship between SED-TOX scores and mean Sediment Quality Guideline (SQG) quotients. Scores varying between 0.1 and 0.9 represent a marginal hazard; scores between 1.0 and 1.9 indicate a moderate hazard; and scores greater or equal to 2.0 represent a high hazard potential.

The association between mean SQG quotients and toxic effects measured in the test batteries were not as clear for sites that yielded moderate SED-TOX scores. Indeed, mean SQG quotients ranged from 0.21 to 76.1 at these sites, indicating a wide range of contamination levels in the moderately toxic sediments. A variety of scenarios may explain this lack of correspondence for sediments falling into the moderate class of hazard potential, for example: (1) toxicity may have occurred in sediments with low mean SQG quotients due to the potential effects of unmeasured contaminants or potentially toxic substances such as ammonia, butyl tin, and sulphur, which are not accounted for with SQGs; (2) some contaminants may not have been present in bio-available forms (*e.g.* complexation with sediment particles, inorganic complexes or organic matter) or the test species composing the battery may have been less sensitive than anticipated to the mixtures of contaminants present in the sediments. A cautious approach is therefore advised when using solely chemical measures or bioassays alone to assess sediment quality.

The incidence on benthos, as suggested by benthic community metrics, was also examined in relation to SED-TOX scores. Figure 2 presents plots of the SED-TOX scores in relation to ascending values of benthic community structure indices. The ICI-SL showed the best relationship with the SED-TOX scores. With two exceptions, relative hazard potential tended to increase with decreasing values of ICI-SL (which suggests benthos alteration). Indeed, high SED-TOX scores (≥ 2.0) were always associated with lower ICI-SL scores (< 8). Of all sites investigated, the reference site yielded the lowest SED-TOX score and the highest ICI-SL value, indicating a low level of stress on the biota. By contrast, three of the highly contaminated sites were of poor quality and two were moderately degraded, based on taxa richness. All of these sites showed high hazard potential in the bioassays. The H' and EPT indices were the least effective at discriminating among sites on the basis of benthos alteration. With one exception, all sites were considered degraded according to the EPT index, while two of the highly contaminated sites were deemed to have "clean sediments" according to the H' classification scheme. It is not surprising to find a certain dichotomy between measures of sediment toxicity and benthic infaunal communities. Indeed, the presence or absence of particular taxa may depend more on environmental characteristics, such as current velocity, physical disturbance, and substrate, for example, than on the degree of contamination. The limited size of the data set may also contribute to the lack of a clear relationship between toxicity and benthic community composition. Other possible reasons for the lack of agreement between bioassay results and benthic indices are that: (1) laboratory bioassays were conducted under controlled conditions and on a limited range of species and life stages which may not accurately mimic *in situ* conditions; (2) laboratory toxicity and chemistry analyses were conducted on composite samples while benthic measures relied on individual grab samples; or (3) the indices used to assess benthic communities, except perhaps for the ICI-SL, are not adequate discriminators of the degree of sediment contamination.

8. Conclusions

Initiatives to develop new methodologies for the assessment of sediment and soil toxicity are ongoing. While there is a need for more sensitive and cost-effective bioassays, the question remains: what concept should be used to integrate as much information as possible from a diverse set of test species and assessment endpoints? It has been demonstrated that it is possible to summarize the results of multiple toxicity assays conducted on a solid matrix such as freshwater sediments by means of a single index. Although there is still little experience with this sediment toxicity index, it is expected to be of great utility in the assessment of sediment and soil quality as a tool for evaluating their toxic potential and tracking their condition over time. Information generated from the SED-TOX index may also prove helpful for investigators who wish to explore causality in future investigations. It might, for instance, be employed to pinpoint sites of concern where investigative strategies involving bioanalytical assessment of spiked sediments or toxicity identification evaluations (TIEs) might be appropriate.

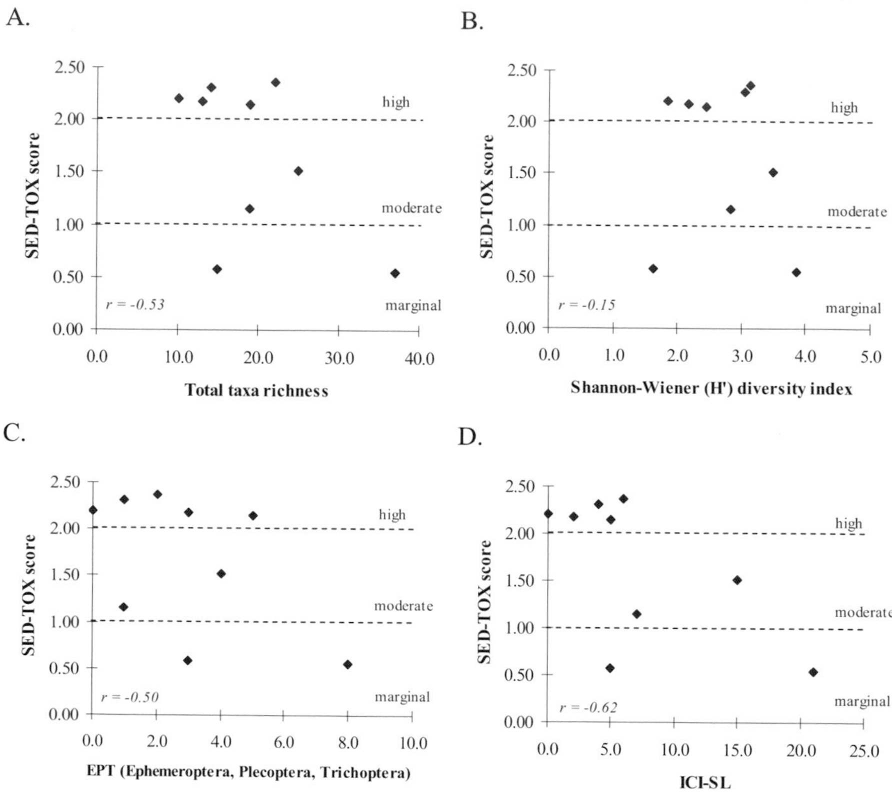

Figure 2. Relationship between SED-TOX scores and benthic community metrics. Scores for each metric are classified as follows: A) Richness < 17 = degraded, 17–23 = moderately degraded, 24–32 = fairly clean, > 33 = clean (taken from Willsie, 1993a, b); B) H' < 1 = degraded, 1–3 = moderately degraded > 3 = clean (Wilhm, 1967); C) EPT < 6 = degraded, 7–13 = fairly clean (taken from U.S. EPA, 1996a); D) 0 ≤ICI-SL < 8 = degraded, 8 ≤ICI-SL < 16 = fairly clean, ≥ 16 = clean (adapted from Willsie, 1993a, b). EPT: taxa richness in the orders Ephemeroptera, Plecoptera and Trichoptera; ICI-SL: Invertebrate community Index for the St. Lawrence River.

The results of the comparison of SED-TOX scores with chemical concentrations and benthic indices in the case-study reported herein underscore the caution with which previous authors have regarded the use of single lines of evidence as indicators of sediment quality (*e.g.*, Chapman, 1989; Luoma and Ho, 1993; Canfield et al., 1996; Ingersoll et al., 1997). We agree with other investigators (Chapman, 1992, 1995; Clements and Kiffney, 1994; Day et al., 1995; Canfield et al., 1996) that the best approach to the assessment of sediment quality is still an integrative approach wherein more than one generic tool, such as a battery of toxicity tests, is combined with chemical analyses and an evaluation of benthic communities structure. In other words, the SED-TOX index and other approaches described herein should be used in an integrative, not a fragmented, manner.

In summary, the SED-TOX index's incorporation of multiple toxicity responses offers a scientifically sound and ecologically meaningful tool to discriminate among sites based on toxicity hazard potential. However, it is recognized that the SED-TOX scores are inherently dependent on the bioassays composing the battery. It is, therefore, of prime importance to employ the same test battery when attempting to compare various sites, although this may not always be feasible. It is also important to carefully select sensitive test species, toxicity endpoints and test protocols to ensure the environmental relevance of the tests and hence, the SED-TOX scores generated from the toxicity results. Additionally, environmental assessors may be enticed to take decisions based solely on a single toxicity hazard value because of the evident convenience it offers. They should nevertheless not lose track of the relevant information that individual toxicity responses also contribute; such information is always easily accessible to users of the SED-TOX approach.

Acknowledgements

Special thanks are given to the editors of this book for their invitation to contribute this chapter. The invaluable technical contribution of Manon Harwood, Lucie Ménard, Christine Girard and Sylvain Trottier from the St. Lawrence Centre is acknowledged for their conduct of the bioassays. Data on chemistry, toxicity to macro-invertebrates and benthos were kindly provided by the Environmental Protection Branch of Environment Canada (Quebec Region), and BEAK International Incorporated.

References

Adams, W.J., Kimerle, R.A. and Mosher, R.G. (1985) Aquatic safety assessment of chemicals sorbed to sediments, in R.D. Cardwell, R. Purdy, and R.C. Bahner (eds.), *Aquatic Toxicology and Hazard Assessment: Seventh Symposium*, STP 854, American Society for Testing Materials, PA, pp. 429-453.

Adams, W.J. (1987) Bioavailability of neutral, lipophilic organic chemicals contained on sediments: a review, in K.L. Dickson, A.W. Maki, and W.A. Brungs (eds.), *Fate and Effects of Sediment-bound Chemicals in Aquatic Systems*, Pergamon Press, NY, pp. 219-245.

Allen, H.E., Fu, G. and Deng, B. (1993) Analysis of acid-volatile sulfide (AVS) and simultaneously extracted metals (SEM) for the estimation of potential toxicity in aquatic sediments, *Environmental Toxicology and Chemistry* **12**, 1441-1453.

Ankley, G.T., Mattson, V.R., Leonard, E.N., West, C.W. and Bennett, J.L. (1993) Predicting the acute toxicity of copper in freshwater sediments: Evaluation of the role of acid volatile sulphide, *Environmental Toxicology and Chemistry* **12**, 315-320.

Alzieu, C. (1999) Gestion des sédiments portuaires, in Ifremer (ed.), *Dragages et Environnement Marin*, France, pp. 167-186.

Alzieu, C. (coord.) (2001) Géodrisk: logiciel d'évaluation des risques liés à l'immersion des déblais de dragage des ports maritimes, Ifremer (ed.), Cederoms.

Alzieu, C., Quiniou, F. and Delesmont, R. (2003) L'approche globale: scores de risques, tests de toxicité, indices biotiques, in C. Alzieu (coord.), *Bioévaluation de la qualité des sédiments portuaires et des zones d'immersion*, Ifremer (ed.), pp. 177-190.

Blaise, C., Sergy, G., Wells, P.G., Bermingham, N. and Van Coillie, R. (1988) Biological testing – development, application and trends in Canadian Environmental Protection laboratories, Toxicity *Assessment* **3**, 385-406.

Blaise, C., Forghani, R., Legault, R., Guzzo, J. and Dubow, M.S. (1994) A bacterial toxicity assay performed with microplates, microluminometry and Microtox^{TM} reagent, *Biotechniques* **16**, 932-937.

Blaise, C. and Ménard, L. (1998) A micro-algal solid-phase test to assess the toxic potential of freshwater sediments, *Water Quality Research Journal of Canada* **33**, 133-151.

Bombardier, M. and Bermingham, N. (1999) The SED-TOX index: Toxicity-directed management tool to assess and rank sediments based on their hazard – Concept and application, *Environmental Toxicology and Chemistry* **18**, 685-698.

Bombardier, M. and Blaise, C. (2000) Comparative study of the Sediment-Toxicity Index, benthic community metrics and contaminant concentrations, *Water Quality Research Journal of Canada* **35**, 753-780.

Bulich, A.A., Green, M.W. and Underwood, S.R. (1992) Measurement of soil and sediment toxicity to bioluminescent bacteria when in direct contact for a fixed time period. Abstract at Water Environment Federation, 65th Annual Conference and Exposition, September 20-24, New Orleans, LA.

Canfield, T.J., Dwyer, F.J., Fairchild, J.F., Haverland, P.S., Ingersoll, C.G., Kemble, N.E., Mount, D.R., La Point, T.W., Burton, G.A., and Swift, M.C. (1996) Assessing contamination in Great Lakes sediments using benthic invertebrate communities and the sediment quality triad approach, *Journal of Great Lakes Research* **22**, 565-583.

CCME (Canadian Council of Ministers of the Environment) (1995) Protocol for the Derivation of Canadian Sediment Quality Guidelines for the Protection of Aquatic Life, Prepared by the Technical Secretariat of the CCME Task Group on Water Quality Guidelines, Ottawa, Canada.

Chapman, P.M. (1989) Current approaches to developing sediment quality criteria, *Environmental Toxicology and Chemistry* **8**, 589-599.

Chapman, P.M. (1992) Pollution status of north Sea sediments – an international integrative study, *Marine Ecology Progress Series* **91**, 313-322.

Chapman, P.M. (1995) Sediment quality assessment: status and outlook, *Journal of Aquatic Ecosystem Health* **4**, 183-194.

Chapman, P.M. (2000) The Sediment Quality Triad: then, now and tomorrow, *International Journal of Environmental Pollution***13**, 351-356.

Chapman, P.M., Wang, F., Germano, J.D. and Batley, G. (2002) Pore water testing and analysis: the good, the bad, and the ugly, *Marine Pollution Bulletin* **44**, 359-366.

Clements, W.H. and Kiffney, P.M. (1994) Integrated laboratory and field approach for assessing impacts of heavy metals at the Arkansas River, Colorado, *Environmental Toxicology and Chemistry* **13**, 397-404.

Cook, N.H. and Wells, P.G. (1996) Toxicity of Halifax Harbour sediments: an evaluation of the Microtox® solid-phase test, *Water Quality Research Journal of Canada* **31**, 673-708.

Costan, G., Bermingham N., Blaise C. and Férard, J.F. (1993) Potential ecotoxic effects probe (PEEP): a novel index to assess and compare the toxic potential of industrial effluents, *Environmental Toxicology and Water Quality* **8**, 115-140.

Côté, C., Blaise C., Schroeder J., Douville M. and Michaud, J.R. (1998a) Investigating the adequacy of selected micro-scale bioassays to predict the toxic potential of freshwater sediments through a tier process, *Water Quality Research. Journal of Canada* **33**, 253-277.

Côté, C., Blaise C., Michaud J.R., Ménard, L., Trottier, S., Gagné, F., Lifshitz, R. (1998b) Comparisons between microscale and whole-sediment assays for freshwater sediment toxicity assessment, *Environmental Toxicology and Water Quality* **13**, 93-110.

Creative Selling (1992) Thamnotoxkit^{TM} F, Crustacean Toxicity Test for Freshwater, Standard Operational Procedure, Creative Selling Ltd., Deinze, Belgium.

Day, K.E., Dutka, B.J., Kwan, K.K. and Batista, N. (1995) Correlations between solid-phase microbial screening assays, whole-sediment toxicity tests with macroinvertebrates and *in situ* benthic community structure, *Journal of Great Lakes Research* **21**, 192-206.

Di Toro, D.M., Mahony, J.D., Hansen, D.J., Scott, K.J., Hicks, M.B., Mayr, S.M. and Redmond, M.S. (1990) Toxicity of cadmium in sediments: The role of acid volatile sulfide, *Environmental Toxicology and Chem.istry* **9**, 1487-1502.

Di Toro, D.M., Zarba, C.S., Hansen, D.J., Berry, W.J., Swartz, R.C., Cowan, C.E., Pavlou, S.P., Allen, H.E., Thomas, N.A. and Paquin, P.R. (1991) Technical basis for establishing sediment quality criteria for non-ionic organic chemicals using equilibrium partitioning, *Environmental Toxicology and Chemistry* **10**, 1541-1583.

Environment Canada (1993) Genotoxicity Using the *Escherichia coli* PQ37 Bacterium (SOS Chromotest), St. Lawrence Centre, Environmental Protection, Conservation and Protection, Environment Canada, Montreal, QC.

Environment Canada (1994) Guidance Document on collection and Preparation of Sediments for Physicochemical Characterization and Biological Testing, Report EPS 1/RM/29, Environment Canada, Ottawa, ON.

Environment Canada (1997a) Biological Test Method: Test for Survival and Growth in Sediment Using the Freshwater *A*mphipod *Hyalella azteca*, Report EPS 1/RM/33, Environmental Protection Publications, Ottawa, ON.

Environment Canada (1997b) Biological Test Method: Test for Survival and Growth in Sediment Using Larvae of Freshwater Midges (*Chironomus tentans* or *Chironomus riparius*), Report EPS 1/RM/32, Environmental Protection Publications, Ottawa, ON.

Hare, L., Tessier, A. and Warren, L. (2001) Cadmium accumulation by invertebrates living at the sediment-water interface, *Environmental Toxicology and Chemistry* **20**, 880-889.

Ingersoll, C.G., Ankley, G.T., Baudo, R., Burton, G.A., Lick, W., Luoma, S.N., MacDonald, D.D., Reynoldson, T.F., Solomon, K.R., Swartz, R.C. and Warren-Hicks, W. (1997) Work group summary report on an uncertainty evaluation of measurement endpoints used in sediment ecological risk assessments, Chapter 18, EPA/600/A-96/097, U.S. Environmental Protection Agency, National Biological Service, Corvallis, OR, USA.

ISO (1998) Water quality - Determination of the inhibitory effect of water samples on light emission of *Vibrio fischeri* (Luminescent bacteria test), Part 3 - Method using freeze-dried bacteria. International Organization for Standardization, TC 147/SC 5.

ISO (1999) Water quality - Determination of acute lethal toxicity to marine copepods (Copepoda, Crustacea). International Organization for Standardization, TC 147/SC 5.

Lamberson, J.O. and Swartz, R.C. (1992) Spiked-sediment toxicity test approach, in *Sediment Classification Compendium*, EPA 823-R-92-006, U.S. EPA, Office of Water, Washington, DC.

Long, E.R. and Chapman, P.M. (1985) A sediment quality triad measures of sediment contamination, toxicity and infaunal community composition in Puget Sound, *Marine Pollution Bulletin* **16**, 405-415.

Long, E.R. and Morgan, L.G. (1990) The potential for biological effects of sediment-sorbed contaminants tested in the National Status and Trends Program, NOAA Technical Memorandum NOS OMA 52, Seattle, WA.

Luoma, S.N. and Ho, K.T. (1993) Appropriate uses of marine and estuarine sediment bioassays, in P. Calow (ed.), *Handbook of Ecotoxicology*, Blackwell Scientific Publications, London, UK, pp. 193-226.

Macdonald, D.D., Smith, S.L., Wong, M.P. and Mudroch, P. (1992) The Development of Canadian Marine Environmental Quality Guidelines, Ecosystem Sciences and Evaluation Directorate, Conservation and Protection, Environment Canada, Ottawa, ON.

Microbics (1992) Microtox® Manual. A Toxicity Testing Handbook, Vol.2 -Detailed Protocols, Vol. 3 - Condensed Protocols, Microbics, Carlsbad, CA, USA.

Nendza, M. (2002) Inventory of marine biotest methods for the evaluation of dredged material and sediments, *Chemosphere* **48**, 865-883.

Novak, L. and Scroggins, R. (2003). Toxicity test methodologies for assessing contaminant mixtures in Canadian soils – workshop overview. Abstract presented at the 24th Annual Meeting of the Society of Environmental Toxicology and Chemistry (SETAC), Austin, Texas, U.S.A., 9-13 November, 2003.

Ohio-EPA (Environmental Protection Agency) (1988) Biological Criteria for the Protection of Aquatic Life, Vol. 1, The role of biological data in water quality assessment, Division of Water Quality Monitoring and Assessment, Surface Water Section, Columbus, Ohio, USA.

Oslo and Paris Commissions (1995) Protocols on Methods for Testing of Chemicals Used in the Offshore Oil Industry. Part A. A Sediment Bioassay Using an Amphipod *Corophium* sp. Reference number: 1995-07.

Quiniou, F. and Alzieu, C. (1999) L'analyse des risques chimiques appliqués au dragage, in C. Alzieu, (coord.), Ifremer (ed.), *Dragages et Environnement Marin: État des connaissances*, pp. 109-125.

SAS Institute Inc. (1988) SAS/STAT® User's Guide, Release 6.03, Cary, N.C., USA.

Sergy, GA. (1987) Recommendations on aquatic biological tests and procedures for environmental protection, C & P DOE, Environment Canada, Environmental Protection, Technology Development and Technical Services Branch, Edmonton, AB.

Sprague, J.B. and Ramsay, B.A. (1965) Lethal levels of mixed copper-zinc solutions for juvenile salmon. *Journal of the Fisheries Research Board of Canada* **22**, 425-432.

Trottier, S., Blaise, C., Kusui, T. and Johnson, E.M. (1997) Acute toxicity assessment of aqueous samples using a microplate-based *Hydra attenuata* assay, *Environment Toxicology and Water Quality* **12**, 265-271.

U.S. EPA (U.S. Environmental Protection Agency) (1983) Methods for Chemical Analysis of Water and Wastes, 3rd edition, U.S. EPA, Environmental Monitoring and Support Laboratory, Cincinnati, OH, USA.

U.S. EPA (U. S. Environmental Protection Agency) (1986) Test Methods for Evaluating Solid Waste. SW-846, U.S. EPA, Office of Solid Waste, Washington, DC, USA.

U.S. EPA (U.S. Environmental Protection Agency) (1996a) Biological criteria - Technical Guidance for Streams and Small Rivers, Revised edition, EPA 822-B96-001, Office of Science and Technology, Health and Ecological Criteria, U.S. EPA, Washington, DC, USA.

U.S. EPA (U. S. Environmental Protection Agency) (1996b) Ecological Effects Test Guidelines. OPPTS 850.1055. Bivalve Acute Toxicity Text (Embryo-Larval). U.S. EPA, Prevention, Pesticides, and Toxic Substances. EPA 712-C-96-160.

U.S EPA (Environmental Protection agency) and U.S. ACE (Army Corps of Engineers). (1998) Evaluation of Dredged Material Proposed for Discharge in Waters of the US – Testing Manual, EPA/823-B-94/002, Washington, D.C.

Warren, L.A., Tessier, A. and Hare, L. (1998) Modeling cadmium accumulation by benthic invertebrates in situ: the relative contributions of sediment and overlying water reservoirs in organism cadmium concentrations, *Limnology and Oceanography* **43**, 1442-1454.

Wells, P.G., Lee, K. and Blaise, C. (eds.) (1998) Microscale Testing in Aquatic Toxicology: Advances, Techniques and Practice, CRC Press, Boca Raton, FL.

Wilhm, J.L. (1967) Comparison of some diversity indices applied to populations of benthic macroinvertebrates in a stream receiving organic wastes, *Journal of the Water Pollution Control Federation* **39**, 1673-1683.

Willsie, A. (1993a) Adaptation of the invertebrate community index (ICI) to the St. Lawrence River, in E.G. Baddaloo, S. Ramamoorthy, and J.W. Moore (eds.), Proc. 19th Annual Aquatic Toxicity Workshop, Edmonton, 4-5 October 1992.

Willsie, A. (1993b) L'Indice des Communautés d'Invertébrés: Définition, Choix et Validation des Métriques en Fonction des Données Historiques, Research Report, St. Lawrence Centre, Conservation and Protection, Environment Canada, QC.

Winger, P.V. and Lasier, P.J. (1995) Sediment toxicity in Savannah Harbour, *Archives of Environmental Contamination and Toxicology* **28**, 357-365.

Abbreviations

BA	Battery Approach
CAPT	Cumulative Average of Phase Toxicity
CCME	Canadian Council of Ministers of the Environment
DL	Detection Limit
EPT	Ephemeroptera, Plecoptera and Trichoptera
ERA	Ecotoxicological Risk Assessment
ICI-SL	Invertebrate Community Index modified for the St. Lawrence River
IFREMER	Institut français de recherche pour l'exploitation de la mer
LOEC	Lowest Observed Effect Concentration
n	Number of phases eliciting toxic effects
NOEC	No Observed Effect Concentration
NTSP	National Status and Trends Program
OE	Organic Extract
PEEP	Potential Ecotoxic Effects Probe
PW	Pore Water
QA/QC	Quality Assurance/Quality Control
SQG	Sediment Quality Guidelines
SQT	Sediment Quality Triad
SSTT	Spiked-Sediment Toxicity Test
TEC	Threshold Effect Concentration

TIF	Toxicity Incremental Factor
TU	Toxic Unit
TU_{dw}	Toxic Unit expressed on a dry weight basis
U.S. ACE	United States Army Corps of Engineers
U.S. EPA	United States Environmental Protection Agency
WAPT	Weighed Average of Phase Toxicity
WS	Whole Sediment
WT	Wet Sediment.

Appendix. Steps in the SED-TOX index calculations[a]

Sediment compartment determination

Study	Site	Moisture content (tm)	Moisture content (rm)	Sediment compartment proportions (x)	Sediment compartment proportions (y)	Sediment compartment proportions (z)
BA	1-5B	0.71	0.25	0.61	0.10	0.29
Mean				0.49	0.13	0.38

Calculation of the detection limit of the assays

	Pore water			Whole sediments	
	Microtox® *EC50* % v/v	*Thamnotoxkit®* *LC50* % v/v	*Hydra* *TEC* % v/v	*Amphipod* % survival %	*Chironomid* % survival %
Cmax, Rmax	100	100	100	100	100
D	1	1	1	—	—
Rmin	—	—	—	69	69
DL	1.3	1.3	1.3	—	—

Equations

$$x = tm - \frac{rm(1-tm)}{1-rm}$$

$$y = \frac{rm \times z}{1-rm}$$

$$z = 1-(x+y)$$

$$DL = \frac{D \times \bar{p}}{\bar{z}} \quad \text{for PW: } \bar{p} = \bar{x} \quad \text{for WS: } \bar{p} = 1$$

$$TU_{fw} = \frac{Cmax \times D}{C_{tox}}$$

$$TU_{dw} = TU_{fw} \times \frac{p}{z}$$

Appendix (continued). Steps in the SED-TOX index calculations[a]

Toxicity data conversion

Study	*Site*	*Porewater*							
		Microtox® *EC50*				*Hydra* *NOEC*			
		%v/v	*TUfw*	*TUdw*	*TIF*	*%v/v*	*TUfw*	*TUdw*	*TIF*
BA	1-5B	35	2.9	6.0	4.6	>100	1.0	2.1	1.6

Study	*Site*	*Whole sediments*			
		Amphipod		*Chironomid*	
		% survival	*TIF*	*% survival*	*TIF*
BA	1-5B	80	0	80	0

Toxicity data integration

Study	*Site*	*PW*	*WS*	*CAPT*	*n*	*SED-TOX score*
BA	1-5B	1	0	1.8	1	0.5

Equations

$$TIF = \frac{TU_{dw}}{DL} \quad \text{or}$$

$$TIF = (R_{\min} - R_{tox}) \times \log(R_{\max} - R_{tox})$$

$$CAPT = \frac{a \dfrac{\sum TIF_{PW}}{N_{PW}} + d \dfrac{\sum TIF_{WS}}{N_{WS}}}{r}$$

$$SED\text{-}TOX = \log\left[1 + n(CAPT)\right]$$

[a] Consult Bombardier and Bermingham (1999) for a detailed description of the calculations.
EC50: concentration causing 50% reduction in the biological function of interest;
LC50: concentration causing 50% lethality;
TEC, threshold effect concentration, calculated as the geometric mean of the no-observable effect concentration (NOEC) and the lowest-observable effect concentration (LOEC); BA: Battery Approach; PW: porewater; WS: whole sediments.

Appendix (continued). Steps in the SED-TOX index calculations[a]

Equation variables:

a,d = sensitivity weighting factor for PW and WS, resp.
$CAPT$ = cumulative average of phase toxicity
$Cmax$ = maximum tested concn
D = dilution factor
DL = test detection limit
p = sum of x, y and/or z proportions
$\bar{p}$, $\bar{x}$, $\bar{z}$ = mean values of all samples
r = redundancy factor
rm = residual moisture
$Rmax$ = maximum response
$Rmin$ = minimum nontoxic response
$Rtox$ = measured response
TIF = toxicity incremental factor
tm = total moisture
x = centrifugeable water fraction
y = uncentrifugeable water fraction
z = dry sediment fraction
$TUdw$ = Toxic Unit expressed on a dry weight basis
$TUfw$ = Toxic Unit expressed on a fresh weight basis

9. THE PT-METHOD AS A HAZARD ASSESSMENT SCHEME FOR SEDIMENTS AND DREDGED MATERIAL

FALK KREBS
German Federal Institute of Hydrology (BfG)
Am Mainzer Tor 1, 56068 Koblenz, Germany
krebs@bafg.de

1. Objective and scope of the pT-method

This method is used to assess the quality of solid media such as sediments and dredged material. It appraises the toxicity of different compartments and phases of solids (*e.g.*, whole sediment, porewater, elutriates, chemical extracts) with standardized bioassays, using a dilution series in geometric sequence with a dilution factor of two. The measured endpoint of toxicity corresponds to the first dilution stage at which the test material is no longer toxic to test organisms. Numerically, toxicity is reported as a pT-value related to the negative binary logarithm of the first non-toxic dilution factor identified. Determined pT-values permit easy and clear identification of toxic effects with different bioassays that are representative of different exposure pathways. Because of its capacity to discriminate grades of toxic hazards, the pT-value is particularly suitable for the ecotoxicological classification of complex environmental samples. The pT-value of the most sensitive organism within a test battery is the pT_{max}-value, and it determines the toxicity class of test samples based on a pT-index. With this scheme, results of different bioassays performed with whole sediment, porewater, or elutriates are considered equal in rank. For example, if the highest pT-value obtained is 8, the test material is then designated as a pT-index value of VIII and is also assigned to toxicity class VIII.

In the case of dredged material classification, the normally open-ended pT-ecotoxicity scale is restricted to seven classes (class 0 and classes I to VI). All pT_{max} –values higher than 6 are included in Class VI. In the context of dredged material management, which may include its relocation within the water body, the seven toxicity classes determined by the pT-method are allocated to three management categories designated as "unproblematic", "problematic" and "hazardous". The pT-ranking system permits comparative studies with results of different test systems and sampling sites. It can also provide simple graphic representations of toxic sediment loading along the course of a river or of a whole river basin. The method does not require river-basin specific classifications.

C. Blaise and J.-F. Férard (eds.), Small-scale Freshwater Toxicity Investigations, Vol. 2, 281-304.

An application that might use the pT-method is the toxicity assessment of dredged material to decide if it can be relocated within a water body provided it meets certain quality criteria. This chapter also describes aspects of the pT-method that influenced the application of an effects-based assessment system in the German Federal Waterways and Shipping Administration (WSV). The pT-method proved useful to the WSV in the development of guidelines and categories for the management of dredged material. This system is now used along with chemical-numerical assessment for the characterization of dredged material. The pT-method has become an element of the guidelines of the German Federal Ministry of Transport for dredged material management in inland and coastal waters.

The pT-method can also be applied to assess liquids (untreated and treated wastewater, surface waters and groundwater). All data from aquatic toxicity tests used to detect pollutants can be integrated into this method. A general description of the pT-method is given in Chapter 3 of this volume. The present chapter specifically addresses the application of the pT-method to sediments and dredged material in order to classify and categorize the hazard associated with the degree of contamination of these matrices.

2. Summary of the pT-method procedure

The toxicity of a solid-medium sample (*e.g.*, whole sediment) or of its associated compartments (*e.g.* porewater, elutriate), is classified by the pT-index, which is based on the pT-value of the most sensitive organism within a given test battery. Thus, pT-values and pT-indices are numerical designations on an open scale to characterize the degree of hazard represented by solid media. Sensitive aquatic ecosystems can be protected using the pT-index as a guide for ensuring sound management decisions and environmental protection. Table 1 provides synthetic information on the pT-method concept related to solid-media assessment.

Table 1. Summary table of the pT-index for sediments and dredged material.

pT-Index **(pT = *potentia Toxicologiae* = toxicological exponent)**
Purpose The pT-index was developed as a management tool to incorporate bioassay data into the decision-making process for assessing and comparing the relative toxic hazards of sediments and dredged material. Once pT-values and pT-indices have been determined, the dredged material can be allocated to a management category.

Table 1 (continued). Summary table of the pT-index for sediments and dredged material.

pT-Index
Principle
The pT-index allows the assessment and comparison of the toxic potential of sediments and dredged material. It is one example of an integrated bioassay-battery approach developed for the purpose of environmental management. This sediment assessment index relies on the use of an appropriate battery of bioassays at different trophic levels (decomposers, primary producers, and consumers) allowing the measurement of various types (acute, chronic) and levels (lethal, sublethal) of toxicity. The pT-values and pT-indices used in tandem provide an ideal system to describe different toxic effects in sediments. The pT-method can be used as an ecotoxicological discriminator in mapping out sediment quality along the course of a river.
Bioassays employed
The Freshwater Test Battery used in inland waters (HABAB-WSV, 2000) is comprised of the following organisms: • *Vibrio fischeri*, bacterial luminescence inhibition test or Microtox® assay (DEV L34, 1988); • *Desmodesmus subspicatus*, micro-algal growth inhibition assay (DEV L33, 1991); • *Daphnia magna*, cladoceran acute immobilisation test (DEV L30, 1989); Additional test systems, including sediment contact tests, are currently being researched. These systems are comprised of the following organisms: • *Lemna minor*, plant growth inhibition assay (Feiler and Krebs, 2001); • *Caenorhabditis elegans*, nematode reproduction test (Höss and Krebs, 2003); • *Danio rerio*, zebra fish, fish egg test (DEV-T6, 2003; Hollert et al., 2003); • *Salmonella typhimurium*, bacterial genotoxicity test or umu-test (DEV T3); The Marine Test Battery used in marine and brackish waters (HABAK-WSV, 1999) is comprised of the following organisms: • *Vibrio fischeri*, bacterial luminescence inhibition test or Microtox® assay (DEV L34, 1988); • *Phaeodactylum tricornutum*, micro-algal growth inhibition assay (DEV L45, 1988); • *Corophium volutator*, amphipod acute immobilisation test (ISO/DIS 16712, 2003).

Table 1 (continued). Summary table of the pT-index for sediments and dredged material.

pT-Index
Determination of sediment hazard potential with the pT-method **1. pT-value for the numerical designation of aquatic toxicity measured in a single bioassay** Hazard potential is determined with standardized aquatic toxicity tests, using 2-fold serial dilutions. The toxicity endpoint corresponds to the first dilution stage that does not produce any toxic effects to the target organisms. The numerical designation of toxicity is the pT-value. The pT-value is the negative binary logarithm of the first non-toxic dilution factor in a dilution series in geometric sequence with a dilution factor of two. The pT-value indicates the number of times a sample must be diluted in a 1:2 ratio until test organisms no longer exhibit toxic effects. The toxic potential of any aqueous sample can be readily quantified in an easily understandable way. The pT-scale is unlimited, as values can theoretically range from 0 to ∞. **2. pT-index for the numerical classification of aquatic toxicity measured within a test battery** The pT-value of the most sensitive organism within a test battery determines the toxicity class of the sample. Different bioassays are considered equal in rank. Roman numerals are assigned to the toxicity classes. For example, if the highest pT-value is 9, then the tested material is assigned to toxicity class IX (*i.e.* the pT-index is IX).
Management categories of dredged material For management of dredged material, the toxicity scale is restricted to seven classes which are allocated to three management categories: "Case 1: unproblematic", "Case 2: problematic" and "Case 3: hazardous". Management of dredged material in Germany is regulated in two Federal guidelines: the Guideline for the Management of Dredged Material in Inland Waters (HABAB-WSV, 2000) and the Guideline for the Management of Dredged Material in Coastal Waters (HABAK-WSV, 1999).
Documented applications of the pT-method The selected case study (Section 7) demonstrates the hazard potential of river sediments. In numerous investigations, pT-values were generated for sediments and dredged material of the rivers Rhine (with its tributaries Moselle and Saar), Ems, Weser, Elbe, Oder and their estuaries, as well as the North and Baltic Seas. Several pT-values were also generated for sediments in the Sepetiba Bay (Federal State of Rio de Janeiro, Brazil) to identify toxic areas. Again, the pT-method was used as an ecotoxicological discriminator to map out sediment quality in polluted zones (Soares and de Freitas, 2000).

3. Historical overview and applications reported with the pT-method

The solid material removed from the bed of a river, canal or harbor in dredging projects must be relocated. In the Federal Republic of Germany, all of the larger waterways are the property of the Federal Government, whereas a number of smaller waterways are the property of Federal States. In the practice of the WSV, dredged material is usually relocated within the water body from which it was removed. This relocation may be performed either by dumping the material directly into flowing water, by hydrodynamic dredging (*e.g.* suction dredge, water jet), or by confined disposal. Sediment removal and its relocation are considered as one continuous process performed under the sovereign administrative activity of the Federal government (Köthe and Bertsch, 1999; Köthe, 2003).

3.1 MANAGEMENT OF DREDGED MATERIAL

Management of dredged material is regulated in Germany in two Federal guidelines: the Guideline for Management of Dredged Material in Inland Waters (HABAB-WSV, 2000) and the Guideline for Management of Dredged Material in Coastal Waters (HABAK-WSV, 1999). For disposal of dredged material on land, only the directive for inland waters (HABAB-WSV, 2000) is applicable.

The quality of dredged material must satisfy environmental protection standards prior to obtaining a permit for material relocation. Its quality is tracked by physical, sedimentological, chemical, biochemical (including oxygen and nutrient balances), including ecotoxicological criteria, and is classified according to definitions of the guidelines. Three categories are eventually identified: Case 1: "The dredged material can be relocated". Case 2: "Further study is required for decision-making". Case 3: "The material must not be relocated".

If a permit for relocation within a Federal waterway cannot be obtained, the material can be used for direct or indirect beneficial uses, upland disposal or disposal in waters other than Federal waterways. For these options, the guidelines provide only a general orientation, because the approval procedures fall under the jurisdiction of the Federal States.

3.2 ASSESSMENT STRATEGY: CHEMICAL AND ECOTOXICOLOGICAL INVESTIGATIONS

A general hazard assessment is based on chemical analyses and a series of established toxicity tests. While chemical analysis permits assessment of sediment by criteria which account for the concentrations of individual substances, the ecotoxicological method uses assessment criteria based on parameters that integrate toxic effects to organisms. Biological test methods can detect the ecotoxicological effects of individual substances as well as the combined effects of all contaminants in a dredged material sample (*e.g.* additive, antagonistic, or synergistic effects). Because they integrate the sum of contaminant effects, ecotoxicological examinations are an indispensable decision-making tool in dredged-material management.

3.3 SELECTION OF ECOTOXICOLOGICAL TESTS FOR A REPRESENTATIVE TEST BATTERY

The selection of ecotoxicological test methods must consider the prevailing ecological situation. Bioassays in a test battery are selected according to the aquatic ecosystem requiring protection (limnic, brackish, or marine) and the assessment endpoints of the tests (*e.g.* survival, growth, reproduction). Test organisms should be representative of at least three trophic levels including producers (*e.g.*, algae), consumers (*e.g.*, crustaceans), and reducers (*e.g.*, bacteria). Depending on indicator requirements, the battery comprises tests reflecting short-term and/or long-term exposures.

3.4 APPLICATION OF THE pT-PROCEDURE

The general pT-procedure can be applied to the assessment of sediments and dredged material in support of monitoring, decision-making, and regulatory needs (Krebs, 2000, 2001). The first publication was that of Krebs (1992).

The pT-procedure has been adopted by other countries. In a project of the GTZ (German government-owned corporation for international technical co-operation) the pT-procedure was applied successfully by the Environmental Agency (FEEMA) of the Federal State of Rio de Janeiro, Brazil, to identify toxic sediments in the Sepetiba Bay (Soares and de Freitas, 2000).

4. Advantages of the pT-method

The advantages of the pT-method as applied to solid media are briefly recalled below:

- easy to use and to interpret (Section 2),
- employs a cost-effective battery of bioassays (Sections 3.3, 5.2),
- universal and flexible in application (Sections 5.1, 5.2),
- simple sample dilution method for toxicity appraisal (Sections 5.3, 5.4),
- good discriminatory potential for sample toxicity (Section 5.4),
- no special software required for data reduction and pT-calculations (Section 5.9),
- ease of technology transfer (Section 3.4),
- based on a sound scientific conceptual framework (Section 5.4),
- user-friendly management tool (Sections 5.6, 5.7),
- enables sediment quality mapping of rivers (Section 7).

5. Description of the pT-procedure for solid-media assessment

5.1 OBJECTIVE, PRINCIPLE, AND SCOPE OF APPLICATION

The general objective, principle, and scope of application of the pT-method are succinctly described in Section 1 and also reported elsewhere in this book (see Chapter 3 of this volume, Section 5.1), where readers will appreciate that this hazard assessment scheme is adaptable to both liquid and solid media. Briefly recalled here in the context of solid-media samples such as dredged material, the pT-value, which relates to a single bioassay, and the pT-index, derived from the most sensitive organism in a test battery, permit a numerical classification of environmental samples on the basis of ecotoxicological principles. Sediment from any aquatic ecosystem (freshwater, brackish, marine) and from any of its phases (whole sediment, porewaters, elutriates or organic extracts) can be appraised provided that the proper standardized toxicity tests are available. There are whole-sediment test protocols standardized for many agencies (*e.g.*, Environment Canada, ASTM).

In the case of whole-sediment toxicity determination, the necessary dilutions can be made with reference sediment material. A standardized method whereby polluted sediments can be diluted with unpolluted sediments for sediment-contact tests is currently being researched (Höss and Krebs, 2003). Hence, the pT-method is capable of capturing the toxic effects of both soluble and adsorbed contaminants in a given sample, assuming that appropriate toxicity tests (*i.e.*, solid-phase contact tests on whole sediment and tests on porewater or elutriates) are used.

Utmost importance must be given to sampling, storing, and processing procedures of collected sediments to ensure the validity of subsequent bioassay results. In Germany, the DEV L1 - DIN EN ISO 5667-16 (1998) guideline is applicable to all pT-method bioassays.

5.2 TOXICITY TESTS RECOMMENDED IN APPLYING THE pT-METHOD

Due to the lack of standardized test methods for direct-contact sediment toxicity with integrated dilutions series, the pT-method described in this chapter focuses on the use of indirect toxicity assessment of solid-media samples using porewater and elutriates. Whole-sediment tests (direct sediment contact tests) have been developed more recently and only a few have reached a recognized level of standardization (see, for example, Chapters 2, 12 and 13, volume 1 of this book). Routine applications of the pT-method with dredged material have thus far been conducted with validated tests with porewaters and elutriates. These latter assays are able to detect effects resulting from sediment-bound contaminants that can easily dissolve into an aqueous phase (Winger et al., 2003). In this sense, pT-values reflect sediment hazard potential for water-column organisms more than they do for sediment dwellers. For the time being, tests based on bulk sediment without dilution series may be used as a supplementary weight of evidence of adverse effects, provided these methods have been proven to be reliable (Den Besten et al., 2003). Standardized whole-sediment tests will eventually become an intrinsic part of the pT-method bioassay battery for assessing solid media toxicity.

Because test results must be suitable as evidence in legal proceedings, only standardized methods are applied. These methods, described in Box 1, include an algal test, a luminescent bacteria test, and a microcrustacean test (Krebs, 1999). Reasons for selecting these toxicity tests are recalled in Chapter 3 of this volume, Sections 3.3 and 5.6. Reference standards are always used as positive controls. The tests are performed according to German Standard Methods (DEV/DIN), which are always based on guidelines detailed in DEV L1 - DIN EN ISO 5667-16 (1998). Additional test systems for a freshwater test battery, including sediment contact tests, are currently being researched and listed in Table 1.

Similarly, standardized tests recommended for investigating marine sediments are listed in Box 2 (Pfitzner and Krebs, 2001). These are conducted according to guidelines prescribed in HABAK-WSV (1999).

Box 1. Recommended freshwater test battery according to HABAB-WSV (2000).

Algal test

Desmodesmus subspicatus HEGEWALD and SCHMIDT, 2000, formerly known as *Scenedesmus subspicatus* CHODAT, 1926.
Taxonomy: Chlorophyta, Chlorophyceae, Chlorococcales.
Test performed according to DEV L 33 - DIN 38 412 Part 33 (1991).
Toxicity endpoint: cell growth inhibition.
Test duration: 72 h.
Number of test organisms per dilution step: 10^4 cells per mL.
Threshold value for the determination of the pT-value: IC < 20%.

Luminescent bacteria test

Vibrio fischeri BEIJERINCK, 1889; LEHMANN and NEUMANN, 1896, formerly known as *Photobacterium phosphoreum* COHN, 1878.
Taxonomy: Bacteria; Proteobacteria; Gammaproteobacteria; Vibrionales; Vibrionaceae.
Test performed according to DEV-L 34 - DIN EN ISO 11348-3 (1998).
Microtox® bacteria
Toxicity endpoint: luminescence inhibition.
Test duration: 30 min.
Number of test organisms per dilution step: about 10^6 cells per mL.
Threshold value for the determination of the pT-value: IC < 20%.

***Daphnia* test**

Daphnia magna STRAUS, 1820, water-flea.
Taxonomy: Crustacea, Branchiopoda, Cladocera, Daphniidae.
Test performed according to DEV L30 - DIN 38 412 Part 30 (1989).
Toxicity endpoint: microcrustacean acute immobilization.
Test duration: 24 h.
Number of test animals per dilution step: 10.
Threshold value for the determination of the pT-value: 90% survival.

Box 2. Recommended marine test battery according to HABAK-WSV (1999).

Algal test

Phaeodactylum tricornutum BOHLIN, 1897.
Taxonomy: Bacillariophyta (Diatoms), Bacillariophyceae, Naviculales.
Test performed according to DEV 45-DIN EN ISO 10253 (1998).
Toxicity endpoint: cell growth inhibition.
Test duration: 72 h.
Number of test organisms per dilution step: 10^4 cells per mL.
Threshold value for the determination of the pT-value: IC < 20%.

Luminescent bacteria test

Vibrio fischeri BEIJERINCK, 1889; LEHMANN and NEUMANN, 1896, formerly known as *Photobacterium phosphoreum* COHN, 1878.
Taxonomy: Bacteria; Proteobacteria; Gammaproteobacteria; Vibrionales; Vibrionaceae.
Test performed according to DEV L 34 - DIN EN ISO 11348-3 (1998).
Microtox® bacteria.
Toxicity endpoint: luminescence inhibition; test duration: 30 min.
Number of test organisms per dilution step: about 10^6 cells per mL.
Threshold value for the determination of the pT-value: IC < 20%.

Amphipod test

Corophium volutator PALLAS, 1766.
Taxonomy: Crustacea, Amphipoda, Corophiidae.
Test performed according to ISO/DIS 16712 (2003) as a liquid-phase and sediment-contact test.
Toxicity endpoint: microcrustacean acute lethal toxicity.
Test duration: 10 d.
Number of test animals per dilution step: 10.
Threshold value for the determination of the pT-value: 90% survival.

5.3 PROCESSING SEDIMENT SAMPLES PRIOR TO TOXICITY TESTING

Appraising the toxic potential of biologically available contaminants in sediment should include three compartments: the whole sediment (with standardized direct contact assays when these are available), the porewater, and the elutriate (aqueous extract). Additional hazard information can also be obtained from toxicity testing conducted on organic extracts using methanol or acetone.

For porewater extraction, whole wet sediment is centrifuged at 17,000 x *g* for 20 min. An elutriate (aqueous extract) is obtained by shaking the wet sediment sample in an aerobic milieu for 24 h. Soluble substances are extracted under such conditions. The elution process is performed with the original wet sediment and the elution ratio is adjusted by adding a standardized (test medium) water to the sediment sample's porewater. This ratio is fixed at 1:3 (1 part fresh sediment to 3 parts of water by weight). The elutriate is subsequently separated from the solid phase by centrifugation at 17,000 x *g* for 20 min (Krebs, 2000).

The organic extract is prepared by mixing 5 g of freeze-dried sediment with 15 mL of methanol. This ratio is also fixed at 1:3 to get a dilution series which is comparable with the dilution steps of the elutriate test. After sonication (ultrasonic bath for 20 min) and shaking (30 min), centrifugation follows to recover the extract, as described for the elutriate. Because of the inherent toxicity of methanol, the extract must be pre-diluted in standardized test water to a concentration known to be non-toxic to the toxicity test organisms. Once this non-toxic concentration of the methanol extract has been prepared, toxicity testing can proceed with the recommended test battery (*e.g.*, algal, bacteria, and microcrustacean bioassays). This sediment organic extract technique is presently being validated.

5.4 THE pT-VALUE AS AN ECOTOXICOLOGICAL DISCRIMINATOR

The sample dilution principle employed in the pT-scale improves the determination of toxic hazard potentials of sediment samples over that of effect percentages reported for undiluted test samples. For instance, a 100% effect measured with a specific endpoint and produced by undiluted samples may become undetectable after a dilution of 1:2 or after much higher dilutions. Clearly, it is essential to know the dilution level at which a whole sediment (or one of its liquid phases) ceases to be toxic[1]. In this respect, the pT-dilution approach offers valuable information allowing sediments to be more accurately classified in terms of the magnitude of toxicity.

This is illustrated in Figures 1 and 2, which show multiple bioassay percentage effect values as a function of dilution found in porewaters and elutriates extracted from sediment samples collected in the Saar River. In Figure 1, for example, sediment porewater produces 100% mortality effects toward *Daphnia* when undiluted and yields a pT-value of 2 after dilution. In contrast, the same undiluted porewater elicits a 62% growth inhibition effect toward algae, but requires increased dilution to reach a non-toxic level (*i.e.*, pT-value = 4). In Figure 2, the "AW 2" core sediment porewater essentially produces 100% effects in all three bioassays tested in its undiluted stage. When diluted, however, *Daphnia*, algae and Microtox®assays, respectively, yield pT-values of 3, 6, and 7, indicating that hazard potential can be markedly different in relation to the level of biological organization being considered. Thus far, the highest pT-value measured with sediment porewaters or elutriates is 11, corresponding to a dilution level of 1:2,048. This toxicity was detected with the luminescent bacteria test (Krebs, 1992).

---

[1] How the "non-toxic" dilution is determined is described in Chapter 3 of this volume, Sections 5.1, 5.3 and 5.4. A short description is given in Boxes 1 and 2 of this chapter.

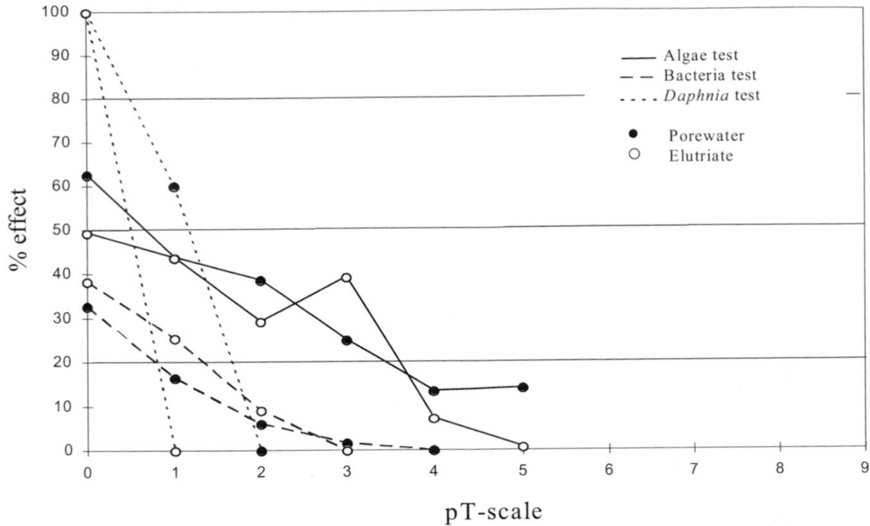

Figure 1. Multiple bioassay examination of porewater and elutriate of sediment core sample "AS 3" collected from the River Saar. The pT-scale is based on a geometric dilution series with the factor 2 (Krebs, 1999).

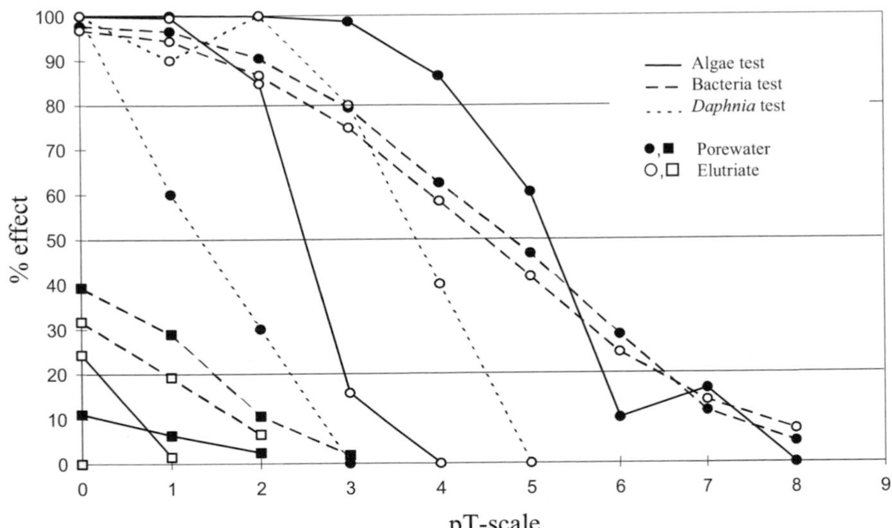

Figure 2. Multiple bioassay examination of porewaters and elutriates of two sediment samples collected from the River Saar. The pT scale is based on a geometric dilution series with the factor 2. Bed surface sample "AW 1" (■, □) and core sample "AW 2" (●, ○), (Krebs, 1999).

5.5 NUMERICAL CLASSIFICATION OF SEDIMENTS WITH THE pT-INDEX

The numerical classification of sediment samples follows the general ecotoxicological classification of aquatic environmental samples with the pT-index (see Chapter 3 of this volume, Section 5.7). In brief, the toxicity class is determined numerically by the pT-value of the most sensitive organism within a battery of bioassays, the pT_{max}-value. The number of bioindicators affected by sediment contaminants is not reflected by the pT-index. Hence, the multi-specificity of the toxic impact is not taken into account. There is no summation of pT-values generated by all bioassays or any other calculation with a mathematical formula integrating the product of sediment toxicity. The same rule should be applied to toxicity tests with multiple measurement endpoints. The most sensitive endpoint for a single-species test should be used to determine the pT-value. The most vulnerable group of test organisms is used as the standard measure with the intent of protecting the most sensitive organisms. This procedure guarantees that a high toxicity result measured in one test cannot be lowered by test results of less sensitive organisms, *i.e.* tests without toxic effects cannot "soften" the assessment of an effect that was measured within another test system. Each toxicity class is therefore defined by the highest pT-value generated and designated by a Roman numeral. Numerically-defined toxicity classes are always related to the undiluted test material (*e.g.* whole sediment, porewater or elutriate). If, for example, a sediment sample like "AW 2" of Figure 2 yields pT-values of 3 (elutriate) and 6 (porewater) with the algal assay, 7 (elutriate) and 7 (porewater) with the bacterial assay, as well as 5 (elutriate) and 3 (porewater) with the daphnid assay, its toxicity class corresponds to VII (*i.e.*, pT-index = VII), based on the most sensitive response obtained (in this example with the bacterial assay). These pT-values are those of real toxicity data obtained with the Wadgassen core sample discussed later on in the context of river sediment quality (Tab. 5).

5.6 ECOTOXICOLOGICAL CLASSIFICATION OF DREDGED MATERIAL

For classification of dredged material, the pT-scale is restricted to seven classes (Krebs, 1999). Responses to toxicity (*i.e.*, pT-values determined with bioassays) of porewater and elutriate to different test organisms are considered equal in rank, and the highest pT-value is used to characterize the sample. Toxicity classes are listed in Table 2. These designations were arbitrarily chosen and their toxic properties are characterized by the pT- and pT_{max}-values (Chapter 3 of this volume, Section 5.7 and 5.8). Each description is related to a toxicity class number up to a maximum of VI which corresponds to a pT_{max}-value of 6. Since sediment toxicity classes at the level VI are considered hazardous, there is no need to designate higher numbered classes. For management considerations, reporting both the toxicity class in form of verbal designation and the corresponding pT-value of the sediment sample investigated is essential in order to properly categorize its hazard potential (*e.g.*, "hazardous sediment, highly toxic, toxicity class V" or "hazardous sediment, extremely toxic, toxicity class VI, pT_{max} = 7": see Tables 2, 4 and 5). For general procedural steps for dredged-material handling, refer to Figure 3.

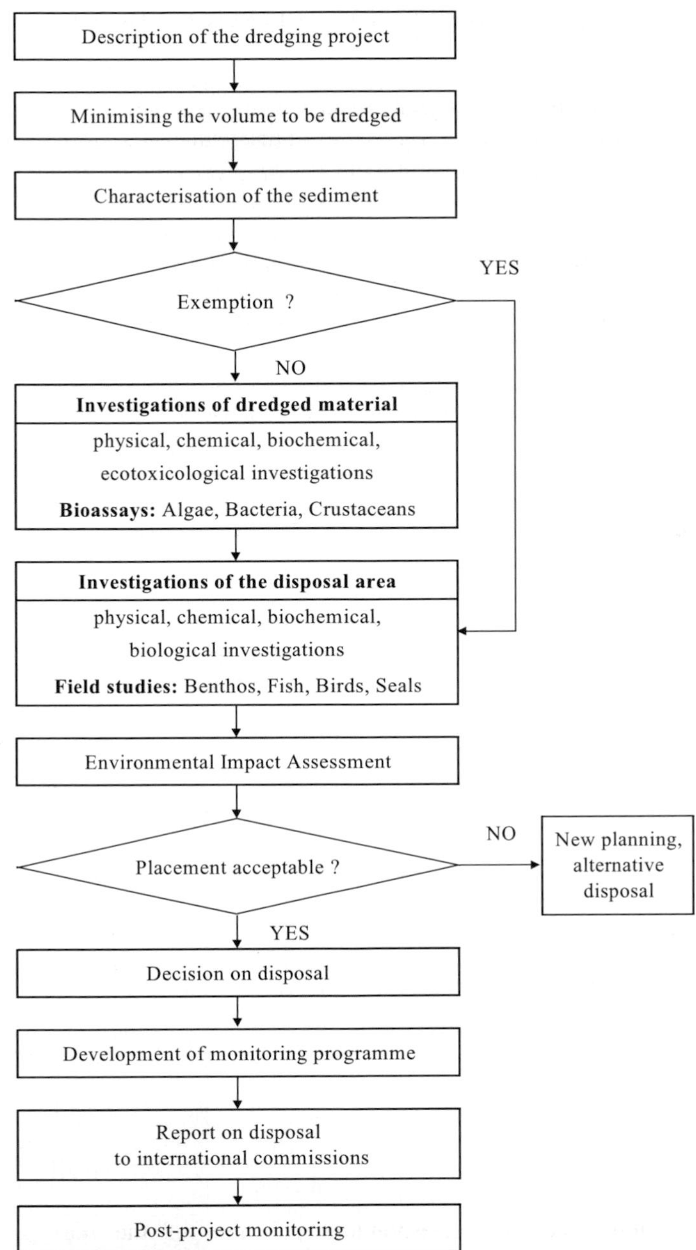

Figure 3. Dredged material handling in coastal waters. Procedural steps for the decision-making on dredged material relocation according to HABAK-WSV (1999).

5.7 MANAGEMENT CATEGORIES FOR THE RELOCATION OF DREDGED MATERIAL

In the context of dredged-material management, the seven toxicity classes established by pT-values are assigned to management categories labeled as "unpolluted", "unproblematic", "problematic" and "hazardous" (Tab. 2). These categories then define "cases" by which dredged material can (or cannot) be relocated, as recalled below:

- **Case 1:** Pursuant to the guideline for handling of dredged material from inland waterways (HABAB-WSV, 2000) and its counterpart for coastal waterways (HABAK-WSV, 1999), dredged material up to toxicity class II can be relocated without restriction.
- **Case 2:** The relocation of dredged material of toxicity classes III and IV must be decided on a case-by-case basis. An impact hypothesis (prediction of potential impacts) is mandatory.
- **Case 3:** Dredged material of the two highest classification levels (toxicity classes V and VI) must not be relocated in inland waterways according to HABAB-WSV (2000) and should not be relocated in coastal areas according to HABAK-WSV (1999).

Table 2. Toxicity classes defined by the German Federal Institute of Hydrology for sediment assessment and ecotoxicological management categories for dredged material relocation. Ecotoxicological characterization is based on porewater and elutriate bioassay responses. For shade codes, refer to Table 3.

Highest dilution level without effect	*Dilution factor*	*pT-value*	*Toxicity classes*		*Management categories*	
			7-level system	*Designation*	*4-level assessment*	*Color coding*
Original sample	2^{0}	0	0	toxicity not detected	unpolluted	0
1:2	2^{-1}	1	I	very slightly toxic	unproblematic	I
1:4	2^{-2}	2	II	slightly toxic		II
1:8	2^{-3}	3	III	moderately toxic	problematic	III
1:16	2^{-4}	4	IV	distinctly toxic		IV
1:32	2^{-5}	5	V	highly toxic	hazardous	V
≤ (1:64)	$\leq 2^{-6}$	≥ 6	VI	extremely toxic		VI

As seen in Table 3, the degree of contamination of dredged material follows matching color codes for ecotoxicological (bioassay-based) and chemical assessments, respectively, set with pT-indices and measured concentrations of specific contaminants. Hence, management categories relating to the degree of hazard of sediment material intended for dredging correspond to cases 1 to 3, and in parallel to the color codes (green, yellow, red) that signal decisions to be taken with respect to the relocation of dredged material.

Table 3. Management categories for dredged-material relocation used by the German Federal Institute of Hydrology. Chemical and ecotoxicological criteria are those of HABAB-WSV (2000) and HABAK-WSV (1999).

Legend: RW_1 and RW_2 = chemical orientation values 1 and 2 (explanation in Section 5.8); c = contaminant concentration.

Shade codes for management categories for dredged material relocation:

Toxicity not detected	*Case 1*	***light***	*toxicity class 0*
Unproblematic pollution	*Case 1*	***dark***	*toxicity classes I and II*
Problematical pollution	*Case 2*	***darker***	*toxicity classes III and IV*
Hazardous pollution	*Case 3*	***darkest***	*toxicity classes V and VI.*

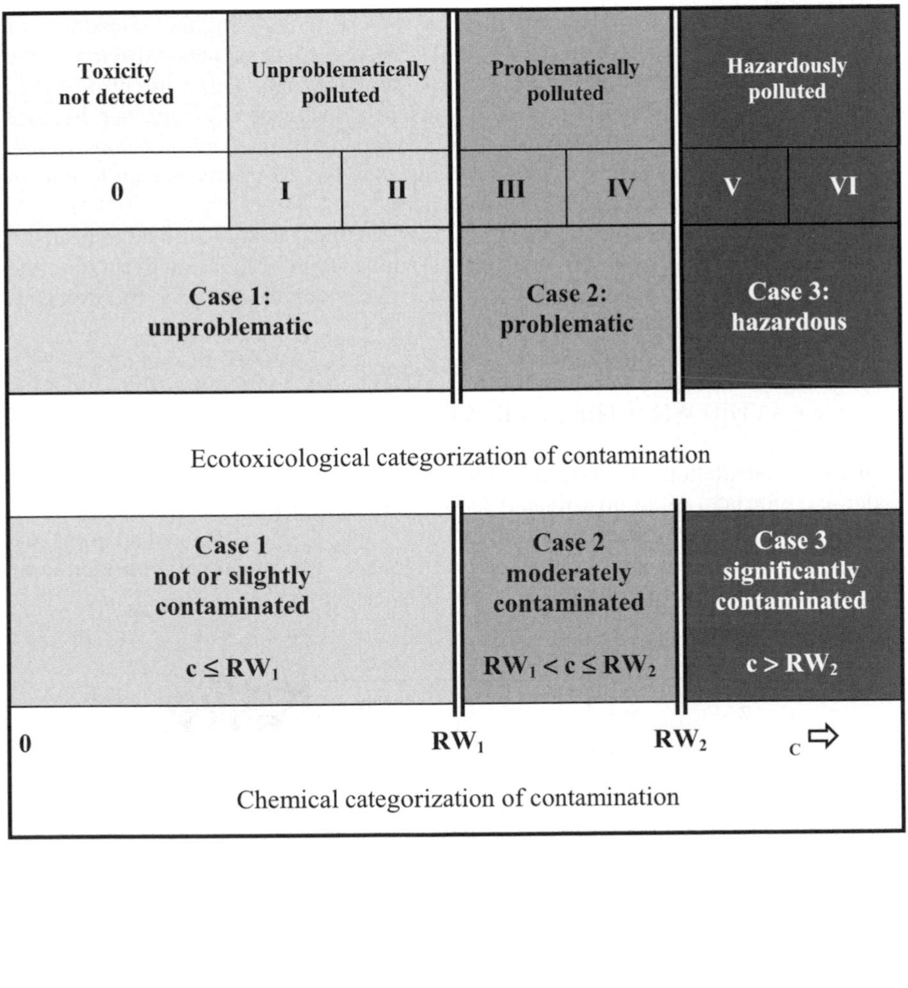

Toxicity not detected	**Unproblematically polluted**		**Problematically polluted**		**Hazardously polluted**	
0	**I**	**II**	**III**	**IV**	**V**	**VI**
Case 1: unproblematic			**Case 2: problematic**		**Case 3: hazardous**	
Ecotoxicological categorization of contamination						
Case 1 not or slightly contaminated $c \leq RW_1$			**Case 2 moderately contaminated** $RW_1 < c \leq RW_2$		**Case 3 significantly contaminated** $c > RW_2$	
0			**RW_1**		**RW_2**	c ⇨
Chemical categorization of contamination						

5.8 CHEMICAL ANALYSES PERFORMED IN PARALLEL TO THE pT-METHOD

Identical samples are used for chemical analysis and toxicity tests. Chemical analyses are an indispensable complement to effects measurements (*i.e.*, toxicity tests) to ensure environmental protection. To define sediment and water quality objectives and standards, the assessment of chemical substances is normally governed by a set of criteria, which involve the determination of threshold (NOEC) values derived from ecotoxicological data. Chemical orientation values derived for determining the relocation status of sediment material intended for dredging are not governed by the same criteria. The chemical orientation values used in the HABAB-WSV and HABAK-WSV guidelines for dredged-material management (RW1 and RW2 in Table 3) are based on the current contamination status, *i.e.*, the prevailing contaminant concentrations in the waters under consideration. The concentrations in suspended matter in river water (average values of a three year period) are the basis for calculations according to HABAB-WSV. In the marine area it is the average concentration found in the mud-flat sediments (Wadden sea) which gives the orientation value according to HABAK-WSV. The actual *in-situ* background concentrations serve as a reference (in terms of orientation) for the assessment of chemical concentrations in dredged material. The procedure is standardized and the chemical concentrations in the 20 μm fraction are compared. This concept promotes management of dredging operations in an acceptable and responsible manner, because it follows the principle that the quality of the environment must not be compromised. These values are a necessary decision-making tool for environmental protection in waterway maintenance and development practices.

In all cases, results of ecotoxicological and chemical analytical data are ranked equally. Generally, the most environmentally-conservative response will serve as the yardstick to assign a sediment sample to a management category for dredged-material relocation (Schubert et al., 2000).

5.9 STATISTICS, CALCULATIONS AND EXAMPLES OF DATA ASSOCIATED WITH THE pT-PROCEDURE

A real-life calculation of pT-values obtained with different compartments (pore-water and elutriate) of an investigated sediment sample is given in Sections 5.4 and 5.5. The dilution steps used are described in Table 4. No other calculations are needed or associated with the pT-scale. Neither are statistical considerations necessary to get the result of a single test or of a test battery.

Table 4. Geometric dilution series, pT-values, and pT-indices for sediment investigations. The pT-values determined to date are marked by the sign +.

Dilution factor as cardinal fraction	*Dilution factor as decimal fraction*	*Dilution factor as exponential fraction*	*pT-value*[a]	*Toxicity class*		*Measured ecotoxicity in sediments*	
				pT-index[b]	*Designation*	*Pore-water*	*Elutriates*
Original sample	1	2^{0}	0	0	toxicity not detected	+	+
1:1.25	0.8	$2^{-0,3}$	0	0	toxicity not detected	+	+
1:2	0.5	2^{-1}	1	I	very slightly toxic	+	+
1:4	0.25	2^{-2}	2	II	slightly toxic	+	+
1:8	0.125	2^{-3}	3	III	moderately toxic	+	+
1:16	0.0625	2^{-4}	4	IV	distinctly toxic	+	+
1:32	0.0313	2^{-5}	5	V	highly toxic	+	+
1:64	0.0156	2^{-6}	6	VI	extremely toxic "Mega toxic"	+	+
1:128	0.00781	2^{-7}	7	VII		+	+
1:256	0.00391	2^{-8}	8	VIII		+	-
1:512	0.00195	2^{-9}	9	IX	"Giga toxic"	+	-
1:1024	0.000977	2^{-10}	10	X		+	-
1:2048	0.000488	2^{-11}	11	XI		+	+
1:4096	0.000244	2^{-12}	12	XII	"Tera toxic"	-	-
1:8192	0.000122	2^{-13}	13	XIII		-	-
1:16384	0.0000610	2^{-14}	14	XIV		-	-

[a] **pT-value:** the highest dilution level devoid of adverse effects is used for the numerical designation of toxicity with regard to a single test organism. The pT-value (potentia Toxicologiae = toxicological exponent) is the negative binary logarithm of the first non-toxic dilution factor in a dilution series in geometric sequence with a dilution factor of 2.

[b] **pT-index:** the numerical toxicological classification of an environmental sample attained with a test battery. The pT-value of the most sensitive organism within a test battery determines the toxicity class of the tested material. Roman numerals are assigned to each toxicity class. If the highest pT-value is 9, for instance, the tested material is then assigned as toxicity class IX (*i.e.*, the pT-index is IX).

6. Factors influencing the interpretation of pT-procedures; quality criteria for sediments and dredged material

Sediments can be relocated within water bodies only if they meet certain quality requirements. In Europe, this is known as the eco-political objective. However, a generally accepted concept to adequately define sediment quality criteria is still lacking, because effects-based scientific knowledge still needs to grow to ameliorate management guidelines and regulations.

Thus far, quality objectives for chemical substances are derived from the most sensitive organisms in acute and chronic toxicity test batteries that determine NOEC values for different trophic levels. The pT-method similarly determines specific sample dilution levels that are devoid of adverse effects toward (micro)organisms of a standardized test battery. Common to both approaches is the more frequent use of water-column test organisms as opposed to benthic-dwelling organism that reflect more intimate contact with sediment. This practice is primarily based on the fact that standardized bioassays capable of appraising sediment porewaters and elutriates are presently more numerous than solid-phase tests for whole-sediment assessment. As more of these latter tests become developed and standardized (see Chapters 12 and 13, volume 1 of this book on amphipod and chironomid tests), their more frequent use will contribute to a better understand of the toxic effects of sediment-bound contaminants.

As a result, the concept of toxicity classes and management categories linked to dredged-material relocation, as presented in Tables 2 and 3, are presently based on sediment porewater and elutriate testing and may have to be adjusted in the long term. Incorporation of solid-phase tests in routine test batteries should also serve to improve sediment quality guidelines and ensure better protection of aquatic ecosystems.

Another consideration, which can influence ecotoxicological criteria for dredged-material management, depends on the chosen toxicity test battery, as sediment toxicity classes are determined by the pT-value of the most sensitive organism. As representative test batteries for porewater, elutriate, and whole sediment assessment become better standardized and less prone to changes with time, ecotoxicological characterization of sediments based on the pT method should increase in environmental relevance. Other considerations will be involved when using direct-contact tests (*e.g.*, using reference sediments for dilution series). Matrix effects and experimental variability have to be considered. The author addresses the concerns that were identified by Den Besten et al. (2003).

The simultaneous coupling of ecotoxicological and chemical criteria in the decision-making process regarding sediment-quality appraisals should also ensure environmentally sound management practices for dredged-material relocation. At present, both criteria are given equal importance to assess such hazards and the most conservative information prevails for recommending the course of action to be taken. Sediment-quality guidelines, it stands to reason, are periodically updated to reflect scientific progress in the field of sediment contamination studies.

7. Applications / case study reported with the pT-scale

During an ecotoxicological study designed to map out sediment quality, surface sediment samples and cores were collected from old arms, harbors, and marinas associated with the rivers Saar and Moselle (Tab. 5). Bioassays conducted according to the pT-scheme indicated that surface sediments (3-cm top layer) displayed some toxic effects, but these never exceeded the toxicity class II level. In contrast, some sediment core samples with documented levels of chemical contamination tended to show more pronounced effects. One particular sample (pT-value of 7 for the AW2 Wadgassen core) was categorized at the highest level of hazard (toxicity class VI). In general, the majority of other surface and core samples from river sites failed to elicit adverse effects on test organism (toxicity class 0) or induced only weak responses (toxicity classes I and II) (Krebs, 2000).

When toxicity-class results of sediment samples collected and analyzed from different localities are mapped (Fig. 4), a comprehensive view of sediment quality can be discerned for the Saar River, a tributary of the Moselle River in the Rhine river basin. Based on the color codes which reflect the different degrees of toxicity, the pollution hot spot, which clearly stands out, is located in the vicinity of the main industry of the Saarland (Völklingen area). Such "red-labeled" sediments, indicative of high hazard (see Tables 2 and 3), could not be relocated in the same river body should they be dredged and would necessitate confinement elsewhere (*e.g.*, upland placement, storage and if possible remediation treatment). This particular case study illustrates the usefulness of applying the pT-scale strategy to facilitate the decision-making process in the management of sediment dredging and disposal.

8. Information related to present and future applications with the pT-method

Owing to its characteristics, the pT-index is distinct from other schemes which also aim to evaluate sediment quality. In contrast to a strategy where multitrophic specificity of toxic impact is taken into account (Den Besten et al., 2003; Henschel et al., 2003a,b; SED-TOX Chapter 8 of this volume), the pT-index does not directly reflect an overall average for the number and types of test organisms affected by sediment contaminants. It simply strives to protect all levels of aquatic organisms on the basis of the response of the most sensitive organism within a test battery. Classifying sediment samples based on the highest pT-value generated is sound in principle and should, in theory, be environmentally protective.

As more bioassays for assessment of the liquid (porewater, elutriate) and solid (whole sediment) compartments of sediments reach standardization, future investigations will, however, have to determine their performance and adequacy. While the present battery employed in the work presented herein is undeniably useful, it may yet be optimized with other liquid-phase tests and likely be supplemented with direct-contact bioassays capable of estimating the toxic potential of contaminants more closely bound to sediments.

Table 5. Sediment quality in waters associated with the rivers Saar and Moselle (old arms, harbors, and marinas). The ecotoxicological characterization is based on porewater and elutriate bioassay responses generated with algae, bacteria, and daphnids. Bioassays conducted according to HABAB-WSV (2000), refer to Box 1. For shade code information of toxicity classes, refer to Table 3.

No.	Location	Sample	Algal Test				Luminescent Bacteria Test				*Daphnia* Test				Toxicity class[e]
			Porewater		Elutriate		Porewater		Elutriate		Porewater		Elutriate		
			%[a]	pT[d]	%[a]	pT[d]	%[b]	pT[d]	%[b]	pT[d]	%[c]	pT[d]	%[c]	pT[d]	
HS 1	Saarbrücken	Bed Surface Sample	-3	0	-19	0	22	1	30	2	0	0	0	0	II
AW 1	Wadgassen	Bed Surface Sample	11	0	24	1	39	2	32	1	0	0	0	0	II
HD 1	Dillingen	Bed Surface Sample	-39	0	-25	0	3	0	-6	0	10	0	0	0	0
BM 1	Merzig	Bed Surface Sample	28	2	-2	0	23	1	36	2	0	0	0	0	II
BT 1	Trier-Monaise	Bed Surface Sample	11	0	-23	0	9	0	9	0	0	0	10	0	0
HS 2	Saarbrücken	Sediment Core Sample	-29	0	-20	0	9	0	26	1	0	0	0	0	I
AW 2	Wadgassen	Sediment Core Sample	100	6	100	3	100	7	100	7	100	3	100	5	VI
BM 2	Merzig	Sediment Core Sample	-16	0	-10	0	-3	0	-1	0	0	0	0	0	0
AS 2	Schwemlingen	Sediment Core Sample	43	2	46	2	30	1	18	0	100	2	70	1	II
AS 3	Schwemlingen	Sediment Core Sample	63	4	49	4	33	1	39	2	100	2	100	1	IV
BT 2	Trier-Monaise	Sediment Core Sample	-22	0	-17	0	7	0	19	0	0	0	0	0	0

[a] Percent growth inhibition for the test alga, *Desmodesmus subspicatus*, in the undiluted test material (negative % values indicate stimulation) (DEV L33).

[b] Percent light inhibition for the test bacterium, *Vibrio fischeri*, in the undiluted test material (negative % values indicate stimulation) (DEV L34, Microtox® bacteria).

[c] Ppercent of immobilized test animals, *Daphnia magna*, in the undiluted test material (DEV L30).

[d] pT-value characterizing the potential toxicity of each sediment sample compartment for a specific test organism (test-specific value), see Section 5.5.

[e] The pT-value of the most sensitive organism in the test battery determines the toxicity class of the dredged material, see Section 5.6.

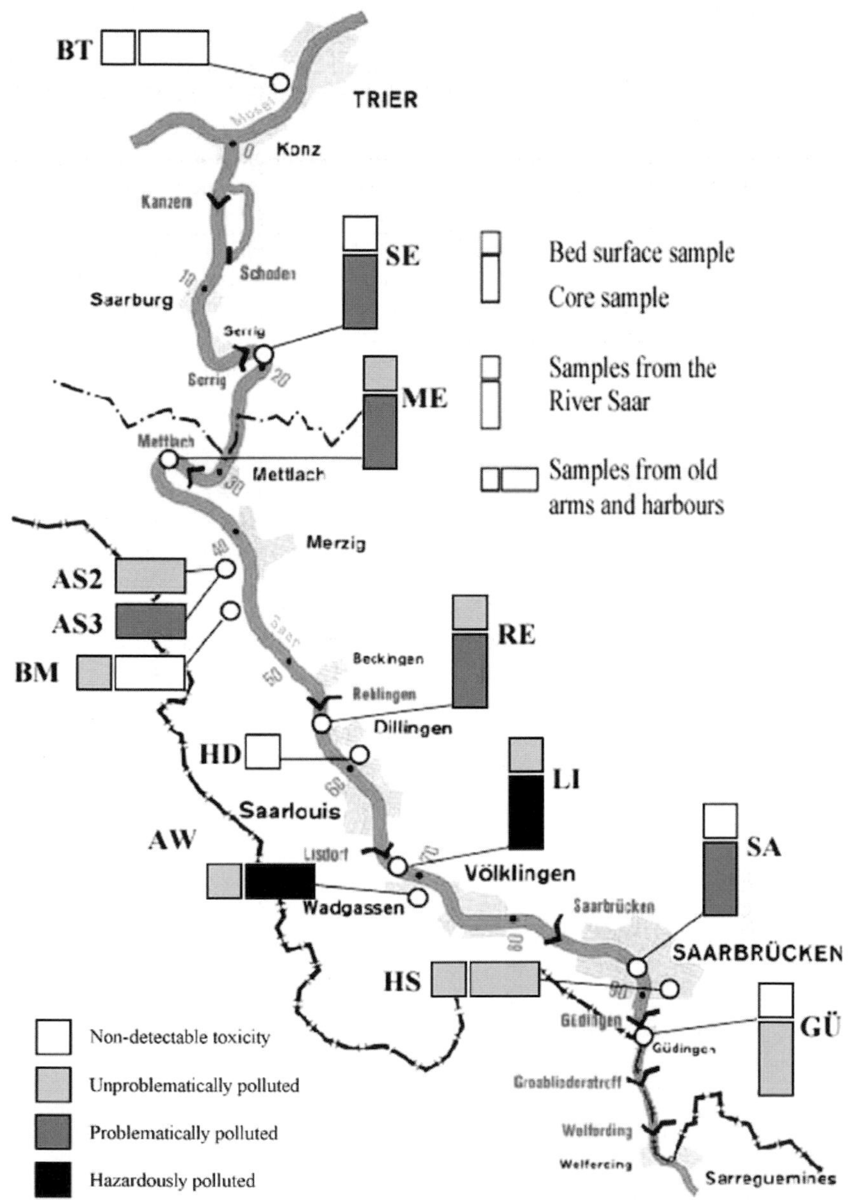

Figure 4. Sediment mapping of the River Saar based on pT-scale ecotoxicological investigations conducted with algal, bacterial and microcrustacean bioassays. This classification is derived from results reported in Table 5. For color coding information of toxicity classes see Table 3.

9. Conclusions/prospects

This simple pT-hazard assessment system designed to determine sediment quality for the management of freshwater and marine ecosystems was developed in the German Federal Institute of Hydrology (BfG). While sediment categorization into several toxicity classes is supported by scientific methodology, evaluation and routine application are based on a convention outlined in specific guidelines. The rationale defining sediment categories based on ecotoxicological principles (*i.e.*, via application of a toxicity test battery) is still very much in its infancy. As the pT-database increases and toxicity test batteries are optimized, sediment toxicity categories will likely be better defined to the benefit of management practices for the protection of inland and coastal waters.

Acknowledgements

I am thankful to Bernd Uebelmann, Dierk-Steffen Wahrendorf, Shannon McDowell and Derek McDowell for their support in writing this paper.

The author wishes to thank the copyright holder for permission to reproduce the following: Figures 1 and 2, ATV-DVWK Deutsche Vereinigung für Wasserwirtschaft, Abwasser und Abfall e.V., Hennef, Germany. Reproduced from: Krebs, F. (1999) Ökotoxikologische Klassifizierung von Sedimenten mit Hilfe der pT-Wert-Methode (Ecotoxicological classification of sediments by the pT-value method), in U. Kern and B. Westrich (eds.): *Methoden zur Erkundung, Untersuchung und Bewertung von Sedimentablagerungen und Schwebstoffen in Gewässern, Schriftenreihe des Deutschen Verbandes für Wasserwirtschaft und Kulturbau (DVWK)* **128**, 297-303 (Figures 1 and 2 , unaltered).

References

Den Besten, P.J., De Deckere, E., Babut, M.P., Power, B., DelValls, T.A., Zago, C. Oen, A.M.P. and Heise, S. (2003) Biological effects-based sediment quality in ecological risk assessment for European waters, *Journal of Soils and Sediments* **3,** 144-162.

DEV L1 - DIN EN ISO 5667-16 (1998) Water quality – Sampling – Part 16: Guidance on biotesting of samples (ISO 5667-16: 1998). Edited by Wasserchemische Gesellschaft - Fachgruppe in der Gesellschaft Deutscher Chemiker in Gemeinschaft mit dem Normenausschuß Wasserwesen (NAW) im DIN Deutsches Institut für Normung e.V., Beuth Verlag, Berlin, Germany.

DEV L30 - DIN 38 412 Part 30 (1989) German standard methods for the examination of water, waste water and sludge – Determination of the non-poisonous effect of waste water to *Daphnia* by dilution limits, Beuth Verlag, Berlin.

DEV L33 - DIN 38 412 Part 33 (1991) German standard methods for the examination of water, waste water and sludge – Determination of the non-poisonous effect of waste water to green algae (*Scenedesmus* chlorophyll fluorescence test) by dilution limits, Beuth Verlag, Berlin.

DEV L34 - DIN EN ISO 11348-3 (1998) Water quality – Determination of the inhibitory effect of water samples on the light emission of *Vibrio fischeri* (Luminescent bacteria test) - Part 3: Method using freeze-dried bacteria, Beuth Verlag, Berlin.

DEV L45 - DIN EN ISO 10253 (1998) Water quality – Marine algae growth inhibition test with *Skeletonema costatum* and *Phaeodactylum tricornutum* (ISO 10253: 1995), Beuth Verlag, Berlin.

DEV T3 - DIN 38415-3 (1996) German standard methods for the examination of water, waste water and sludge – Sub-animal testing – Part 3: Determination of the genotype potential of water with the umu-test, Beuth Verlag, Berlin.

DEV T6 - DIN 38415-6 (2003) German standard methods for the examination of water, waste water and sludge – Sub-animal testing – Part 6: Toxicity to fish; Determination of the non-acute-poisonous effect of waste water to fish eggs by dilution limits, Beuth Verlag, Berlin.

Feiler, U. and Krebs, F. (2001) Entwicklung und Anwendung pflanzlicher Biotestverfahren für ökotoxikologische Untersuchungen von Gewässersedimenten (Aquatic plant bioassays used in the ecotoxicological assessment of sediments), Bundesanstalt für Gewässerkunde (Federal Institute of Hydrology), BfG-**1336**.

HABAB-WSV (2000) Handlungsanweisung für den Umgang mit Baggergut im Binnenbereich (Directive for the Management of Dredged Material in Inland Waters), 2nd ed., Bundesanstalt für Gewässerkunde (Federal Institute of Hydrology), Koblenz, Germany, BfG-**1251**.

HABAK-WSV (1999) Handlungsanweisung für den Umgang mit Baggergut im Küstenbereich (Directive for the Management of Dredged Material in Coastal Waters).- 2nd ed. Bundesanstalt für Gewässerkunde (Federal Institute of Hydrology), Koblenz, Germany, BfG-**1100**.

Henschel, T., Calmano, W., Maaß, V. and Krebs, F. (2003a) Bewertung der Sedimentqualität: Empfehlungen für eine integrierte Gefährdungsabschätzung (Evaluation of sediment quality: a guidance for an integrated approch), *Wasser & Boden* **55**, 89-92.

Henschel, T., Calmano, W., Maaß, V. and Krebs, F. (2003b) Gefährdungsabschätzung von Gewässersedimenten – Handlungsempfehlungen und Bewertungsvorschläge für eine integrierte Bewertung (Risk assessment of aquatic sediments – recommendations for action and proposals for integrated assessments*), Handbuch Angewandte Limnologie* (16. Erg.Lfg. 7/03) **VIII-8.1**, 1-27.

Höss, S. and Krebs, F. (2003) Dilution of toxic sediments with unpolluted artificial and natural sediments – Effects on *Caenorhabditis elegans* (*Nematoda*) in a whole sediment bioassay, Proceedings 13th Annual meeting, Society of Environmental Toxicology and Chemistry – Europe Branch (SETAC-Europe), Hamburg, 28.04.-01.05. 2003.

Hollert, H., Keiter, S., König, N., Rudolf, M., Ulrich, M. and Braunbeck, T. (2003) A new sediment contact assay to assess particle-bound pollutants using zebrafish (*Danio rerio*) embryos, *Journal of Soils and Sediments* **3,** 197-207.

ISO/DIS 16712 (2003) Water quality - Determination of acute toxicity of marine or estuarine sediment to amphipods, Geneve, Switzerland.

Köthe, H. (2003) Existing sediment management guidelines: An overview, *Journal of Soils and Sediments* **3,** 139-143.

Köthe, H. and Bertsch, W. (1999) Der Umgang mit Baggergut in Deutschland (Dredged material handling in Germany), in Gesellschaft für Umweltgeowissenschaften (ed.), *Ressourcen - Umwelt - Management: Boden - Wasser - Sedimente.* Springer Verlag, Berlin, Heidelberg, pp. 155-171.

Krebs, F. (1992) Über die Notwendigkeit ökotoxikologischer Untersuchungen an Sedimenten (On the necessity of ecotoxicological assessments of aquatic sediments), *Deutsche Gewässerkundliche Mitteilungen* **36**,165-169.

Krebs, F. (1999) Ökotoxikologische Klassifizierung von Sedimenten mit Hilfe der pT-Wert-Methode (Ecotoxicological classification of sediments by the pT-value method), in U. Kern and B. Westrich (eds.), *Methoden zur Erkundung, Untersuchung und Bewertung von Sedimentablagerungen und Schwebstoffen in Gewässern, Schriftenreihe des Deutschen Verbandes für Wasserwirtschaft und Kulturbau (DVWK)* **128**, 297-303.

Krebs, F. (2000) Ökotoxikologische Bewertung von Baggergut aus Bundeswasserstraßen mit Hilfe der pT-Wert-Methode (Ecotoxicological assessment of dredged material from federal waterways by the pT-value method), *Hydrologie und Wasserbewirtschaftung* **44**, 301-307.

Krebs, F. (2001) Ökotoxikologische Baggerguntuntersuchung, Baggergutklassifizierung und Handhabungskategorien für Baggergutumlagerungen (Ecotoxicological studies on dredged material: classification and categories for management of dredged material), in W. Calmano (ed.), *Untersuchung und Bewertung von Sedimenten - ökotoxikologische und chemische Testmethoden.* Springer-Verlag, Berlin, Heidelberg, pp. 333-352.

Pfitzner, S. and Krebs, F. (2001) Ökotoxikologische Tests in Küstengewässern (Ecotoxicological tests for coastal waters).- 2. Ostsee-Worksshop "Sedimentuntersuchungen in Ostseeküstengewässern und Schlussfolgerungen für Ausbau- und Unterhaltungsmaßnahmen in der WSV", *BfG-Veranstaltungen* **4/2001**, 19-25.

Schubert, B., Krebs, F. and Bergmann, H. (2000) Federal regulations for the disposal of dredged material in German coastal areas - Experiences with chemical and biological criteria, in J. Gandrass, W. Salomons and U. Foerstner (eds.), *River sediments and related dredged material in Europe – Scientific background from the viewpoints of chemistry, ecotoxicology and regulations*, GKSS Research Centre, Geesthacht, Germany, pp. 41-44.

Soares, F. and de Freitas, L. (2000) Avaliação da Qualidade de Água e Sedimentos da Baía de Sepetiba, Anexo 5: Ecotoxicologia, 1995 a 1999, Projeto de Cooperação Técnica Brasil/Alemanha - FEEMA/GTZ. FEEMA Fundação Estadual de Engenharia do Meio Ambiente, Rio de Janeiro, RJ, Brasil, (Water and Sediment Quality Assessment of Sepetiba Bay, Annex 5: Ecotoxicology, 1995 to 1999, Brazil/Germany Cooperation Project - FEEMA /GTZ. Report (CD-ROM), FEEMA Environmental Agency of the Federal State of Rio de Janeiro, Rio de Janeiro, Brazil).

Winger, P.V., Albrecht, B., Anderson, B.S., Bay, S.M., Bona, F. and Stephenson, G.L. (2003) Comparison of porewater and solid-phase sediment toxicity tests, in R.S. Carr and M. Nipper (eds.), *Porewater toxicity testing: biological, chemical, and ecological considerations*, SETAC Press, Pensacola, Florida, USA, pp. 37-61.

Abbreviations

BfG	German Federal Institute of Hydrology
DEV	German Standard Methods for Examination of Water, Wastewater and Sludge
DIN	German Organization for Standardization
EN	European Organization for Standardization
HAS	Hazard Assessment Scheme
ISO	International Organization for Standardization
NOEC	No-Observed Effect Concentration
OECD	Organization for Economic Cooperation and Development
pH	*potential Hydrogenii*
pT	*potentia Toxicologiae*
RW	Chemical Orientation Value (from German *"Richtwert"*)
WSV	German Federal Waterways and Shipping Administration.

10. USING THE SEDIMENT QUALITY TRIAD (SQT) IN ECOLOGICAL RISK ASSESSMENT

PETER M. CHAPMAN
& BLAIR G. MCDONALD
EVS Environment Consultants
195 Pemberton Avenue
North Vancouver
B.C. V7P 2R4, Canada
pchapman@attglobal.net

1. Objective and scope

The Sediment Quality Triad (SQT) is an effects-based conceptual approach that can be used to assess and determine the status of contaminated sediments based on biology (laboratory and/or *in situ* toxicity tests), chemistry (chemical identification and quantification), and ecology (community structure and/or function). It provides a means for comparing three different lines of evidence (LOE) and arriving at a weight of evidence (WOE) determination regarding the risk posed by contaminated sediments. Effectively, each LOE comprises an independent assessment of hazard; combined and integrated, they provide an assessment of risk.

The SQT is not restricted to only three LOE and can incorporate additional LOE, for instance biomagnification and sediment stability. It can also incorporate additional investigative studies such as toxicity identification evaluation (TIE) and contaminant body residue (CBR) analyses. Moreover, the concept is not restricted to sediments but can also be applied to soils in the terrestrial environment, as well as to groundwater discharges entering the aquatic receiving environment, and to the water column above the sediments. It can be applied to any and all contaminants and to other stressors, for example, physical habitat changes. It can be implemented in a tiered manner to maximize its utility and cost-effectiveness and provides, when fully implemented, the most complete information possible, in a wide variety of media, on biological effects to biota due to physical, chemical or biological stressors.

2. Summary of SQT procedure

An SQT provides for a WOE evaluation of different LOE. It consists, at a minimum, of the following determinations: (1) concentrations of sediment contaminants of

C. Blaise and J.-F. Férard (eds.), Small-scale Freshwater Toxicity Investigations, Vol. 2, 305-329.

potential concern (COPCs); (2) toxicity of those sediments: (3) composition of the benthic community in those sediments. It should also include measures of sediment stability, and can also include measures of biomagnification for those organic chemicals whose concentrations can increase, via feeding, up three or more trophic levels (e.g., DDT, PCBs, methyl mercury) as well as toxicity identification evaluations (TIEs) and/or contaminant body residues (CBRs). Determination of COPCs depends on sophisticated chemical analyses of the sediments as well as of ancillary parameters such as grain-size and total organic carbon (TOC). Measurements of sediment toxicity require at least one and preferably a minimum of three tests using appropriate species and measuring both acute and chronic endpoints. To determine benthic community structure, appropriately sieved sediment samples are sorted and organisms are identified. Sediment stability is determined based on laboratory and/or field measurements. Biomagnification is determined based on measured and modelled body burdens of specific contaminants. Sediment COPC concentrations are compared to benchmarks (sediment quality guidelines [SQGs] and/or concentrations in reference areas); stability measurements (*e.g.*, shear stress) are compared to worst case flow / event (*e.g.*, 100-year storm) conditions. Sediment toxicity, benthic community and biomagnification data from impacted areas are typically compared to data from reference areas. Comparisons to reference areas typically involve univariate and/or multivariate statistical analyses. Integration of the different LOE typically involves sophisticated statistical analyses (Chapman, 1996; Hunt et al., 2001; Burton et al., 2002a,b; Hollert et al. 2002a,b; Beiras et al., 2003), which can then be summarized relatively simply in a decision framework such as that shown in Figure 1.

3. Historical overview and reported applications

The SQT procedure for contaminated sediments was originally proposed by Long and Chapman (1985), and subsequently refined by Chapman et al. (1987), Chapman (1990, 1996), Chapman et al. (1997, 2002) and Grapentine et al. (2002). Chapman (2000) summarises published SQT studies in marine, freshwater and estuarine environments in the following geographic areas: North America, Europe, and Antarctica. Subsequent SQTs have included presently unpublished studies in Australasia and South America (Chapman, pers. comm.), and published studies in Europe (Hollert et al., 2002a,b; Beiras et al., 2003; Lahr et al., 2003; Riba et al., 2004) and North America (Anderson et al., 2001; Balthis et al., 2002; Schmidt et al., 2002).

The SQT provides, as recommended by Suter (1997), a WOE analysis based on *a priori*, not *ad hoc* logic, as exampled by Table 1. Alternative approaches for interpretation of SQT components are summarized by Chapman et al. (2002); a subsequent refinement to one of these approaches is suggested by Forbes and Calow (2004). Hollert et al. (2002b) evaluated small streams in the Netherlands using the SQT, then assessed and integrated the resulting data using both Hasse diagrams and fuzzy logic to provide relatively easy-to-understand visual representations of different sites. In all cases, the objective is to provide a comprehensive overview of the findings of both individual LOE and of the overall WOE.

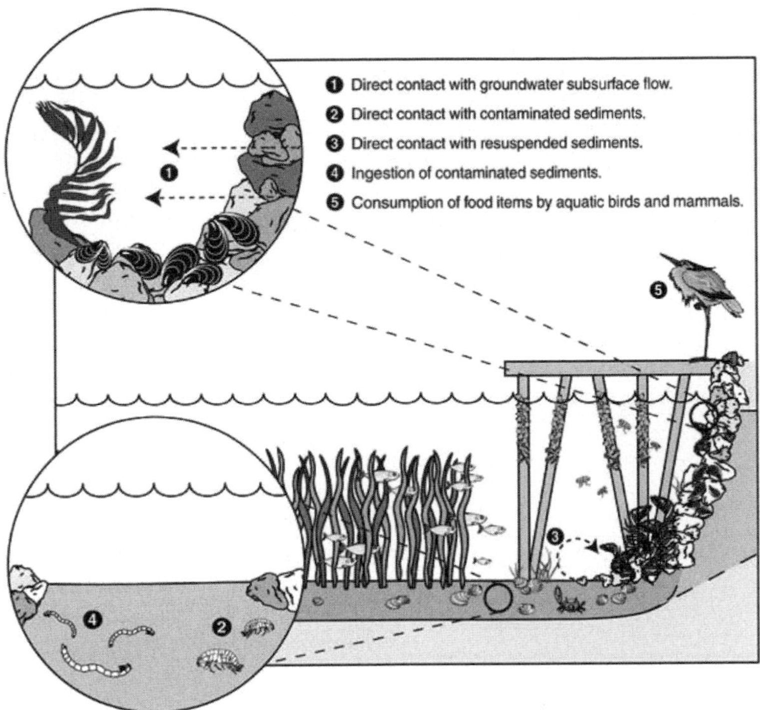

Figure 1. Conceptual model of sediment exposure pathways potentially evaluated using a SQT approach (see case study example).

The SQT approach, though derived explicitly for use with contaminated aquatic sediments, can also be adapted for use with terrestrial soils, groundwater discharges, or the water column (similar LOE apply, albeit designed for water or soil rather than sediment). For instance, the SQT approach implicitly comprises key components of the framework proposed by Bartell (2003) for holistically estimating ecological risks from nutrients and trace elements in a river. Groundwater discharges to the marine environment have been evaluated using LOEs for groundwater chemistry, groundwater toxicity to bivalves, kelp and larval fish, and alteration to the intertidal community in the vicinity of the discharge (McDonald, pers. comm.). Hollert et al. (2002b) applied the SQT to both sediments and surface waters. The approach also provides data useful for assessing human and wildlife health with the inclusion of a biomagnification LOE. Bacterial contamination can be addressed by modification of the LOE: bacterial contamination in water, sediments, or soils; bacterial diseases or concentrations in tissues; literature data (or testing) quantifying the relationships between environmental concentrations and effects.

Table 1. Possible conclusions from different outcomes of the SQT + biomagnification[a].

Outcome	***C[b]***	***T[b]***	***Be[b]***	***Bm[b]***	***Possible conclusions***	***Possible actions / decisions***
1	-	-	-	-	Measured sediment contaminants either not present or not bio-available	None needed
2	+	-	-	-		None needed
3*	-	-	-	+		Determine source(s) of biomagnification
4	-	+	-	-		Determine reason(s) for toxicity
5	-	-	+	-		Determine reason(s) for benthos alteration
6	+	-	+	-		
7	-	-	+	+		Determine reason(s) for benthos alteration and biomagnification
8	+	+	-	-	Sediment contaminants toxic but not causing *in situ* effects	Determine reason(s) for toxicity but no *in situ* effects
9*	-	+	+	-	Adverse biological effects unrelated to measured sediment contaminants	Assess other stressor(s) / source(s) that may be causing the observed responses
10*	-	+	+	+		
11*	-	+	-	+		
12*	+	-	-	+	Biomagnification may be due to sediment contamination	Risk management (can include further investigations)
13*	+	-	+	+	Biomagnification and benthos alteration may be due to sediment contamination	
14*	+	+	-	+	Sediment contaminants toxic but not causing benthos alteration; may be causing biomagnification	
15*	+	+	+	-	Adverse biological effects occurring, apparently related to measured sediment contaminants	
16*	+	+	+	+		

[a] Simplified tabular representation that does not consider differences within LOE, which commonly occur and which can influence the outcomes. Adapted from Chapman (1990, 1996), Grapentine et al. (2002).
[b] C: chemistry; T: toxicity; Be: benthos; Bm: biomagnification.
* Unacceptable risk (though not always necessarily related to sediment contamination).

4. Advantages of applying the SQT

The SQT minimizes uncertainties and provides information to address all three primary objectives of any sediment assessment: (1) identification of problem areas where sediment contamination (or other stressors) are causing adverse ecological effects; (2) prioritization and ranking of such areas and their environmental significance; and, (3) identification of areas where sediment contamination (or other stressors) are not causing adverse ecological effects, including determining the quality of reference areas.

Its advantages include:

- Allows for (1) interactions between contaminants in complex sediment mixtures (additivity, antagonism, synergism); (2) actions of unmeasured toxic chemicals; (3) actions of other stressors (physical, biological).
- Based on well-developed, standardized techniques for determination of individual LOE.
- Does not require *a priori* assumptions concerning specific mechanisms of interaction between biota and contaminants.
- Can be used to develop site-specific SQGs for measured contaminants.
- Provides empirical evidence of sediment quality.
- Combines separate hazard-based LOE to provide a risk-based WOE determination.
- Allows site- and situation-specific flexibility in LOE, including the possibility of tiering and/or adding additional studies (Chapman, 1996).
- Though developed for sediments (can be applied to any sediment type), can be applied to other media (*e.g.*, soils, water column).
- Though developed for use with chemical contaminants, can be applied to other stressors (physical, biological).
- Provides excellent discrimination potential; requires no or limited follow-up when a complete study is conducted.
- Although not inexpensive in monetary terms, extremely cost-effective in terms of providing risk-based information for decision-making; provides information that would otherwise not be available.
- Provides a means by which detailed, sophisticated technical information can be presented in a user-friendly, readily understandable format.

The SQT addresses the two components comprising the logic of the sciences as put forth by Francis Bacon in 1620: a determination based on past experiences coupled with a prediction for the future (Root, 2003). The first component is derived from observation. In the case of the SQT, this component is addressed by chemistry measurements and community structure analyses – to determine present status. The second component is derived from experimentation, which can include predictive modelling. In the case of the SQT, this second component is addressed by sediment toxicity testing and biomagnification analyses coupled with food chain modelling – to both determine present status and provide information for predicting future status based on changes to contaminant loadings.

Future predictions are improved by the inclusion of TIE and CBR analyses. TIEs have been and continue to be used to establish causality based on the toxicity of sediment interstitial pore waters (Ankley and Schubauer-Berigan, 1995; Stronkhorst et al., 2003). However, because interstitial water testing may overestimate toxicity of non-persistent, readily water soluble substances (*e.g.*, ammonia) and underestimate toxicity of persistent, poorly water soluble substances, the focus of TIEs is shifting to studies of whole sediments (Burgess et al., 2000, 2003; Ho et al., 2002). TIEs have been used as part of the SQT to determine causation (Hunt et al., 2001). The information provided regarding specific contaminants responsible for observed toxicity provides additional information for predictions related to changes in loadings of contaminants such as metals, which are not metabolized.

The dose or contaminant body residue in organisms is a key determinant of potential risk from contaminant uptake. For instance, metal body concentrations in small exposed organisms can be related to metal-induced effects at the organism and population levels of organization (Chapman et al., 2003). CBRs have been used, similar to TIEs, to establish causation as part of the SQT (Borgmann et al., 2001), and to provide additional information for predictions related to changes in contaminant loadings.

5. SQT procedure description

The SQT can involve tiered studies for the different components (Chapman, 1996), or all components. A common tiering strategy involves collection of sufficient sediment for chemistry and multiple toxicity tests, as well as benthic community samples, but conducting only chemistry and a limited number of short-term toxicity tests (*e.g.*, 48-h bivalve larval development and 10-d amphipod survival) as an initial tier. Toxicity test data are typically available within two weeks of test initiation—additional toxicity tests can be then implemented within the minimum 6-week sample-holding period if required. Sediment benthic samples are archived and analysed only if toxicity data are insufficient to make informed site management decisions. In a tiered approach, the SQT conforms to the screening-level risk assessment model as opposed to a higher-tier or detailed-level risk assessment (Hill et al., 2000; Pittinger et al., 2003).

It is recommended that personnel conducting all aspects of the SQT have a very high level of technical expertise, from the initial study design, through to field sampling, data analyses and interpretation. Such high levels of expertise, as well as the necessary costly and sophisticated equipment, are commonly required for sediment contamination investigations irrespective of whether or not the SQT is explicitly employed.

Sediment samples are collected synoptically from both exposed and reference areas (*e.g.*, by the use of remote grab samplers operated from a vessel, or using divers for locations where grab sampling is impractical). Spatial heterogeneity in sediment contamination and toxicity render coincident sampling extremely difficult to interpret, and such sampling is not recommended. Appropriate quality assurance /

quality control (QA/QC) procedures should be followed in all aspects of SQT studies, from field collections through to data analyses.

Sediment samples used for toxicity testing are subsampled for chemical analyses (following compositing). Samples for benthic community analyses are collected at the same time and from the same areas. Field replicates (typically n = 5 unless a different value is determined by power analyses) are collected for toxicity testing and for benthic community analyses, but generally not for chemical analyses due to cost considerations.

Measurement endpoints for each of the LOE (chemistry, toxicity, benthic community, biomagnification) should adequately represent the respective assessment endpoints (*e.g.*, chemicals measured include toxicants actually present; testing involves appropriate fauna that can be affected by those toxicants). In some cases direct measurements are possible (*e.g.*, number of species, types of species). In other cases inferences are necessary (*e.g.*, between toxicity tests and alterations in benthic communities; between tissue body burdens and biomagnification potential).

COPCs are determined based on local concerns and existing data, and typically include but are not necessarily restricted to some or all of: metals and metalloids, polyaromatic hydrocarbons, organochlorine pesticides, PCBs, butyltins, phthalate esters. TOC and grain size measurements provide a basis for normalizing the data to different types of sediments. Acid volatile sulfides (AVS) and simultaneously extracted metals (SEM) provide information for determining whether certain metals are not bioavailable, or may be bioavailable and thus potentially toxic. Petrographic analysis has also been used as qualitative evidence that high PAH concentrations are not necessarily bioavailable due to the presence of large quantities of coal particles.

The toxicity testing LOE may involve sample collection and / or *in situ* testing. The focus is typically on whole sediments, however testing and analysis can also be done on pore water or sediment extracts, provided that the ecological relevance of such testing has been established. A small number of tests are conducted, attempting to cover as wide a range as possible of organism type, life cycle, exposure route, and feeding type. Ideally, at least three different tests are conducted, reflective of the ecologically most relevant receptor groups present at the site under investigation. Species selection should reflect physical parameters, such as interstitial salinity and grain size. Toxicity tests should involve at least three different taxonomic groups (*e.g.*, amphipods, polychaetes, bivalves) comprising both acute and chronic endpoints (ideally survival, growth and reproduction). Sediment and porewater toxicity testing involving subchronic endpoints (*e.g.*, enzyme induction and/or inhibition) have also been incorporated — however, the ecological relevance of such tests is questionable. Examples of possible sediment toxicity tests involving bioindicators (whole organisms) are shown in Table 2. Generally both ammonia and sulphide concentrations are measured in the toxicity test containers since these substances can cause toxicity at elevated concentrations and thus mask the effects of COPCs.

The LOE involving assessment of the benthic community determines whether or not alterations have occurred, generally compared to reference areas, though comparisons to baseline (*e.g.*, pre-discharge) conditions are preferable. The benthic infaunal community is generally measured in preference to water column or epibenthic organisms because these organisms are relatively sessile and location-specific. Area-wide comparisons can use more mobile organisms such as fish; for instance, Chapman (1986) measured bottom fish histopathology. Key variables measured for the benthos include: numbers of taxa, numerical dominance, total abundance, and percentage composition of major taxonomic groupings. In the marine environment, this last category includes polychaetes, crustaceans (*e.g.*, amphipods), molluscs, and echinoderms. In the freshwater environment, oligochaetes and insect larvae (*e.g.*, chironomids) fit into this last category. Indices of biotic integrity have been used in watersheds where the benthic community is well characterized relative to habitat (Llanso et al., 2003).

The biomagnification LOE involves determining concentrations of contaminants such as DDT, PCBs, dioxins, TBT, and methyl mercury in benthic invertebrates or fish. This LOE only applies to those few organic contaminants that actually biomagnify. Concentrations are compared to reference areas or literature-based toxicity reference values (TRVs) and assessed via food chain bioaccumulation models (Grapentine et al., 2002).

Table 2. Examples of bioindicator sediment toxicity tests[a].

Toxicity tests	*Duration (days)*	*Primary endpoints*	*Litres of sediment required*
Marine/estuarine waters:			
Amphipod	~10-28	Survival, growth, reproduction	1.5
Bivalve/Echinoderm larvae	~2-5	Survival, development	0.5
Polychaete worm	~20	Survival, growth	2.0
Fresh waters:			
Amphipod	~10-28	Survival, growth, reproduction	1.5
Water flea	~10-21	Survival, reproduction	0.5
Insect larvae	~20	Survival, growth	1.5
Oligochaete worm	~20	Survival, growth, reproduction	1.5

[a] Different species are used / applicable to different geographic areas. Biomarkers exampled by bacteria (Microtox®) toxicity tests can be conducted in all cases; endpoint for this test is enzyme function, duration is hours, and amount of sediment required is minimal (~ 0.1 L).

TIE and CBR analyses are conducted as appropriate. CBR analyses can be conducted, depending on the contaminant, as part of the biomagnification LOE (measuring contaminants in tissues of field collected organisms, *e.g.*, Hg), or as part of the toxicity LOE (measuring contaminant concentrations in tissues of toxicity test organisms, *e.g.*, other metals). TIE analyses are typically conducted following a determination of toxicity, in that LOE.

Assessment of sediment stability serves to determine whether or not SQT studies need to be limited to superficial sediments (Chapman et al., 2002). Such assessments usually involve determining current flows required to resuspend sediments compared to worst case current flows or events such as 100-year storms at that site.

Data analyses are a means, not an end. The specific data analyses used will depend on the hypotheses being tested, and on site- and situation-specific circumstances. There is no "best" method for conducting SQT analyses, though general guidance is provided by Chapman (1996). A summary of that guidance and of additional useful analyses is provided below.

Univariate and/or multivariate analyses are conducted of the four primary LOE (sediment chemistry, toxicity, benthic community, biomagnification) individually. Typically, ANOVA and a set of *a priori* contrasts are used to determine spatial trends and statistically significant differences among stations or groupings of stations. ANCOVA can be used when effects may be due to non-contaminant factors such as grain size. COPC data are normalized to dry weight and TOC, and possibly also to AVS-SEM; means, standard deviations and ranges are summarized where replicate data are available. Exposed and reference areas are compared to themselves and to SQGs; principal components analyses (PCA) can be conducted for sub-groups of COPCs (*e.g.*, organics versus inorganics) and differences between locations can be tested by ANOVAs. For toxicity, data calculations can include between stations differences in mean response, ANOVA, and multiple comparison tests. For the benthic community data, univariate analyses can include abundance, richness and dominance; multivariate analyses can include: cluster analyses; ordination analyses; analyses of similarities; two- and three-way ANOVAs to test for differences among locations, habitats, and stations. Biomagnification data are treated similarly to those for COPCs except that normalizations are to percent lipids, and the results of food chain models are presented.

Relationships between the individual LOE can be examined via principal components analysis (PCA). Correlations among principal components for individual LOE indicate concordance or agreement. Relationships between different SQT LOE can also be assessed using other methods including: Mantel's test (Legendre and Fortin, 1989) coupled with a measure of similarity or ordination; canonical discriminant (or correspondence) analyses; multidimensional scaling (MDS).

More innovative methods for examining relationships between individual LOE for the SQT include quantitative estimation of probability derived from odds ratio (Smith et al., 2002) and meta-analysis resulting in pooled, empirically derived P-values (Bailer et al., 2002). Comparison of odds ratio and meta-analysis with PCA for clustering sites into groups of similar impact (Reynoldson et al., 2002a) revealed similarities and differences. The differences between the three methods (PCA, odds ratio and meta-analysis) were ascribed to three factors, which almost certainly apply to all integrations: the variables selected; the manner in which information is combined within a LOE; and, the statistical methodology employed.

6. Factors capable of influencing interpretation of the SQT

The SQT can be influenced by the following factors:

- It is assumed that the COPCs are selected appropriately, such that they are the appropriate indicators of overall chemical contamination. However, restrictions on chemical analyses include funds, equipment and facilities. Unmeasured chemicals can include degradation products.
- It is assumed that the measurement endpoints for biological effects (toxicity tests, benthic community structure) are appropriate relative to the assessment endpoints.
- Correlations between measured COPCs and biological effects (toxicity, alterations to the benthic community) could be strongly influenced by the presence of unmeasured toxicants that may or may not vary with the measured contaminants.
- Interpretation of the SQT depends strongly on the correct choice of individual LOE, in particular: COPCs; sediment toxicity tests and endpoints.
- Interpretation of the SQT also depends strongly on the choice of reference site(s), which may not always be available.
- The SQT does not explicitly incorporate variance in the quality of the different LOE. For instance, it is assumed that given adequate QA/QC, analytical methods provide accurate data; it is assumed that the toxicity tests chosen are sufficiently sensitive to detect the effects of any toxicants that are present. This assumption is not always correct and may require testing (cf. McPherson and Chapman, 2000).

7. Application of the SQT: case study

This example describes the approach and interpretative methods used in a SQT conducted in 1998 for a former shipyard located within a busy urban harbour. Site facilities included metal-smithing shops, electrical equipment shops, sandblasting and painting berths, and fuel storage facilities, which led to the introduction of numerous contaminants to the environment. This particular case study was chosen as illustrative due to its complexity coupled with the fact that it has neither been previously published, nor is it subject to client confidentiality (which is the case for other, more recent SQT studies). While a marine case study is presented, the SQT is equally applicable to freshwater environments.

7.1 PROBLEM FORMULATION/SAMPLING AND ANALYSIS PLAN (PF/SAP)

A PF/SAP document was prepared in order to guide the implementation and interpretation of the overall sediment quality triad. Objectives in a PF/SAP are identical to that of a problem formulation following a traditional ERA approach (U.S. EPA, 1998). PF/SAP documents are often submitted for regulatory approval prior to any sample collection in order to facilitate dialogue between interested parties. Our experience suggests that this dialogue during the problem formulation frequently eliminates (or at least, substantially reduces) the need for additional sampling at a later date, and facilitates a more rapid review and regulatory acceptance of the final SQT report.

Table 3. Assessment/measurement endpoints and impact hypotheses (see case study example).

Assessment endpoint	***Impact hypothesis***	***Measurement endpoints***
Maintenance of soft-bottom benthic invertebrate communities capable of supporting local fish populations.	COPC concentrations in sediment will not result in toxicity to the soft-bottom benthic community.	Assess the effect of COPCs in sediment on the survival of two amphipod species (*Eohaustorius estuarius* and *Rhepoxynius abronius*). Assess the effect of COPCs in sediment on survival and growth of the polychaete, *Neanthes arenaceodentata.* Assess *in situ* changes in benthic community structure (diversity and abundance).
Maintenance of hard-bottom benthic community capable of providing fish habitat and supporting wildlife populations.	Resuspension of COPC-contaminated sediment will not result in toxicity to the hard-bottom benthic community.	Assess the effects of suspended COPCs on the normal development of larval bivalves (*Mytilus galloprovincialis*). Assess the mobility and bioavailability of sediment COPCs through measurement of AVS-SEM and porewater COPC concentrations.
Maintenance of bottom fish and marine mollusk communities.	Bioaccumulation in the marine food chain will not result in unacceptable tissue COPC concentrations.	Measure COPC concentrations in tissue samples and compare to literature-based toxicity reference values.

Contaminants of potential concern (COPCs) included metals, polycyclic aromatic hydrocarbons (PAHs), polychlorinated biphenyls (PCBs) and tributyltin (TBT). Ecological resources in the intertidal and subtidal portions of the study area were typical for a site with similar depth and orientation. Immobile species in or near the study area (based on available habitat surveys) included barnacles, kelp, brown algae, and anemones. Common invertebrate species included shrimp and various species of crabs and seastars. Fish species included sculpins and perch. As a result of the observed biology of the study area, a series of receptors and assessment/measurement endpoints were defined (Tab. 3), and a conceptual model was prepared (Fig. 1).

A preliminary WOE assessment was conducted on five surface samples as part of the PF/SAP. Our experience is that such an initial reconnaissance provides valuable information about the spatial extent of contamination (including the potential influence of off-site or background conditions), which allows refinement of sample numbers and sample locations in the PF/SAP. A total of five surface sediment samples were collected from a boat using a van Veen grab and analyzed for selected COPCs (*e.g.*, metals, PAHs, PCBs, TBT) in order to estimate the potential magnitude of the sediment contamination. All chemical analyses were conducted according to well-established analytical protocols (*e.g.*, U.S. EPA, 1986). Preliminary toxicity testing was conducted using two test methods: a 10-d amphipod (*Eohaustorius estuarius*) survival test (Environment Canada, 1992) using whole-sediment, and a 48-h bivalve normal development toxicity test (PSEP, 1995) using sediment elutriate.

Data from the reconnaissance survey indicated that bulk sediment concentrations of all COPCs were greater than apparent effect levels (AELs; Environment Canada, 1995), indicating that contamination was widespread. A 20% reduction in toxicity test endpoint performance (relative to the negative control) was used to evaluate toxicity data. Such a reduction typically indicates real differences from the control. All samples demonstrated greater than a 20% reduction in bivalve normal development, however, similar reductions in amphipod survival were not observed, with the exception of one marginal "hit" for a single sample. As a result of the bivalve toxicity and elevated COPC concentrations observed in the reconnaissance survey, a full SQT was considered necessary for the site.

7.2 STUDY DESIGN

The following LOE were incorporated in the SQT:

- Whole-sediment chemistry (metals, PAHs, PCBs, total organic carbon [TOC] grain size, acid-volatile sulphides/simultaneously extractable metals [AVS/SEM]).
- Porewater chemistry (TBT).
- A 10-d *Eohaustorius estuarius* (amphipod) survival and avoidance test using whole-sediment.
- A 10-d *Rhepoxynius abronius* (amphipod) survival and avoidance test using whole-sediment.
- A 20-d *Neanthes arenaceodentata* (polychaete) survival and growth test using whole-sediment.
- A 48-h *Mytilus galloprovincialis* (bivalve) larval development test using sediment elutriate.
- Benthic community diversity and abundance.
- Tissue chemistry (PCBs, TBT) on fish and mussel samples.

A total of 34 stations were included in the SQT assessment in a "near-shore/far-shore" study design. Sample locations were selected to avoid known areas of rocky substrates, and to provide a higher sample density in near-shore areas (in which the reconnaissance survey found higher COPC concentrations than far-shore areas).

7.3 SAMPLE COLLECTION AND ANALYSIS

Surface sediment samples for sediment chemistry and toxicity were collected from all 34 stations using a van Veen grab. Grab samples were considered acceptable if proper penetration and complete jaw closure was obtained; sediment from the upper 10 cm of the grab was removed using a stainless steel spoon. Multiple grabs from the same location were composited in a stainless steel mixing bowl to achieve necessary sample volume. Porewater for chemical analysis was extracted from the surface sediment samples using centrifugation under oxygen-free conditions. Subsurface sediment samples were collected from 1-m and 3-m depth horizons using a diver-assisted coring device at eighteen locations spaced throughout the study area. Chemical analysis of sediment and porewater was conducted using approved methods (*e.g.*, U.S. EPA, 1986). Toxicity testing was conducted using approved methods for each species listed above (*e.g.*, Environment Canada, 1992; PSEP, 1995; ASTM, 1997). Appropriate quality assurance/quality control (QA/QC) measures were included (e.g., field duplicates, spike recovery, field swipes for chemistry; negative controls and reference toxicant tests for toxicity).

One major deviation from the typical toxicity test methodologies was that field replicates were used for toxicity testing in lieu of the standard laboratory replication (*i.e.*, one replicate per station rather than 5). This modification was made *a priori* in order to maximize the number of stations at the expense of statistical power within the available project resources.

Sediment benthos were sampled from 32 of 34 surface sediment stations using a Ponar grab with a total surface area of 0.05 m^2. Sample acceptance criteria were the same as for sediment chemistry and toxicity sampling. Two samples were excluded due to an inability to get proper grab penetration. Grab contents were gently rinsed through a 0.5 mm mesh and the remaining material transferred to a 1-L plastic container and preserved with 10% buffered formalin. Samples were transferred to 70% ethanol after 24-h in formalin and sent to the taxonomy laboratory where organisms were counted and identified to the lowest practical level (usually to species).

A total of 24 fish were collected from the site using gillnets. Species collected included sculpins (*Leptocottus armatus*), greenling (*Hexagrammos decagrammus* and *H. octogrammus*), sole (*Parophrys vetulas*), flounder (*Platichthys stellatus*) and sanddab (*Citharichthys sordidus*). Whole individual fish were composited into three samples, consisting of sculpins (Sample 1); sole, sanddab and flounder (Sample 2); and greenling (Sample 3). A total of three composite mussel samples were also collected from the shoreline of the site. Fish and mussel tissue samples were analyzed for PCBs, mercury, methylmercury and TBT using approved methods.

Table 4. Example WOE integration of chemistry, toxicity, and benthic community data for seven representative stations from the case study example. (□ = negligible effects; ▣ = moderate effects; ■ = severe effects).

Station	Chemistry		Toxicity				Benthic community Richness-Abundance)	Overall
	Sediment (Metal-PAH-PCB-SEM:AVS)	Porewater (TBT)	Amphipod (Rhepox) (Survival-Avoidance)	Amphipod (Eohaustorius) (Survival-Avoidance)	Bivalve (Normal-Survival-Combined Normal Survival)	Polychaete (Survival-Growth)		
STN-1	■-▣-▣-□	▣	□-□	□-□	□-□-□	□-□	□-□	□
STN-2	▣-▣-■-□	■	■-□	□-□	□-□-▣	□-□	□-□	□
STN-3	▣-▣-▣-□	■	■-□	□-□	□-□-□	□-□	□-□	□
STN-4	▣-□-□-■	▣	□-□	□-□	▣-□-▣	▣-□	Not available	▣
STN-5	▣-■-▣-□	■	■-□	□-■	■-□-■	□-▣	▣-□	▣
STN-6	■-▣-▣-□	▣	▣-□	□-□	▣-□-▣	□-□	▣-▣	▣
STN-7	▣-▣-▣-□	▣	▣-□	□-□	□-□-□	■-▣	□-▣	▣

Table 4 (continued).

Metals-PAH-PCB

□ All COPC concentrations > numerical standards

▣ One or more COPC concentrations < numerical standard

■ One or more COPC concentrations < 10 x numerical standard

SEM:AVS

□ SEM:AVS ratio < 1

■ SEM:AVS ratio > 1

Porewater TBT

□ TBT concentration < ambient water quality criterion (0.001 μg/L)

▣ TBT concentration > ambient water quality criterion (0.001 μg/L)

■ TBT concentration > toxicity reference value (1 μg/L)

Toxicity

□ Endpoint performance not reduced more than 20% relative to negative control

▣ Endpoint performance reduced more than 20% relative to negative control

■ Endpoint performance reduced more than 50% relative to negative control

Benthic community

□ Taxonomic richness/abundance similar to historical background

▣ Taxonomic richness/abundance significantly reduced relative to historical background

Overall – see text.

7.4 INTERPRETATION OF INDIVIDUAL LINES OF EVIDENCE

Sediment chemistry, porewater chemistry, sediment toxicity and benthic community data were interpreted as follows, and as summarized in Table 4 for seven representative stations:

7.4.1 Surface sediment chemistry

COPC concentrations greater than the AEL were considered indicative of moderate effects, while concentrations greater than 10 x AEL were considered indicative of severe effects. Contamination by metals, PAHs and PCBs was widespread in the study area. Concentrations of copper, mercury and zinc were greater than AELs for all surface sediment stations. Arsenic and lead concentrations exceeded the AEL for all but one station, while nickel and silver concentrations exceeded the AEL at approximately half of the stations. PAH and PCB concentrations exceeded their respective AEL benchmark at all stations except one.

AVS:SEM data were also considered as a separate LOE. AVS:SEM ratios of greater than 1 were considered to indicate negligible effects (for the metals Ag Cd, Cu, Pb, Hg and Zn), and AVS:SEM ratios greater than one were considered to indicate severe effects (*i.e.*, the measured metals would tend be more bioavailable). Nearly all stations had AVS-SEM ratios greater than 1.

7.4.2 Porewater chemistry

Porewater TBT concentrations greater than the provincial marine ambient water quality guideline (AWQG) value of 0.001 μg/L (Nagpal, 1995) were considered indicative of moderate effects, while concentrations greater than a literature-based TRV of 1 μg/L (Garrett and Shrimpton, 1997) were considered indicative of a severe effect. This TRV represented a concentration that corresponded to increased abnormal development and mortality in bivalve larvae and, as such, was considered more indicative of adverse ecological effects than an AWQG value, which was never intended for interpreting sediment porewater concentrations. Porewater TBT concentration exceeded the 1 μg/L TRV at the majority of stations, and exceeded the 0.001 μg/L AWQG value at all stations.

7.4.3 Sediment toxicity

Reductions in toxicity test endpoint performance (*e.g.*, amphipod survival, polychaete growth) of greater than 20% (relative to the negative control) were considered indicative of moderate effects, while reductions greater than 50% were considered indicative of severe effects. Results of the toxicity testing were as follows:

- *R. abronius* survival was reduced by greater than 20% in 15 samples, and was reduced by greater than 50% in 10 samples. No effects on *R. abronius* avoidance were observed.

- *E. estuarius* survival was unimpaired in all samples, although greater than 20% increase in sediment avoidance was observed in two samples.

- *N. arenaceodentata* survival was unimpaired in the majority of samples, with the exception of one sample that demonstrated a greater than 20% reduction, and one sample that demonstrated greater than 50% reduction. *N. arenaceodentata* growth was similarly unimpaired, with the exception of four samples that demonstrated greater than a 20% reduction.

- Normal development and survival of *M. galloprovincialis* larvae was unimpaired in 16 samples. All remaining stations demonstrated greater than a 20% reduction in normal development and/or survival; one station demonstrated greater than a 50% reduction in normal development.

7.4.4 Benthic community structure

Historical data from the surrounding harbour indicated that species richness was typically above 30 taxa — a 10% reduction in the background species richness was permitted to compensate for differences in the sieve size used in this study (0.5 mm) versus historical investigations (0.3 mm). As a result, species richness values of less than 27 taxa were considered to indicate a moderate effect. Three stations demonstrated a reduction in benthic community diversity, with species richness values ranging from 7 to 20 taxa. All other stations had species richness values that were greater than 27.

Historical abundance data were not sufficient to make meaningful comparisons. As a result, the average abundance data from the samples included in this SQT were used to make qualitative observations about the degree of impairment in individual stations. Two stations had abundances that were considerably reduced relative to other stations at the site and, therefore, were considered indicative of moderate effects.

7.4.5 Tissue chemistry

Data from the three composite fish and three mussel tissue samples were compared using a hazard quotient approach to the lowest available TRVs derived from the literature. TRVs were derived for dioxin-like PCB congeners, *i.e.*, using a toxic equivalent quotient (TEQ) approach (one TRV for fish and mussels), total PCBs (separate TRVs for fish and mussels), mercury (separate TRVs) and TBT (separate TRVs). All tissue concentrations were less than their respective TRVs, with the exception of total PCBs in the sculpin and greenling tissue samples and TBT in the sculpin tissue sample. These exceedances were interpreted based on their magnitude and the relevance of the available data used to develop the TRV as either indicative of potential effects (for PCBs) or moderate effects (for TBT).

7.5 OVERALL WEIGHT-OF-EVIDENCE ASSESSMENT

Table 4 provides the overall WOE assessment for a subset of seven representative (of 34) stations included in the SQT. Assessment of bioaccumulation was conducted on an area-wide basis rather than for individual stations, and as a result, bioaccumulation LOE are not included in this table. Different LOE were not assigned equal weight in the WOE nor was a mathematical approach used to generate the WOE assessment. Rather, best professional judgement was used to evaluate each LOE depending on its inherent conservatism, degree of uncertainty, and relevance to site-specific conditions. LOE were weighted in the following order, from highest to lowest priority:

- Benthic community LOE received the highest weight since it provided direct measurement of an ecosystem-level endpoint (*i.e.*, diversity and abundance) and was highly relevant to the stated assessment endpoint (*i.e.*, soft-bottom benthic community).
- Toxicity LOE received medium weight, since they measure adverse biological effects on receptor species intended to represent the overall assessment endpoint, and incorporate a measurement of COPC bioavailability. However, toxicity LOE are ranked lower than the benthic community LOE due to the need to extrapolate laboratory results as indicative of field effects.
- Chemistry LOE received the lowest weight since elevated concentrations of bulk sediment and porewater COPC concentrations do not necessarily translate into adverse effects in the other two LOE.

Additionally, *R. abronius* survival and growth was assigned a much lower weight than *E. estuarius* survival and growth. *R. abronius* typically inhabits coarse-grained sediments, however, samples collected from the site had grain-size distributions that were predominantly silts and clays. Fine-grained sediments can result in significant mortality to *R. abronius* (DeWitt et al., 1988); as a result, it was impossible to separate COPC-related effects from potential grain size-related effects. *R. abronius* was an inappropriate test species for this study area—instead, an amphipod that prefers fine-grained sediment should have been selected as the second amphipod species (*e.g.*, *Ampelisca abdita*). In addition to the station-specific WOE assessment, the overall risk management/remediation plan for the site considered the fact that bioaccumulation of TBT and total PCBs may result in unacceptable effects to demersal fish.

When all available LOE were integrated using a WOE assessment, severe adverse ecological effects were not predicted for any station, although a total of six stations (out of 34) demonstrated a potential for moderate effects. These effects were anticipated to be limited in spatial scale (*i.e.*, stations with moderate effects were surrounded by stations with negligible effects) and limited in magnitude (*i.e.*, adverse effects were not demonstrated in all LOE simultaneously). A summary of the different LOE and overall WOE assessment are provided for representative stations in Table 4. Results from the sub-surface sediment coring indicated an overall vertical profile where surface sediment contamination is moderate, shallow sub-surface sediment contamination is severe, and deep sub-surface sediment contamination is low — suggesting that the contamination from historical site operation is being buried through the aggregation of new sediment layers.

Site management recommendations were made based on the results of the SQT. An overall assessment of negligible effects resulted in a recommendation for *in situ* management of contaminated sediments (*i.e.*, no dredging). This assessment and recommendation assumed that sub-surface sediment layers remained undisturbed. Additional toxicity testing was recommended for the six stations that demonstrated moderate effects in order further evaluate the spatial extent and magnitude of any adverse effects to sediment benthos. Monitoring TBT concentrations in demersal fish in both the study area as well as adjacent properties was also recommended in order to evaluate the relative contribution of the study area to TBT burdens in aquatic organisms (multiple shipyards are present in the harbour, which made it problematic to assume that all TBT was the result of historical operation in this particular study area).

8. Lessons learned

The SQT approach is not a static methodology—one of its strengths is its flexibility in terms of the inclusion of different LOE to reflect the latest scientific knowledge and practices. Design and implementation of the SQT also reflects the experience of the scientists involved. Consequently, a new SQT designed today for this study area would likely be different than that described above. Potential improvements include five major components as follows. First, explicit consideration would be provided for toxicity and benthic community structure at an appropriate reference site (or sites), rather than the current approach where toxicity data were interpreted relative to negative controls, and benthic community data were interpreted relative to average values for the site obtained from historical data. Second, selection of toxicity test species would likely include tests with longer duration (*e.g.*, 28-d *Leptocheirus plumulosus* survival, growth and reproduction test). Third, field replicates in the toxicity tests would be unlikely, given the difficulty in reliably detecting a 20% difference in endpoint performance based on a single replicate (which only contains 20 amphipods or 5 polychaetes). Fourth, literature-based TRVs for tissue concentrations would require updating to reflect recent advances. Fifth, a formal *a priori* framework for integrating the results of multiple toxicity tests would be implemented, rather than relying solely on best professional judgment. An example of such an *a priori* framework is provided in Table 5. Results from multiple LOE are then integrated into an overall assessment using the approach outlined in Table 1.

9. Future prospects

Areas for future development of the SQT include improvements to reference comparisons, to both individual LOE and to the integrated WOE. As recommended by Chapman (2000), elaboration of reference comparisons is encouraged including appropriate use of formulated sediments and regional reference-envelope comparisons. Reference envelopes have been developed by Reynoldson et al. (1995) for the North American Great Lakes; an example of their application is provided by Reynoldson et al. (2002b).

Table 5. Weight of evidence framework for integrating multiple toxicity lines of evidence.

Observed pattern in toxicity data	***Overall toxicity***	
	Symbol	***Narrative statement***
Greater than a 50% reduction in one or more acute endpoints (*i.e.*, survival)	●	Adverse impacts associated with toxicity are probable
Greater than a 20% reduction in two or more acute endpoints (*i.e.*, survival) and the differences are statistically significant.	●	Adverse impacts associated with toxicity are probable
Greater than a 50% reduction in two or more non-acute endpoints (*e.g.*, growth, reproduction), and the differences are statistically significant.	●	Adverse impacts associated with toxicity are probable
Greater than a 50% reduction in one non-acute endpoint (*e.g.*, growth, reproduction), and the differences are statistically significant.	⦿	Adverse impacts associated with toxicity are possible
Greater than a 20% reduction in one or more acute endpoints (*i.e.*, survival) and the differences are statistically significant	⦿	Adverse impacts associated with toxicity are possible
Greater than a 50% reduction in one non-acute endpoint (*e.g.*, growth, reproduction), but the differences are not statistically significant.	○#	Adverse impacts associated with toxicity are possible, but likely limited in magnitude
Greater than a 20% reduction in two or more acute endpoints (*i.e.*, survival) but the differences are not statistically significant	○#	Adverse impacts associated with toxicity are possible, but likely limited in magnitude
Greater than a 20% reduction in one non-acute endpoint (*e.g.*, growth, reproduction), and the differences are statistically significant.	○#	Adverse impacts associated with toxicity are possible, but likely limited in magnitude
Greater than a 20% reduction in one non-acute endpoint (*e.g.*, growth, reproduction), but the differences are not statistically significant.	○	No adverse impacts associated with toxicity anticipated
No reduction in endpoint performance	○	No adverse impacts associated with toxicity anticipated

Chemical measurements should be more relevant to biological availability of contaminants. CBR measurements would be more useful if the equivalent of SQGs could be developed for a wide range of contaminants in a wide range of organisms.

TIE analyses are, as noted above, changing focus from interstitial waters to whole sediments to increase the relevance of the results. Development of whole sediment TIEs for all contaminant classes is desirable.

Biomarkers are, at present, effectively only indicators of exposure, not of effects. Development of effects-based biomarkers that provide predictive information for higher levels of biological organization is highly desirable.

Toxicity measurements should incorporate tests that reflect what will actually occur in the environment and, in some cases (*e.g.*, herbicide contamination), should include plants in the suite of tests. Consideration should also be given to including *in situ* tests; such tests have been developed, primarily for low-energy, shallow systems. Comparisons between these toxicity tests and laboratory toxicity tests using field-collected sediments have shown both similarities and differences (Sasson-Brickson and Burton, 1991; Hatch and Burton, 1999; DeWitt et al., 1999; Castro et al., 2003). Although "In situ tests provide a potentially powerful and important means for validating and extending information from laboratory toxicity tests under more realistically variable environmental conditions", experimental artifacts can and do occur (DeWitt et al., 1999). Chemical conditions affecting COPC bioavailability are changed during collection of field sediments but can also change under field conditions. Organisms in the field may experience other stressors, including toxicity from contaminants in overlying waters (Sasson-Brickson and Burton, 1991).

Benthic community analyses need to determine whether measurements of structure are adequate to protect community function. They also need to allow for extrapolation from the level of individual organisms and small spatial scales to higher levels of biological organization and larger spatial scales.

As noted above, the SQT originally focused on sediment chemistry, toxicity, and community structure. It has been expanded to include bioaccumulation, sediment stability, CBR and TIE analyses. Schmidt et al. (2002) have provided a means for inclusion of data on habitat. Habitat is a key factor controlling the distributions of biota, which has been implicitly but not explicitly considered to date in the SQT (*e.g.*, via measurements of sediment grain size and TOC). Habitat should be explicitly considered as part of the SQT in future.

10. Conclusions

The SQT has been accepted internationally as the most comprehensive approach available for assessing contaminated sediments. It has been widely used, not just in North America, but in Europe, Australasia, South America, and the Antarctic. It is extremely cost-effective for the level of information provided when applied in a tiered and iterative fashion. It can be used with all sediment types and can be adapted for use with soils and the water column. It provides information on potential effects of biomagnifying contaminants to the health of humans and wildlife, and can be adapted for use with bacterial contaminants and for human health assessments. It is a framework, not a formula, and thus will continue to be improved and possibly expanded by subsequent investigators.

Acknowledgements

We thank Christian Blaise for the invitation to write this book chapter and would like to acknowledge the contributions of the EVS scientists who conducted the SQT provided as an example: Gary Mann, Gary Lawrence, Ryan Stevenson, Ryan Hill as well as field staff and laboratory biologists.

References

Anderson, B.S., Hunt, J.W., Phillips, B.M., Fairey, R., Roberts, C.A., Oakden, J.M., Puckett, H.M., Stephenson, M., Tjeerdema, R.S., Long, E.R., Wilson, C.J. and Lyons, M. (2001) Sediment quality in Los Angeles Harbor, USA: A Triad assessment, *Environmental Toxicology and Chemistry* **20**, 359-370.

Ankley, G.T. and Schubauer-Berigan, M.K. (1995) Background and overview of current sediment toxicity identification evaluation procedures, *Journal of Ecosystem Health* **4**, 133-149.

ASTM (American Society for Testing and Materials) (1997) *Standard Guide for Conducting Static Acute Toxicity Tests Starting with Embryos of Four Species of Saltwater Bivalve Mollusks*, Method E724-94, Vol. 11, American Society for Testing and Materials, Philadelphia, PA, USA.

Bailer, A.J., Hughes, M.P., See, K., Noble, R. and Schaefer, P. (2002) A pooled response strategy for combining multiple lines of evidence to quantitatively estimate impact, *Human and Ecological Risk Assessment* **8**, 1597-1611.

Balthis, W.L., Hyland, J.L., Scott, G.I., Fulton, M.H., Bearden, D.W. and Greene, M.D. (2002) Sediment quality of the Neuse River estuary, North Carolina: an integrated assessment of sediment contamination, toxicity, and condition of benthic fauna, *Journal of Aquatic Ecosystem Stress and Recovery* **9**, 213-225.

Bartell, S.M. (2003) A framework for estimating ecological risks posed by nutrients and trace elements in the Patuxent River, *Estuaries* **26**, 385-397.

Beiras, R., Fernández, N., Bellas, J., Besada, V., González-Quijano, A. and Nunes, T. (2003) Integrative assessment of marine pollution in Galician estuaries using sediment chemistry, mussel bioaccumulation, and embryo-larval toxicity bioassays, *Chemosphere* **52**, 1209-1224.

Borgmann, U., Norwood W. P., Reynoldson, T.B. and Rosa, F. (2001) Identifying cause in sediment assessments: Bioavailability and the Sediment Quality Triad, *Canadian Journal of Fisheries and Aquatic Sciences* **58**, 950-960.

Burgess, R.M., Cantwell, M.G., Pelletier, M.C., Ho, K.T., Serbst, J.R., Cook, H.F. and Kuhn, A. (2000) Development of a toxicity identification evaluation procedure for characterizing metal toxicity in marine sediments, *Environmental Toxicology and Chemistry* **19**, 982-991.

Burgess, R.M., Pelletier, M.C., Ho, K.T., Serbst, J.R., Ryba, S.A., Kuhn, A., Perron, M.M., Raczelowski, P. and Cantwell, M.G. (2003) Removal of ammonia toxicity in marine sediment TIEs: a comparison of *Ulva lactuca*, zeolite and aeration methods, *Marine Pollution Bulletin* **46**, 607-618.

Burton, G.A. Jr., Chapman, P.M. and Smith, E.P. (2002a) Weight-of-evidence approaches for assessing ecosystem impairment, *Human and Ecological Risk Assessment* **8**, 1657-1673.

Burton, G.A. Jr., Batley, G.E., Chapman, P.M., Forbes, V.E., Smith, E.P., Reynoldson, T., Schlekat, C., den Besten, P.J., Bailer, A.J., Green, A.S. and Dwyer, R.L. (2002b) A weight-of-evidence framework for assessing sediment (or other) contamination: improving certainty in the decision-making process, *Human and Ecological Risk Assessment* **8**, 1675-1696.

Castro, B.B., Guilhermo, L. and Ribeiro, R. (2003) *In situ* bioassay chambers and procedures for assessment of sediment toxicity with *Chironomus riparius*, *Environmental Pollution* **125**, 325-335.

Chapman, P.M. (1986) Sediment quality criteria from the Sediment Quality Triad – an example, *Environmental Toxicology and Chemistry* **5**, 957-964.

Chapman, P.M. (1990) The Sediment Quality Triad approach to determining pollution-induced degradation, *Science of the Total Environment* **97/98**, 815-825.

Chapman, P.M. (1996) Presentation and interpretation of Sediment Quality Triad data, *Ecotoxicology* **5**, 327-339.

Chapman, P.M. (2000) The Sediment Quality Triad: then, now and tomorrow, *International Journal of Environment and Pollution* **13**, 351-356.

Chapman, P.M., Dexter, R.N. and Long, E.R. (1987) Synoptic measures of sediment contamination, toxicity and infaunal community structure (the Sediment Quality Triad), *Marine Ecology Progress Series* **37**, 75-96.

Chapman, P.M., Anderson, B., Carr, S., Engle, V., Green, R., Hameedi, J., Harmon, M., Haverland, P., Hyland, J., Ingersoll, C., Long, E., Rodgers J. Jr., Salazar, M., Sibley, P.K., Smith, P.J., Swartz, R.C., Thompson, B. and Windom, H. (1997) General guidelines for using the Sediment Quality Triad, *Marine Pollution Bulletin* **34**, 368-372.

Chapman, P.M., McDonald, B.G. and Lawrence, G.S. (2002) Weight of evidence issues and frameworks for sediment quality (and other) assessments, *Human and Ecological Risk Assessment* **8**, 1489-1515.

Chapman, P.M., Wang, F., Janssen, C., Goulet, R.R. and Kamunde, C.N. (2003) Conducting ecological risk assessments of inorganic metals and metalloids – Current status, *Human and Ecological Risk Assessment* **9**, 641-697.

DeWitt, T.H., Ditsworth G.H. and Swartz, R.C. (1988) Effects of natural sediment features on survival of the phoxocephalid amphipod, *Rhepoxynius abronius*, *Marine Environmental Research* **25**, 99-124.

DeWitt, T.H., Hickey, C.W., Morrisey, D.J., Nipper, M.G., Roper, D.S., Williamson, B., Van Dam, L. and Williams, E.K. (1999) Do amphipods have the same concentration-response to contaminated sediment in situ as in vitro?, *Environmental Contamination and Toxicology* **18**, 1026-1037.

Environment Canada (1992) *Biological Test Method: Acute Test for Sediment Toxicity Using Marine or Estuarine Amphipods*, EPS 1/RM/26, Environment Canada, Conservation and Protection Branch, Ottawa, ON, Canada.

Environment Canada (1995) *Interim Sediment Quality Guidelines*, Ecosystem Conservation Directorate, Guidelines Branch, Ottawa, ON, Canada.

Forbes, V.E. and Calow, P. (2004) A systematic approach to weight of evidence in sediment quality assessments: challenges and opportunities, *Journal of Aquatic Ecosystem Health and Management 7, 339-350.*

Garrett, C.L. and Shrimpton, J.A. (1997) *Organotin Compounds in the British Columbia Environment*, Regional Program Report No. 98-03, Environment Canada, Environmental Protection Service, Pacific and Yukon Region, Vancouver, BC, Canada.

Grapentine, L., Anderson, J., Boyd, D., Burton, G.A., DeBarros, C., Johnson, G., Marvin, C., Milani, D., Painter, S., Pascoe, T., Reynoldson, T., Richman, L., Solomon, K. and Chapman, P.M. (2002) A decision making framework for sediment assessment developed for the Great Lakes, *Human and Ecological Risk Assessment* **8**, 1641-1655.

Hatch, A.C. and Burton, G.A. Jr. (1999) Sediment toxicity and stormwater runoff in a contaminated receiving system: consideration of different bioassays in the laboratory and field, *Chemosphere* **39**, 1001-1007.

Hill, R.A., Chapman, P.M., Mann, G.L. and Lawrence, G.S. (2000) Level of detail in ecological risk assessments, *Marine Pollution Bulletin* **40**, 471-477.

Ho, K.T., Burgess, R.M., Pelletier, C., Serbst, J.R., Ryba, S.A., Cantwell, M.G., Kuhn, A. and Raczelowski, P. (2002) An overview of toxicant identification in sediments and dredged materials, *Marine Pollution Bulletin* **44**, 286-293.

Hollert, H., Dürr, M., Olsman, H., Halldin, K., van Bavel, B., Brack, W., Tysklind, M., Engwall, M. and Braunbeck, T. (2002a) Biological and chemical determination of dioxin-like compounds in sediments by means of a Sediment Triad approach in the catchment area of the River Neckar, *Ecotoxicology* **11**, 323-336.

Hollert, H., Heise, S., Pudenz, S., Brüggemann, R., Ahlf, W. and Braunbeck, T. (2002b) Application of a Sediment Quality Triad and different statistical approaches (Hasse diagrams and fuzzy logic) for the comparative evaluation of small streams, *Ecotoxicology* **11**, 311-321.

Hunt, J. W., Anderson, B.S., Phillips, B.M., Tjeerdema, R.S., Taberski, K.M., Wilson, C.J., Puckett, H.M., Stephenson, M., Fairey, R. and Oakden, J. (2001) A large-scale categorization of sites in San Francisco Bay, USA, based on the Sediment Quality Triad, toxicity identification evaluations, and gradient studies, *Environmental Toxicology and Chemistry* **20**, 1252-1265.

Lahr, J., Maas-Diepeveen, J.L., Stuijfzand, S.C., Leonards, P.E. G., Drüke, J.M., Lüker, S., Espeldoorn, A., Kerkum, L.C.M., van Stee, L.L.P. and Hendriks, A.J. (2003) Responses in sediment bioassays used in the Netherlands: can observed toxicity be explained by routinely monitored priority pollutants?, *Water Research* **37**, 1691-1710.

Legendre, P. and Fortin, M.J. (1989) Spatial pattern and ecological analysis, *Vegetatio* **80**, 107-138.

Llanso, R.J., Dauer, D.M., Vølstad, J. and Scott, L. (2003) Application of the benthic index of biotic integrity to environmental monitoring in Chesapeake Bay, *Environmental Monitoring and Assessment* **81**, 164-174.

Long, E.R. and Chapman, P.M. (1985) A sediment quality triad: measures of sediment contamination, toxicity and infaunal community composition in Puget Sound, *Marine Pollution Bulletin* **16**, 405-415.

McPherson, C.A. and Chapman, P.M. (2000) Copper effects on potential sediment test organisms: the importance of appropriate sensitivity, *Marine Pollution Bulletin* **40**, 656-665.

Nagpal, N.K. (1995) *Approved and Working Criteria for Water Quality, 1995*, BC Ministry of Environment, Lands and Parks, Water Quality Branch, Victoria, BC, Canada.

Pittinger, C.A., Brennan, T.H., Badger, D.A., Hakkinen, P.J. and Fehrenbacker, M.C. (2003) Aligning chemical assessment tools across the hazard-risk continuum, *Risk Analysis* **23**, 529-535.

PSEP (Puget Sound Estuary Program) (1995) *Recommended Guidelines for Conducting Laboratory Bioassays on Puget Sound Sediments*, Puget Sound Estuary Program, Olympia, WA, USA.

Reynoldson, T.B., Day, K.E. and Norris, R.H. (1995) Biological guidelines for freshwater sediment based on BEnthic Assessment of SedimenT (BEAST) using a multivariate approach for predicting biological state, *Australian Journal of Ecology* **20**, 198-219.

Reynoldson, T.B., Smith, E. and Bailer, A.J. (2002a) Comparison of three weight of evidence approaches for integrating sediment contamination data within and across lines of evidence, *Human and Ecological Risk Assessment* **8**, 1613-1624.

Reynoldson, T.B., Thompson, S.P. and Milani, D. (2002b) Integrating multiple toxicological endpoints in a decision-making framework for contaminated sediments, *Human and Ecological Risk Assessment* **8**, 1569-1584.

Riba, I.R., Forja, J.M., Gómez-Parra, A. and DelValls, A. (2004) Sediment quality in littoral ecosystems from the Gulf of Cádiz: A Triad approach to address influence of mining activities, *Environmental Pollution* **132**, 341-353.

Root, D.H. (2003) Bacon, Boole, the EPA, and scientific standards, *Risk Analysis* **23**, 663-668.

Sasson-Brickson, G. and Burton, G.A. Jr. (1991) *In situ* and laboratory sediment toxicity testing with *Ceriodaphnia dubia*, *Environmental Toxicology and Chemistry* **10**, 201-207.

Schmidt, T.S., Soucek, D.J. and Cherry, D.S. (2002) Modification of an ecotoxicological rating to bioassess small acid mine drainage-impacted watersheds exclusive of benthic macroinvertebrate analysis, *Environmental Toxicology and Chemistry* **21**, 1091-1097.

Smith, E., Lipkovich, P.I. and Ye, K. (2002) Weight of evidence: quantitative estimation of probability derived from odds ratio, *Human and Ecological Risk Assessment* **8**, 1585-1596.

Stronkhorst, J., Schot, M.E., Dubbeldam, M.C. and Ho, K.T. (2003) A toxicity identification evaluation of silty marine harbor sediments to characterize persistent and non-persistent constituents, *Marine Pollution Bulletin* **46**, 56-64.

Suter, G.W. III. (1997) Overview of the ecological risk assessment framework, in C.G. Ingersoll, T. Dillon and G.R. Biddinger (eds.), *Ecological Risk Assessment of Contaminated Sediments*, SETAC Press, Pensacola, FL, USA, pp. 1-6.

U.S. EPA (U.S. Environmental Protection Agency) (1986) *Test Methods for Evaluating Solid Waste – Physical/Chemical Methods*, SQ-846, 3rd Edition, Washington, DC, USA.

U.S. EPA (U.S. Environmental Protection Agency) (1998) *Guidelines for ecological risk assessment,* EPA/600/R-95/002F, Washington DC, USA.

Abbreviations

AEL	Apparent Effect Levels
AVS	Acid Volatile Sulphide
AWQG	Ambient Water Quality Guideline
CBR	Contaminant Body Residue

COPC	Contaminant Of Potential Concern
ERA	Ecological Risk Assessment
DDT	dichlorodiphenyltrichloroethane
LOE	Line Of Evidence
MDS	MultiDimensional Scaling
PAH	polycyclic aromatic hydrocarbon
PCA	Principal Components Analysis
PCB	polychlorinated biphenyl
PF/SAP	Problem Formulation/Sampling and Analysis Plan
QA/QC	Quality Assurance / Quality Control
SEM	Simultaneously Extracted Metal
SQG	Sediment Quality Guideline
SQT	Sediment Quality Triad
TBT	tributyltin
TEQ	Toxic Equivalent Quotient
TIE	Toxicity Identification Evaluation
TOC	Total Organic Carbon
TRV	Toxicity Reference Value
WOE	Weight of Evidence.

11. WASTOXHAS: A BIOANALYTICAL STRATEGY FOR SOLID WASTES ASSESSMENT

JEAN FRANCOIS-FÉRARD
Université Paul Verlaine
Laboratoire Ecotoxicité et Santé Environnementale
CNRS FRE 2635, Campus Bridoux,
rue du Général Delestraint
57070 METZ, France
ferard@sciences.univ-metz.fr

BENOIT FERRARI
Université de Genève, Institut F.A. Forel
10, route de Suisse, CH-1290 VERSOIX, Suisse
benoit.ferrari@terre.unige.ch

1. Objective and scope of WASTOXHAS

WASTOXHAS is the acronym for WASte ecoTOXic Hazard Assessment Scheme. This method was developed to ensure that unacceptable adverse effects would not arise from landfilled or re-used waste disposal. It is dedicated to assess the long-term leaching hazardous impact of any solid waste containing potentially hazardous substances (*e.g.*, bulk, stabilized, solidified, or vitrified wastes as well as contaminated soils or sediments intended for soil disposal).

As stated by Johnson (1993), "*hazardous waste becomes a problem when it moves*". Because water is the main vector for transporting pollutants from wastes towards receiving ecosystems, WASTOXHAS, presented hereinafter, is only focused on ecotoxicological assessment of different leachates with aquatic bioassays.

2. Summary of WASTOXHAS procedure

WASTOXHAS is a part of a tiered approach (see Figure 1, and also Figure 2 in Section 4) for conducting long-term impact assessment of leachates produced by solid wastes. It takes place after a prerequisite step consisting in a classical batch shaking leaching test (*e.g.*, EN 12457/1 to 4 (2002) or equivalent) followed by application of a large bioassay battery (examples are given in Section 3.4). This

C. Blaise and J.-F. Férard (eds.), Small-scale Freshwater Toxicity Investigations, Vol. 2, 331-375.

prerequisite step has to be carried out in order to decide to pursue ecotoxic hazard assessment and to select adequate and sensitive bioassays for proper waste assessment.

WASTOXHAS is based on two different dynamic leaching procedures (see Fig.1):

(1) Simulation leaching tests are upward-flow (NEN, 7343, 1995) and downward-flow column leaching tests (Huang et al., 2003). Laboratory leachates are collected on a regular basis, chemically monitored and tested with (at least) two aquatic bioassays (see below).

(2) Field leaching tests, where wastes are stored in big tanks, are also performed. Field leachates are collected on a regular basis, chemically monitored and tested with (at least) the same bioassays as those used in the simulation leaching tests, so as to compare the results with the upper procedure.

A large number of bioassay batteries have been used in waste toxicity testing (see Section 3.4). The following toxicity tests are examples of a wide list of available bioassays (see also other chapters of this book): Microtox™ light inhibition test (*Vibrio fischeri*), micro-algal growth inhibition assay (*Selenastrum capricornutum*), Mutatox™ revertant light test (*Vibrio fischeri* M169 mutant), acute microcrustacean immobilization test (*Daphnia magna*) and chronic microcrustacean reproduction test (*Ceriodaphnia dubia*). Interesting features of the Microtox™ and microalgal tests are rapidity of testing and automation potential respectively.

Measurement endpoints are integrated to calculate a waste PEEP (Potential Ecotoxic Effects Probe) index value (derived from the PEEP index published by Costan et al. (1993) and also presented in the Chapter 1 of this volume). This index can be calculated at any time period with a simple and user-friendly mathematical formula integrating leachate toxicity (*i.e.*, the summation of toxic units generated by all bioassays) and the extent of trophic/specific toxicity (*i.e.*, ratio of the number of bioassays exhibiting a calculable ecotoxic response (n) divided by the total number of bioassays (N) used). For explanations and formulae, see Section 5.6. The waste PEEP index value resulting from the integration of different bioassay responses is reflected by a $\log_{10}$ value that normally varies from 0 to 10. An interesting feature of this index is that it can accommodate any number and type of bioassays to fit particular needs.

Also, measurement endpoints such as inhibition concentration related to x% of effect (ICx) compared to a control, can be plotted versus time for each bioassay. Mathematical models for such "toxicity curves" could generate an infinite time Inhibition Concentration related to a x% of effect (∞ICx) for each bioassay. Such data, equivalent to the incipient lethal concentration defined for example by Giesy and Graney (1989), are valuable in terms of long-term environmental impact assessment.

After carrying out WASTOXHAS, decision-makers can choose to dispose solid wastes in the environment with (or without) treatment (for a review see Conner and Hoeffner, 1998) or continue waste hazard assessment with more sophisticated ecotoxicological approaches like microcosm (Pollard et al., 1999) or mesocosm (Propst et al., 1999) approaches taking into account the waste disposal scenario. A summary of the complete procedure is provided in Table 1.

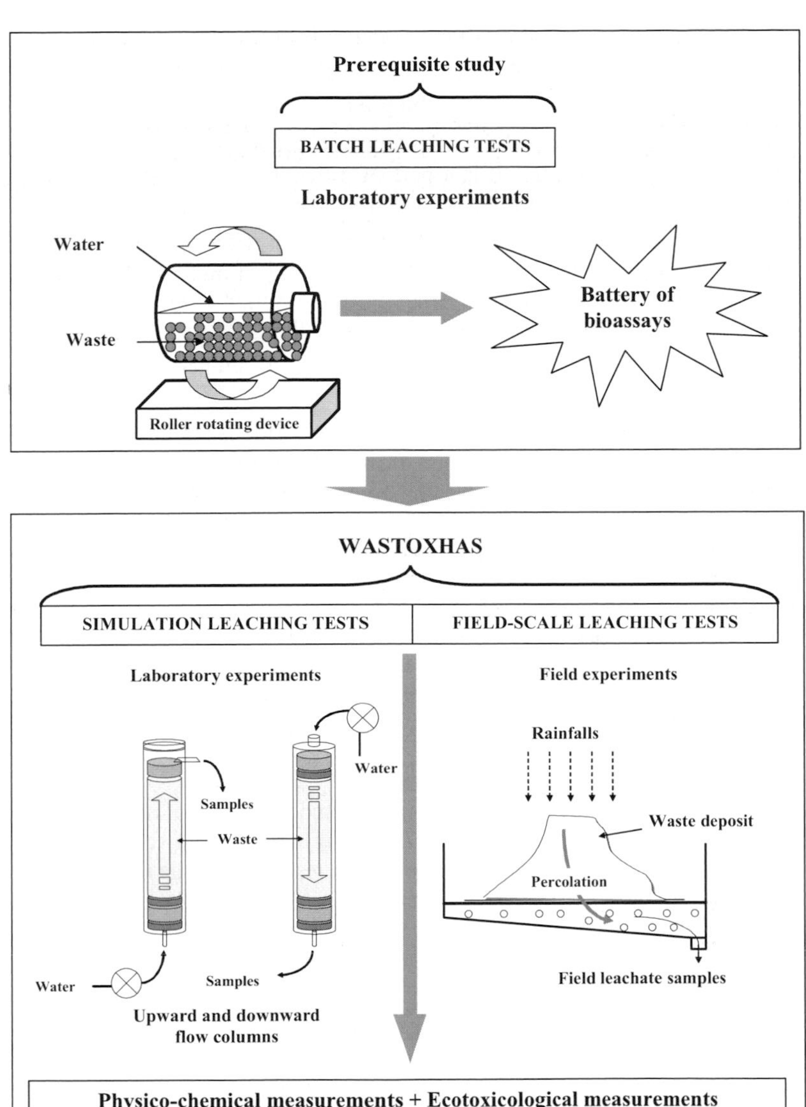

Figure 1. Place of WASTOXHAS in a general assessment of wastes.

Table 1. Summary table of WASTOXHAS.

Purpose
WASTOXHAS was developed for assessing leaching hazardous impact of wastes in laboratory and field situations. It is a part of a tiered approach (Fig. 1). It assumes that classical batch leaching tests (see for example Sahuquillo et al., 2003), followed by application of relevant bioassays, have been initially undertaken for i) deciding to continue ecotoxic hazard assessment and ii) selecting adequate and sensitive bioassays. It can be eventually followed by a more complex and elaborate hazard assessment scheme based on microcosms or mesocosms.
Principle
WASTOXHAS relies on the use of: 1- pilot-scale (or simulation) and large-scale (or field) leaching tests (see Fig. 1), chosen for their simplicity, practicality and standardization, 2- selected small-scale bioassays (see below and Section 5.4) conducted in parallel with physico-chemical measurements.
Bioassays employed
In principle, any battery of bioassays can be employed, but small-scale toxicity tests are preferred because of their performance output (Wells et al., 1998). It is highly desirable that bioassays used were part of the initial bioassay battery (prerequisite step) that proved to be sufficiently sensitive in the WASTOXHAS approach. Examples include the Microtox^{TM} light inhibition test (*Vibrio fischeri*) and the microalgal growth inhibition assay (*Selenastrum capricornutum**) that were found suitable for two tested wastes (see Section 7). *This alga, *S. capricornutum* (Printz), has undergone two recent taxonomical changes. It was first renamed *Raphidocelis subcapitata* and later became *Pseudokirchneriella subcapitata*. Following these taxonomical tribulations, the correct appellation is purported to be *Pseudokirchneriella subcapitata* (Korshikov) Hindak (for explanations see Chapter 3, volume 1 of this book).
Operational results
Measurement endpoints result in a waste PEEP index value (similar to the PEEP index presented in the Chapter 1 of this volume) resulting from the integration of different bioassay responses.
Notes of interest
WASTOXHAS needs essential prerequisites for its optimal operation, namely the initial application of a large bioassay battery conducted after classical small-scale batch shaking leaching test (for example EN 12457-1 to 4, 2002 or equivalent). The final choice of bioassays selected for the battery is not self-evident, but some examples of possible bioassays are given in Table 2 and some concepts are given in Table 3 and Section 3.4. Before undertaking some bioassays, waste pre-treatment may also be required (see Section 7).

Table 1 (continued). Summary table of WASTOXHAS.

Notes of interest (continued)
The waste PEEP index formula can be employed with any appropriate number and type of tests depending on laboratory expertise and means (any bioassay can be replaced by another or added to the proposed list). In theory, waste PEEP index values can vary from 0 to infinity. In practice, it has been shown to produce values ranging from 0 to 10, thereby simulating a readily-understandable "waste scale" indicative of ecotoxic impact. WASTOXHAS is straightforward to apply but requires one trained technician during two months to carry out a full assessment on one or two waste samples.
Applications of WASTOXHAS
WASTOXHAS has been applied to bottom ash from municipal solid waste incineration (MSWI) and matured slags from the second smelting of lead (Ferrari and Férard, 1999; Ferrari, 2000).

3. Historical overview and applications reported with the WASTOXHAS procedure

3.1 A BRIEF HISTORY OF WASTES

Human activity and civilizations have always left behind some kind of waste. But wastes are not necessarily "evil". For example, beneath our modern towns are buried layers, including building foundations and huge quantities of domestic rubbish and waste, that can give information about life in the past (*e.g.*, McCorriston and Weisberg, 2002). In this context, wastes reflect each type of civilization, particularly its degree of social, artistic, and technical development. Moreover, archaeologists can study the food people ate, the illnesses they suffered and died from, the tools they made and used, the goods they traded, the coinage they spent, and the buildings they lived and worshipped in.

Dumping has been the most popular means of garbage disposal from prehistory to the present. In prehistoric times, men left waste where they ate. In the Bronze Age, inhabitants of Troy (approximately 3000 to 1000 BC) simply covered trash in their homes with layers of dirt or clay. As more and more people began to live in cities, the problem of waste disposal grew acute. It seems that the first garbage removal management appeared in antic Greece. In the Middle Ages, streets and alleys were often filled with garbage, and rain would turn them into open sewers. In France, King Philippe Auguste ordered to pave, in 1185, Paris's streets to avoid mud and odors! But it was only in 1870 that it was strictly forbidden to dump wastewater, urine, faeces and other garbage out of windows. In 1884, a city prefect named Eugène Poubelle demanded that landlords provide containers for such refuse. They responded by naming them “poubelle” in his honor. The word has stuck and is now used by French people to signify a garbage bin.

3.2 WHERE ARE WE NOW?

Nowadays the nature of solid waste has evolved. Every home contains hazardous products, or products that can harm human health or the environment if improperly handled. They are potential sources of hazardous contaminants in municipal solid waste (MSW). As an example, Americans generate 1.6 million tons of household hazardous waste (HHW) per year (http://www.epa.gov/epaoswer/non-hw/househld/hhw.htm). Compared to the MSW waste produced by U.S. residents, businesses, and institutions before recycling (http://www.epa.gov/epaoswer/non-hw/muncpl/facts.htm), HHW is around 0.7 % of MSW.

Industry is by far the largest source of hazardous wastes. It is assumed that 10-15 % of wastes produced by industry in most developing countries are likely to be hazardous (Chaaban, 2001). Over five million tons of hazardous wastes were produced in England and Wales in 2000 (http://www.environment-agency.gov.uk/yourenv/eff/ resources_waste/213982/203145/?lang=_e) and in 1996, the U.S. EPA (http://www.enviroliteracy.org/article.php/56.html) reported that 279 million tons of hazardous wastes were generated annually.

In today's world, generation, storage, treatment, transport, recovery, transboundary movement, and disposal of hazardous wastes pose formidable problems for society and represent a serious threat for human health and the environment. Great concern exists for the future if this issue is not properly addressed (Rummel-Bulska, 1993) and solid waste production management is clearly a necessity. Different institutions, in fact, have now developed Integrated Solid Waste Management Plans (ISWMP). For instance, the U.S. Army has developed a technical guide (TG 197), which is an interesting framework document.

Moreover, waste management is now moving from a "ways of dealing-based approach" (disposal, incineration and/or treatment of wastes) to an "objectives-based approach" (waste reduction, re-use, recycling, valorization, stabilization or solidification, vitrification, risk assessment, ecocompatibility). In Agenda 21 (http://www.un.org/esa/sustdev/documents/agenda21/index.htm), for example, the framework for required action is based on a hierarchy of objectives and focused on the four following major waste-related program areas:

- Minimizing wastes.
- Maximizing environmentally sound waste reuse and recycling.
- Promoting environmentally sound waste disposal and treatment.
- Extending waste service coverage.

3.3 DEFINITION/CLASSIFICATION OF HAZARDOUS WASTES

Different definitions of hazardous wastes exist in the literature (see for example Chaaban, 2001). The United Nations Environment Program (UNEP) hazardous wastes definition is as follows: "*Wastes other than radioactive wastes which, by reason of their chemical activity or toxic, explosive, corrosive or other characteristics causing danger or likely to cause danger to health of the environment, whether alone or coming into contact with other wastes, are legally defined hazardous in the state in which they are generated or in which they are disposed of or through which they are transported*". According to the U.S. EPA,

hazardous wastes usually have one or more of the 4 following characteristics (http://www.safety.vanderbilt.edu/resources/hazard_charwaste.htm):
- Ignitability refers to wastes that can readily catch fire and sustain combustion. Many paints, cleaners, and other industrial wastes pose such a hazard.
- Corrosiveness concerns wastes that are acidic or alkaline and can readily corrode or dissolve flesh, metal, or other materials. An example is waste sulfuric acid from automotive batteries.
- Reactivity relates to wastes that readily explode or undergo violent reactions. Common examples are discarded munitions or explosives.
- Toxicity ascribes to hazardous waste disposed of in a land disposal unit (at the best) and refers to toxic compounds or elements, which can leach into underground drinking water supplies and expose water users to hazardous chemicals and constituents.

In Europe, the hazardous character of wastes depends on 14 criteria (H1 to H14) distributed among 4 types: H1 to H3 = physical hazard; H4 to H12 = hazard for human health; H13 = hazard following elimination of waste; H14 = environmental hazard.

3.4 REVIEW OF THE BATTERY OF BIOASSAYS USED FOR EVALUATING ECOTOXICITY OF SOLID WASTES

The ecotoxicity of wastes has to be evaluated after application of bioassays to raw wastes and to their leachates (French Ministry of Environment, 1998). As emphasized several times in this book, bioassays give a direct and comprehensive estimate of environmental toxicity. When confronted with complex mixtures of chemicals, responses of biological tests integrate different factors such as antagonism, synergism, and bioavailability of pollutants.

In undertaking our search of the literature linked to bioanalytical assessment of solid waste leachates (Tab. 2), we circumscribed it to small-scale toxicity testing performed on leachates. Furthermore, we did not exclude marine bioassays, but we exclusively selected literature references involving test battery approaches (TBAs) on solid wastes (or their elutriates). As defined previously in the first chapter of this book, a TBA represents a study conducted with two or more tests representing at least two biotic levels. As also pointed out in Section 2 of this chapter, TBAs are suitable to assess hazard at different levels so as not to underestimate ecotoxicity. Nevertheless, we have not excluded from this review publications describing other types of bioassays (*e.g.*, terrestrial bioassays, sub-cellular bioassays or those carried out with recombinant DNA (micro)organisms and biosensors), when those were part of the TBA.

It has to be noted that this review was undertaken to illustrate the large number of bioassays that have been used for waste assessment (45 different species and more than 45 tests if different test methods dedicated for each species are taken in account) and to give some general remarks on the 37 different test batteries presented in Table 2. Although a selected test battery is proposed in Section 5.4, a unique test battery cannot be recommended because each waste is specific in its composition and may thus require the use of a particular test battery.

Table 2. Application of bioassays to assess ecotoxicity of (solid) waste leachates: test batteries are listed in chronological order.

Number and type of bioassays employed	**Waste origin / *leaching procedure specifics***
Atwater et al., 1983	
Three bioassays: a) 48h and 96h acute crustacean test (*Daphnia pulex*) b) 96h acute fish test (*Oncorhynchus mykiss* and *Oncorhynchus nerka*) c) residual oxygen fish test *(Oncorhynchus mykiss and Oncorhynchus nerka)*	Municipal solid waste (MSW) leachates originating from: a) Landfills b) Laboratory lysimeters (downward) c) Field lysimeters (downward) *Unspecified pretreatment*
Plotkin and Ram, 1984	
Four bioassays : a) 5 min acute bacterial test *(Vibrio fischeri)* b) 11 to 21d algal test *(Pseudokirchneriella subcapitata)* c) 48h acute crustacean test *(Daphnia magna)* d) 96h acute fish test (*Pimephales promelas)*	Municipal and industrial solid waste leachates originating from a sanitary landfill *Filtered (glass fiber and 0.45 μm membrane filters for algal test) and unfiltered*
Calleja et al., 1986	
Two bioassays : a) 5, 15 and 30 min acute bacterial test *(Vibrio fischeri)* b) 24h acute crustacean test *(Daphnia magna)*	Leachates from pesticide industry and from electroplating sludges (metals) *12 different extraction procedures*
Mezzanotte et al., 1988	
Two bioassays : a) 48h plate incorporation AMES test with *Salmonella typhimurium* his- (TA 98, 100, 1535, 1537 et 1538) b) 24h acute crustacean test *(Daphnia magna)*	MSW landfill leachates; Ashes and slags from MSW incinerator leachates *Centrifuged and 0.45 μm filtered only for AMES test*
Peterson et al., 1990	
Four bioassays : a) 30 min acute bacterial test *(Vibrio fischeri)* b) 96h algal test *(Pseudokirchneriella subcapitata)* c) 120h lettuce seed root elongation test (*Lactuca sativa*) d) 24h acute crustacean test *(Daphnia magna)*	Municipal and industrial solid waste Leachates originating from landfills Simulated wastes *Two different extraction procedures* *pH adjusted according to tolerance of organisms* *Centrifuged and 0.45 μm or glass fiber filtered*

Table 2 (continued). Application of bioassays to assess ecotoxicity of (solid) waste leachates: test batteries are listed in chronological order.

Number and type of bioassays employed	**Waste origin /** ***leaching procedure specifics***
Day et al., 1993	
Nine bioassays : a) 5 min electron transfer test (Beef Heart mitochondria) b) 15 min acute bacterial test *(Vibrio fischeri)* c) acute bacterial test *(Spirillum volutans)* d) SOS Chromotest (*Escherichia coli PQ 37*) e) TOXI Chromotest (*Escherichia coli PQ 37*) f) mortality/mutagenicity worm test (*Panagrellus redivivus*) g) 48h acute crustacean test *(Daphnia magna)* h) 96h acute fish test (*Pimephales promelas*) i) 96h acute fish test (*Oncorhynchus mykiss)*	Leachates of immersed whole tires *Unspecified pretreatment*
Hamilton et al., 1993	
Two bioassays : a) 7d chronic shrimp test *(Mysidopsis bahia)* b) 7d chronic fish test *(Menidia beryllina)*	Waste-to-energy (WTE) ash concrete leachates *Sea-water leaching with stabilized wastes; sea-water elutriation with crushed wastes; decantation and/or filtration*
Schrab et al., 1993	
Four bioassays: a) 5, 15 and 30 min acute bacterial test *(Vibrio fischeri)* b) 48h plate incorporation AMES test (TA98 *Salmonella typhimurium* his-) c) chromosomal aberration fungus assay (*Aspergillus nidulans*) d) 24h DNA repair bacterial assay (*Bacillus subtilis*)	MSW landfill leachates *Organic extraction*
Devare and Bahadir, 1994a	
Four bioassays : a) 30 min acute bacterial test *(Vibrio fischeri)* b) 8d growth rate aquatic plant test (*Lemna minor*) c) 72h germination plant test (*Lepidium sativum*) and 8d root elongation plant test (*Lepidium sativum*) d) 72h germination plant test (*Brassica rapa*) and 8d root elongation plant test (*Brassica rapa*)	Municipal and industrial solid waste leachates originating from landfills *Unspecified pretreatment*

Table 2 (continued). Application of bioassays to assess ecotoxicity of (solid) waste leachates: test batteries are listed in chronological order.

Number and type of bioassays employed	**Waste origin / *leaching procedure specifics***
Devare and Bahadir, 1994b	
Three bioassays : a) 8d growth rate aquatic plant test (*Lemna minor*) b) 72h germination plant test (*Lepidium sativum*) and 8d root elongation plant test (*Lepidium sativum*) c) 72h germination plant test (*Brassica rapa*) and 8d root elongation plant test (*Brassica rapa*)	Different industrial solid waste leachates *Saline extraction*
Ernst et al., 1994	
Three bioassays : a) 48h acute crustacean test (*Daphnia magna*) b) 96h acute fish test (*Oncorhynchus mykiss*) c) 96h acute fish test (*Salvelinus fontinalis*)	Municipal solid waste landfill leachates *Unspecified pretreatment*
Hjelmar et al., 1994	
Three bioassays : a) 48h acute rotifer test (*Brachionus plicatilis*) b) 120d chronic sea anemone test (*Aiptasia pallida*) c) Acute (48h) and chronic (19d) marine copepod test (*Acartia tonsa*)	MSWI bottom and fly ash leachates *Column and batch leaching tests with ocean water and acidic artificial rain*
Kampke-Thiel et al., 1994	
Four bioassays : a) Sapromat (non identified bacteria coming from a municipal sewage treatment plant) b) 30 min acute bacterial test *(Vibrio fischeri)* c) 72h algal test *(Pseudokirchneriella subcapitata)* d) 24h acute crustacean test *(Daphnia magna)*	Different industrial solid waste leachates *Saline extraction*
Lambolez et al., 1994	
Four bioassays : a) 30 min acute bacterial test *(Vibrio fischeri)* b) 72h algal test *(Pseudokirchneriella subcapitata)* c) Acute (24h) and chronic (28d) crustacean test *(Daphnia magna)* d) 48h plate incorporation AMES test *(Salmonella typhimurium* his-with TA 97a, 98, 100, 102)	Industrial solid waste leachates *Batch leaching test with demineralized water followed by paper filtration (crude leachate), Liquid/liquid extraction (organic extract), lyophilization (lyophilized extract.* *pH adjusted according to tolerance of organisms and 0.22 µm filtration for AMES test*
Griest et al., 1995	
Two bioassays : a) 48h plate incorporation AMES test *(Salmonella typhimurium* his- (TA 98, 100) b) 7d chronic crustacean test (*Ceriodaphnia dubia*)	Windrow composts from explosives-contaminated sediment leachates *Aqueous and organic extraction*

Table 2 (continued). Application of bioassays to assess ecotoxicity of (solid) waste leachates: test batteries are listed in chronological order.

Number and type of bioassays employed	**Waste origin /** ***leaching procedure specifics***
Nimmo et al., 1995	
Four bioassays : a) 120h lettuce germination test (*Lactuca sativa*) b) 48h acute crustacean test (*Ceriodaphnia dubia*) c) 96h acute amphipod test (*Hyalella azteca*) d) 96h acute fish test (*Pimephales promelas*)	Waters collected from wells dug in an urban landfill, sediments collected in an adjacent river *Unspecified pretreatment*
Ortiz et al., 1995	
Two bioassays: a) 15 min acute bacterial test *(Vibrio fischeri)* b) 24h acute crustacean test *(Daphnia magna)*	Leachates from solid waste sludges generated from metal finishing wastewater treatment processes *2 different extraction procedures* *Filtration (not precisely defined)*
Clément et al.,1996	
Eight bioassays : a) 30 min acute bacterial test *(Vibrio fischeri)* b) 120h algal test *(Pseudokirchneriella subcapitata)* c) 120h growth rate aquatic plant test (*Lemna minor*) d) 24h acute ciliate test (*Spirostomum ambiguum*) e) 24h acute rotifer test (*Brachionus calyciflorus*) f) 24h acute crustacean test *(Thamnocephalus platyurus)* g) 7d chronic crustacean test (*Ceriodaphnia dubia*) h) 24h acute crustacean test *(Daphnia magna)*	Domestic, pure industrial and mixed landfill leachates *Decantation*
Font et al., 1998	
Two bioassays : a) 15 min acute bacterial test *(Vibrio fischeri)* b) 24h and 48h acute crustacean test *(Daphnia magna)*	Tannery waste leachates *0.45 μm filtration* *pH adjusted between 6 and 8 (for bacterial test)*
Hartwell et al., 1998	
Two bioassays : a) acute (96h) and short-term (7d) shrimp test *(Palaemonetes pugio)* b) acute (96h) and short-term (7d) fish test *(Cyprinodon variegatus)*	Scrap tire leachates *Saline extraction* *37 μm and 10 μm mesh screen filtrations followed by decantation*
Kahru et al., 1998	
Three bioassays : a) 15 min acute bacterial test *(Vibrio fischeri)* b) 24h acute rotifer test (*Brachionus calyciflorus*) c) 48h acute crustacean test *(Daphnia magna)*	Ash-heap leachates from oil shale industrial dumps *Paper filtration*

Table 2 (continued). Application of bioassays to assess ecotoxicity of (solid) waste leachates: test batteries are listed in chronological order.

Number and type of bioassays employed	Waste origin / *leaching procedure specifics*
Miadokova et al., 1998	
Two bioassays : a) 48h plate incorporation AMES test (TA 97, 98, 100 and 102 *Salmonella typhimurium* his-) b) 72h chromosomal aberration test (*Vicia sativa*)	Waste acid mine drainage waters *Unspecified pretreatment*
Rojickova-Padrtova et al., 1998	
Eight bioassays : a) 15 min acute bacterial test *(Vibrio fischeri)* b) 96h microplate and flask algal test *(Pseudokirchneriella subcapitata)* c) 24h acute ciliate test (*Spirostomum ambiguum*) d) 24h acute Rotoxkit test (*Brachionus calyciflorus*) e) 24h Thamnotoxkit test *(Thamnocephalus platyurus)* f) 24h Ceriodaphtoxkit (*Ceriodaphnia dubia*) g) 24h acute crustacean test *(Daphnia magna)* h) 48h acute fish test *(Poecilia reticulata)*	Foundry dust and fly-ash leachates *Paper filtration followed by 1.5 μm filtration*
Ferrari et al., 1999	
Five bioassays : a) 60 min acute bacterial test *(Vibrio fischeri)* b) 72h algal test *(Pseudokirchneriella subcapitata)* c) 7d chronic crustacean test (*Ceriodaphnia dubia*) d) 21d chronic crustacean test *(Daphnia magna)* e) 10d germination and biomass plant test *(Avena sativa, Brassica campestris and Lactuca sativa* and 4 stress oxidant enzyme activities (Glutathion reductase, Superoxide dismutase, Peroxidase, Catalase)	MSWI bottom ash leachates *Paper filtration followed by pH adjustment or not*
Czerniawska-Kusza and Ebis, 2000	
Four bioassays : a) 72h Algaltoxkit *(Pseudokirchneriella subcapitata)* b) 24h Protoxkit test (*Tetrahymena thermophila*) c) 24h Thamnotoxkit test *(Thamnocephalus platyurus)* d) 24-48h Daphtoxkit *(Daphnia magna)*	Leachates from the drainage system of municipal waste dumps *Unspecified pretreatment*

Table 2 (continued). Application of bioassays to assess ecotoxicity of (solid) waste leachates: test batteries are listed in chronological order.

Number and type of bioassays employed	**Waste origin / *leaching procedure specifics***
	Joutti et al., 2000
Seven bioassays : a) 20 min electron transfer test (Beef Heart mitochondria) b) 24h genotoxicity Mutatox™ test *(Vibrio fischeri* mutant) c) 6h Toxi-Chromotest (*Escherichia coli PQ 37*) d) 4-5 h MetPLATE bacterial test (*Escherichia coli* mutant) e) 4-5h Met PAD bacterial test (*Escherichia coli* mutant) f) 3-5d growth rate duckweed test (*Lemna minor*) g) 4-7d germination plant test (*Hordeum vulgare, Spinacia oleracea, Trifolium pratense*)	Leachates of different solid industrial wastes *pH adjustment (range 5-8) followed by 0.2 μm filtration*
	Latif and Zach, 2000
Six bioassays : a) 30 min acute bacterial test *(Vibrio fischeri)* b) 72h Algaltoxkit *(Pseudokirchneriella subcapitata)* c) 72h algal test *(Pseudokirchneriella subcapitata)* d) 24h Daphtoxkit *(Daphnia magna)* e) 24h acute crustacean test *(Daphnia magna)* f) 48h root elongation plant test *(Lepidium sativum)*	Treated (mechanically and biologically) residual wastes leachates *Two successive centrifugations followed by 0.45 μm filtration (except for daphnid test)*
	Mala et al., 2000
Three bioassays : a) Toxichromopad (*Escherichia coli,* mutant) b) 72h Algaltoxkit *(Pseudokirchneriella subcapitata)* c) 24h Thamnotoxkit test *(Thamnocephalus platyurus)*	Cement, fly-ash and slag leachates *Acidic extraction and 7.8 pH adjustment*
	Vaajasaari et al., 2000
Four bioassays : a) 30 min acute bacterial test *(Vibrio fischeri)* b) 48h chronic bacterial test *(Pseudomonas putida)* c) 72h algal test *(Pseudokirchneriella subcapitata)* d) 48h acute crustacean test *(Daphnia magna)*	Leachates from forest and metal solid industry wastes *Six different extraction procedures 7.0 pH adjustment*

Table 2 (continued). Application of bioassays to assess ecotoxicity of (solid) waste leachates: test batteries are listed in chronological order.

Number and type of bioassays employed	**Waste origin** / ***leaching procedure specifics***
Sekkat et al., 2001	
Eight bioassays : a) 30 min acute bacterial test *(Vibrio fischeri)* b) 2h SOS chromotest (*Escherichia coli PQ 37*) c) 72h algal test *(Scenedesmus subspicatus)* c) 72h algal test *(Chlamydomonas reinhardtii)* d) biomass plant test (*Vicia lens*) e) acute ciliate test (*Colpidium campylum*) f) 24h acute brine shrimp test *(Artemia salina)* g) 24h acute crustacean test *(Daphnia magna)*	Urban and industrial wastes *Unspecified pretreatment*
Lapa et al., 2002a	
Three bioassays : a) 15-30 min acute bacterial test *(Vibrio fischeri)* b) 120h algal test *(Pseudokirchneriella subcapitata)* c) 7d germination plant test *(Lactuca sativa)*	Pyrolised/vitrified material leachates *0.45 µm filtration with or without pH adjustment*
Lapa et al., 2002b	
Four bioassays : a) 15-30 min acute bacterial test *(Vibrio fischeri)* b) 120h algal test *(Pseudokirchneriella subcapitata)* c) 7d germination plant test *(Lactuca sativa)* d) 48h acute crustacean test *(Daphnia magna)*	Leachates from MSWI bottom ashes *0.45 µm filtration with or without pH adjustment*
Schultz et al., 2002	
Five bioassays : a) electron transfer test (Beef heart mitochondria) b) 30 min acute bacterial test *(Vibrio fischeri)* c) ToxiChromopad (*Escherichia coli*, mutant) d) 4d root growth plant test (*Allium cepa*) e) 2d germination plant test *(Lactuca sativa)*	Furniture (varnishing and organic solvent contaminated) and resin industry waste leachates *pH adjustment (range 6-8)*
Ward et al., 2002	
Three bioassays : a) 15 min acute bacterial test *(Vibrio fischeri)* b) 96h algal test *(Pseudokirchneriella subcapitata)* c) 48h acute crustacean test (*Ceriodaphnia dubia*)	MSW landfill leachates *Glass fiber B filtration*

Table 2 (continued). Application of bioassays to assess ecotoxicity of (solid) waste leachates: test batteries are listed in chronological order.

Number and type of bioassays employed	Waste origin / *leaching procedure specifics*
Aït-Aïssa et al., 2003	
Four bioassays : a) 30 min acute bacterial test *(Vibrio fischeri)* b) 72h algal test *(Pseudokirchneriella subcapitata)* c) 7d chronic crustacean test (*Ceriodaphnia dubia*) d) 48h acute crustacean test *(Daphnia magna)* and *in vitro* induction stress protein test (*HELA-hsp-CAT cells*)	Different industrial waste leachates *Decantation, 100 µm filtration followed by 5.5-8.5 pH adjustment (for daphnid test), followed by 0.45 µm filtration (for other tests)*
Birkholz et al., 2003	
Seven bioassays : a) acute bacterial test *(Vibrio fischeri)* b) algal test *(Pseudokirchneriella subcapitata)* c) acute crustacean test *(Daphnia magna)* d) acute fish test *(Pimephales promelas)* e) SOS chromotest (*Escherichia coli PQ 37*) f) genotoxicity Mutatox™ test *(Vibrio fischeri mutant)* g) fluctuation AMES test *(Salmonella typhimurium* his- TA 98, 100, 1535, 1537)	Tire crumbs *1) Unspecified filtration of water extracts for the first four tests* *2) Soxhlet extraction for the three last tests*

During 20 years (1983-2003), 37 test batteries were applied to solid wastes, comprising 2 to 9 bioassays. Test batteries with four bioassays were most frequently used. The three most frequently used bioassays were the acute *Vibrio fischeri* test (26 times), the acute *Daphnia magna* test (22 times) and the chronic *Pseudokirchneriella subcapitata* test (16 times), confirming that acute tests are used more frequently than chronic ones. Plant as well as genotoxic (including mutagenic) tests are characterized by a large variety of species and methods. As concerns the Ames test, it is interesting to note that some authors did not observe mutagenic activity on leachates (*e.g.*, Lambolez et al., 1994), who nevertheless tested organic or lyophilised concentrated fractions to improve assay sensitivity. Moreover, most genotoxic substances are lipophilic and it is obvious that they are not recovered in the leachates considering the design of the leaching procedure.

A total of 45 different species were employed, but authors did not always specify their choice of species. Ideally, bioassays should have some basic characteristics, as defined by Giesy and Hoke (1989). An adequate battery of bioassays needs in principle to measure various types (acute, chronic, genotoxic) and levels (lethal, sublethal) of ecotoxicity, without any redundancy, with test species belonging to different trophic levels or characterized by different ecological and biological traits (Ducrot et al., 2005). Another important aspect in the selection of bioassays for a test battery is clearly linked to cost-effectiveness. Care must also be taken to avoid redundancy in toxic responses when using different bioassays in test batteries (Blanck, 1984). Some characteristics of an ideal test battery are highlighted in Table 3.

3.5 LEACHING TESTS

As stated by Sahuquillo et al. (2003), there is an increased use of widely different leaching tests. Again, selection of appropriate and cost-effective methods used to assess environmental impact must provide a basis for long-term prediction of hazard and ecocompatibility.

Table 3. Characteristics of an ideal test battery for solid waste toxicity assessment.

- Cost-effectiveness
- Avoidance of redundancy
- Waste discriminating potency
- Large (or selected) number of trophic levels, toxicity measurements, endpoints and exposure routes
- Composed of :
    - rapid tests
    - simple tests
    - reliable tests (*i.e.*, reproducibility of toxic responses)
    - standardized tests
    - sensitive tests
    - ecologically relevant tests (*i.e.*, related to field effects)
    - non-vertebrate tests

The rationale for choosing both upward- and downward-flow column leaching tests is based on the fact that both simulation tests are representative of different water-contact scenarios. The downward process reflects all kinds of precipitation and subsequent influent seepage (*i.e.*, the gravity movement of water in the zone of aeration from the ground surface toward the water table) of water through a non-saturated zone. Several authors have used downward-flow columns for different purposes (Kaschl et al., 2002; Núñez-Delgado et al., 2002; Chen et al., 2003; Hofstee et al., 2003; Sajwan et al., 2003). The upward process is meant to simulate flooding (*i.e.*, overflowing by water of the normal confines of a stream or other body of water), a rarer event, where water circulates through a saturated zone. Such a process is the object of the Dutch standard (NEN 7343, 1995). Upward flow simulation has been used, for example, on contaminated soils (Masfaraud et al., 1999).

The scientific basis for using simulation tests must ultimately depend on their degree of accuracy in predicting results of full-scale field tests. Simultaneous undertaking of laboratory simulation tests and field tests are therefore of interest for validation purposes, as ecological factors can clearly influence field situations.

3.6 APPLICATIONS REPORTED WITH THE WASTOXHAS PROCEDURE

So far, the WASTOXHAS procedure was applied to bottom ash from municipal solid waste incineration (BA) and matured slags from the second smelting of lead (2SL) (Ferrari and Férard, 1999). Detailed results are given in Section 7.

4. Advantages of applying the WASTOHAS procedure

WASTOXHAS is a valuable, cost-effective and sound framework for decision-makers that require relevant and precise waste hazard information. Minimizing cost and increasing effectiveness of the hazard assessment scheme have driven the choice of the simulation and field tests. Thus simplicity, practicality and predictive accuracy were important criteria for elaborating WASTOXHAS. It is an essential part of a decision tree illustrated in Fig. 2. Application of the WASTOXHAS concept can be beneficial for general protection of aquatic systems by providing environmental managers with relevant information. Indeed, ecotoxicological assessment of waste leachates simply performed with a test battery of small-scale tests can provide data on the ecocompatibility of wastes without having to perform more elaborate and costly assessment.

One of the most important concerns in solid waste management is the long-term behavior of such residues once disposed of (Crawford and Neretnieks, 1999). In this regard, WASTOXHAS appears to be a robust and reliable instrument for improving knowledge of the long-term fate and ecocompatibility of solid wastes. This scheme was part of a French research program designed to evaluate the ecocompatibility of two different wastes (BA and 2SL), as described in Section 7.

WASTOXHAS leaching tests are also interesting research tools offering possibilities for long-term predictions by studying the influence of various factors on leaching and subsequent toxicity endpoints. Obvious factors include the L/S ratio (*i.e.*, volume of liquid to mass of solid ratio), composition of leaching medium, sample preparation (*e.g.*, particle size) and disposition (*e.g.*, mode of compaction). Other factors to be considered, related to on-going research on different types of materials, are ageing of wastes (reviewed by Alexander, 2000), which can be studied by different temperature cycles, biodeterioration and biodegradation of solidified/stabilized materials (Gourdon et al., 1999; Knight et al., 1999), as well as air purging with CO_2 (Mizutani et al., 1999) and episodic (or intermittent) leaching events (Crane et al., 2001). All of these effects must be studied, by integrating both physico-chemical and ecotoxicological measurements.

5. WASTOXHAS procedure description

5.1 OBJECTIVE, PRINCIPLE AND SCOPE OF APPLICATION

As reported briefly at the beginning of this chapter, WASTOXHAS is an HAS approach for predicting leachate toxicity on biota of aquatic systems by any solid wastes. WASTOXHAS is part of a tiered approach for conducting long-term and ecocompatibility assessment of leachates produced by different leaching tests.

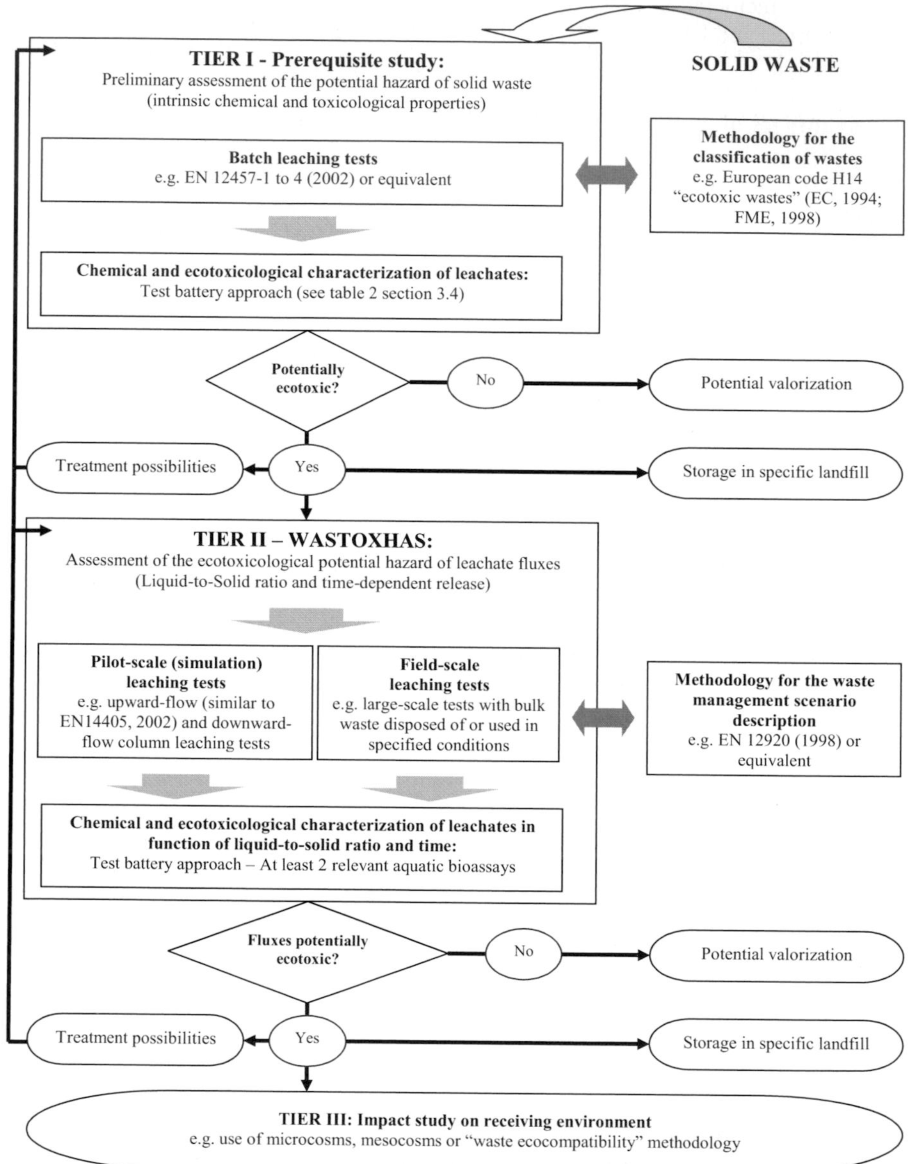

Figure 2. Decision scheme integrating the WASTOXHAS procedure with examples of current waste management guidelines (EC = European Communities; FME = French Ministry of Environment).

Batch leaching tests followed by standard physico-chemical and ecotoxicological measurements (normally part of any prerequisite study) may not be sufficient to adequately assess hazards for aquatic biota. Moreover, production of large amount of wastes, that are recycled at low cost, pleads for a more detailed evaluation. Indeed, 3-4 million metric tons of MSWI bottom ashes are produced in France each year from the incineration of MSW (Clozel-Leloup et al., 1999) and they are currently used in road construction (Crignon et al., 1999). Testing different MSWI bottom ashes with a four-test battery recently showed that leaching samples (obtained with L/S = 10) were classified as ecotoxic and hazardous according to both French guidelines and German regulations (Lapa et al., 2002b). In order to ensure that adverse effects toaquatic life does not arise from waste disposal, assessment of potential time-induced ecotoxic fluxes in both pilot (or simulation) and full-scale (or field) tests must be performed. Periodic (semi-continuous) sampling with time was chosen for simplicity followed by hazard assessment on samples taken with different L/S ratios.

While different kinds of leaching simulation tests (*e.g.*, involving columns, lysimeters) have been described in the literature (Kylefors et al., 2003), WASTOXHAS leaching methods were selected in order to:

- favor the use of standard leaching methods,
- study leaching processes in both saturated and non-saturated zones,
- apply simple laboratory tools as much as possible.

5.2 ENVIRONMENTAL MEDIUM APPRAISED BY THE HAS PROCEDURE

WASTOXHAS is designed to assess all solid wastes likely to contain potentially hazardous contaminants, such as landfilled, re-used and solidified/stabilized wastes. Besides solid wastes *per se*, this approach can also be applied to a wide range of other solid media including:

- contaminated soil or soil undergoing bioremediation treatment,
- deposited sewage sludge, compost or sediment,
- construction or demolition material,
- chemically-impregnated wood,
- miscellaneous products that are recycled via environmental disposal (rubber, plastic, metal products, etc.).

5.3 OTHER CONSIDERATIONS FOR PROPER WASTOXHAS APPLICATION

It is highly preferable to use standard methods for sampling and storage (Lapa et al., 2002b) and conditioning (EN 12457-1 to 4, 2002) of solid wastes. It is also important to know, before any leaching experiments are conducted, relevant properties of the waste, such as:

- the nature and origin of the material,
- its physical properties (density, porosity, permeability, particle size distribution),
- its morphology (granular or monolithic).

It is important that pre-treatment and preservation of the waste samples are in harmony with pre-requisite classical leaching batch tests (Tier I of Figure 2).

Within a period of a few days after sampling, wastes have to be submitted (or not) to a crushing procedure with the aim of obtaining fragmented material having a particle size lower than 4 mm (EN 12457/1-3) or 10 mm (EN 12457/4). There are specificities, however, and this step is necessary for BA type waste but not for SL type waste (see Section 7).

It is also necessary to determine the moisture content of waste (by drying a small portion of each sample at 105 ± 5°C until constant weight is reached) in order to define an L/S ratio on a dry weight basis.

It is also recommended to sample a sufficient quantity of waste so as to be able to undertake Tier I, II and III. Cost-effective conditions of storage must therefore be defined (for example, at ambient temperature, in darkness and inside well-closed containers) to prevent air contact prior to initiating analyses. Chemical and ecotoxicological stability of a waste sample should also be verified via time-sequential testing of sub-samples.

With respect to leachate pre-treatment, different methods have been reported in the literature involving steps such as decantation, centrifugation, filtration and pH adjustment. Specific pre-treatment may well have some influence on chemical or ecotoxicological results (Isidori et al., 2003). Moreover, some bioassays can only be run on 0.45 μm filtered leachates (*e.g.* Ames test) or those having undergone pH adjustment (*e.g.*, Microtox™ test).

Clearly, each type of solid waste is specific and may require a particular pre-treatment. Because of the complexity associated with waste characteristics, laboratory personnel must keep abreast of the scientific literature linked to this field and of the evolution of waste pre-treatment methodology.

5.4 TOXICITY TESTS AND ENDPOINTS EMPLOYED

As pointed out previously, any bioassay can be employed (see Section 2), but micro-scale tests are preferred because of their small sample volume requirements. This ensures a sufficient quantity of leachate for subsequent laboratory bioassays. Table 4 lists some basic features of five small-scale bioassays that can be used for WASTOXHAS applications.

The Microtox assay which measures light inhibition with the bacterium *Vibrio fisheri* is a well known and useful aquatic toxicity test (see Chapter 1, volume 1 of this book). As previously reported (Blaise et al., 1994) and based on our own experience, it appears more appropriate to determine 60 min IC50 for waste leachates, as opposed to 15 min or 30 min endpoints. IC50s measured after 60 min on MIOM leachates were clearly more sensitive and reproducible than those measured at 30 min and 15 min (Ferrari et al., 1999). Since WASTOXHAS was applied on (poly)metallic matrices in this study, we also found it more suitable to use zinc sulphate as a reference toxicant to periodically verify the sensitivity of the Microtox bacterial light reagent.

Table 4. Features of some small-scale bioassays useful for WASTOXHAS applications.

Common (or commercial) test name	***Organism***	***Reference of standardized protocol***	***Assessment endpoint***	***Type of toxicity***
(Microtox™ test)	*Vibrio fisheri*	ISO 11348-3 (1999)	Inhibition of bioluminescence	Acute
Microplate algal test	*Selenastrum capricornutum*	Environment Canada SPE 1/RM/25 (1992)	Inhibition of growth	Chronic
Microcrustacean test	*Daphnia magna*	ISO 6341 (1996)	Immobilisation	Acute
Microcrustacean test	*Ceriodaphnia dubia*	U.S. EPA (2002)	Inhibition of reproduction	Chronic
(Mutatox™)	*Vibrio fischeri* M169 mutant	No standard	Genotoxic effects	Chronic

The Microplate algal toxicity test, another popular small-scale bioassay, is equally employed for WASTOXHAS applications (see Chapter 3, volume 1 of this book). In our case, rapid endpoint determinations of growth (72h-IC50s) are made with a microplate fluorescence reader, where algal biomass is indirectly measured via chlorophyll *a* fluorescence (excitation filter: 440 nm – emission filter: 640 nm).

Another commonly used test is the *Daphnia* immobilization assay (see Chapter 10, volume 1 of this book). Results are expressed as 24 and 48h-EC 50s.

Chronic exposure micro-crustacean toxicity tests (7-d *Ceriodaphnia dubia* test; 21-d *Daphnia magna* test) are relevant as well for evaluation of waste leachates. The major differences between these two assays have been discussed elsewhere (Férard and Ferrari, 1997). For WASTOXHAS applications, we tend to favor the former over the latter for the following reasons:

(1) Daily renewal of medium guarantees less pH modification inside test vessels.

(2) Shorter test duration has less repercussion on potential (bio)transformation or degradation processes inside the leachate samples.

(3) Breeding conditions of adult ceriodaphnids are more precisely defined in the U.S. EPA standard than for daphnids in the ISO one. Moreover, individual cultures (between stock cultures and tests) generates juveniles in the same conditions as those required for the test;

(4) Quality and quantity of food is more precisely described in the U.S. EPA standard.

(5) No EDTA is added in the U.S. EPA standard, thereby curtailing any potential underestimation of metal toxicity.

Finally, the Mutatox™ test is generally less common in use, although it is a commercial assay with several interesting features. The Mutatox procedure has been employed for some 15 years after its initial publication made it known to the scientific community (Kwan et al. 1990). The Mutatox™ test, now commercialized by SDI (http://www.sdix.com/ProductSpecs.asp?nProductID=7), is based on the use of a dark variant (named M169) of the luminescent bacterium *Vibrio fischeri*. It is used to screen for genotoxic effects in aqueous samples. A very large range of genotoxic (primary DNA) damages can induce the recovery of luminescence. In the Mutatox assay, SOS system activation of bacteria leads to the formation of a protease that breaks down a repressor protein of the *lux*-pathway thereby leading to luminescence that serves as a measure for genotoxicity. Test exposure lasts for 16–24 h and enables different toxicity parameters to be reported, such as:

- LOEC: the Lowest Observed Effect Concentration where induced luminescence is at least twice higher than that of controls;
- HEC: the Highest Effect Concentration at which induced luminescence is highest;
- ECx: the Effective Concentration where there is an x% of induced luminescence increase calculated as a percentage of the difference between controls mean response and HEC (= 100 %).
- IF: Induction factors for any of the previous concentrations calculated as follows:

$$IF = \frac{\textit{Observed bioluminescence at one concentration (LOEC, HEC or ECx)}}{\textit{Mean control luminescence}} \qquad (1)$$

The test is routinely undertaken with and without metabolic activation (S9), which allows detection of direct genotoxic agents, and those requiring metabolic activation to express their genotoxic potency (indirect genotoxicants).

The procedure follows the supplier's protocol. Glass cuvettes are normally used, but light readings can also be performed in microplate wells after appropriate micropipette transfers. At the start of the test, 1:1 dilution series (ten successive dilutions) with test sample and Mutatox™ medium containing nutrients, salts and antibiotics are prepared. After addition of bacteria (rehydrated from a lyophilized powder), the mixture is incubated at 27°C. Light levels are recorded after 16, 20 and 24 h of exposure with a bioluminometer (Microbics M5000). Direct and indirect positive controls are respectively 9-aminoacridine (10 and 0.02 µg/mL) and benzo(a)pyrene (5 and 0.25 µg/mL). The latter also needs a DMSO control. With this procedure, a positive genotoxicity test result is recorded if:

(1) Light levels increase to at least twice the average (negative or solvent) light level of controls;

(2) The latter light level induction is observed in at least 2 successive sample concentrations.

Additionally, the test does not require strict aseptic conditions, and can be purchased as a test kit with all test materials included except the bioluminometer. Other advantages include small sample volume requirement (10 mL) and increased bioanalytical output (particularly when the assay is run in a microplate that can be subsequently placed in a light reader).

5.5 OTHER TYPES OF ANALYSES EMPLOYED WITHIN THE WASTOXHAS PROCEDURE

Along with the toxicity test battery, physico-chemical analysis of waste and each leachate produced either in the prerequisite study (*i.e.* Tier I in Figure 2) or in the WASTOXHAS procedure are useful to understand the main processes that can influence release (and rate of release) of pollutants from a solid matrix. In this sense, ecotoxicological and physico-chemical approaches are complementary to ensure a sound and reliable assessment of the potential environmental impacts of solid wastes.

All test methods adopted or under development by national or international standard organizations (*e.g.*, OECD, ISO...) are suitable for WASTOXHAS. However, the choice of endpoints to be evaluated, among a large number now available, and their quantification methods, are directly related to criteria defining limit values. Examples of different chemical limits can be found in Lapa et al. (2002b).

5.6 FORMULAE/FLOWCHARTS ASSOCIATED WITH THE WASTOXHAS PROCEDURE

Results obtained with bioassays on each leachate sample in the prerequisite study or in the WASTOXHAS procedure can be integrated through a waste toxicity scale system indicative of a specific level of ecotoxic impact. The aim of such a system is to convert individual endpoint values of different tests into a unique hazard index, representing the overall toxicity of the tested leachate.

This index is built on the model of the PEEP (Potential Ecotoxic Effects Probe) index developed by Costan et al. (1993), which is largely described in Chapter 1 of this volume. Briefly, at the time of its conception, the PEEP index was developed to compare, and classify hazard of effluents, within a set group discharging to a common aquatic system, by integrating the results of selected small-scale screening bioassays (acute, chronic and genotoxicity tests), while also considering biodegradation and effluent flow.

To facilitate the integration of measurement endpoints from different bioassays into a single hazard index value, data need to be expressed on the same scale of measurement. Therefore, prior to calculating each effluent PEEP index value, the measurement endpoint of each bioassay is converted to toxic units (TU), by means of the following equation:

$$TU = \frac{1}{MEV} \times 100 \qquad (2)$$

where MEV is the measurement endpoint value determined with each bioassay. TU corresponds to the dilution ratio of tested leachate producing a given effect: the higher the value, the higher the toxicity. Afterwards, each PEEP index value is calculated with the following formula:

$$PEEP = \log 10\left[1 + n\left(\frac{\sum_{i=1}^{N} Ti}{N}\right)Q\right] \quad (3)$$

Where n is the number of bioassays exhibiting calculable (geno)toxic response, N is the total number of bioassays carried out, Ti is (geno)toxicity expressed in TU of each test before or after samples have been submitted to a biodegradation procedure and Q is the effluent flow. PEEP values reflect an index varying from 0 *ad infinitum* in theory, but which in practice does not exceed 10 for effluents.

Lambolez (1994) and Bispo (1998) showed that this kind of index can be used for integrating different results obtained from solid waste leachates without having to take into account the MEVs obtained after biodegradation nor the flow (Q in the PEEP formula) specific to effluents. The Waste PEEP formula defining a PEEP index for waste then becomes:

$$WastePEEP = \log 10\left[1 + n\left(\frac{\sum_{i=1}^{N} Ti}{N}\right)\right] \quad (4)$$

Where n is the number of bioassays exhibiting calculable (geno)toxicity responses, N is the total number of bioassays carried out, Ti is (geno)toxicity expressed in TU of each test. In quantifying the toxicity of leachates produced in the prerequisite study or in the WASTOXHAS procedure, the waste PEEP index allows clear identification of the most problematic wastes requiring priority in terms of clean-up action or attention.

5.7 STATISTICS/CALCULATIONS/EXAMPLE OF DATA ASSOCIATED WITH THE HAS PROCEDURE

In the WASTOXHAS procedure, ecotoxicity testing of leachate samples obtained at different liquid-to-solid ratios (or at different times of release) aims at measuring effects on species representing various levels of biological organization (see Section 5.4) as a function of dilution rate while controls without leachate are used as reference. In order to express results in a synthetic form, raw data obtained from concentration-response curves are transformed into a summary criterion corresponding to a specific measurement endpoint (*e.g.,* EC_{50}, EC_x, NOEC, LOEC, etc.) for each test (Fig. 3).

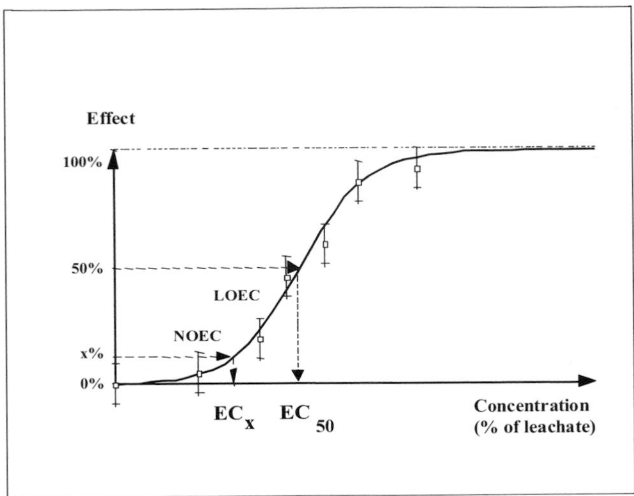

Figure 3. Example of a typical concentration-response curve obtained from a bioassay and the associated measurement endpoints (NOEC, LOEC and ECx) calculated by statistical methods.

Although based on the same statistical grounds, two types of statistical methods allow calculating different measurement endpoints (Isnard et al., 2001):

(1) ANOVA-type data analyses (or hypothesis testing) are used to generate NOEC and LOEC from an assessment endpoint (*e.g.* growth, reproduction, mortality, etc…).

(2) Regression methods (linear or non-linear) are used to determine an EC value that produces a specific percent reduction (*e.g.* 10, 20 or 50%) in an assessment endpoint.

Among the large set of available hypothesis tests and regression models that can be suitable to calculate different assessment endpoints, an example of flowchart that can be used to guide statistical analysis of the *C. dubia* reproduction assay is presented in Figure 4.

Following this flowchart, the LOEC, which differed significantly ($p < 0.05$) from control, can be determined by Dunnett's test after verifying Shapiro-Wilk's test for normality and Bartlett's test for homogeneity of variance. If unequal numbers of replicates occurred among the concentration levels tested, a *t*-test with Bonferroni adjustment must be used. Then, the NOEC, which corresponds to the next lowest concentration in relation to the LOEC, can be deduced. The principle of hypothesis tests is adequately described in the standard U.S. EPA 821/R-02/013 (2002) and all statistical calculations can be easily performed using adapted software in statistical analysis. However, readers must be informed that a commercial software package called TOXSTAT™ including all procedures used for estimating NOEC/LOEC endpoints can be purchased from Western Ecosystem Technology, Inc. (address via http://www.west-inc.com).

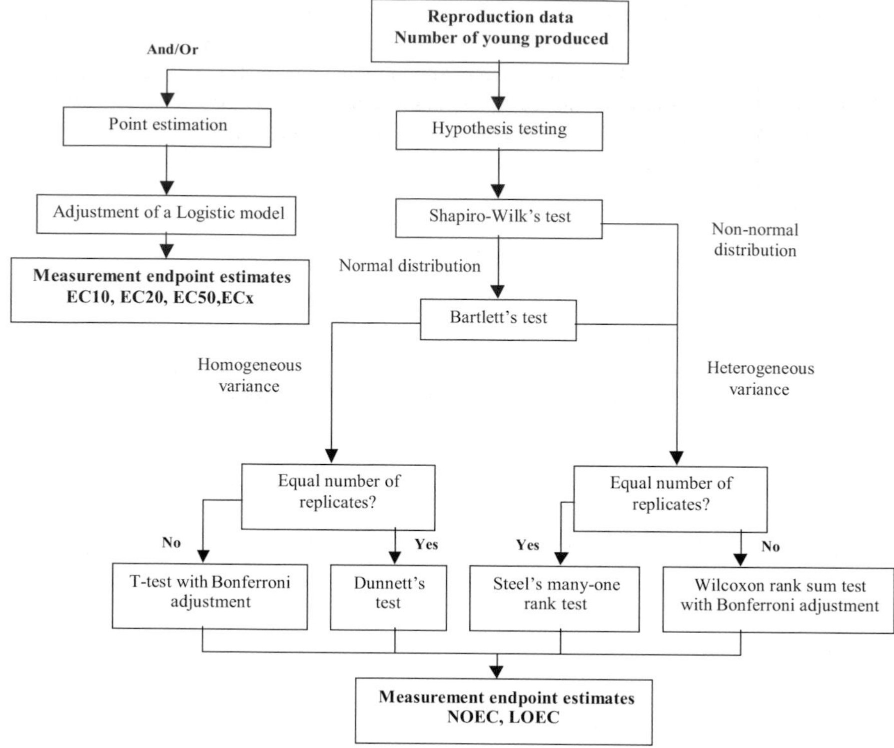

Figure 4. Flowchart for statistical analysis of Ceriodaphnia dubia reproduction assay data (adapted from the standard U.S. EPA 821/R-02/013, 2002).

In contrast, an ECx that produces a specific percent reduction (*e.g.*, x = 10, 20 or 50%) in ceriodaphnid reproduction can be calculated by adjusting a logistic model derived from the Hill equation to the test results (Vindimian et al., 1983). This model is characterized by the following equation:

$$Y = \frac{Y\max}{1 + \left(\frac{C}{EC50}\right)^{Hn}} \tag{5}$$

Where Y is the observed total number of live young ceriodaphnids per replicate, C is the concentration being tested, Ymax is the adjusted value of live young ceriodaphnids expected in the control, EC50 is the estimated concentration which causes 50% of reproduction inhibition and Hn is the estimated Hill number corresponding to the slope of the sigmoid curve. Then, the ECx can be estimated by an equation directly derived from the logistic model, as follows:

$$ECx = EC50 \times \left(\frac{x}{100-x}\right)^{\frac{1}{Hn}} \quad (6)$$

Where x corresponds to the x% level of effect compared to the control. Estimation of each parameter of such a logistic model and their confidence intervals associated can be performed with any statistical analysis software. Calculations can be programmed into Microsoft Excel® solver. An Excel® Macro called REGTOX has been elaborated and can be downloaded from the Internet (http://eric.vindimian.9online.fr/en_index.html). REGTOX is freely distributed with the moral obligation to cite its source whenever its use results in publications. The advantage of REGTOX is that other regression models, such as Weibull and Log-Normal models, can also be used. In addition to REGTOX, TOXSTAT™ also includes also a regression method, which is the ICp method based on a linear interpolation of means.

6. Factors capable of influencing WASTOXHAS interpretation potential

As stated by Van der Sloot (1998), several factors can influence the release of contaminants from both granular and monolithic materials. These include major element chemistry, pH, redox status of the system, presence of complexants, humic substances or other dissolved organic compounds, liquid to solid ratio, and biological activity.

For interpretation issues related to toxicity testing, general caveats are mentioned in several chapters of this book or in Environment Canada (1999). Attention must also be paid to atmospheric deposition (wash-out or fall-out) in field tests. It is recommended to implement such tests in the vicinity of ambient air monitoring stations. If this is not possible, passive or active biomonitoring could be carried out (Fernández et al., 2000).

7. Application of the WASTOXHAS procedure in a case study

The WASTOXHAS procedure was applied in a case study involving two kinds of solid waste, a municipal solid waste incinerator Bottom Ash (BA) and a slag from a second Smelting of Lead (2SL). This case study describes the ecotoxicological portion of a multidisciplinary French national research program on the "Waste

Ecocompatibility" concept (Perrodin et al., 1996) that the authors of this chapter were entrusted with. This integrated program sought to define a reliable methodology for assessing a situation where pollutant flux from waste, either disposed of or reused under specific conditions (physical, hydrogeological, chemical and biological conditions) would be ecocompatible with a receiving ecosystem (Grelier-Volatier et al., 2002; Perrodin et al., 2002). In presenting this case study, we focus only on some of the ecotoxicological results obtained with aquatic bioassays and the minimum physico-chemical parameters required to illustrate the WASTOXHAS procedure.

Prior to applying the WASTOXHAS approach, a prerequisite study is necessary that consists in testing leachate obtained from standardized batch leaching test with a large battery of bioassays. For this purpose, samples BA and 2SL were leached according to part 1 of the draft European standard EN 12457 (2002) using a liquid-to-solid ratio (L/S) of 2, and according to part 2 using a L/S ratio of 10. The leachates were then appraised with a battery of bioassays consisting of the 60-min Microtox™ test, the 72-h green alga, *P. subcapitata*, growth inhibition test, the 48-h daphnid, *D. magna*, immobilization test, the 7-d daphnid, *C. dubia*, reproduction inhibition test and the 24-h Mutatox™ test (see Section 5.4).

Before performing the leaching procedure, the two wastes were pre-treated. Briefly, the samples of bulk BA were previously submitted to a crushing procedure with the aim of obtaining fragmented material with a particle size lower than 4 mm, as required by the leaching procedure. Because samples of waste 2SL consisted of a powder in which particle size was lower than 4 mm, no crushing treatment was applied. The moisture content was also determined for both wastes by drying a small portion of each sample at 105 ± 5°C, until constant weight was reached. Values obtained were then taken into account for adjustment of the L/S ratio expressed in dry-weight of waste in the leaching procedure.

Afterwards, a portion of each pre-treated waste sample was submitted to the leaching methodology while the other part was stored at ambient temperature inside well-closed containers to prevent air contact prior to use for laboratory experiments called for in the WASTOXHAS procedure (see Fig. 1). For the leaching procedure, sub-samples prepared from each pre-treated sample were brought into contact with demineralized water in the defined L/S ratio for a 24-h duration under constant agitation at 20 ± 2°C. Leaching took place in capped 1L polyethylene bottles and the extraction process was performed in a roller-rotating device working at 100 rpm. After 24 h of leaching, each mixture was allowed to settle during 15 min and centrifuged during 10 min at 3500 rpm in order to remove suspended matter from the leachates. pH and conductivity were then measured. Finally, the ecotoxic potential of the leaching supernatants was assessed immediately without filtration and pH adjustment.

Table 5 summarizes pH, conductivity and ecotoxicological results obtained for each waste and each L/S ratio. For the Microtox™ test, a preliminary study (results not shown) had demonstrated that a 30-min exposure time was sufficient for testing 2SL leachates while 60 min was a more optimal time for testing the BA leachates. pH and conductivity values, obtained after water extraction (L/S of 2 and 10) of both wastes, were relatively high. Toxicity responses obtained from BA leachates varied from 4.2 TU (ceriodaphnid test, L/S 2) to 65.3 TU (algal test, L/S 2), whereas they varied from 231.5 TU (algal test) to 6106 TU (Microtox™ test) for the 2SL leachates.

Table 5. Prerequisite study - Ecotoxicity data of leachates of a municipal solid waste incinerator bottom ash (BA) and a slag from a second smelting of lead (2SL) obtained after following the draft standard EN 12457 (2002) using liquid-to-solid ratios (L/S) of 2 and 10.

Test	Measurement endpoints	BA waste leachate		2SL waste leachate	
		L/S 2	L/S 10	L/S 2	L/S 10
pH		9.6	10	12.1	11.9
Conductivity (µS/cm at 20°C)		5940	1575	152500	52750
Microtox™ (30/60 min)[a]	TU[b] (100/EC50)	5.5	5.5	6106	1448
P. subcapitata (72h)	TU (100/IC50)	65.3	47.2	1276	231.5
D. magna (48h)	TU (100/EC50)	17.4	26.7	1080	251
C. dubia (7d)	TU (100/EC50)	4.2	4.4	2077	360
Mutatox™ (24h)	TU (100/LOEC) S9-	NG[c]	NG	NG	NG
	TU (100/LOEC)S9+	NG	NG	NG	NG

[a] 30 min for 2SL leachates and 60 min for BA leachates; [b] Toxic Unit (see Section 5.6); [c] Non-Genotoxic.

No genotoxic effects were observed with either waste leachate. Although the various bioassay measurement endpoints clearly do not have the same ecotoxicological significance (*e.g.* ceriodaphnid EC50 *versus* Mutatox™ LOEC), they nevertheless allow ranking each waste and/or each L/S ratio as a function of their sensitivity. For the BA leachate at both L/S ratios, the sequence in decreasing order of sensitivity was as follows: algal test > daphnid test > Microtox™ test > ceriodaphnid test > Mutatox™ test. This information clearly identifies the algal test as a good candidate to assess BA toxicity fluxes in the WASTOXHAS procedure. Lambolez et al. (1994) and Ferrari et al. (1999) had previously showed the sensitivity of algae for this type of waste.

In contrast, the decreasing sensitivity sequence for the 2SL leachates was the following: Microtox™ test > ceriodaphnid test > algal test ≈ daphnid test > Mutatox™ test. This indicates that the Microtox™ test should be the bioassay of choice to assess 2SL toxicity fluxes in the WASTOXHAS procedure.

After integrating results shown in Table 5, a waste PEEP index value was calculated for each waste and each experimental L/S ratio (Fig. 5). These values allowed ranking the BA and 2SL leachates, along with the L/S ratios used, according to their increasing level of ecotoxicity. The increasing sequence of sensitivity obtained was BA (L/S 10) < BA (L/S 2) < 2SL (L/S 10) < 2SL (L/S 2). Classifying wastes with this index shows that leachates prepared from a leaching procedure using a L/S 10 generated less toxicity than those prepared from a procedure using a L/S 2, whatever the waste studied. Moreover, such a classification points out that the 2SL leachate possesses a higher level of hazard potential than the BA leachate. While priority attention is required for the 2SL waste in terms of clean-up action, the hazard potential of the BA waste should nevertheless not be neglected. In this light, waste BA leachate for the L/S ratio of 10 generates 26.7 TU with the *D. magna* immobilization assay (Tab. 5) and clearly exceeds the minimum limits (≥ 10 TU) imposed by the French Ministry of Environment (FME, 1998) to classify a waste as ecotoxic for the daphnid test under this ratio (Tab. 6). Ultimately, BA and 2SL wastes were considered as adequate material to implement the WASTOXHAS procedure.

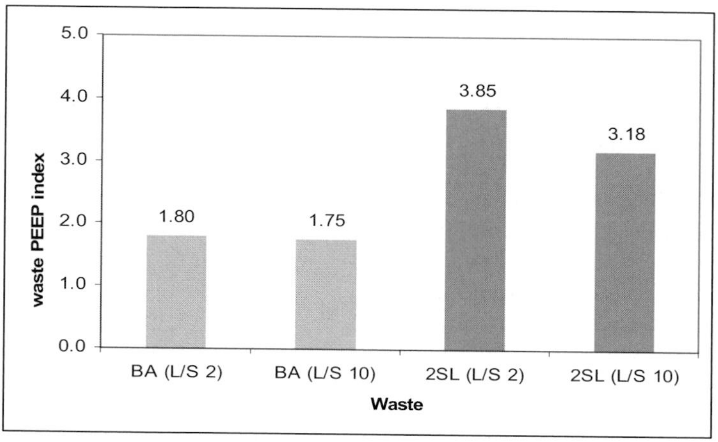

Figure 5. Prerequisite study - Comparison of the waste PEEP index values calculated for a municipal solid waste incinerator bottom ash (BA) and a slag from a second smelting of lead (2SL) as a function of each experimental liquid-to-solid ratios (L/S). Each waste PEEP index value was calculated using ecotoxicological results presented in Table 5 (see Section 5.6 for details of calculations).

Table 6. Ecotoxicological limits defined in the French proposal for criterion and evaluation methods of waste ecotoxicity based on an L/S ratio of 10 (FME, 1998).

Biological indicator	***Limit value[a] expressed in toxic unit (TU)***
Microtox™ - IC50 (30 min exposure)	10
P. subcapitata – IC20 (72h exposure)	1000
D. magna – EC50 (48h exposure)	10
C. dubia – EC20 (7d exposure)	1000

[a] Maximum limits for non-ecotoxic wastes.

Briefly recalled, the WASTOXHAS approach consists in characterizing the ecotoxicological hazard potential of contaminant fluxes from waste leachate obtained under defined conditions with two different dynamic leaching procedures: laboratory simulated leaching tests and field leaching tests. The approach developed below considered a specific scenario that simulates a waste deposit receiving rain or run-off water (Perrodin et al., 2002).

To illustrate the laboratory simulation leaching procedure, only the studies carried out in columns with gravitating percolation (Fig. 1, downward-flow column) are presented here. This leaching process is conducted under conditions of non-saturation of water in order to simulate the defined scenario adopted for the assessment of ecotoxic fluxes from the BA and 2SL wastes. For each waste, three cylindrical Plexiglas columns, with an internal diameter of 50 mm and a height of 350 mm, were filled with 480 g (in dry weight) of pre-treated waste materials used for the prerequisite study. During the filling procedure, each 2 cm layer of the material was slightly compacted by dropping an 80 g piston ten consecutive times from a constant height (10 cm). Afterwards, the columns were vertically aligned in darkness in a temperature-controlled chamber (20 ± 2°C) and were continuously flushed from top to bottom with demineralized water. Under these conditions, the leachate flow rate was close to 40 mL/h at the exit of each column. Experiments were stopped when a final L/S ratio of 30 (expressed in volume of obtained leachate by dry-weight of waste) was obtained, which corresponded approximately to 360 h of percolation for either waste. During this leaching preparation step, leachates were collected in fractions of 5 to 10 mL corresponding to different L/S ratios as they were reached (*i.e.*, 0, 0.25, 0.5, 1, 2, 4, 10, 24, 30 for BA waste, and 0, 1, 2, 4, 6, 8, 10, 30 for 2SL waste). Each leachate was then immediately treated as those obtained in the prerequisite study (*i.e.*, centrifugation during 10 min at 3500 rpm, followed by pH and conductivity measurements of each supernatant). Finally, the ecotoxicity of leachates was immediately assessed without filtration and pH adjustment by undertaking the 72-h algal test and the Microtox™ test (60-min exposure for BA leachates and 30-min exposure for 2SL leachates). These bioassays were used because they proved to be the most sensitive in the prerequisite study toward the BA and 2SL leachates, respectively.

Figure 6a summarizes the pH measured in different fractions of BA and 2SL leachates collected with different L/S ratios during the downward-flow column experiments. For the BA waste, whatever the L/S ratio reached, pH values did not vary more than 0.5 units and remained between 9 and 9.5 units. For the 2SL waste, values increased from 11.3 to 12.2 units in the first fractions collected (until L/S 2), then decreased irregularly down to 10.4 units at the end of the experiment (L/S of 30). Despite such variations, pH values were respectively of the same order of magnitude as those observed in the prerequisite study for BA and 2SL leachates (Tab. 5).

Similarly, Figure 6b summarizes conductivity results. In contrast with pH, only conductivity measured in the first fractions (up to L/S 0.5 for BA and L/S 2 for 2SL) was of the same order of magnitude as that observed in the prerequisite study (Tab. 5). Moreover, conductivity measured in BA leachates, as well as in 2SL leachates, depicted a hyperbolic relationship with L/S ratio and showed marked decreases at the highest L/S ratio. These results likely indicate that the major portion of soluble salts (*e.g.*, chlorides, sulfates) is easily leached from waste during the initial extractions and less so afterwards because of their diminishing quantities.

Results of the ecotoxic potential of leachate fluxes from BA waste are presented in Figure 7. Similarly results for the 2SL waste are presented in Figure 8. For each waste tested, bioassay responses showed adequate repeatability for the three replicate extraction columns (n = 3) employed. Indeed, averaged coefficients of variation determined with the Microtox™ test and the algal test were respectively 14.1% and 15.6% for columns filled with BA waste, and 19.4% and 27.3%, respectively, for columns filled with 2SL waste.

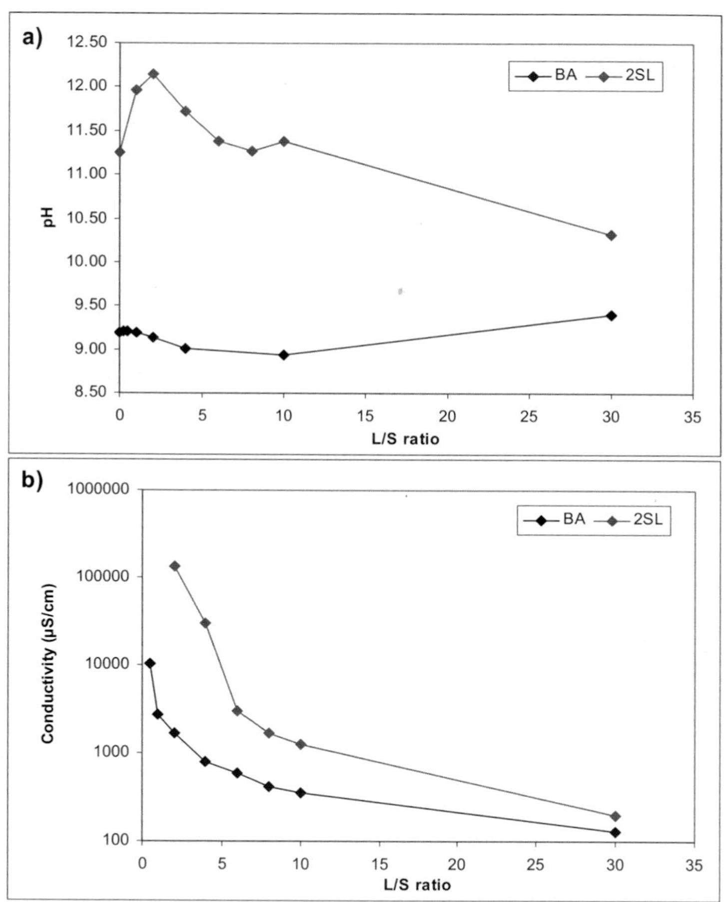

Figure 6. Downward-flow column study - Variation of pH (a) and conductivity (b) in leachate fluxes obtained from a municipal solid waste incinerator bottom ash (BA) and a slag from a second smelting of lead (2SL) as a function of the liquid-to-solid ratio (L/S).

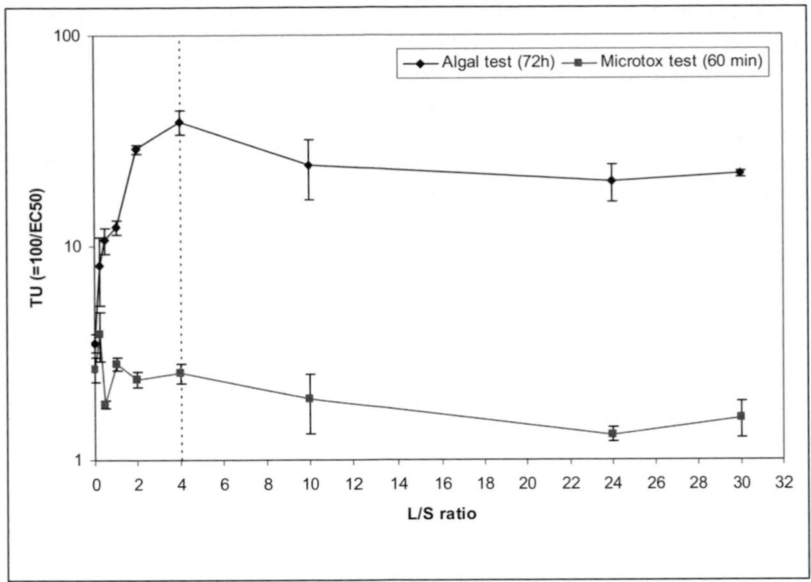

Figure 7. Downward-flow column study - Toxicity responses of the Microtox™ test and the algal test on BA leachates in relation to the liquid-to-solid ratio (L/S).

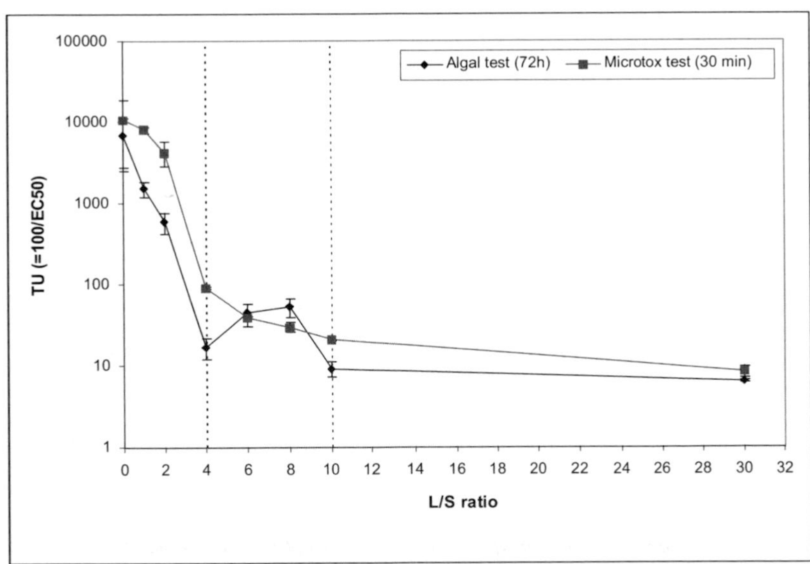

Figure 8. Downward-flow column study – Toxicity responses of the Microtox™ test and the algal test on 2SL leachates as a function of the liquid-to-solid ratio (L/S).

As expected, based on the prerequisite study, algae were more sensitive to tested fractions of BA leachates than bacteria (Microtox™ test). In contrast, bacteria proved to be more sensitive to most of 2SL leachate fractions, although algae were more sensitive to the L/S 6 and L/S 8 fractions.

With BA leachates (Fig. 7), phytotoxicity significantly increased from 3.5 TU at the start of the experiment up to a maximum of 38.9 TU reached for the L/S 4 leachate. It then significantly decreased to 21.8 TU at the final L/S 30 leachate (t-Student test, $p < 0.05$). For the bacterial test, no clear L/S ratio toxicity relationship was observed from L/S 0 to L/S 4. Its toxicity responses did not vary by more than 2 TU and remained between 1.8 and 3.8 TU. A significant decrease (t-Student test, $p < 0.05$) was observed, however, from 2.5 TU at L/S 4 ratio to 1.6 TU at the final L/S 30 ratio.

With the 2SL leachates (Fig. 8), toxicity significantly decreased ($p < 0.05$) between the start and end of the experiment, falling from 6759 TU to 6.4 TU for algae, and from 10609 TU to 8.3 TU for bacteria. Although a global decreasing tendency was observed, algal toxicity significantly rebounded between L/S 4 and L/S 8 (from 16.9 TU up to 52.5 TU), then significantly fell once more between L/8 and L/S 10 (from 52.5 TU down to 9.1 TU). These responses highlight the fact that bioassays can integrate toxicity of any variation associated with pollutant release from a waste. This makes them relevant and necessary tools for long-term hazard assessment of wastes carried out according to a specific simulator leaching procedure.

Based on the different results obtained for the algal and Microtox™ tests (Figures 7 and 8), a waste PEEP index value was calculated for each waste and each L/S ratio assessed. Each waste index value was then plotted as a function of the corresponding L/S ratio (Fig. 9) and a simple non-linear regression fit (Power model, $y = ax^b$) was applied to predict the ecotoxicological hazard potential of leachate fluxes between L/S 4 and L/S 30 ratios.

Assessing waste PEEP index values in relation to leachate fluxes indicates tendencies for the long-term ecotoxicological hazard potential of BA and 2SL wastes (Fig. 9). These values varied from 4.24 at L/S 0 down to 1.20 at L/S 30 for the 2SL waste, whereas they increased from 0.86 at L/S 0 to 1.63 at L/S 4, then decreased to 1.39 at L/S 30 for the BA waste. Ultimately, even if a general decrease in waste PEEP values is observed, a residual hazard persists for the two wastes at high L/S ratios. Based on the relatively good relationship of waste PEEP index values with L/S ratios between L/S 4 and L/S 30 (Fig. 9), there is a cut-off L/S ratio of approximately 17 above which BA leachate fluxes appear to be more hazardous than 2SL leachate fluxes. Consequently, although the 2SL waste leachate was identified as more hazardous than the BA waste leachate in the prerequisite study, which is supported by results obtained with the column leaching procedure for small L/S ratios, it seems to be less hazardous than its BA counterpart for the long term when its leachates surpass L/S ratios of 17.

These waste simulation trials based on laboratory column experiments still needed to be validated using a field approach. Two field leaching tests were thus built on an experimental site (CERED, Vernon, France) to simulate real conditions of a waste deposit site receiving rain or run-off water (Perrodin et al., 2002). The first field leaching test consisted of a large tank where the BA waste received water leading to the production of 2 m^3 of percolates per ton of dry BA waste every four months. The second field-leaching test consisted of a smaller tank where the 2SL waste received water leading to the production of 7.5 m^3 of percolates per ton of dry 2SL waste every four months. While detailed descriptions of the two field trial installations are not given herein, some characteristics concerning the conditions of their application are indicated in Table 7.

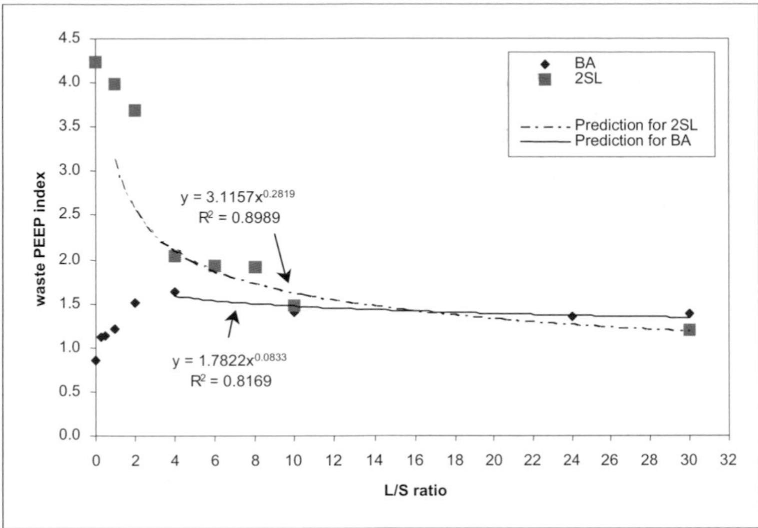

Figure 9. Downward-flow column study – Comparison of waste PEEP index values calculated for a municipal solid waste incinerator bottom ash (BA) and a slag from a second smelting of lead (2SL) as a function of each liquid-to-solid ratio (L/S). Each waste PEEP index value was calculated using ecotoxicological results presented in Figs. 7 and 8 (see Section 5.6 for the detail of calculations).

Watering the two tanks was made by rainwater, but required the addition of tap water to satisfy expected L/S ratios. Based on a 4-month time period for experiments, additional inflow of water was calculated according to the total volume required and on average rainfall normally associated over this period. Total water addition was meant to produce 2 m^3 of percolates per ton of dry BA and 7.5 m^3 of percolates per ton of dry 2SL. On the whole, three "P" fractions were recovered from each rack *in situ* and were analyzed for pH and conductivity before being sent to the laboratory. These "P" fractions were defined as follows:

- P0.5, P1 and P2 corresponded to accumulated quantities of final percolates based on L/S ratios of 0.5, 1 and 2 for the BA waste (expressed as the accumulated volume of leachate obtained at the exit of the rack by dry weight of waste).
- P2, P2.5 and P7.5 corresponded to accumulated quantities of final percolates based on L/S ratios of 2.5, 5 and 7.5 for the 2SL waste.

Table 7. Characteristics of application of the field leaching tests for the municipal solid waste incinerator bottom ash (BA) and the slag from a second smelting of lead (2SL).

		BA waste	2SL waste
Waste	- Weight (in tons)	39.5	0.450
	- Moisture (in %)	22	8.6
	- Dry density	1.55	-
	- Dry weight (in tons)	30.8	0.411
	- Height of the waste layer (in cm)	40 to 45	20 to 22
	- Compacted or not?	Yes	No
Total quantity of percolate (in m^3) produced after the 4 month experiment		61.5	3.1
Bottom area of the tank (in m^2)		54.4	2

Table 8. Field scale study - Ecotoxicity of accumulated percolates of a municipal solid waste incinerator bottom ash (BA) and a slag from a second smelting of lead (2SL) from field experiments and their corresponding waste PEEP index values (see Section 5.6 for the detail of calculations).

Test	Measurement endpoints	BA waste			2SL waste		
		P0.5	P1	P2	P2.5	P5	P7.5
pH		7.0 (6.9)[a]	7.6 (10.3)[a]	7.8 (9.8)[a]	12.1 (12.5)[a]	12.2 (12.6)[a]	11.5 (12.3)[a]
Conductivity (µS/cm at 20°C)		13000 (9500)	3075 (2740)	1975 (1460)	91500 (93900)	54750 (84100)	21500 (21500)
Microtox™ (30/60 min)[b]	TU[c]	3	1.4	1.5	7353	2182	23.2
P. subcapitata (72h)	TU	20	6.5	5.7	1423	363	75.3
D. magna (48h)	TU	1.2	NT[d]	NT	375	187	16.4
C. dubia (7d)	TU	7.7	< 3	< 3	955	266	23.2
Mutatox™ (24h)	TU S9-	NG[e]	NG	NG	NG	NG	NG
	TU S9+	NG	NG	NG	NG	NG	NG
Waste PEEP index value		1.35	0.81	0.78	3.83	3.30	1.97

[a] Measured *in situ* before sending the samples to the laboratory; [b] 30 min for 2SL leachates and 60 min for BA leachates; [c] Toxic Unit (see Section 5.6 and Table 5); [d] Non-Toxic; [e] Non-Genotoxic.

As an example, fraction P2 of the BA waste experiment corresponded to that volume (= 1 m^3) of percolate recovered between the time when the L/S 1 ratio was reached and that when the L/S 2 was reached. Received in the laboratory after a maximum time of 48 h, percolate samples were immediately treated in a manner similar to leachates obtained in the prerequisite study (*i.e.*, centrifugation during 10 min at 3500 rpm, pH and conductivity measurements of each supernatant). The ecotoxicity of the different fractions was then assessed without filtration and pH adjustment with the same battery of bioassays used for the prerequisite study.

Table 8 summarizes pH, conductivity, ecotoxicological results and the corresponding waste PEEP index values obtained for the accumulated percolates recovered *in situ* for each waste. Measurements of pH remained relatively constant, but conductivity decreased in relation to L/S ratio, as observed before in the column study.

For both BA and LS2 wastes, PEEP index values consistently indicated a reduction in ecotoxicity of the percolates as a function of the L/S ratios. At the end of experimentation, no apparent threshold devoid of ecotoxic effects appeared to have been reached. Comparison of these index values revealed that BA waste leachate fluxes appeared to be less hazardous than 2SL waste leachate fluxes at all L/S ratios investigated, even the L/S ratio range are different in both cases. As with the prerequisite study, the different bioassays can be ranked in terms of sensitivity for each waste and each percolate. For the BA waste and all tested percolates, decreasing order of sensitivity responses is as follows: algal test > ceriodaphnid test > Microtox™ test > daphnid test > Mutatox™ test. For the 2SL waste, sensitivity responses were the following: Microtox™ test > algal test > ceriodaphnid test > daphnid test > Mutatox™ test for percolates P2.5 and P5; algal test > Microtox™ test ≈ ceriodaphnid test > daphnid test > Mutatox™ test for percolate P7.5. This sensitivity classification shows good agreement with the prerequisite study and reinforces the assumption that the algal and Microtox™ tests are adequate tools to assess leachate fluxes from BA and 2SL wastes using the column procedure.

Table 9 gives an overview of results obtained from the column tests and the field tests for BA and 2SL wastes up to an L/S ratio of 2 and an L/S ratio of 8, respectively. This outlook allows a comparison of the two procedures based on the sensitivity responses of the algal and Microtox™ tests and on the waste PEEP index values for L/S ratios experimented in the field. For the purposes of this comparison, waste PEEP index values for the field percolates were recalculated using results of both tests presented in Table 8.

For the BA waste, even if sensitivity ranking of ecotoxicity tests is similar, the evolution of ecotoxic hazard potential of the leachate fluxes was different between the two approaches. In the field, the ecotoxic hazard potential of leachate fluxes decreased, whereas it increased for the laboratory column study. The column approach tends to overestimate the long-term ecotoxic hazard potential of BA leachate fluxes generated in the field. This overestimation of the long-term ecotoxic hazard potential of leachate fluxes generated in the field may indicate that relevant factors such as 1) residence time of water in the waste, 2) the continuous or discontinuous watering of the waste and/or 3) the physico-chemical characteristics of the water used for obtaining leachates were not considered in the column approach.

In such a case, the ecotoxic hazard potential assessment of BA leachate fluxes must be refined before definitive conclusions on its hazard can be reached.

In contrast, field and laboratory results showed good agreement for the 2SL waste. Again, sensitivity ranking of the ecotoxicity tests was similar. The waste PEEP index values closely corresponded and decreased as a function of the L/S ratios. Considering the column results, it is assumed that the field waste PEEP index values continue to decrease between 7.5 and 30 L/S ratio, then the column approach tends to give a realistic estimation of the long term ecotoxic hazard potential of 2SL waste leachate fluxes generated in the field. Because a residual ecotoxic hazard potential of 2SL waste leachate fluxes in the long term is assumed, this waste needs to be treated, or stored in a specific landfill, or evaluated in a higher tier (see Fig. 2).

Table 9. Downward-flow column study and field scale study - Comparison of the two procedures based on the sensitivity responses of ecotoxicity tests and on the corresponding waste PEEP index values (see Section 5.6 for the detail of calculations) for a municipal solid waste incinerator bottom ash (BA) and a slag from a second smelting of lead (2SL) at different liquid-to-solid ratios (L/S).

Waste	***Procedure***	***L/S ratio***	***Sensitivity of ecotoxicity tests***	***Waste PEEP index value***	***Waste PEEP trend***
BA	Field scale leaching test	0 to 0.5 (P0.5)	Algae > Microtox™	1.38	↘
		0.5 to 1 (P1)	Algae > Microtox™	0.95	
		1 to 2 (P2)	Algae > Microtox™	0.91	
	Downward flow column leaching test	0	Algae > Microtox™	0.86	↗
		0.5	Algae > Microtox™	1.13	
		1	Algae > Microtox™	1.21	
		2	Algae > Microtox™	1.51	
2SL	Field scale leaching test	0 to 2.5 (P2.5)	Microtox™ >Algae	3.94	↘
		2.5 to 5 (P5)	Microtox™ >Algae	3.40	
		5 to 7.5 (P7.5)	Algae > Microtox™	1.99	
	Downward flow column leaching test	0	Microtox™ >Algae	4.24	↘
		2	Microtox™ >Algae	3.68	
		4	Microtox™ >Algae	2.03	
		6	Algae > Microtox™	1.93	
		8	Algae > Microtox™	1.92	

8. Accessory/miscellaneous WASTOXHAS procedure information

To ensure proper undertaking of this HAS procedure, personnel should be knowledgeable in the field of ecotoxicology and be specifically trained in toxicity testing and analytical methods. Moreover, careful attention to sampling methods and quality control will ensure the successful undertaking of the WASTOXHAS procedure.

Safety measures must be in place prior to carrying out bioassays. Consulting manufacturer data sheets on the potential effects that reference substances could exert on human health is a necessary precaution, as is that of using care when dealing with specific types of complex liquid wastes. In all cases, proper safety measures must be applied before testing. Among others, this will include protective wear (laboratory coat, gloves, eye-glasses) and other suitable means of defense (*e.g.*, respirator and fume hoods to guard against volatile toxicants). It stands to reason that a laboratory should also possess an approved safety plan describing ways of handling hazardous chemicals, as well as one for their disposal and that of other contaminated material (*e.g.*, pipettes, gloves, spent solvent).

Sampling strategies must be carried out with an important emphasis placed on safety. Safety considerations include the safety of personnel conducting the work, the surrounding community, and the environment. Occupational Safety and Health Administration standards and regulations should be followed, and appropriate monitoring equipment should be used during sampling operations.

In terms of applicability, the waste used in the two scales of tests must come from the same origin, must be sampled at the same date and must follow the same pre-treatments and procedures. The sampling method used must ensure that a representative sample of the waste to be assessed will be obtained.

WASTOXHAS leaching tests must be carried out on columns (preferentially in polyethylene) with an internal diameter of at least 5 cm and a filling height of at least 4 times the internal diameter (NEN 7343, 1995). Filling test material into the columns must approximate field density, without altering either the waste physical or chemical properties or creating too much water preferential ways.

As pointed out in Table 1 (part 2), any type of bioassay can be used in the WASTOXHAS procedure. Alternative choices can be found in Table 2.

9. Conclusions/prospects

There is a requirement for bioanalytical methods (integrating biological and leaching strategies) that provide information on the mobility of toxicants and their fluxes from solid wastes. Presently, WASTOXHAS is still in its infancy. It needs to be applied to different wastes by research institutions in order to evolve and to be optimized. Along with other types of leaching tests and bioassays, it comprises one choice for laboratories desirous of ranking and prioritizing actions for protecting the aquatic environment. It offers a universal and flexible framework for going in this direction. New environmental issues and prospects were presented in Section 4 and future studies are planned in the area of solid waste assessments.

Acknowlegments

This study was carried out within the framework of a three year research program: The French National program on waste ecocompatibility provided funding through ADEME (French Environmental and Energy Management Agency). The authors would like to acknowledge all team members that were part of this research network. Our special thanks go to Dr. Christian Blaise and to Dr. P.T. Johnson for useful suggestions for the improvement of the manuscript.

References

Aït-Aïssa, S., Pandard, P., Magaud, H., Arrigo, A.P., Thybaud, E. and Porcher, J.M. (2003) Evaluation of *in vitro* hsp70 induction test for toxicity assessment of complex mixtures: comparison with chemical analyses and ecotoxicity tests, *Ecotoxicology and Environmental Safety* **54,** 92-104.

Alexander, M. (2000) Ageing, bioavailability and overestimation of risk from environmental pollutants, *Environmental Science and Technology* **34**, 4259-4264.

Atwater, J.W., Jasper, S., Mavinic, D.S. and Koch, F.A. (1983) Experiments using *Daphnia* to measure landfill leachate toxicity, *Water Research* **17**, 1855-1861.

Birkholz, D., Belton, K. and Guidotti, T. (2003) Toxicological evaluation for the hazard assessment of tire crumb for use in public playgrounds, *Journal of the Air & Waste Management Association* **53**, 903-907.

Bispo, A. (1998) Contribution à l'élaboration d'une méthodologie pour évaluer les dangers liés aux matériaux solides contaminés, Ph. D. thesis, INPL Nancy, France, 138 pp. (in French).

Blaise, C., Forghani, R., Legault, R., Guzzo, J. and Dubow, M.S. (1994) A bacterial toxicity assay performed with microplates, microluminometry and Microtox, *Biotechniques* **16**, 932-937.

Blanck, H. (1984) Species dependant variation among aquatic organisms in their sensitivity to chemicals, *Ecological Bulletins* **36**, 107-119.

Calleja, A., Baldasano, J.M. and Mulet, A. (1986) Toxicity analysis of leachates from hazardous wastes via Microtox and *Daphnia magna*, *Toxicity Assessment* **1**, 73-83.

Chaaban, M.A. (2001) Hazardous waste source reduction in materials and processing technologies, *Journal of Materials Processing Technology* **119**, 336-343.

Chen, Y.X., Zhu, G.W., Tian, G.M. and Chen, H.L. (2003) Phosphorus and copper leaching from dredged sediment applied on a sandy loam soil: column study, *Chemosphere* **53**, 1179-1187.

Clément, B., Persoone, G., Janssen, C. and Le Dû-Delepierre, A. (1996) Estimation of the hazard of landfills through toxicity testing of leachates: I. Determination of leachate toxicity with a battery of acute tests, *Chemosphere* **33,** 2303-2320.

Clozel-Leloup, B., Bodénan, F. and Piantone, P. (1999) Bottom ash from municipal solid waste incineration: Mineralogy and distribution of metals, in J. Méhu et al. (Eds.), *Waste Stabilization and Environment 99*, Villeurbanne, France, 13-16 April 1999, Proceedings of conference presentation, Société alpine de publications, Grenoble, France, pp. 45-51.

Conner, J.R. and Hoeffner, S.L. (1998) Critical review of stabilization/solidification technology, *Critical Reviews in Environmental Science and Technology* **28**, 397-462.

Costan, G., Bermingham, N., Blaise, C. and Férard, J.F. (1993) Potential Ecotoxic Effects Probe (PEEP): A Novel Index to Assess and Compare the Toxic Potential of Industrial Effluents, *Environmental Toxicology and Water Quality* **8**, 115-140.

Crane, M., Newmann, M.C., Chapman, P.F. and Fenlon, J.S. (2001) Risk assessment with time to event models, CRC Press/Lewis Publishers, Boca Raton, Florida, USA, 192 pp.

Crawford, J. and Neretnieks, I. (1999) Development of methods to assess the long term leaching behavior of solid wastes, in J. Méhu et al. (Eds.), *Waste Stabilization and Environment 99*, Villeurbanne, France, 13-16 April 1999, Proceedings of conference presentation, Société alpine de publications, Grenoble, France, pp. 411-414.

Crignon, C., Pecqueur, G., Garcia Diaz, E., Germaneau, B. and Siwak, J.M. (1999) Study of cement-treated MSWI bottom ash expansion, in J. Méhu et al. (eds.), *Waste Stabilization and Environment 99*, Villeurbanne, France, 13-16 April 1999, Proceedings of conference presentation, Société alpine de publications, Grenoble, France, pp. 64-68.

Czerniawska-Kusza, I. and Ebis, M. (2000) Toxicity of waste dump leachates and sugar factory effluents and their impact on ground water and surface water quality in the Opole province in Poland, in G.Persoone *et al.* (Eds.), New microbiotests for routine toxicity screening and biomonitoring, Kluwer Academic Plenum Publishers, New-York, USA, pp. 319-322.

Day, K.E., Holtze, K.E., Metcalf-Smith, R., Bishop, C.T. and Dutka, B.J. (1993) Toxicity of leachate from automobile tires to aquatic biota, *Chemosphere* **27**, 665-675.

Devare, M. and Bahadir, M. (1994a) Biological monitoring of landfill leachate using plants and luminescent bacteria, *Chemosphere* **28**, 261-271.

Devare, M. and Bahadir, M. (1994b) Ecotoxicological assessment of inorganic waste disposal in salt mines, part II: phytotoxicity tests, *Fresenius Environ. Bull.* **3**, 119-126.

Ducrot, V., Usseglio-Polatera, P., Péry, A.R.R., Mouthon, M., Laffont, M., Roger, M.C., Garric, J. and Férard, J.F. (2005) Functional grouping based on species traits: contribution to improvement of relevance in the selection of test species for ecotoxicology, *Environmental Toxicology and Chemistry* (in press).

EC (1994) European Catalogue of Hazardous Wastes, Decision 94/904/EC, *Official Journal* L 356, 31.12.1994.

EN 12457-1 (2003) Characterization of waste – Leaching – Compliance test for leaching of granular waste materials and sludges – Part 1: One-stage batch test at a liquid to solids ratio of 2 L/kg for materials with high solids content and with a particle size below 4 mm (with or without size reduction), CEN/TC292/WG2, European Committee for Standardization, Brussels.

EN 12457-2 (2002) Characterization of waste – Leaching – Compliance test for leaching of granular waste materials and sludges – Part 2: One-stage batch test at a liquid to solids ratio of 10 L/kg for materials with a particle size below 4 mm (with or without size reduction), CEN/TC292/WG2, European Committee for Standardization, Brussels.

EN 12457-3 (2002) Characterization of waste – Leaching – Compliance test for leaching of granular waste materials and sludges – Part 3: One-stage batch test at a liquid to solids ratio of 2 L/kg and 8 L/kg for materials with high solids content and with a particle size below 4 mm (with or without size reduction), CEN/TC292/WG2, European Committee for Standardization, Brussels.

EN 12457-4 (2002) Characterization of waste – Leaching – Compliance test for leaching of granular waste materials and sludges – Part 4: One-stage batch test at a liquid to solids ratio of 10 L/kg for materials with a particle size below 10 mm (with or without size reduction), CEN/TC292/WG2, European Committee for Standardization, Brussels.

EN 12920 (1998) Characterisation of waste - Methodology guideline for the determination of leaching behaviour of waste under specific conditions, CEN/TC292/WG6, European Committee for Standardization, Brussels.

EN 14405 (2002) Characterization of waste - Leaching behavior test - Up-flow percolation test, CEN/TC292/WG6, European Committee for Standardization, Brussels.

Environment Canada (1992) Biological test method: growth inhibition test using the freshwater alga *Selenastrum capricornutum*, Report EPS 1/RM/25, Environmental Protection Series, Environment Canada, Ottawa, Ontario, 43 pp.

Environment Canada (1999) Guidance document on application and interpretation of single-species tests in environmental toxicology, Report EPS 1/RM/34, Environmental Protection Series, Environment Canada, Ottawa, Ontario, 203 pp.

Ernst, W.R., Hennigar, P., Doe, K., Wade, S. and Julien, G. (1994) Characterization of chemical constituents and toxicity to aquatic organisms of a municipal landfill leachate, *Water Pollution Research Journal of Canada* **29**, 89-101.

Férard, J.F. and Ferrari, B. (1997) Quel test de toxicité chronique faut-il choisir pour l'évaluation de la dangerosité des déchets ? *Déchets, Sciences et Techniques*, **8**, 44-47 (in French).

Fernández, J.A., Aboal, J.R. and Carballeira, A. (2000) Use of native and transplanted mosses as complementary techniques for biomonitoring mercury around an industrial facility, *The Science of The Total Environment* **256**, (2-3), 151-161.

Ferrari, B. (2000) Contribution à l'évaluation de l'écocompatibilité des déchets : approche écotoxicologique, Ph. D. thesis, University of Metz, France, 151 pp. (in French).

Ferrari, B. and Férard, J.F. (1999) Ecotoxicologie des déchets : evaluation des flux de polluants par des essais de perco-lixiviation en laboratoire, in J. Mehu et al. (Eds.) *Waste Stabilization and Environment 99*, Villeurbanne, France, 13-16 april 1999, Proceedings of poster presentation, pp. 192-194.

Ferrari, B., Radetski, C.M., Veber, A.M. and Férard, J.F. (1999) Ecotoxicological assessment of solid wastes: a combined liquid- and solid-phase testing approach using a battery of bioassays and biomarkers, *Environmental Toxicology and Chemistry* **18**, 1195-1202.

Font, R., Gomis, V., Fernandez, J. and Sabater, M.C. (1998) Physico-chemical characterization and leaching of tannery wastes, *Waste Management Research* **16,** 139-149.

French Ministry of Environment (1998) Critères et méthodes d'évaluation de l'écotoxicité des déchets, Ministère de l'Environnement/Direction de la prévention des pollutions et des risques, Paris, France.

Giesy, J.P. and Graney, R.L. (1989) Recent developments in and intercomparisons of acute and chronic bioassays and bioindicators, *Hydrobiologia* **188/189**, 21-60.

Giesy, J.P. and Hoke, R.A. (1989) Freshwater sediment toxicity bioassessment: rationale for species selection and test design, *Journal of Great Lakes Research* **15**, 539-569.

Gourdon, R., Ménoret, C. and Bayard, R. (1999) Effect of algal growth on the leaching behavior of APC residues solidified with hydraulic binders, in J. Mehu et al. (Eds.), *Waste Stabilization and Environment 99*, Villeurbanne, France, 13-16 April 1999, Proceedings of poster presentation, pp. 171-173.

Grelier-Volatier, L., Hugrel, C., Perrodin,Y. and Chateau, L. (2002) Evaluation de l'écocompatibilité de déchets mis en depôts ou valorisés en travaux publics : une méthode pluridisciplinaire pour une approche "en scénario", *Revue Française des Sciences de l' Eau* **15**(special), 57-66 (in French).

Griest, W.H., Tyndall, R.L., Stewart, A.J., Caton, J.E., Vass, A.A., Ho, C.H. and Caldwell, W.M. (1995) Chemical characterization and toxicological testing of windrow composts from explosives-contaminated sediments, *Environmental Toxicology and Chemistry* **14,** 51-59.

Hamilton, K.L., Nelson, W.G. and Curley, J.L. (1993) Toxicological assessment of effects of waste-to-energy ash-concrete on two marine species, *Environmental Toxicology and Chemistry* **12**, 1919-1930.

Hartwell, S.I., Jordahl, D.M., Dawson, C.E.O. and Ives, A.S. (1998) Toxicity of scrap tire leachates in estuarine salinities: Are tires Acceptable for Artificial Reefs? *Transactions of the American Fisheries Society* **127,** 796-806.

Hjelmar, O., Hansen, E.A., Andersen, K.J., Andersen, J.B. and Bjørnestad, E. (1994) An approach to the assessment of environmental impacts of marine applications of municipal solid waste combustion residues, in J.J.J.M. Goumans et al. (Eds.), Environmental Aspects of Construction with Waste Materials, Elsevier Sciences B.V., Amsterdam, NL, pp. 137-16.

Hofstee, C., Gutiérrez-Ziegler, C., Trötschler, O. and Braun, J. (2003) Removal of DNAPL contamination from the saturated zone by the combined effect of vertical upward flushing and density reduction, *Journal of Contaminant Hydrology* **67**, 61-78.

Huang, H.C., Liu, C.W., Chen, S.K. and Chen, J.S. (2003) Analysis of percolation and seepage through paddy bunds, *Journal of Hydrology* **284**, 13-25.

Isidori, M., Lavorgna, M., Nardelli, A. and Parrella, A. (2003) Toxicity identification evaluation of leachates from municipal solid waste landfills: a multispecies approach, *Chemosphere* **52**, 85-94.

Isnard, P., Flammarion, P., Roman, G., Babut, M., Bastien, P., Bintein, S., Esserméant, L., Férard, J.F., Gallotti-Schmitt, S., Saouter, E., Saroli, M., Thiébaud, H., Tomassone, R. and Vindimian, E. (2001) Statistical analysis of regulatory ecotoxicity tests, *Chemosphere* **45**, 659-669.

ISO 6341 (1996) Water quality – Determination of the inhibition of the mobility of *Daphnia magna* Straus (Cladocera, Crustacea) – Acute toxicity test, International Organization for Standardization, Geneva.

ISO 11348-3 (1999) Water quality – Determination of the inhibitory effect of water samples on the light emission of *Vibrio fisheri* (luminescent bacteria test) – Part 3. Method using freeze-dried bacteria, International Organization for Standardization, Geneva.

Johnson, B.L. (1993) Commentary, in Proceedings of *Hazardous Waste Conference*, 3-6 mai 1993, Atlanta, USA, (http://www.atsdr.cdc.gov/cx3a.html)

Joutti, A., Schultz, E., Tuukkanen, E. and Vaajasaari, K. (2000) Industrial waste leachates: toxicity detection with microbiotests and biochemical tests, in G. Persoone *et al.* (Eds.), New microbiotests for routine toxicity screening and biomonitoring, Kluwer Academic Plenum Publishers, New-York, USA, pp. 347-355.

Kahru, A., Reiman, R. and Rätsep, A. (1998) The efficiency of different phenol-degrading bacteria and activated sludges in detoxification of phenolic leachates, *Chemosphere* **37,** 301-318.

Kampke-Thiel, K., Freitag, D., Kettrup, A. and Bahadir, M. (1994) Ecotoxicological assessment of inorganic waste disposal in salt mines, part I: tests with aquatic organisms, *Fresenius Environmental Bulletin* **3,** 113-118.

Kaschl, A., Römheld, V. and Chen, Y. (2002) The influence of soluble organic matter from municipal solid waste compost on trace metal leaching in calcareous, *The Science of the Total Environment* **291**, 45-57.

Knight, J., Cheeseman, C. and Rogers, R. (1999) Microbial influenced degradation of solidified wastes, in J. Méhu et al. (Eds.), *Waste Stabilization and Environment 99*, Villeurbanne, France, 13-16 April 1999, Proceedings of conference presentation, Société alpine de publications, Grenoble, France, pp. 143-148.

Kwan, K.K., Dutka, B.J., Rao, S.S. and Liu, D. (1990) Mutatox test: A new test for monitoring environmental genotoxic agents, *Environmental Pollution* **65**, 323-332.

Kylefors, K., Andreas, L. and Lagerkvist, A. (2003) A comparison of small-scale, pilot-scale and large-scale tests for predicting leaching behavior of landfilled wastes, *Waste Management* **23**, 45-59.

Lambolez, L. (1994) Etude des relations mobilité-biodisponibilité-toxicité des micropolluants présents dans les déchets industriels. Application à la gestion des centres d'enfouissement technique de classe I, Ph. D. thesis, University of Metz, France, 100 pp., (in French).

Lambolez, L., Vasseur, P., Férard, J.F. and Giesbert, T. (1994) The environmental risks of industrial waste disposal: an experimental approach including acute and chronic toxicities studies, *Ecotoxicology and Environmental Safety* **28,** 317-328.

Lapa, N., Santos Oliveira, J.F., Camacho, S.D.L. and Circeo, L.J. (2002a) An ecotoxic risk assessment of residual materials produced by the plasma pyrolysis/vitrification (PP/V) process, *Waste management* **22**, 335-342.

Lapa, N., Barbosa, R., Morais, J., Mendes, B., Méhu, J. and Santos Oliveira, J.F. (2002b) Ecotoxicological assessment of leachates from MSWI bottom ashes, *Waste Management* **22**, 583-593.

Latif, M. and Zach, A. (2000) Toxicity studies of treated residual wastes in Austria using different types of conventional assays and cost-effective microbiotests, in G. Persoone, C. Janssen and W. de Coen (Eds.), New microbiotests for routine toxicity screening and biomonitoring, Kluwer Academic Plenum Publishers, New-York, USA, pp. 367-383.

Mala, J., Marsalkova, E. and Rovnanikova, P. (2000) Toxicity testing of solidified waste leachates with microbiotests, in G. Persoone *et al.* (Eds.), New microbiotests for routine toxicity screening and biomonitoring, Kluwer Academic Plenum Publishers, New-York, USA, pp. 385-390.

Masfaraud, J.F., Beaunoir, V., Perrodin, Y., Thybaud, E., Savanne, D. and Férard, J.F. (1999) Assessment of soil contamination and ecotoxicity by the use of an ascending flow percolation - Bioassay approach, in J. Méhu et al. (Eds.), *Waste Stabilization and Environment 99*, Villeurbanne, France, 13-16 April 1999, Proceedings of poster presentation, pp. 198-200.

McCorriston, J. and Weisberg, S. (2002) Spatial and temporal variation in Mesopotamiam agricultural practices in the Khabur Basin, Syrian Jazira, *Journal of Archaeological Science* **29**, 485-498.

Mezzanotte, V., Sora, S., Vigano, L. and Vismara, R. (1988) Using bioassays to evaluate the toxic and mutagenic properties of landfill leachate, in L. Anderson and J. Moller (eds.), *ISWA '88 Proceedings*, **1,** Academic Press, London, UK, pp. 131-136.

Miadokova, E., Duhova, V., Rezna, M., Kralikova, A., Sucha, V. and Vlcek, D. (1998) Genotoxicological research on the waste mine drainage water, *Journal of Trace and Microprobe Techniques* **16**, 453-463.

Mizutani, S., Sakai, S.I. and Takatsuki, H. (1999) Effects of CO_2 in the air on leaching behavior of heavy metal from alkaline residues, in J. Méhu et al. (Eds.), *Waste Stabilization and Environment 99*, Villeurbanne, France, 13-16 April 1999, Proceedings of poster presentation, pp. 125-127.

NEN 7343 (1995) Leaching Characteristics of Solid Earthy and Stony Building and Waste Materials. Leaching Tests. Determination of the Leaching of Inorganic Components from Granular Materials with the Column Test, Netherlands Normalization Institute, Delft, Netherlands.

Nimmo, D.W.R., Willox, M.J., Karish, J.F., Tessari, J.D., Craig, T.L., Gasser, E.G. and Self, J.R. (1995) Non-availability of metals from an urban landfill in Virginia, *Chemical Speciation and Bioavailability* **7,** 65-72.

Núñez-Delgado, A., López-Períago, E. and Díaz-Fierros-Viqueira, F. (2002) Pollution attenuation by soils receiving cattle slurry after passage of a slurry-like feed solution: Column experiments, *Bioresource Technology* **84**, 229-236.

Ortiz, M.I., Ibanez, R., Andrés, A. and Irabien, A. (1995) Ecotoxicological characterization of metal finishing wastes, *Fresenius Environmental Bulletin* **4**, 189-194.

Perrodin, Y., Cavéglia, V., Barna, R., Moszkowicz, P., Gourdon, R., Férard, J.F., Ferrari, B., Fruget, J.F., Plenet, S., Jocteur-Monrozier, L., Poly, F., Texier, C., Cluzeau, D., Lambolez, L., and Billard, H. (1996) Programme de recherche sur l'écocompatibilité des déchets: étude bibliographique, Report, ADEME, France (in French).

Perrodin, Y., Gobbey, A., Grelier-Volatier, L., Canivet, V., Fruget, J.F., Gibert, J., Texier, C., Cluzeau, D., Gros, R., Poly, F. and Jocteur-Monrozier, L. (2002) Waste ecocompatibility in storage and reuse scenarios: global methodology and detailed presentation of the impact study on the recipient environments, *Waste Management* **22**, 215-228.

Peterson, S.A., Greene, J.C. and Miller, W.E. (1990) Toxicological evaluation of hazardous waste samples extracted with deionized water or sodium acetate (TCLP) leaching media. in D. Friedman (ed.), *Waste Testing and Quality Assurance*, ASTM STP 1062, Philadelphia, USA, **2**, 107-129.

Plotkin, S. and Ram, N.M. (1984) Multiple bioassays to assess the toxicity of a sanitary landfill leachate, *Archives of Environmental Contamination and Toxicology* **13**, 197-206.

Pollard, S.J.T., Whittaker, M. and Risden, G. C. (1999) The fate of heavy oil wastes in soil microcosms I: a performance assessment of biotransformation indices, *The Science of The Total Environment* **226**, 1-22.

Propst, T.L., Lochmiller, R.L., Qualls, C.W.Jr. and McBee, K. (1999) *In situ* (mesocosm) assessment of immunotoxicity risks to small mammals inhabiting petrochemical waste sites, *Chemosphere* **38**, 1049-1067.

Rojickova-Padrtova, R., Marsalek, B. and Holoubek, I. (1998) Evaluation of alternative and standard toxicity assays for screening of environmental samples: selection of an optimal test battery, *Chemosphere* **37**, 495-507.

Rummel-Bulska, I. (1993) The basel convention: A global approach for the management of hazardous wastes, in *Proceedings of Hazardous Waste Conference*, 3-6 may 1993, Atlanta, USA.

Sahuquillo, A., Rigol, A. and Rauret, G. (2003) Overview of the use of leaching/extraction tests for risk assessment of trace metals in contaminated soils and sediments, *Trends in Analytical Chemistry* **22**, 152-159.

Sajwan, K.S., Paramasivam, S., Alva, A.K., Adriano, D.C. and Hooda, P. S. (2003) Assessing the feasibility of land application of fly ash, sewage sludge and their mixtures, *Advances in Environmental Research* **8**, 77-91.

Schrab, G.E., Brown, K.W. and Donnelly, K.C. (1993) Acute and genetic toxicity of municipal landfill leachate, *Water, Air, and Soil Pollution* **69**, 99-112.

Schultz, E., Vaajasaari, K., Joutti, A. and Ahtiainen, J. (2002) Toxicity of industrial wastes and waste leaching eluates containing organic compounds, *Ecotoxicology and Environmental Safety* **52**, 248-255.

Sekkat, N., Guerbet, M. and Jouany, J.M. (2001) Etude comparative de huit bioessais à court terme pour l'évaluation de la toxicité de lixiviats de déchets urbains et industriels, *Revue des Scinces de l'Eau* **14**, 63-72 (in French).

U.S. EPA (2002) Short-term methods for estimating the chronic toxicity of effluents and receiving waters to freshwater organisms, EPA 821/R_02/013, Environmental Systems Laboratory, Cincinnati, OH, USA.

Vaajasaari, K., Ahtiainen, J., Nakari, T. and Dahlbo, H. (2000) Hazard assessment of industrial waste leachability: chemical characterization and biotesting by routine effluent tests, in G. Persoone et al. (eds.), New microbiotests for routine toxicity screening and biomonitoring, Kluwer Academic Plenum Publishers, New-York, USA, pp. 413-423.

Van der Sloot, H.A. (1998) Quick techniques for evaluating the leaching properties of waste materials: their relation to decisions on utilization and disposal, *Trends in Analytical Chemistry* **17**, 298-310.

Vindimian, E., Robaut, C. and Fillion, G. (1983) A method for co-operative and non co-operative binding studies using non linear regression analysis on microcomputer, *Journal of Applied Biochemistry* **5**, 261-268.

Ward, M.L., Bitton, G., Townsend, T. and Booth, M. (2002) Determining toxicity of leachates from Florida municipal solid waste landfills using a battery-of-tests approach, *Environmental Toxicology* **17**, 258-266.

Wells, P.G., Lee, K. and Blaise, C. (1998) Microscale testing in aquatic toxicology: Advances, Techniques, and Practice, CRC Presss, Boca Raton, Florida, USA, 679 pp.

Abbreviations

2SL	matured slags from the second smelting of lead
BA	bottom ash from municipal solid waste incineration
CERED	centre d'études et de recherches sur l'élimination des déchets
d	day(s)
DMSO	dimethyl sulfoxide
EC	European Community
ECx	x% effective concentration
EC50	50% effective concentration
EDTA	ethylenediamine tetraacetate (C10H14O8N2)
EN	European Normalization
FMN	French Ministry of Environment
h	hour(s)
HEC	highest effect concentration
HSW	hazardous solid waste
ICx	x% effect inhibitory concentration
IC50	50% effect inhibitory concentration
IF	Induction factor
ISO	International organization for standardization
ISWMP	Integrated Solid Waste Management Plans
LOEC	Lowest observed effect concentration
L/S	liquid to solid ratio
min	minute(s)
MEV	Measurement endpoint value
MSW	municipal solid waste
MSWI	municipal solid waste incineration
NOEC	No observed effect concentration
OECD	Organization for Economic Co-operation and Development
PEEP	Potential Ecotoxic Effects Probe
SL	Smelting Lead
TBA	Test Battery Approach
TU	Toxic Unit
UNEP	United Nations Environment Program
WTE	Waste-to-energy.

Glossary

Note to readers: Volume and chapter number(s) indicated after each **Glossary** term are those for which authors contributed a definition. They may also be found in other chapters of both volumes.

Acclimation	Adaptation to environmental conditions (usually controlled laboratory conditions). Acclimation is generally conducted over a specified period of time. Volume 1(10).
Acid volatile sulfides (AVS)	Chemical analysis that quantifies the amount of sulfides present in a sample that are assumed to be capable of forming insoluble precipitates with divalent metals. See also AVS:SEM ratio and SEM. Volume 2(10).
Acid-washing	Procedure in which laboratory articles are soaked overnight in 4% detergent (*e.g.*, Contrad 70) and rinsed five times in reverse osmosis water, soaked overnight in 10% HCl and rinsed five times in Milli-Q water and oven dried (58°C). Volume 1(6).
Activated sludge	Product that results when primary effluent is mixed with bacteria-laden sludge and then agitated and aerated to promote biological treatment, speeding the breakdown of organic matter in raw sewage undergoing secondary waste treatment (U.S. EPA, 2004). Volume 2(7).
Active biomonitoring	Use of transplanted living organisms (or part of) to assess water, air, sediment or soil quality. See also Passive biomonitoring and Biomonitoring. Volume 2(11).
Acute	Lasting a short time (test or exposure), severe enough to induce a response rapidly (stress or stimulus), having a sudden onset (effect) as opposed to chronic. Volume 1(1,2,3,5,10), Volume 2(5,8,11).
Acute effect	Overt adverse effect (lethal or sublethal) induced in test organisms within a short period of exposure to a test material. Acute effects often induce highly toxic responses (*e.g.*, mortality or assessment endpoints related to mortality). See also Acute exposure and Acute toxicity. Volume 1(1,2,3,5,10), Volume 2(5,8,11).

Acute exposure	Short period of exposure (minutes, hours, or a few days) relative to the life span of the organism (usually set at < 10% of an organism's life span). See also Acute effect and Acute toxicity. Volume 1(1,2,3,5,10), Volume 2(5,8,11). For *Selenastrum capricornutum*, whose cell numbers double every 12 h at 24°C, a contact time of 1-4 h with a test sample would correspond to an acute exposure allowing for the determination of corresponding acute toxicity effects. Measuring esterase inhibition in *S. capricornutum* after a 1-h exposure to a test chemical is another example of an acute exposure toxicity bioassay (Snell et al., 1996). Volume 1(3).
Acute toxicity	Inherent potential or capacity of a material to cause acute effects that occur rapidly as a result of a short exposure time. See also Acute effect and exposure. Volume 1(1,2,3,5,10), Volume 2(5,8,11).
Additive effect	Effect of a mixture of chemicals whereby the summation of the known effects of individual chemicals is essentially additive. For example, if individual aqueous solutions of chemical A and chemical B each yield an IC50 = 50% v/v (or 2 toxic units) for a particular biotest, their combined toxicity will correspond to an IC50 = 25% v/v (or 4 toxic units). See also Additivity. Volume 2(1,10).
Additivity	Toxicity of a contaminant mixture equal to the sum of toxic effects of the individual contaminants. See also Additive effects, Antagonism and Synergy. Volume 2(1,10).
Ad libitum	Literally means "at one's pleasure". This term is generally used with respect to feeding (see below). Volume 1(11,13).
Ad libitum feeding	Feeding with more food than the organisms are able to ingest during a period *i.e.,* until the fed organisms no longer consume food feeding or until satiation occurs. Volume 1(11,13).
Aeration of medium	Operation during which air from a compressor, which is passed through a particle and moisture filter followed by activated carbon, is directed into the aqueous solution through a Pasteur pipette to bubble the solution. The aeration period stabilizes the carbonate system (bicarbonate and CO_2) so that it is in equilibrium with air, thus preventing pH drift. Volume 1(6).
Algal fluorescence	Re-emission of light initially absorbed by chlorophyll *a* pigments in algal cells. In algal toxicity testing, it can be employed as an indirect measure of algal biomass. Volume 1(3, 6).

Algal symbiotes	Freshwater green algae which live inside the tissue of green *Hydra* in a similar symbiotic relationship to marine zooanthellae algae and corals. They are also named Zoochlorellae algae. They provide additional nutrients to *Hydra* in the form of carbohydrates via photosynthesis, while *Hydra* provides them with a protected environment. Both *Hydra* and corals can experience 'bleaching' where the symbiotic algae are expelled from the organism following some significant environmental stress, particularly increased water temperature. Volume 1(11).
Algicidal	Property of killing algae. The algicidal concentration is the lowest concentration tested which allows no net growth of the population of test organisms during either exposure to the test material or during the recovery period in the absence of test material. See also Algistatic. Volume 1(4).
Algistatic	Property of inhibiting algal growth. The algistatic concentration is the highest concentration tested which allows no net growth of the population of test organisms during exposure to the test material but permits re-growth during the recovery period in the absence of test material. See also Algicide and Algistatic effect. Volume 1(4).
Algistatic effect	Effect caused by a chemical agent which inhibits algal growth. Volume 1(3).
Algorithm	Detailed sequence of actions required for accomplishing a specific task. Volume 2(2).
Alternative assay	Biological-based assay destined to reduce the sacrifice of organisms (usually vertebrates), to reduce the cost and to replace old tests by more rapid and efficient ones. The use of fish cells is an example of an alternative for fish. Volume 1(14).
Analysis of covariance (ANCOVA)	Used to test the main and interactive effects of categorical variables on a continuous dependent variable, controlling for the effects of other selected continuous variables that covary with the dependent variable. Volume 2(4).
Antagonism	Interaction of several agents resulting in a lower effect than the one expected by addition of the individual effects. See also antagonistic effect. Volume 2(2,10).
Antagonistic effect	Toxicity of a mixture of chemicals whereby the summation of the known toxicities of individual chemicals is less than that expected from a simple summation of the toxicities of the individual chemicals comprising the mixture. Volume 2(1).

Antilogarithm	Number to which a given logarithm belongs. If $b^x = a$, then a is called the antilogarithm of x to the base b. Finding an antilogarithm is, in a sense, the inverse of finding a logarithm. Volume 2(3).
Aposymbiotic	Lacking a symbiotic organism (*e.g.*, pink *Hydra*). Volume 1(11).
Artificial sediment	Mixture of materials used to mimic the physical components of a natural sediment. See also Reference sediment. Volume 1(13).
Assessment endpoint	Effect criterion by which toxicity is estimated (*e.g.*, mortality, growth, reproduction). Volume 1(3,10).
Autolysis	Dissolution or destruction (self-digestion) of cell. Volume 1(8).
Auxinic effect	Chemical substance capable of stimulating the growth of phototrophic (micro-)organisms. Phosphorus and nitrogen are two examples of common nutrients capable of enhancing micro-algal growth. Volume 1(3).
AVS:SEM ratio	Surrogate measure of bio-availability. An AVS:SEM ratio > 1 (*i.e.*, more AVS than SEM) indicates a sample where Cd, Cu, Hg, Ni, Zn, and Pb are unlikely to be bio-available (*i.e.*, have formed an insoluble metal precipitate with sulfides). It is expressed in terms of molar differences (*e.g.*, AVS - SEM). See also AVS and SEM. Volume 2(10).
Axenic culture	A mono-specific culture of a test organism (*e.g.*, a single micro-algal species) which is devoid of other species of micro-organisms (*e.g.* other types of algae) and also free of bacterial contamination. Volume 1(3,7,8).
Bacteria	Large group of organisms that do not have organelles enclosed in cell membranes and have DNA in both a chromosome and circular plasmids. They have a protein and complex carbohydrate cell wall over a plasma membrane. Although eukaryotic and prokaryotic cells are structurally different, their basic biochemical processes are similar. Volume 1(1, 2), Volume 2(3).
Bacterial bioluminescence	Production of light by certain marine bacteria. The general consensus is that light is produced when bacterial luciferase catalyzes the bioluminescent oxidation of $FMNH_2$ and a long chain aldehyde by molecular oxygen. Volume 1(1,2).

Bacterial lyophilization	Procedure conducted under vacuum in which water is removed from bacteria (also known as freeze-drying). If the vacuum seal of the container is maintained and the bacteria are stored in the dark, they will remain viable indefinitely. Viable bacteria are activated by rehydration. Volume 1(1,2).
Bacterial reagent	In the Microtox® test, it is a standard culture of freeze-dried (lyophilized) *Vibrio fischeri*, stored in small, sealed vials which each contain about 100 million cells. Volume 1(1,2).
Basal cytotoxicity	Impairment of one or more cellular activities common to all cells. Volume 1(15).
Basal medium	In cultured fish cells assay, it is an aqueous solution of nutrients and buffering agents, such as Leibovitz's L-15, that contains a hexose, bulk ions, trace elements, amino acids, and vitamins. Volume 1(15).
Battery of (toxicity) tests	Use of several laboratory toxicity tests (at least two), usually representative of different levels of biological organization (*e.g.*, a battery composed of a bacterial, algal, micro-invertebrate and fish test) to attempt to circumscribe the full toxicity potential of a liquid or solid matrix sample. Volume 2(1,8).
Bioassay	Biological test in which the severity of the toxic effect caused by a test material is measured by the response of living organisms. Synonyms: biotest, toxicity test, toxicity assay. Volume 2(8).
Bioassay battery approach	Use of several laboratory toxicity tests (at least two), usually representative of different levels of biological organization (*e.g.*, a battery composed of a bacterial, algal, micro-invertebrate and fish test) to attempt to circumscribe the full toxicity potential of a liquid or solid matrix sample. See also battery of (toxicity) tests. Volume 2(1,8).
Biodegradability	Ability of a substance to be broken down into simpler substances by organisms such as bacteria. Volume 2(1).
Biodegradation	Process (*e.g.*, enzymatic breakdown) whereby an organic compound is transformed to a simpler carbon entity (*e.g.*, glucose to carbon dioxide). Volume 2(1,7).
Biodeterioration	Process caused by activities of living organisms whereby properties of a material are modified. Volume 2(11).

Bioindicator (biological indicator)	Measure, index of measures, or model that characterizes an ecosystem or one of its critical components. It may reflect biological, chemical or physical attributes of ecological condition. The primary uses of an indicator are to characterize current status and to track or predict significant change. With a foundation of diagnostic research, an ecological indicator can also be used to identify major ecosystem stress. Volume 2(10).
Biomagnification	Cumulative increase in contaminant body burdens up three or more trophic levels in a food chain. Biomagnification occurs when the intake of a contaminant exceeds the capacity of an organism to excrete and/or metabolize the contaminant in question. Volume 2(10).
Biomarker	Any one of a series of physiological, biochemical, behavioural or metrics measurements reflecting an interaction between a living system (tissue, organ, cell, etc.) and an environmental agent, which may be chemical, physical or biological. For example, the induction of metallothionein, a heavy metal biomarker of defense, is activated in fish hepatic tissue exposed to metals such as cadmium or mercury. Volume 1(14), Volume 2(1,10).
Biomass	Dry or wet weight of living matter. In algal tests, for example, it can be expressed in terms of mg of algae per liter. Because dry weight is difficult to measure accurately, however, surrogate measures of biomass, such as cell counts, are typically used in algal toxicity testing. Volume 1(4).
Biomonitoring	Use of resident or transplanted living organisms (or parts of) to assess water, air, sediment or soil quality. See also Passive biomonitoring and Biomonitoring. Volume 2(11).
Biosolid	Treated sewage sludge that meets US EPA regulations for land application. Volume 2(7).
Biotransformation	Ability of biological tissues to transform chemical compounds. Transformations can involve, for instance, oxidation reactions. Volume 1(14).
Bootstrap method	Re-sampling method that randomly chooses new datasets among experimental data. Volume 2(2).
Brackish	Low salinity exemplified by freshwater and seawater that are mixed near the estuary of a river flowing into the sea. Tidal flats and lagoons of low salinity are also considered as brackish areas (PIANC, 2000). Volume 2(9).

Bray-Curtis Index — Distance coefficient (*e.g.*, linked to fish and benthos field surveys) that reaches a maximum value of 1 for two sites that are entirely different and a minimum value of 0 for two sites that possess identical descriptors. It measures the amount of association between sites. Volume 2(4).

Cell line — Cells obtained from a tissue that are transferred (or passaged) to a new culture vessel and that divide readily in the culture vessel. They can be propagated *in vitro* by repeating the cycle through cell proliferation followed by transferring an aliquot of the cell population into new culture vessels, usually flasks. Volume 1(14,15).

Cell viability assay — Determined on the basis of loss of cell membrane permeability in response to a deliberate modification in culture conditions. Cell viability can be determined by either neutral red or fluorescein diacetate retention assays. Volume 1(14,15).

Chain of custody — Documented and traceable transfer of a sample from the point of collection to reception at the testing laboratory. Volume 1(10).

Chironomus — Non-biting midge with an aquatic larval stage (order Diptera). Volume 1(12,13).

Chronic — Lasting a long time (test or exposure); it can involve a stimulus or stress that is lingering or continues for a long time; it has a light onset (effect) as opposed to an acute one. Volume 1(3,5), Volume 2(2,5,11).

Chronic effects — Subtle adverse effects (lethal or sublethal) induced in the test organisms within a long period of exposure to a test material. Chronic effects often relate to growth or reproduction impairments. See also Acute exposure and Acute toxicity. Volume 1(3,5), Volume 2(2,5,11).

Chronic exposure — Long period of exposure (days, weeks or months) relative to the life span of the test organism (*i.e.*, > 10% of an organism's life span) and also relative to several life-cycle phases (*e.g.*, development, reproduction) (Férard et al., 1992). See also Chronic effects and Chronic toxicity. Volume 1(3,5), Volume 2(2,5,11).

Chronic toxicity	Inherent potential or capacity of a material to cause chronic effects that occur following long exposure times. For *S. capricornutum*, whose cell numbers double every 12 h at 24°C, a 3-d contact time with a test sample corresponds to a chronic exposure period allowing for the determination of corresponding chronic toxicity effects. See also Chronic effects and Chronic exposure. Volume 1(3,5), Volume 2(2,5,11).
CYP1A1	Gene producing cytochrome P4501A1 that biotransforms coplanar aromatic hydrocarbons (*e.g.*, benzo(a)pyrene). Volume 1 (14).
Coefficient of determination (r^2)	Measure of the closeness of fit of a scatter graph to its regression line where $r^2 = 1$ is a perfect fit. Volume 1(6).
Coefficient of variation	A statistical index of precision calculated as ([standard deviation × 100] ÷ mean). The CV is a measure of the variability in a group of measurements. Since the CV is unitless, it can be used to compare CVs from different "experiments". It is also a quality control tool. For example, in the algal microplate toxicity test, algal cell density in control wells at the end of the test exposure period must have a CV not exceeding 20% to meet test acceptability criteria. Volume 1(1,2,3,10).
Coincident Sampling	Sampling at the same location but at different times. Volume 2(10).
Collagenase	A protease (*i.e.*, a protein enzyme that degrades other proteins) specific to collagen which is the main protein matrix that holds liver cells together. Volume 1(14).
Concordance	Total number of correct predictions (*i.e.*, presence or absence of toxic effects) between two bioassays over the total number of test samples. Volume 1(14).
Confidence interval	A range of values estimated by a sample within which the true population value is expected to fall. For example, if an LC50 and its 95% confidence intervals are estimated from a toxicity test, the true population LC50 is expected to fall within the interval 95% of the time. Volume 1(10), Volume 2(5).
Confidence limits	Upper and lower boundaries of the confidence interval. Volume 1(10), Volume 2(20).

Confined disposal	Placement of dredged material within diked nearshore or upland confined placement facilities that enclose and isolate the dredged material from adjacent waters. Confined dredged-material placement does not refer to sub-aqueous capping or contained aquatic dredged-material placement (PIANC, 2000). Volume 2(9).
Confluent monolayer	Animal cells completely covering the surface of a culture vessel. Volume 1(15).
Conspecific	Belonging to the same species. Volume 1(14).
Consumer (primary and secondary)	Heterotrophic organisms which consume other organisms and/or particulate organic matter. Primary consumers are herbivores (*e.g.*, daphnids eating micro-algae) whereas secondary consumers are carnivores (*e.g.*, hydras eating daphnids). Volume 2(1).
Contaminated dredged material	Sediments or materials having unacceptable levels of contaminant(s) that have been demonstrated to cause an unacceptable adverse effect on human health or the environment (PIANC, 2000). Volume 2(9).
Control	Treatment in an investigation or study that duplicates all the conditions and factors that might affect the results of the investigation, except the specific condition that is being studied. In an aquatic toxicity test, the control must duplicate all the conditions of the exposure treatment(s), but must contain no added test material or substance. The control is used to determine the absence of measurable toxicity due to basic test conditions (*e.g.*, temperature, health of test organisms, or effects due to their handling or manipulation). Volume 1(2), Volume 2(5).
Control Chart	Graphical plot of test results with respect to time or sequence of measurement upon which control and warning limits are set to guide in detecting whether the system is in a state of control. Volume 1(10).
Control limits	Limits or combination of limits which, when exceeded, trigger analyst intervention. These limits may be defined statistically or based on test method requirements. Control limits may be assigned to method blanks, check standards, spike recoveries, duplicates and reference samples. Most control limits for toxicity tests are based on thrice the standard deviation of the mean (*i.e.*, one in every 100 tests would be expected to exceed the control limits due to chance alone). Volume 1(10).

Corer — Hollow tubes or casings that are used to collect soil or sediment samples. Small soil corers are normally pushed into the soil or sediment by hand-held tools. See also Sediment core sample. (PIANC, 1997). Volume 2(9).

Correlation analysis — Statistical analysis that calculates the coefficient of correlation (*i.e.*, covariance divided by the product of variances) for a set of variables. Volume 2(2).

Cryovial — A two mL capacity polypropylene container with sealable screw-top lid and "V" shaped bottom. Ideal for storing dried organisms (*e.g.*, amphipods) and water samples and good for digesting small tissue samples because small acid volumes remain in contact with tissue samples. Volume 1(12).

Cryptobiotic — Relating to the dormant stage of a particular micro-organism or organism. Examples include cyst formation in micro-invertebrates such as water fleas (*e.g.*, *Daphnia magna*) or the embedding of physiologically-active algal cells (*e.g.*, *Selenastrum capricornutum*) in an alginate matrix to produce algal beads. Water fleas can later be hatched "on demand" to conduct biological testing, as can be algal cells once they are removed from their beaded matrix. Volume 1(3).

Culture — As a noun, stock of plants or animals raised under defined and controlled conditions to produce healthy test organisms. As a verb, it means to conduct the procedure of raising organisms (Environment Canada, 1999). Volume 1(7,10).

Cytogram — Bi-parametric plot of data from a flow cytometer. Each axis of the plot displays one parameter (light scatter and/or fluorescence). Data from each event (particle) analysed is represented as a dot (particle) on the cytogram. Volume 1(5).

Decomposer — Organism (*e.g.*, a bacterium) that feeds on dead or decaying plants and animals, transforming them chemically, thereby contributing to recycling (in)organic materials to the environment. Volume 2(1).

Dialysis — Removal of a small molecule from a solution with macromolecule(s) by allowing it to diffuse through a semipermeable membrane into a solvent. Volume 1(1).

Diapause — Period during which an organism does not grow, while it awaits necessary environmental conditions. Volume 1(13).

Diluent	In the Microtox® test, it is a solution of 3.5% sodium chloride in distilled or deionized water, which is prepared using reagent-grade salt. Diluent comprised of 3.5% NaCl may be used with samples of marine, estuarine, or freshwater sediment. See also "distilled water" and "deionized water". Volume 1(2).
Dilution water	Solution used to prepare the reference toxicant or effluent dilutions required for toxicity testing. Volume 1(6).
Discriminant analysis	Multivariate statistical analysis using classes of variables and calculating discriminant functions as linear combinations of the variables that maximize the inter-class variance and minimize the intra-class variance. Volume 2(2).
Dispersant	Chemical substance that reduces the surface tension between water and a hydrophobic substance (*e.g.*, oil), and thereby facilitates its dispersal via a water emulsion. Volume 1(3,7).
Dose (or concentration) response model	Function of dose (or concentration) of a chemical able to link a toxicity response to any dose (or concentration) value. Volume 2(2).
Dredged material	Material excavated from waters. The term "dredged material" refers to that which has been dredged usually from the bed of a water body, while the term "sediment" refers to material in a water body prior to the dredging process (PIANC 2000). Volume 2(9).
Dredging	Loosening and lifting earth and sand from the bottom of water bodies. Dredging is often carried out to widen the stream of a river, deepen a harbor or navigational channel, or collect earth and sand for landfill; it is also carried out to remove contaminated bottom deposit or sludge to improve water quality (PIANC, 2000). Volume 2(9).
Dredging process	A process consisting of the following three elements: 1) Excavation: this process involves the dislodgment and removal of sediments (soils) and/or rocks from the bed of the water body. A special machine – the dredger – is used to excavate the material either mechanically, hydraulically or by combined action. 2) Transport of excavated material: transporting materials from the dredging area to the site of utilization, disposal or intermediate treatment, is generally achieved by one of the following methods: in self-containing hoppers of the dredgers; in barges; pumping through pipelines; and using natural forces such as waves and currents. 3) Other, rarely used transport methods are truck and conveyer belt transport. The method of transport is generally linked to the type of dredger being used. Volume 2(9).

Duplicate	Quality control sample, often chosen randomly, from a batch of samples and undergoing separate, but identical sample preparation and analysis whose purpose is to monitor method precision and sample homogeneity. Duplicate testing also aids in the evaluation of analyst proficiency. Volume 1(10).
EC50	See ECx.
ECx	Effective concentration of a test material in the test matrix (*e.g.*, growth medium) that is calculated to exhibit a specified non-lethal or lethal effect to x% of a group of test organisms during exposure over a specified period of time. The ECx and its 95% confidence limits are usually derived by statistical analysis of responses in several test concentrations. The particular effect must be specified as well as the exposure time (*e.g.*, 48-h EC50 for immobilization). Volume 1(1,4,10).
Ecocompatibility	Situation where pollutant release from waste, when deposited in a specific physical, hydrogeological, physico-chemical and biological context, is in keeping with the acceptable pollutant level of receiving environments (Perrodin et al., 1996). Volume 2(11).
Effluent	Any liquid, gaseous or aerosolic waste discharged in the environment. Generally, it is a complex mixture. For example, wastewaters include: mine water effluent, mill process effluent, tailings impoundment area effluent, treatment pond or treatment facility effluent, seepage and surface drainage. Volume 1(9,10,14), Volume 2(2,5).
Electrophiles	Compounds representing a non reversible mode of action. Electrophilic interactions involve substitution or conjugation of electron-rich groups to nucleophilic sites in cellular macromolecules. Volume 1(8).
Elutriate	Aqueous solution obtained after adding a fixed volume of water to a solid medium (*e.g.*, waste, soil or sediment), shaking of the mixture, then centrifuging, or filtering it or decanting the supernatant. Volume 1(3,9), Volume 2(8,9).
Emulsifier	Substance that aids the fine mixing (in the form of small droplets) within water of an otherwise hydrophobic substance (Environment Canada, 1999). Volume 1(7).
Endocrine disruption	Any one of a series of effects caused by hormonally-active agents that alter the homeostatic function of hormone or physiological system under the control of hormone(s). Volume 1(14), Volume 2(1).

Endocrine disruptors	Exogenous chemicals which cause adverse health effects in organisms or their progeny as a result of changes in endocrine function. Volume 1(9, 13,14), Volume 2(1).
Endpoint	Measurement(s) or value(s) that characterize the results of a test (*e.g.*, LC50, ICp). This term might also mean the reaction of the test organisms to show the effect which is measured upon completion of the test (*e.g.*, inhibition of light production). Volume 1(2,10).
EPT Index	Total number of distinct taxa within the orders Ephemeroptera, Plecoptera and Trichoptera compared to total taxa present. Volume 2(4).
Ephippium (s.), *ephippia* (pl.)	Egg case that develops under the postero-dorsal part of the adult *Daphnia* female carapace in response to unfavorable environmental conditions. Ephippia eggs are the outcome of sexual reproduction. Volume 1(10).
Epibenthic	Characteristic of organisms that have regular contact with sediment and live just above the sediment/water interface. Volume 2(8).
Equitox parameter	Toxic unit used by the French Water Agencies. See also Toxic unit. Volume 2(2).
Esterases	Group of enzymes involved in phospholipid turnover in cell membranes. Esterase activity in algae has been shown to relate well to metabolic activity and cell viability. Volume 1(5).
Estuarine water	Coastal body of ocean water that is measurably diluted with fresh water derived from land drainage. Volume 1(2).
Eukaryotes	All organisms except viruses, bacteria and archaea. See eukaryotic cell. Volume 1(3,8).
Eukaryotic cell	Advanced cell type with a nuclear membrane surrounding genetic material and numerous membrane-bound organelles dispersed in a complex cellular structure see Eukaryotes. Volume 1(8).
Eutrophication	Excessive enrichment of waters with nutrients (essentially nitrate and phosphate), including the associated adverse biological effects (*i.e.*, aquatic plant blooms). Volume 1(3).
Exuvium (s.), *exuviae* (pl.)	Remains of the pupa, which is discarded when an insect has emerged. Volume 1(13).

Term	Definition
Far-far field	Receiving water near an industry's effluent discharge that is more distant from the effluent outfall than the far-field and in which the effluent concentration is lower than that of the far-field. Volume 2(4).
Far-field	Receiving water near an industry's effluent discharge and located along a dilution gradient in which effluent concentration is less than or equal to 1%. Volume 2(4).
Field swipes for chemistry	Check on the quality of equipment decontamination procedures involving the "swiping" of sterile filter paper over sampling equipment after decontamination has occurred, followed by chemical analysis of the field swipe and an unused filter paper. Volume 2(10).
Fines	Sediment or soil particles which are $\leq$ 63 µm in size. Measurements of % fines include all particles defined as silt (*i.e.*, particles $\leq$ 63 µm but $\geq$ 4 µm) or clay (*i.e.*, particles $<$ 4 µm). Volume 1(2).
Flow cytometer	Instrument that is capable of rapid and quantitative measurements of individual cells in a moving fluid. Thousands of cells pass through a light source (usually a laser, 488 nm) and measurements of cell density, light scatter (two parameters) and fluorescence (three or more parameters) are collected simultaneously. Volume 1(5).
Flow-through	Tests in which solutions in test vessels are renewed continuously by the constant inflow of a fresh solution, or by a frequent intermittent inflow. Synonymous term is "dynamic". Volume 1(10).
Fluorescent unit (FU)	Arbitrary unit of measurement by fluorescent plate reader. Volume 1(15).
Fluorometer	Instrument that measures the fluorescence properties of solutions. It is composed of a high-energy lamp for excitation and a phototube for emission readings. Instruments are available in either tube or microplate formats. Volume 1(14).
Foot-candle	One of several units of illumination based on units per square meter. One foot-candle = 10.76 lux. Volume 1(3).
Formulated sediment	Artificial sediment formulated from constituents such as silica sand and peat moss according to standardized recipes, intended to match the physical characteristics (*e.g.*, grain size, TOC) of the site under investigation. Volume 2(10).

Frond — Individual leaf-like structure of a duckweed plant. It is the smallest unit (*i.e.*, individual) capable of reproducing (Environment Canada, 1999). Volume 1(7).

Gamma — In the Microtox® test, it is a measure of light loss used in calculating the IC50 or ICp. It is calculated individually for each cuvette containing a filtrate of a particular test concentration. Gamma (Γ) is calculated based on the ratio between the amount of light emitted by a test filtrate and that emitted by the control solutions, as follows: $\Gamma = (I_c/I_t) - 1$, where I_c = the average light reading of filtrates of the control solutions, and I_t = the light reading of a filtrate of a particular concentration of the test material. When Gamma equals unity ($\Gamma = 1$), half of the light production has been lost. Vol. 1(2).

Genomics — Branch of genetics that studies organisms in terms of their genomes (*i.e.*, full DNA sequences). Volume 1(14).

Genotoxicity — Inherent potential or capacity of a chemical, biological or physical agent to damage the hereditary material of cells (DNA) or organ tissues (*i.e.*, causing DNA damage or alterations that can give rise to mutations, tumors and/or cancer). Volume 2(1,2).

Geometric mean — Mean of repeated measurements, calculated on a logarithmic basis. It has the advantage that extreme values do not influence the mean as is the case for an arithmetic mean. It can be calculated as the nth root of the product of the n values, and it can also be calculated as the antilogarithm of the mean of the logarithms of the n values. Volume 1(2).

Gibbosity — Fronds exhibiting a humped or swollen appearance (Environment Canada, 1999). Volume 1(7).

Groundwater — Source of water that is found below ground level. Volume 1(14).

Growth — Increase in size or weight as the result of proliferation of new tissues in a specified period of time. For example, in the duckweed test, it refers to an increase in frond number over the test period as well as the dry weight of fronds at the end of the test. Volume 1(4,7).

Growth medium — Medium promoting growth. For example, for culturing cells, basal medium plus a supplement of fetal bovine serum (FBS). Volume 1(15).

Growth rate — Rate at which the biomass increases (Environment Canada, 1999). Volume 1(7).

Hardness	Concentration of cations in water that will react with a sodium soap to precipitate an insoluble residue. Total hardness is a measure of the concentration of calcium and magnesium ions in water, usually expressed as mg/L $CaCO_3$. Volume 1(10).
Hazard	Potential for adverse effect(s) that might result from exposure to a chemical, biological or physical agent. Volume 2(8,10).
Hazard assessment	Process that evaluates the type and magnitude of adverse effect(s) caused by a stressor (such as chemical contamination). Volume 2(8,10).
Hepatocyte	Main epithelial cell in the liver. Volume 1(14).
Heterotroph	Organism that requires complex nutrient molecules as a source of carbon and energy. Volume 1(1).
Hexagenia	Burrowing mayfly (order Ephemeroptera). Volume 1(12).
Highest effect concentration (HEC)	Concentration related to the highest induced effect. In the Mutatox® test, for example, this effect refers to induced luminescence. Volume 2(11).
Histogram	Single-parameter plot of data. In flow cytometry, the horizontal axis displays the light scatter or fluorescence intensity parameter and the vertical parameter displays the number of events (*e.g.,* cell count). Volume 1(5).
Holding Time	Time elapsed between the end of sample collection or sample preparation and the initiation of analysis. Volume 1(10).
Hyalella	Amphipod crustacean (suborder Gammaridea). Volume 1(12).
Hydraulic dredgers	Dredgers using hydraulic centrifugal pumps to provide the dislodging and lifting force for sediment material removal in a liquid slurry form. Hydraulic dredging and transport methods "slurry the sediment", that is, they add large amounts of process water and thus change the original structure of sediments (PIANC, 2001). Volume 2(9).
Hydrodynamic dredging	See Hydraulic dredgers. Volume 2(9).
Hydroid	Individual *Hydra* including any attached buds. Volume 1(11).
Hydrophobic	Molecules or molecular groups that mix poorly with water (*e.g.*, hydrocarbons and fats are hydrophobic). Volume 1(3).
Hydroponic cultures	Methods of culturing plants by growing them, for example, in gravel, through which water containing dissolved inorganic nutrient salts is pumped. Volume 2(6).
IC25 or IC50	See ICp. Volume 1(4), Volume 2(4,5).

ICp (or ICx) Inhibiting concentration for a (specified) percentage effect. It relates to a point estimate of a test sample concentration that causes a designated percent inhibition (*p*) compared to the control, *e.g.*, a corresponding percent reduction in a quantitative assessment endpoint such as algal growth inhibition. The ICp and its 95% confidence limits are usually derived by statistical analysis of responses in several test concentrations. Examples of frequently-reported ICps are IC50 (50% effect in relation to control organisms) or IC20 (20% effect in relation to control organisms). This term should be used for any bioassay which measures a continuously-variable effect, such as light production, reproduction, respiration, or dry weight at test end. Volume 1(2, 3, 4), Volume 2(4,5,8).

Imhoff settling cone Cone-shaped container (1 L capacity) for measuring the volume of suspended matter in liquids. Also used as toxicity test chambers because their shape results in adequate sediment depth when using small volumes of sediment and large volumes of water. Volume 1(12).

Immobility In the daphnid test, inability to swim during the 15 seconds following gentle agitation of the test solution, even if the daphnids can still move their antennae. Volume 1(10).

Immunotoxicity Inherent potential or capacity of a chemical agent which specifically affects cells having immune functions (*e.g.*, heavy metals can intoxicate bivalve hemocytes and impede them from ingesting and lysing pathogenic micro-organisms which can lead to either sub-lethal or lethal infections). Volume 2(1).

Index Single parameter summarizing several values while minimizing the loss of information and attempting to be relevant to the notion of interest (*e.g.*, toxicity). Volume 2(2).

Inhibitory concentration (IC) See ICp. Volume 1(6).

Inter-laboratory Among-laboratory activities. For example, inter-laboratory variability evaluates reproducibility of similar analyses by different laboratories. Estimation of inter-laboratory variability addresses a measure of quality assurance of laboratories (Environment Canada, 1999). Volume 1(10).

Interstitial water	Water occupying space between sediment particles. The amount of interstitial water in sediment is calculated and expressed as the percentage ratio of the weight of water in the sediment to the weight of the whole sediment including the pore water. It can be recovered by methods such as squeezing, centrifugation, or suction. Synonymous term is pore water. Volume 1(2,9,14), Volume 2(5,8,9).
Intra-laboratory	Within-laboratory activities. For example, intra-laboratory variability evaluates repeatability of analysis within the same laboratory system. Estimation of intra-laboratory variability of data is a principal quality control measure of a laboratory (Environment Canada, 1999). Volume 1(10).
Isogenic population	Members of a population having similar genetic make-up since they are clones of the original organisms. For example, *Hydra* use asexual budding as their prime form of reproduction, and all buds are genetic clones of the parent *Hydra*. Sexual reproduction in *Hydra* involving the production of testes and ovaries only occurs when environmental conditions become unfavorable: in this case, *Hydra* produce sperm and eggs which result in a resistant fertilized zygote being produced that can withstand dessication (drying out) and freezing. Volume 1(11).
L-15/ex	Simplified version of the basal medium L-15 that contains only galactose, pyruvate and bulk ions Volume 1(15).
Laboratory	Body or part of an organization that is involved in calibration and/or testing. Volume 1(10).
Laboratory accreditation	Formal recognition, by a registered accrediting body, of the competence of a laboratory to conduct specific functions. The process by which a laboratory quality system (*i.e.*, laboratory management system) is evaluated through regular site assessments by the accrediting body, and may include annual or twice-yearly proficiency testing rounds. Volume 1(10).
Lag phase	Stage in the growth cycle when the growth rate is changing. There may be increase or decrease in algal cell mass per unit volume of cell suspension. Volume 1(6).
Larval instar	Period of the life-cycle between molts. Volume 1(13).

LC50 Median lethal concentration of a test material in the test matrix (*e.g.*, growth medium) that is calculated to exhibit a lethal effect to 50% of a group of test organisms during exposure over a specified period of time. The LC50 and its 95% confidence limits are usually derived by statistical analysis of mortalities in several test concentrations. The duration of exposure must be specified (*e.g.*, 48-h LC50). Volume 1(1,4,10), Volume 2(5).

Leachate Water recovered after its percolation through a solid medium (*e.g.*, soil or solid waste). Volume 1(3).

***Lemna* root** Part of the *Lemna* plant that assumes a root-like structure (Environment Canada, 1999). Volume 1(7).

Lentic system Still-water aquatic system, such as a lake, a pond or a swamp. Volume 1(13), Volume 2(1).

Lethal Causing death. Death is defined as the cessation of visible signs of all movement or other activity. For example, death of daphnids is defined as the cessation of all visible signs of movement or other activity, including second antennae, abdominal legs, and heartbeat as observed through a microscope. Volume 1(10,14), Volume 2(8).

Limnic environment Ecological conditions (affecting the life of a plant or animal) related to lakes and other bodies of fresh standing water or (more widely) all inland water. Volume 2(9).

Linear interpolation Statistical method used to determine a precise point estimate of the test sample (*e.g.*, toxicant solution, effluent) that produces a specific percent effect. In algal assays, for example, one would strive to determine a particular reduction (*e.g.*, 20, 25 or 50%) in algal growth by calculating ICps corresponding to IC20, IC25 or IC50. Volume 1(3).

Liquid-phase (toxicity) test Bioassay using a biological system which measures toxic effects of the liquid/aquatic phase of a test material (*e.g.*, porewater, elutriate, leachate) and determines a response (*e.g.*, acute and/or chronic toxicity). See also Solid-phase (toxicity) test. Volume 1(2), Volume 2(9).

LOEC Lowest observed effect concentration, that is the lowest concentration in the tested series at which a biological effect is observed (*i.e.,* where the mean value for the observed response is significantly different from the controls). It is one of the tested concentrations obtained, for example, after analysis of variance and multiple comparison statistical testing (*e.g.*, Dunnett test). Volume 1(3,4), Volume 2(8,11).

Log (logarithmic) phase	Stage in the growth cycle when the mass of microbial cells doubles over each of the successive and equal time intervals. The doubling time and, therefore, the growth rate during the entire log phase is thus constant. Volume 1(6).
Lotic system	Running-water aquatic system including rivers, brooks or streams. Volume 1(13), Volume 2(1).
LT50	Lethal time (period of exposure) estimated to cause 50% mortality in a group of organisms held in a particular test solution. The value can be estimated graphically or by regression. Volume 1(10).
Lumen	One of several units of illumination based on units per square metre. Synonymous term is lux (*i.e.*, 1 lumen = 1 lux). Volume 1(3).
Lux	One of several units of illumination based on units per square metre. One lux = 0.0929 foot-candles, and 1 foot-candle = 10.764 lux. Relationships between lux and $\mu E.m^{-2}.s^{-1}$ is variable and depends on light source, light meter used, geometrical arrangement of the exposure environment and possible light reflections, so one lux ≈ $0.015 \mu E.m^{-2}.s^{-1}$ (range of 0.012 to 0.019). Synonymous term is lumen (*i.e.*, 1 lux = 1 lumen). Volume 1(3,10).
Lyophilization	Process which extracts water from biological products or field samples, so that they remain stable over time. It is carried out using a principle called sublimation, which is the transition of a substance from the solid to the vapour state. Synonymous term is freeze-drying. Volume 1(2,3).
Lyophilized organism	Organisms which have been freeze-dried under vacuum (see above). Some bacteria, for example, can be lyophilized and stored for months at room temperature. They can then be rehydrated on demand and used to conduct bioassays. In the Microtox® test, lyophilized *Vibrio fischeri* are stored in a freezer at -20°C and will be ready for use until the expiration date, which is provided with each batch of Bacterial Reagent. Volume 1(2,3).
Macro	Computer program able to execute sequences of interactive software functions together with instructions using a programming language. Volume 2(2).
Manning sampler	Piece of equipment employed in fluid monitoring as in the collection of specific volumes of wastewater over time and commercialized by Manning Environmental Inc. Volume 2(1).

Marine water	Water coming from or within the ocean, sea, or inshore location where there is no appreciable dilution of water by natural fresh water derived from land drainage. Volume 1(2).
Matrix effect	Phenomenon occurring when toxicants interact with other effluent constituents in ways that change their toxicity. Volume 2(5).
Maximum standing crop	Algal biomass which results after cells have used up all available growth-stimulating nutrients under controlled experimental conditions. Volume 1(3).
Measurement endpoint	Numerical expression of a specific assessment endpoint or effect criterion (*e.g.*, IC50, NOEC, LOEC). Volume 1(3,10), Volume 2(8).
Mechanical dredgers	Dredgers well suited for removing hard-packed sediment material or debris and for working in confined areas (PIANC, 2001). Volume 2(9).
Mesocosm	Experimental system reflecting semi-realistic conditions. Volume 2(2).
Metallothionein	Small molecular weight protein family, rich in cysteine, that binds strongly to divalent heavy metals. The synthesis is under the control of essential metals like zinc and copper. Other metals such as cadmium, mercury and silver can induce its concentration in cells. Volume 1(14).
Milli-Q water	Reverse osmosis water which is passed through a Milli-Q Plus system (Millipore Corp.) to produce water, which meets the American Society of Testing Materials (ASTM) type 1 reagent grade water standard. Volume 1(6).
Model parameter	Constant value in a model that explains its properties. Volume 2(2).
Molting	Shedding of carapace during the growth phase. Volume 1(10).
Monitoring	Act of observing something and sometimes keeping a record of it over space and time. It can refer to the periodic (routine) checking and measurement of certain biological or water-quality variables, or the collection and testing for toxicity of samples of effluent, elutriate, leachate, or receiving water. Volume 1(7,14).
Monotonous response	Response that continuously increases (or decreases) with dose or concentration. Volume 2(2).
Mortality	Ratio of deaths in a population of cells. It is usually expressed in percentage (%).Volume 1(14).

Multiple regression method	Linear regression using several variables. Volume 2(2).
Multitrophic	Use of organisms from several different trophic levels, which can include decomposers, primary producers and (primary, secondary and tertiary) consumers. Volume 2(1).
Near-field	Receiving water adjacent to the point of industry's effluent discharge in which the water or sediment quality is potentially affected by the effluent discharge. Effluent concentration in the receiving water of the near-field will be greater than or equal to 1%. Volume 2(4)
Neat effluent sample	Undiluted or unaltered wastewater sample. Volume 2(1).
Necrosis	It indicates dead (*i.e.*, with brown or white spots) frond tissue, (Environment Canada, 1999). Volume 1(7).
Negative control sediment	Uncontaminated (clean) sediment which does not contain concentrations of one or more contaminants that could affect the performance (*e.g.*, light production) of test organisms. This sediment may be natural, field-collected sediment from an uncontaminated site, or artificial sediment formulated in the laboratory using an appropriate mixture of uncontaminated (clean) sand, silt, and/or clay. This sediment contains no added test material or substance. For example, in the solid-phase test using *V. fischeri*, it must enable an acceptable rate of light production in line with test conditions and procedures. The use of negative control sediment provides a basis for judging the toxicity of coarse-grained (< 20% fines) test sediment. See also Artificial control sediment and Reference sediment. Volume 1(2).
Neonate	Newly born organism (*e.g.*, daphnid). Volume 1(10).
NOEC	No-observed-effect-concentration, that is the highest concentration in the tested series where exposed organisms present no significant effect in relation to control organisms (*i.e.*, where the mean value for the observed response is not significantly different from the controls). It is always the next lowest concentration in the dilution series after the LOEC. Volume 1(3,4), Volume 2(3,8,11).
Non linear regression	Regression where the model is not a linear function of each parameter. Volume 2(2).
Non polar narcotics	Compounds causing baseline toxicity, *i.e.*, reversible state of arrested activity of protoplasmic structures (Bradbury et al., 1989). Volume 1(8).

Organic extract	Organic solution obtained from, for example, Soxhlet extractions, after adding an extractant (*e.g.*, dimethyl sulfoxide) to samples. Volume 2(8).
Orthogonal variables	Variables for which coefficients of correlation are inexistent. Volume 2(2).
Oxidative stress	Stress condition where oxygen (radical) reacts with internal components in cells (*e.g.*, lipids and DNA) and produces damages that eventually kill or destroy tissues. Considered as a universal mechanism of toxic damage in cells. Vol. 1(14).
Parshall flume	Specially-shaped open channel flow section device which may be installed in a canal, lateral, or ditch to measure the flow rate, such as that of an industrial effluent. Volume 2(1).
Passive biomonitoring	Use of resident living organisms (or part of) to assess water, air, sediment or soil quality. See also Active biomonitoring and Biomonitoring. Volume 2(11).
Pelagic	Aquatic organism which remains free-swimming or free-floating. Volume 2(8).
Perfusion	Pumping a liquid into an organ or tissue by way of blood vessels. Volume 1(14).
Permeability	Property of a cell or a material that can be pervaded by a liquid such as by osmosis or diffusion. Volume 1(14).
Persistence	Resistance of an organic molecule to transformation by either chemical or biological processes contributing to its longevity in the environment (*e.g.*, many organochlorine compounds are known to be persistent). Persistent organic compounds, because they are lipid-soluble, tend to accumulate in aquatic biota where they may exert adverse effects. Volume 2(1).
Petrographic analysis	Examination of a sediment sample under a high-powered microscope by trained experts in order to quantify the percentage of coal particles present. Volume 2(10).
pHi	Initial pH of an effluent sample as received by the test laboratory, before any adjustment or manipulation has been performed. Volume 2(5).
Photoperiod	Duration of light and darkness within 24 hours. Volume 1(10).
Phototrophic	Organism which must use sunlight as an energy source for nutritional purposes (*e.g.*, phytoplankton). Volume 1(3).

pH-scale	Logarithmic scale devised by Sørensen for expressing acidity or alkalinity of a solution. It is expressed numerically as the logarithm to the base 10 of the reciprocal of the hydrogen ions activity (in moles per litre). Volume 2(3).
Phytotoxicity	Potential of any agent (physical, biological, chemical) to cause adverse effects toward vegetal systems. Volume 1(3).
PLS regression	Partial least square regression: a regression method that maximizes the co-inertia of a table of independent and a table of dependent variables. Volume 2(2).
Polar narcotics	Aromatic compounds with strong electron releasing amino or hydroxy moieties, which have a narcotic mode of action (Bradbury et al., 1989). Volume 1(8).
"Polluter pays" principle	Principle stating that a polluting entity (*e.g.*, an industrial plant) should be charged the cost of restoration of the environment. Volume 2(2).
Ponar grab	Sampling device operated using a boat-mounted winch that allows collection of a relatively undisturbed surface sediment sample. Essentially, a set of "jaws" with a trigger that closes the sampling device on impact. Volume 2(10).
Pore water	See interstitial water. Volume 1(2,9), Volume 2(5,8,9).
Positive control sediment	Sediment which is known to be contaminated with one or more toxic chemicals, and which causes a predictable toxic response (for instance, inhibition of light production) with the test organisms according to the procedures and conditions of the test. This sediment might be one of the following: a standard contaminated sediment; artificial sediment or reference sediment that has been spiked experimentally with a toxic chemical; or a highly-contaminated sample of field-collected sediment, shown previously to be toxic to a (battery of) bioassay and for which its physicochemical characteristics are known. The use of positive control sediment assists in interpreting data derived from toxicity tests using test sediment. For a reference method, positive control sediment must be used as a reference toxicant when appraising the sensitivity of the test organisms and the precision and reliability of results obtained by the laboratory for that material. See also Standard contaminated sediment, Artificial sediment, Reference sediment, and Reference toxicant. Volume 1(2).
Primary consumer	Animal that eats, for example, green plants or algae in a food chain. Volume 1(8).

Primary cultures — Cells freshly extracted and isolated from an organ or tissue and plated in a defined culture medium (*e.g.*, PBS or L-15 media). During this procedure two parallel processes occur: 1) differentiated cells of the original tissue explants usually do not divide and, with time, will successively lose some of their specialized functions (dedifferentiation); and 2) decrease of number of specialized cells (*e.g.*, fibroblasts divide rapidly, and will eventually outnumber the specialized cells). Volume 1(14).

Producer (primary) — Autotrophic organisms (plants and algae) which synthesize organic matter from inorganic materials (*e.g.*, algae photosynthesize sugars from CO_2). Volume 2(1).

pT-bioassay — Bioassay belonging to a test battery for the determination of the toxicity class of a wastewater effluent. Volume 2(3).

pT-index — Numerical ecotoxicological classification of environmental samples attained with a test battery. The pT-value of the most sensitive organism within a test battery, the pT_{max}-value, determines the toxicity class of an investigated sample. Roman numerals are assigned to each toxicity class which corresponds to a pT-index. Volume 2(3).

pT-method — Procedure in accordance with a particular theory for environmental protection which includes the determination of pT-values and pT-indices. Volume 2(3).

pT-scale — A logarithmic scale for expressing aquatic toxicity with regard to a single test organism, along which distances are proportional to the pT-values. Volume 2(3).

pT-value — Numerical designation of aquatic toxicity: the highest dilution level without effect is used for the numerical designation of the toxicity with regard to a single test organism. The pT-value is the negative binary logarithm of the first non-toxic dilution factor in a dilution series in geometric sequence with a dilution factor of two. Volume 2(3).

Quantal — Toxicity test or effect endpoint for which the result can only be expressed as pass/fail or yes/no (for instance, survival/no survival). Volume 2(8).

Quantal flux — Illumination or irradiance of light in the photosynthetically effective wavelength range (400 - 700 nm), expressed in lux, foot-candles or $\mu E.m^{-2}.s^{-1}$. Volume 1(3).

Quantitative — Toxicity test or effect endpoint for which the result can be anywhere on a numerical scale (for instance, weight gained, number of young produced). Volume 2(8).

Receiving water	Surface water (*e.g*, stream, lake) receiving the effluent of a discharged waste. A representative receiving water sample should be collected upstream from the source of contamination or adjacent to the source but unaffected by it. Volume 1(6), Volume 2(5).
Reconstitution solution	Non-toxic distilled or deionized water that is used to activate a vial of Bacterial Reagent. Volume 1(2).
Reference material	Material that may consist of one or more substances whose properties are sufficiently well established to be used for the calibration of a test system. Volume 1(10).
Reference sediment	Field-collected sample of presumably clean (uncontaminated) sediment, selected for properties (*e.g.*, particle size, compactness, total organic content) representing sediment conditions that closely match those of the sample(s) of test sediment except for the degree of chemical contaminants. It is often selected from a site that is uninfluenced or minimally influenced by the source(s) of anthropogenic contamination but within the general vicinity of the site(s) where samples of test sediment are collected. A reference sediment should not produce a toxic effect (or have a minimum effect) on a test species. A sample of reference sediment should be included in each series of toxicity tests with test sediment(s). See also Artificial sediment, Positive control sediment and Test sediment. Volume 1(2,13), Volume 2(8).
Reference substance (or toxicant)	Selected chemical employed to measure the sensitivity of the test organisms in order to establish confidence in toxicity data obtained for a given test sample (or a batch of test samples). In most instances, a toxicity test with a reference toxicant is performed i) to confirm that test organisms (or cells) are in good physiological health for bioanalytical purposes at the time the test sample is evaluated, and ii) to assess the precision and reliability of results obtained by the laboratory for that reference toxicant. The toxicant selected should meet different properties as defined by Environment Canada, 1990. Volume 1(2,3,6,7,14), Volume 2(11).
Reference toxicant testing	See above. Volume 1(2,7,10).
Reference toxicity test	Test conducted using a reference toxicant in conjunction with a toxicity test to appraise the sensitivity of the organisms and the precision and reliability of results obtained by the laboratory at the time the test material is evaluated. Deviations outside an established normal range indicate that the sensitivity of the test organisms (and/or the performance and precision of the test) are suspect. Volume 1(2,7,10).

Regression	Statistical method to calculate a set of model parameters for which a model best fits the experimental data. Volume 2(2).
Regulatory authorities	Administrative or political authorities in charge of setting-up and enforcing a law or set of rules. For example, regulatory authorities implement rules to protect the aquatic environment from impairment due to the release of toxic effluents. Volume 2(2).
Residue	Difference between a modeled value and an experimental observation. Volume 2(2).
Response ratio (relative to control cells)	Amount of retained dye in cells treated to a test substance divided by the amount of retained dye in control (unexposed) cells. It indicates changes in cell viability. Volume 1(14).
Resting egg	Cyst, dormant organism or organism in a cryptobiotic stage. Volume 1(9).
Rhizosphere	Part of the ground which is located in the immediate environment of plant roots. It is very rich in micro-organisms and biological substances. Volume 2(6).
Richter scale	Logarithmic scale devised by Richter for expressing the magnitude of an earthquake from seismograph oscillations. Volume 2(3).
Risk assessment	Process of estimating the probabilities and magnitude of undesired effects resulting from the release of chemicals, other human actions or natural catastrophes. Volume 2(10).
Root exudates	Low molecular weight metabolites that enter the soil from the roots of plants. Volume 2(6).
Rotifer cyst	Encysted, diapausing embryo capable of remaining dormant for decades. Volume 1(9).
Sample	Portion of a lot or population consisting of one or more single units. Volume 1(10).
Sample preparation	All procedures applied to a sample prior to analysis; may include pre-treatment (*e.g.*, filtration, homogenization). Volume 1(10).
Sample pre-treatment	All procedures applied to a collected sample prior to sample analysis, including removal of unwanted material, removal of moisture, sub-sampling and/or homogenization. Vol. 1(10).

Sediment	Particulate material (*e.g.*, sand, silt, clay) which has been transported and deposited in the bottom of a body of water. Sediment input to a body of water comes from natural sources, such as erosion of soils and weathering of rock, or as the result of anthropogenic activities, such as forest or agricultural practices, or construction activities. The term can also describe a material that has been experimentally prepared (formulated) using selected particulate material (*e.g.*, sand of particular grain size, bentonite clay, etc.). Volume 1(2), Volume 2(5,9).
Sediment core sample	Sediment sample collected with a corer. The advantage of corers is that they preserve the vertical profile of the chemical constituents of the sediment. This allows for sediments to be sub-sampled to specific depths. Volume 2(9).
Sediment quality triad	Effects-based approach for assessing the status of contaminated sediments based on chemistry, biology and ecotoxicology. Volume 2(10).
Sediment reference area	Area with sediment that has similar physical characteristics as the site under investigation, but without elevated contaminant concentrations. Volume 2(10).
Sediment relocation	Aquatic disposal/placement of dredged material in water bodies including navigable and non-navigable waters, small lakes, lagoons and rivers (PIANC, 2000). Volume 2(9).
Sensitivity	1- Ability to detect a toxic effect at a very low concentration of test sample, 2- In quality control, it is the slope of a concentration-response relationship, 3- Number of toxic samples in a test system (*e.g.*, trout hepatocyte culture) divided by the number of toxic samples in another test system (*e.g.*, rainbow trout test). In the context of alternative tests, the sensitivity of fish cell methods is usually compared with the corresponding whole organism response. Volume 1(14).
Sexual dimorphism	Differences between males and females. Volume 1(13).
Sexually immature fish	Young fish that has not started its reproductive cycle with the absence of secondary sexual characteristics. Volume 1(14).
Simpson's Diversity Index	Proportion of individuals for each taxonomic group that contributes to the total individuals in a field site under study. The arithmetic mean (plus or minus the standard error, plus or minus the standard deviation), minimum and maximum for the area are also calculated. Volume 2(4).

Simpson's Evenness Index	Expressing Simpson's Diversity Index, D, as a proportion of the maximum possible value of D_s assumes individuals were completely evenly distributed among the species. Volume 2(4).
Simultaneously Extractable Metals (SEM)	Chemical analysis that quantifies the sum of Cd, Cu, Hg, Ni, Zn and Pb that can be extracted from a sediment sample. Volume 2(10).
Soil	Whole, intact material representative of the terrestrial environment that has had minimal manipulation following collection. It is formed by the physical and chemical disintegration of rocks and the deposition of leaf litter and/or decomposition and recycling of nutrients from plant and animal life. Its physicochemical characteristics are influenced by microbial and invertebrate activities therein, and by anthropogenic activities. Volume 1(2).
Solid-phase (toxicity) test	Bioassay using a biological system which measures toxic effects of solid phase of a test material (*e.g.*, bulk/whole sediment) and determines a response (*e.g.*, acute and/or chronic toxicity). It usually comprises a series of test concentrations prepared using an aliquot of the test material. See also Liquid-phase (toxicity) test. Volume 1(2), Volume 2(9).
Speciation effects	Any of a series of physical, chemical or biological factors that can cause changes in the form, bioavailability, uptake, mobility and toxicity of a chemical substance. Volume 1(3).
Stabilization pond	Relatively shallow body of wastewater contained in an earthen basin used for secondary biological treatment. Volume 2(7).
Standard deviation	Square root of the sample variance. Volume 1(10).
Standard operating procedure (SOP)	Written, authorized and controlled quality document that details instructions for the conduct of laboratory activities; SOPs are developed by laboratories when adopting a standard method or when developing laboratory-specific procedures. Volume 1(10).
Standardization	Imposition of rules permitting to check or validate the accuracy of a test using live organisms. For example, the use of a well-defined experimental procedure and the use of a reference toxicant are important rules to standardize a test. Test standardization also requires that the test be feasible by many laboratories and yield comparable results with the same test substance. Volume 1(14).

Static test	Toxicity test in which test solutions are not renewed during the test period (Environment Canada, 1999). Volume 1(3,7,10).
Static renewal test	Toxicity test in which test solutions are renewed (replaced) periodically (*e.g.*, at specific intervals) during the test period. Synonymous terms are batch replacement, renewed static, renewal, intermittent renewal, static replacement, and semi-static (Environment Canada, 1999). Volume 1(7,10).
Static replacement	See above. Volume 1(10).
Stock	Ongoing laboratory culture of a specific test organism from which individuals are selected and used to set up separate test cultures (Environment Canada, 1999). Volume 1(7).
Strain	Variant group within a species maintained in culture, with more or less distinct morphological, physiological, or cultural characteristics (Environment Canada, 1999). Volume 1(7).
Subculture	1- As a noun, laboratory culture of a specific test organism that has been prepared from a pre-existing culture, such as the stock culture. 2- As a verb, to conduct the procedure of preparing a subculture (Environment Canada, 1999). Volume 1(7).
Sublethal	Stress condition that is not immediately lethal to the organisms or below the level which directly causes death within the test period. Sublethal effects are most of the times reversible in contrast with mortality which is an irreversible condition. Volume 1(10,14).
Sum of squares	Sum of the squared residues. This sum is used as a criterion for goodness of fit in a regression procedure. Volume 2(2).
Surface water	Water column of a given water body (*e.g.*, lake, river, estuary, bay). Volume 1(14).
Surfactant	Surface tension decreasing agent that facilitates dispersion of hydrophobic materials in water. Volume 1(3).
Suspended matter	1- Fine insoluble particles originating from soil erosion, organic debris, urban wastewater or industrial effluent. Excessive levels of suspended matter lead to oxygen deficiencies in water bodies, and may have harmful effects on fauna and flora. 2- Part of the sediment load that is in suspension. Vol. 2(9).
Synergism	Interaction of several agents resulting in a greater effect than the one expected by addition of the individual effects. See also Synergistic effect. Volume 2(10).

Synergistic effect	Toxicity of a mixture of chemicals whereby the summation of the known toxicities of individual chemicals is greater than would be expected from a simple summation of the toxicities of the individual chemicals comprising the mixture. Volume 2(1).
Synoptic Sampling	Sampling at the same location at the same times; ideally, subsampling from the same original or composite sample. Volume 2(10).
Taxation principle	Guideline used to tax economic actors (*e.g.*, as a function of the load of pollutants that their activity generates). Ecological monetary taxes place pressure on polluters to limit pollution provided they are sufficiently substantial to incite clean-up action. Volume 2(2).
Test culture	Culture for providing organisms for use in a toxicity test. It can be established from organisms isolated from a stock culture. In the *Lemna* test, it refers to the 7 to 10-day old *Lemna* cultures maintained in Hoagland's medium that are then transferred to control/dilution water for an 18 to 24-h acclimation period. Volume 1(7).
Test medium	Synthetic culture medium that enables the survival or growth of test organisms during exposure to the test substance. It is prepared with deionized or glass-distilled water (*e.g.*, ASTM type-1 water) to which reagent-grade chemicals have been added. The resultant synthetic test medium is free from contaminants. The test substance will normally be mixed with, or dissolved in, the test medium. Volume 1(7).
Test sediment	Field-collected sample of whole sediment, taken from a marine, estuarine, or freshwater site thought to be contaminated (or potentially so) with one or more chemicals, and intended for use in solid-phase toxicity tests. In some instances, the term also applies to any solid-phase sample (including reference sediment, artificial sediment, negative control sediment, positive control sediment, or dredged material) used in testing. Volume 1(2).
Threshold Effect Concentration (TEC)	Value lying between the NOEC and LOEC derived by calculating the geometric mean of the NOEC and LOEC where $TEC = (NOEC \times LOEC)^{1/2}$. Volume 1(3), Volume 2(8).
Threshold Observed Effect Concentration (TOEC)	See above. Volume 1(3), Volume 2(8).

TIE Blank	During Toxicity Identification Evaluation, performance of a Phase I test on control water to determine if toxicity is added by the effluent manipulation itself. See also Control. Volume 2(5).
Toxic	Poisonous. A toxic chemical or material can cause adverse effects on living organisms, if present in sufficient amount at the right location. "Toxic" is an adjective, and should not be used as a noun, the term "toxicant" being the legitimate noun. Volume 1(2).
Toxic Threshold Effect Concentration (TTEC)	See Threshold Effect Concentration. Volume 1(3).
Toxic Unit (TU)	Inverse of the concentration of the test sample that is toxic calculated to make toxicity data directly proportional to the intensity of toxicity. For example, if a 25% dilution of a municipal wastewater has an effect on organisms, then the sample will have 100% v/v ÷ 25% v/v = 4 toxic units. Volume 1(14), Volume 2(5,8).
Toxicity	Inherent potential or capacity of a material or substance to cause adverse effect(s) on living organisms. The effect(s) can be lethal or sublethal. Volume 1(2,6,10).
Toxicity Identification Evaluation (TIE)	Iterative series of chemical manipulations (*e.g.*, pH adjustment, filtration, aeration) followed by toxicity testing designed to determine the contaminant responsible for the observed toxicity in the original sample. Volume 1(10), Volume 2(5,10).
Toxicity test	Determination of the effect of a material or substance on a group of selected organisms (*e.g.*, *Vibrio fischeri*), under defined conditions. An aquatic toxicity test usually measures either (a) the proportions of organisms affected (quantal); or (b) the degree of effect shown (quantitative or graded), after exposure to specific concentrations of test material or complex mixture (*e.g.*, chemical, effluent, elutriate, leachate, or receiving water). Volume 1(2,10).
Trickling filter	Fixed–film biological process for secondary domestic wastewater treatment. Volume 2(7).
Tubifex	Oligochaete worm, deep burrower and relatively tolerant to anoxia. Volume 1(12).
Ubiquitous	Found everywhere, present in most ecosystems around the world. Volume 1(11).

Viable cells	Cells capable of maintaining membrane permeability which is essential for the maintenance of life processes. A viable cell is able to maintain its membrane integrity to assure proper exchanges with its environment. Volume 1(14).
Vitellogenin (Vg)	Precursor of egg-yolk proteins rich in carbohydrates, lipids, phosphates and calcium. It is the principal energy reserve in oocytes. Vg expression is under the control of estradiol-17β receptors. This protein complex is produced in the liver by oviparous vertebrates and used as a biomarker to detect environmental estrogens. Volume 1(14).
Vortex mixer	Compact laboratory mixer used for stirring small sample volumes in containers (*i.e.*, test tubes, centrifuge tubes, colorimetric tubes, small flasks). Volume 1(1,3).
Warning chart	Graph used to follow changes over time, in the endpoints for a reference toxicant. Date (or number) of the test is on the horizontal axis and the effect-concentration is plotted on the vertical logarithmic scale. Volume 1(2).
Warning limits	Boundary or combination of limits which, when exceeded, may trigger analyst intervention; most toxicity laboratories use 2 X the standard deviation of the mean to create warning limits (*i.e.*, one in every 20 tests would be expected to exceed the warning limits, due to chance alone). Volume 1(2,10).
Wastewater	Water mixed with waste matter usually released by man-made activities, townships, municipal treatment plants and industries. Volume 1(14).
WaterTox Program	International network organized by the IDRC (International Development Research Centre), in collaboration with the National Water Research Institute and the Saint-Lawrence Centre of Environment Canada, to undertake bioanalytical intercalibration exercises with participating laboratories from eight different countries (Argentina, Canada, Chile, Colombia, Costa Rica, India, México and Ukraine). The battery of simple, affordable and robust tests was initially selected to detect the toxic potential of chemical contaminants in drinking water and freshwater sources. Volume 2(7).
Weak acid respiratory uncouplers	Compounds that abolish the link between substrate oxidation and adenosine triphosphate (ATP) synthesis (Cajina-Quezada and Schultz, 1990). They are generally bulky and electronegative. Volume 1(8).

Wet sediment phase	Solid phase obtained after extracting pore (or interstitial) water from whole sediment. Porewater is commonly extracted by centrifugation (*e.g.,* at 3,000 rpm, 30 min, 15°C). Volume 2(8).
Whole-water sample	Sample of water that has not been filtered or extracted. Volume 1(15).
Xenobiotics	Chemicals that have no relevant function for maintenance and reproduction of biological organisms. These compounds are usually produced by anthropogenic activity. Volume 1(14).

Glossary References

Bradbury, S.P., Carlson, R.W. and Henry, T.R. (1989) Polar narcosis in aquatic organisms, in U.M. Cowgill and L.R Williams (eds.), *Aquatic toxicology and hazard assessment: 12th Vol.*, ASTM, Philadelphia, PA, pp. 59-73.

Cajina-Quezada, M. and Schultz, T.W. (1990) Structure – toxicity relationships for selected weak acid respiratory uncouplers, *Aquatic Toxicology* **17**, 239-252.

Environment Canada (1999) Guidance documentation: Control of toxicity test precision using reference toxicants, Environmental Protection Service, Report EPS 1/RM/12, Ottawa, ON, 85 pp.

Environment Canada (1999) Biological test method: Test for measuring the inhibition of growth using the freshwater macrophyte, *Lemna minor*, Environmental Protection Service, Report EPS 1/RM/37, Ottawa, ON, 98 pp.

Férard, J.-F., Costan, G., Bermingham, N. and Blaise, C. (1992) Comparative assessment of effluents with the *Ceriodaphnia dubia* and *Selesnastrum capricornutum* chronic toxicity tests, Communication Meeting of SETAC-Europe, Postdam (Germany).

Perrodin, Y., Cavéglia, V., Barna, R., Moszkowicz, P., Gourdon, R., Férard, J.-F., Ferrari, B., Fruget, J.F., Plenet, S., Jocteur-Monrozier, L., Poly, F., Texier, C., Cluzeau, D., Lambolez, L. and Billard, H. (1996) Programme de recherche sur l'écocompatibilité des déchets: étude bibliographique, Report, ADEME, France (in French).

PIANC (1997) Handling and treatment of contaminated dredged material from ports and inland waterways, Permanent International Association of Navigation Congresses (PIANC), Report of Working Group No. 17, Supplement to Bulletin No. 89.

PIANC (2000) Glossary of selected environmental terms, Permanent International Association of Navigation Congresses (PIANC), Report of Working Group No. 3 of the Permanent Environmental Commission, Supplement to Bulletin No. 104.

PIANC (2001) Dredging: the facts, Permanent International Association of Navigation Congresses (PIANC), ISBN 90-75254-11-3.

Snell, T., Mitchell, J.L. and Burbank, S.E. (1996) Rapid toxicity assessment with microalgae using *in vivo* esterase inhibition, in G.K. Ostrander (Ed.), CRC Press/Lewis Publishers, Boca Raton, Florida, U.S.A., pp. 13-22.

US EPA, 2004, http://www.epa.gov/OCEPAterms/aterms.html.

C

D

E

M

N

O

P

Z